Robot Manipulator Control
Theory and Practice
Second Edition, Revised and Expanded

CONTROL ENGINEERING

A Series of Reference Books and Textbooks

Editors

NEIL MUNRO, PH.D., D.SC.

Professor
Applied Control Engineering
University of Manchester Institute of Science and Technology
Manchester, United Kingdom

FRANK L. LEWIS, PH.D.

Moncrief-O'Donnell Endowed Chair
and Associate Director of Research
Automation & Robotics Research Institute
University of Texas, Arlington

1. Nonlinear Control of Electric Machinery, *Darren M. Dawson, Jun Hu, and Timothy C. Burg*
2. Computational Intelligence in Control Engineering, *Robert E. King*
3. Quantitative Feedback Theory: Fundamentals and Applications, *Constantine H. Houpis and Steven J. Rasmussen*
4. Self-Learning Control of Finite Markov Chains, *A. S. Poznyak, K. Najim, and E. Gómez-Ramírez*
5. Robust Control and Filtering for Time-Delay Systems, *Magdi S. Mahmoud*
6. Classical Feedback Control: With MATLAB, *Boris J. Lurie and Paul J. Enright*
7. Optimal Control of Singularly Perturbed Linear Systems and Applications: High-Accuracy Techniques, *Zoran Gajić and Myo-Taeg Lim*
8. Engineering System Dynamics: A Unified Graph-Centered Approach, *Forbes T. Brown*
9. Advanced Process Identification and Control, *Enso Ikonen and Kaddour Najim*
10. Modern Control Engineering, *P. N. Paraskevopoulos*
11. Sliding Mode Control in Engineering, *edited by Wilfrid Perruquetti and Jean Pierre Barbot*
12. Actuator Saturation Control, *edited by Vikram Kapila and Karolos M. Grigoriadis*

13. Nonlinear Control Systems, *Zoran Vukić, Ljubomir Kuljača, Dali Đonlagić, Sejid Tešnjak*
14. Linear Control System Analysis and Design with MATLAB: Fifth Edition, Revised and Expanded, *John J. D'Azzo, Constantine H. Houpis, and Stuart N. Sheldon*
15. Robot Manipulator Control: Theory and Practice, Second Edition, Revised and Expanded, *Frank L. Lewis, Darren M. Dawson, and Chaouki T. Abdallah*
16. Robust Control System Design: Advanced State Space Techniques, Second Edition, Revised and Expanded, *Chia-Chi Tsui*

Additional Volumes in Preparation

Robot Manipulator Control
Theory and Practice
Second Edition, Revised and Expanded

Frank L. Lewis
University of Texas at Arlington
Arlington, Texas, U.S.A.

Darren M. Dawson
Clemson University
Clemson, South Carolina, U.S.A.

Chaouki T. Abdallah
University of New Mexico
Albuquerque, New Mexico, U.S.A.

MARCEL DEKKER, INC.   NEW YORK • BASEL

First edition: Control of Robot Manipulators, FL Lewis, CT Abdallah, DM Dawson, 1993. This book was previously published by Prentice-Hall, Inc.

Although great care has been taken to provide accurate and current information, neither the author(s) nor the publisher, nor anyone else associated with this publication, shall be liable for any loss, damage, or liability directly or indirectly caused or alleged to be caused by this book. The material contained herein is not intended to provide specific advice or recommendations for any specific situation.

Trademark notice: Product or corporate names may be trademarks or registered trademarks and are used only for identification and explanation without intent to infringe.

Library of Congress Cataloging-in-Publication Data
A catalog record for this book is available from the Library of Congress.

ISBN: 0-8247-4072-6

This book is printed on acid-free paper.

Headquarters
Marcel Dekker, Inc., 270 Madison Avenue, New York, NY 10016, U.S.A.
tel: 212-696-9000; fax: 212-685-4540

Distribution and Customer Service
Marcel Dekker, Inc., Cimarron Road, Monticello, New York 12701, U.S.A.
tel: 800-228-1160; fax: 845-796-1772

Eastern Hemisphere Distribution
Marcel Dekker AG, Hutgasse 4, Postfach 812, CH-4001 Basel, Switzerland
tel: 41-61-260-6300; fax: 41-61-260-6333

World Wide Web
http://www.dekker.com

The publisher offers discounts on this book when ordered in bulk quantities. For more information, write to Special Sales/Professional Marketing at the headquarters address above.

Copyright © 2004 by Marcel Dekker, Inc. All Rights Reserved.

Neither this book nor any part may be reproduced or transmitted in any form or by any means, electronic or mechanical, including photocopying, microfilming, and recording, or by any information storage and retrieval system, without permission in writing from the publisher.

Current printing (last digit):

10 9 8 7 6 5 4 3 2 1

PRINTED IN THE UNITED STATES OF AMERICA

To My Sons Christopher and Roman
F. L. L.

To My Faithful Wife, Dr. Kim Dawson
D. M. D.

To My 3 C's
C. T. A.

Series Introduction

Many textbooks have been written on control engineering, describing new techniques for controlling systems, or new and better ways of mathematically formulating existing methods to solve the ever-increasing complex problems faced by practicing engineers. However, few of these books fully address the applications aspects of control engineering. It is the intention of this new series to redress this situation.

The series will stress applications issues, and not just the mathematics of control engineering. It will provide texts that present not only both new and well-established techniques, but also detailed examples of the application of these methods to the solution of real-world problems. The authors will be drawn from both the academic world and the relevant applications sectors.

There are already many exciting examples of the application of control techniques in the established fields of electrical, mechanical (including aerospace), and chemical engineering. We have only to look around in today's highly automated society to see the use of advanced robotics techniques in the manufacturing industries; the use of automated control and navigation systems in air and surface transport systems; the increasing use of intelligent control systems in the many artifacts available to the domestic consumer market; and the reliable supply of water, gas, and electrical power to the domestic consumer and to industry. However, there are currently many challenging problems that could benefit from wider exposure to the applicability of control methodologies, and the systematic systems-oriented basis inherent in the application of control techniques.

This series presents books that draw on expertise from both the academic world and the applications domains, and will be useful not only as academically recommended course texts but also as handbooks for practitioners in many applications domains. *Nonlinear Control Systems* is another outstanding entry in Dekker's Control Engineering series.

Preface

The word 'robot' was introduced by the Czech playwright Karel Capek in his 1920 play Rossum's Universal Robots. The word 'robota' in Czech means simply 'work'. In spite of such practical beginnings, science fiction writers and early Hollywood movies have given us a romantic notion of robots. The anthropomorphic nature of these machines seems to have introduced into the notion of robot some element of man's search for his own identity.

The word 'automation' was introduced in the 1940's at the Ford Motor Company, a contraction for 'automatic motivation'. The single term 'automation' brings together two ideas: the notion of special purpose robotic machines designed to mechanically perform tasks, and the notion of an automatic control system to direct them.

The history of automatic control systems has deep roots. Most of the feedback controllers of the Greeks and Arabs regulated water clocks for the accurate telling of time; these were made obsolete by the invention of the mechanical clock in Switzerland in the fourteenth century. Automatic control systems only came into their own three hundred years later during the industrial revolution with the advent of machines sophisticated enough to require advanced controllers; we have in mind especially the windmill and the steam engine. On the other hand, though invented by others (e.g. T. Newcomen in 1712) the credit for the steam engine is usually assigned to James Watt, who in 1769 produced his engine which combined mechanical innovations with a control system that allowed automatic regulation. That is, modern complex machines are not useful unless equipped with a suitable control system.

Watt's centrifugal flyball governor in 1788 provided a constant speed controller, allowing efficient use of the steam engine in industry. The motion of the flyball governor is clearly visible even to the untrained eye, and its principle had an exotic flavor that seemed to many to embody the spirit of the new age. Consequently the governor quickly became a sensation throughout Europe.

Master-slave telerobotic mechanisms were used in the mid 1940's at Oak Ridge and Argonne National Laboratories for remote handling of radioactive material. The first commercially available robot was marketed in the late 1950's by Unimation (nearly coincidentally with Sputnik in 1957- thus the space age and the age of robots began simultaneously). Like the flyball governor, the motion of a robot manipulator is evident even for the untrained eye, so that the potential of robotic devices can capture the imagination. However, the high hopes of the 1960's for autonomous robotic automation in industry and unstructured environments have generally failed to materialize. This is because robotics today is at the same stage as the steam engine was shortly after the work of Newcomen in 1712.

Robotics is an interdisciplinary field involving diverse disciplines such as physics, mechanical design, statics and dynamics, electronics, control theory, sensors, vision, signal processing, computer programming, artificial intelligence (AI), and manufacturing. Various specialists study various limited aspects of robotics, but few engineers are able to confront all these areas simultaneously. This further contributes to the romanticized nature of robotics, for the control theorist, for instance, has a quixotic and fanciful notion of AI.

We might break robotics into five major areas: motion control, sensors and vision, planning and coordination, AI and decision-making, and man-machine interface. Without a good control system, a robotic device is useless. The robot arm plus its control system can be encapsulated as a generalized data abstraction; that is, robot-plus-controller is considered a single entity, or 'agent', for interaction with the external world.

The capabilities of the robotic agent are determined by the mechanical precision of motion and force exertion capabilities, the number of degrees of freedom of the arm, the degree of manipulability of the gripper, the sensors, and the sophistication and reliability of the controller. The inputs for a robot arm are simply motor currents and voltages, or hydraulic or pneumatic pressures; however, the inputs for the robot-plus-controller agent can be desired trajectories of motion, or desired exerted forces. Thus, the control system lifts the robot up a level in a hierarchy of abstraction.

This book is intended to provide an in-depth study of control systems for serial-link robot arms. It is a revised and expended version of our 1993 book. Chapters have been added on commercial robot manipulators and devices, neural network intelligent control, and implementation of advanced controllers on actual robotic systems. Chapter 1 places this book in the context of existing commercial robotic systems by describing the robots that are available and their limitations and capabilities, sensors, and controllers.

We wanted this book to be suitable either for the controls engineer or the roboticist. Therefore, Appendix A provides a background in robot kine-

PREFACE

matics and Jacobians, and Chapter 2 a background in control theory and mathematical notions. The intent was to furnish a text for a second course in robotics at the graduate level, but given the background material it is used at UTA as a first year graduate course for electrical engineering students. This course was also listed as part of the undergraduate curriculum, and the undergraduate students quickly digested the material.

Chapter 3 introduces the robot dynamical equations needed as the basis for controls design. In Appendix C and examples throughout the book are given the dynamics of some common arms. Chapter 4 covers the essential topic of computed-torque control, which gives important insight while also bringing together in a unified framework several sorts of classical and modern robot control schemes.

Robust and adaptive control are covered in Chapters 5 and 6 in a parallel fashion to bring out the similarities and the differences of these two approaches to control in the face of uncertainties and disturbances. Chapter 7 addresses some advanced techniques including learning control and arms with flexible joint coupling.

Modern intelligent control techniques based on biological systems have solved many problems in the control of complex systems, including unknown non-parametrizable dynamics and unknown disturbances, backlash, friction, and deadzone. Therefore, we have added a chapter on neural network control systems as Chapter 8. A robot is only useful if it comes in contact with its environment, so that force control issues are treated in Chapter 9.

A key to the verification of successful controller design is computer simulation. Therefore, we address computer simulation of controlled nonlinear systems and illustrate the procedure in examples throughout the text. Simulation software is given in Appendix B. Commercially available packages such as MATLAB make it very easy to simulate robot control systems.

Having designed a robot control system it is necessary to implement it; given today's microprocessors and digital signal processors, it is a short step from computer simulation to implementation, since the controller subroutines needed for simulation, and contained in the book, are virtually identical to those needed in a microprocessor for implementation on an actual arm. In fact, Chapter 10 shows the techniques for implementing the advanced controllers developed in this book on actual robotics systems.

All essential information and controls design algorithms are displayed in tables in the book. This, along with the List of Examples and List of Tables at the beginning of the book make for convenient reference by the student, the academician, or the practicing engineer.

We thank Wei Cheng of Milagro Design for her LaTeX typesetting and figure preparation as well as her scanning in the contents from the first edition into electronic format.

F.L. Lewis, Arlington, Texas
D.M. Dawson, Clemson, South Carolina
C.T. Abdallah, Albuquerque, New Mexico

Contents

Series Introduction		v
Preface		vii
1 Commercial Robot Manipulators		**1**
1.1	Introduction	1
	Flexible Robotic Workcells	2
1.2	Commercial Robot Configurations and Types	3
	Manipulator Performance	3
	Common Kinematic Configurations	4
	Drive Types of Commercial Robots	9
1.3	Commercial Robot Controllers	10
1.4	Sensors	12
	Types of Sensors	13
	Sensor Data Processing	16
References		19
2 Introduction to Control Theory		**21**
2.1	Introduction	21
2.2	Linear State-Variable Systems	22
	Continuous-Time Systems	22
	Discrete-Time Systems	28
2.3	Nonlinear State-Variable Systems	31
	Continuous-Time Systems	31
	Discrete-Time Systems	35
2.4	Nonlinear Systems and Equilibrium Points	36
2.5	Vector Spaces, Norms, and Inner Products	39

		Linear Vector Spaces	39
		Norms of Signals and Systems	40
		Inner Products	48
		Matrix Properties	48
	2.6	Stability Theory	51
	2.7	Lyapunov Stability Theorems	67
		Functions Of Class K	67
		Lyapunov Theorems	69
		The Autonomous Case	72
	2.8	Input/Output Stability	80
	2.9	Advanced Stability Results	82
		Passive Systems	82
		Positive-Real Systems	84
		Lure's Problem	85
		The MKY Lemma	86
	2.10	Useful Theorems and Lemmas	88
		Small-Gain Theorem	88
		Total Stability Theorem	89
	2.11	Linear Controller Design	93
	2.12	Summary and Notes	101
	References		103

3 Robot Dynamics — 107

	3.1	Introduction	107
	3.2	Lagrange-Euler Dynamics	108
		Force, Inertia, and Energy	108
		Lagrange's Equations of Motion	111
		Derivation of Manipulator Dynamics	119
	3.3	Structure and Properties of the Robot Equation	125
		Properties of the Inertia Matrix	126
		Properties of the Coriolis/Centripetal Term	127
		Properties of the Gravity, Friction, and Disturbance	134
		Linearity in the Parameters	136
		Passivity and Conservation of Energy	141
	3.4	State-Variable Representations and Feedback Linearization	142
		Hamiltonian Formulation	143
		Position/Velocity Formulations	145
		Feedback Linearization	145
	3.5	Cartesian and Other Dynamics	148
		Cartesian Arm Dynamics	148

		Structure and Properties of the Cartesian Dynamics	150
	3.6	Actuator Dynamics	152
		Dynamics of a Robot Arm with Actuators	152
		Third-Order Arm-Plus-Actuator Dynamics	154
		Dynamics with Joint Flexibility	155
	3.7	Summary	161
	References		163
	Problems		166

4 Computed-Torque Control — 169

	4.1	Introduction	169
	4.2	Path Generation	170
		Converting Cartesian Trajectories to Joint Space	171
		Polynomial Path Interpolation	173
		Linear Function with Parabolic Blends	176
		Minimum-Time Trajectories	178
	4.3	Computer Simulation of Robotic Systems	181
		Simulation of Robot Dynamics	181
		Simulation of Digital Robot Controllers	182
	4.4	Computed-Torque Control	185
		Derivation of Inner Feedforward Loop	185
		PD Outer-Loop Design	188
		PID Outer-Loop Design	197
		Class of Computed-Torque-Like Controllers	202
		PD-Plus-Gravity Controller	205
		Classical Joint Control	208
	4.5	Digital Robot Control	222
		Guaranteed Performance on Sampling	224
		Discretization of Inner Nonlinear Loop	225
		Joint Velocity Estimates from Position Measurements	226
		Discretization of Outer PD/PID Control Loop	226
		Actuator Saturation and Integrator Antiwindup Compensation	228
	4.6	Optimal Outer-Loop Design	243
		Linear Quadratic Optimal Control	243
		Linear Quadratic Computed-Torque Design	246
	4.7	Cartesian Control	248
		Cartesian Computed-Torque Control	248
		Cartesian Error Computation	250
	4.8	Summary	251

References 253
Problems 257

5 Robust Control of Robotic Manipulators 263
5.1 Introduction 263
5.2 Feedback-Linearization Controllers 265
        Lyapunov Designs 268
        Input-Output Designs 273
5.3 Nonlinear Controllers 293
        Direct Passive Controllers 293
        Variable-Structure Controllers 297
        Saturation-Type Controllers 306
5.4 Dynamics Redesign 316
        Decoupled Designs 316
        Imaginary Robot Concept 318
5.5 Summary 320
References 321
Problems 324

6 Adaptive Control of Robotic Manipulators 329
6.1 Introduction 329
6.2 Adaptive Control by a Computed-Torque Approach 330
        Approximate Computed-Torque Controller 330
        Adaptive Computed-Torque Controller 333
6.3 Adaptive Control by an Inertia-Related Approach 341
        Examination of a PD Plus Gravity Controller . . . 343
        Adaptive Inertia-Related Controller 344
6.4 Adaptive Controllers Based on Passivity 349
        Passive Adaptive Controller 349
        General Adaptive Update Rule 356
6.5 Persistency of Excitation 357
6.6 Composite Adaptive Controller 361
        Torque Filtering 362
        Least-Squares Estimation 365
        Composite Adaptive Controller 368
6.7 Robustness of Adaptive Controllers 371
        Torque-Based Disturbance Rejection Method . . . 372
        Estimator-Based Disturbance Rejection Method . . 375
6.8 Summary 377
References 379
Problems 381

7 Advanced Control Techniques — 383
7.1 Introduction — 383
7.2 Robot Controllers with Reduced On-Line Computation — 384
Desired Compensation Adaptation Law — 384
Repetitive Control Law — 392
7.3 Adaptive Robust Control — 399
7.4 Compensation for Actuator Dynamics — 407
Electrical Dynamics — 408
Joint Flexibilities — 416
7.5 Summary — 426
References — 427
Problems — 429

8 Neural Network Control of Robots — 431
8.1 Introduction — 431
8.2 Background in Neural Networks — 433
Multilayer Neural Networks — 433
Linear-in-the-parameter neural nets — 437
8.3 Tracking Control Using Static Neural Networks — 440
Robot Arm Dynamics and Error System — 440
Adaptive Control — 442
Neural Net Feedback Tracking Controller — 443
8.4 Tuning Algorithms for Linear-in-the-Parameters NN — 445
8.5 Tuning Algorithms for Nonlinear-in-the-Parameters NN — 449
Passivity Properties of NN Controllers — 453
Passivity of the Robot Tracking Error Dynamics — 453
Passivity Properties of 2-layer NN Controllers — 455
Passivity Properties of 1-Layer NN Controllers — 458
8.6 Summary — 458
References — 459

9 Force Control — 463
9.1 Introduction — 463
9.2 Stiffness Control — 464
Stiffness Control of a Single-Degree-of-Freedom Manipulator — 464
The Jacobian Matrix and Environmental Forces — 467
Stiffness Control of an N-Link Manipulator — 474
9.3 Hybrid Position/Force Control — 478
Hybrid Position/Force Control of a Cartesian Two-Link Arm — 479

Hybrid Position/Force Control of an N-Link
Manipulator 482
Implementation Issues 487
9.4 Hybrid Impedance Control 489
Modeling the Environment 490
Position and Force Control Models 492
Impedance Control Formulation 494
Implementation Issues 499
9.5 Reduced State Position/Force Control 501
Effects of Holonomic Constraints on the
Manipulator Dynamics 501
Reduced State Modeling and Control 504
Implementation Issues 509
9.6 Summary 510
References 513
Problems 514

10 Robot Control Implementation and Software 517
10.1 Introduction 518
10.2 Tools and Technologies 520
10.3 Design of the Robotic Platform 523
Overview 523
Core Classes 526
Robot Control Classes 527
External Device Classes 532
Utility Classes 533
Configuration Management 533
Object Manager 534
Concurrency/Communication Model 537
Plotting and Control Tuning Capabilities 538
Math Library 540
Error Management and the Front-End GUI 542
10.4 Operation of the Robotic Platform 543
Scene Viewer and Control Panels 543
Utility Programs for Moving the Robot 544
Writing, Compiling, Linking, and Starting
Control Programs 545
10.5 Programming Examples 548
Comparison of Simulation and Implementation . . 548
Virtual Walls 548
10.6 Summary 550
References 551

A	**Review of Robot Kinematics and Jacobians**	**555**
	A.1 Basic Manipulator Geometries	555
	A.2 Robot Kinematics	558
	A.3 The Manipulator Jacobian	576
	References	589
B	**Software for Controller Simulation**	**591**
	References	597
C	**Dynamics of Some Common Robot Arms**	**599**
	C.1 SCARA ARM	600
	C.2 Stanford Manipulator	601
	C.3 PUMA 560 Manipulator	603
	References	607
Index		**609**

Chapter 1

Commercial Robot Manipulators

This chapter sets the stage for this book by providing an overview of commercially available robotic manipulators, sensors, and controllers. We make the point that if one desires high performance flexible robotic workcells, then it is necessary to design advanced control systems for robot manipulators such as are found in this book.

1.1 Introduction

When studying advanced techniques for robot control, planning, sensors, and human interfacing, it is important to be aware of the systems that are commercially available. This allows one to develop new technology in the context of existing technology, which allows one to implement the new techniques on existing robotic systems.

A National Association of Manufacturer's report [NAM 1998] states that the two most important drivers for US commercial business manufacturing success in the 1990's have been reconfigurable manufacturing workcells and local area networks in the factory. In this chapter we discuss flexible robotic workcells, commercial robot configurations, commercial robot controllers, information integration to the internet, and robot workcell sensors. More information on these topics can be found in the Mechanical Engineering Handbook [Lewis 1998] and the Computer Science Engineering Handbook [Lewis and Fitzgerald 1997].

Flexible Robotic Workcells

In factory automation and elsewhere it was once common to use fixed layouts built around conveyors or other transportation systems in which each robot performed a specific task. These assembly lines had distinct workstations, each performing a dedicated function. Robots have been used at the workstation level to perform operations such as assembly, drilling, surface finishing, welding, palletizing, and so on. In the assembly line, parts are routed sequentially to the workstations by the transport system. Such systems are very expensive to install, require a cadre of engineering experts to design and program, and are extremely difficult to modify or reprogram as needs change. In today's high-mix low-volume (HMLV) manufacturing scenario, these characteristics tolled the death knell for such rigid antiquated designs.

Figure 1.1.1: UTA's Automation and Robotics Test Cell.

In the assembly line, the robot is restricted by placing it into a rigid sequential system. Robots are versatile machines with many capabilities, and their potential can be significantly increased by using them as a basis for *flexible robotic workcells* [Decelle 1988], [Jamshidi et al. 1992], [Pugh 1983] such as the UTA Automation and Robotics Test Cell in Figure 1.1.1. In the flexible robotic workcell, robots are used for part handling, assembly, and other process operations. By reprogramming the robots one changes the entire functionality of the workcell. The workcell is designed to make full use of the workspace of the robots, and components such as milling machines, drilling machines, vibratory part feeders, and so on are placed within the robots' workspaces to allow servicing by the robots. Contrary to the assembly line, the physical layout does not impose a priori a fixed sequencing of the operations or jobs. Thus, as product requirements change, all that is required is to reprogram the workcell in software [Mireles and Lewis 2001]. The workcell is ideally suited to emerging HMLV conditions in manufacturing and elsewhere.

The rising popularity of robotic workcells has taken emphasis away from hardware design and placed new emphasis on innovative software techniques and architectures that include planning, coordination, and control (PC&C) functions. A great deal of research into robot controllers has been required to give robots the flexibility, precision, and functionality needed in modern flexible workcells. The remainder of this book details such advanced control techniques.

1.2 Commercial Robot Configurations and Types

Much of the information in this section was prepared by Mick Fitzgerald, who was then Manager at UTA's Automation and Robotics Research Institute (ARRI).

Robots are highly reliable, dependable and technologically advanced factory equipment. The majority of the world's robots are supplied by established companies using reliable off-the-shelf component technologies. All commercial industrial robots have two physically separate basic elements— the manipulator arm and the controller. The basic architecture of most commercial robots is fundamentally the same, and consists of digital servo-controlled electrical motor drives on serial-link kinematic machines, usually with no more than six axes (degrees of freedom). All are supplied with a proprietary controller. Virtually all robot applications require significant design and implementation effort by engineers and technicians. What makes each robot unique is how the components are put together to achieve performance that yields a competitive product. The most important considerations in the application of an industrial robot center on two issues: manipulation and integration.

Manipulator Performance

The combined effects of kinematic structure, axis drive mechanism design, and real-time motion control determine the major manipulation performance characteristics: reach and dexterity, payload, quickness, and precision. Caution must be used when making decisions and comparisons based on manufacturers' published performance specifications because the methods for measuring and reporting them are not standardized across the industry. Usually motion testing, simulations, or other analysis techniques are used to verify performance for each application.

Reach is characterized by measuring the extent of the *workspace* described by the robot motion and *dexterity* by the angular displacement of the individual joints. Some robots will have unusable spaces such as dead zones, singular poses, and wrist-wrap poses inside of the boundaries of their reach.

Payload weight is specified by the manufacturers of all industrial robots. Some manufacturers also specify inertial loading for rotational wrist axes. It is common for the payload to be given for extreme velocity and reach conditions. Weight and inertia of all tooling, workpieces, cables and hoses must be included as part of the payload.

Quickness is critical in determining throughput but difficult to determine from published robot specifications. Most manufacturers will specify a maximum speed of either individual joints or for a specific kinematic tool point. However, *average speed* in a working cycle is the quickness characteristic of interest.

Precision is usually characterized by measuring *repeatability*. Virtually all robot manufacturers specify static position repeatability. *Accuracy* is rarely specified, but it is likely to be at least four times larger than repeatability. Dynamic precision, or the repeatability and accuracy in tracking position, velocity, and acceleration over a continuous path, is not usually specified.

Common Kinematic Configurations

All common commercial industrial robots are serial-link manipulators, usually with no more than six kinematically coupled axes of motion. By convention, the axes of motion are numbered in sequence as they are encountered from the base on out to the wrist. The first three axes account for the spatial positioning motion of the robot; their configuration determines the shape of the space through which the robot can be positioned. Any subsequent axes in the kinematic chain generally provide rotational motions to orient the end of the robot arm and are referred to as wrist axes. In a robotic wrist, three axes usually intersect to generate true independent positioning in terms of 3-D orientation. See Appendix A for a kinematic analysis of the spherical robot wrist mechanism. Note that in our 3-dimensional space, one requires three degrees of freedom for fully independent spatial positioning and three degrees of freedom for fully independent orientational positioning.

There are two primary types of motion that a robot axis can produce in its driven link– either *revolute* or *prismatic*. Revolute joints are anthropomorphic (e.g. like human joints) while prismatic joints are able to extend and retract like a car radio antenna. It is often useful to classify robots according to the orientation and type of their first three axes. There are four very common commercial robot configurations: Articulated, Type I SCARA, Type II SCARA, and Cartesian. Two other configurations,

1.2 Commercial Robot Configurations and Types

Figure 1.2.1: Articulated Arm. Six-axis CRS A465 arm (courtesy of CRS robotics).

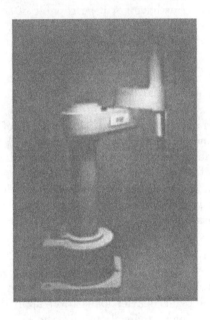

Figure 1.2.2: Type I SCARA Arm. High precision, high speed midsized SCARA I. (courtesy of Adept Technologies, Inc.).

Cylindrical and Spherical, are now much less common.

Appendix C contains the dynamics of some common robot manipulators for use in controls simulation in this book.

Articulated Arms. The variety of commercial articulated arms, most of which have six axes, is very large (Fig. 1.2.1). All of these robots' axes are revolute. The second and third axes are co-planar and work together to produce motion in a vertical plane. The first axis in the base is vertical and revolves the arm to sweep out a large work volume. Many different types of drive mechanisms have been devised to allow wrist and forearm drive motors and gearboxes to be mounted close to the first and second axis of rotation, thus minimizing the extended mass of the arm. The workspace efficiency of well designed articulated arms, which is the degree of quick dexterous reach with respect to arm size, is unsurpassed by other arm configurations when five or more degrees of freedom are needed. A major limiting factor in articulated arm performance is that the second axis has to work to lift both the subsequent arm structure and the payload. Historically, articulated arms have not been capable of achieving accuracy as high as other arm configurations, as all axes have joint angle position errors which are multiplied by link radius and accumulated for the entire arm.

Type I SCARA. The Type I SCARA (selectively compliant assembly robot arm) arm, Figure 1.2.2, uses two parallel revolute joints to produce motion in the horizontal plane. The arm structure is weight-bearing but the first and second axes do no lifting. The third axis of the Type I SCARA provides work volume by adding a vertical or z axis. A fourth revolute axis will add rotation about the z axis to control orientation in the horizontal plane. This type of robot is rarely found with more than four axes. The Type I SCARA is used extensively in the assembly of electronic components and devices, and it is used broadly for the assembly of small- and medium-sized mechanical assemblies.

Type II SCARA. The Type II SCARA, Figure 1.2.3, also a four axis configuration, differs from Type I in that the first axis is a long vertical prismatic z stroke which lifts the two parallel revolute axis and their links. For quickly moving heavier loads (over approximately 75 pounds) over longer distance (more than about three feet), the Type II SCARA configuration is more efficient than the Type I.

Cartesian Coordinate Robots. Cartesian coordinate robots use orthogonal prismatic axes, usually referred to as x, y, and z, to translate their end-effector or payload through their rectangular workspace. One, two, or three revolute wrist axes may be added for orientation. Commercial robot companies supply several types of Cartesian coordinate robots with workspace sizes ranging from a few cubic inches to tens of thousands of cubic feet, and payloads ranging to several hundred pounds. Gantry robots,

1.2 Commercial Robot Configurations and Types

Figure 1.2.3: Type II SCARA (courtesy of Adept Technologies, Inc.).

Figure 1.2.4: Cartesian Robot. Three-axis robot constructed from modular single-axis motion modules (courtesy of Adept Technologies, Inc.).

which have an elevated bridge structure, are the most common Cartesian style and are well suited to material handling applications where large areas and/or large loads must be serviced. They are particularly useful in applications such as arc welding, waterjet cutting, and inspection of large complex precision parts.

Modular Cartesian robots, see Figure 1.2.4, are also commonly available from several commercial sources. Each module is a self-contained completely functional single-axis actuator; the modules may be custom assembled for special-purpose applications.

Spherical and Cylindrical Coordinate Robots. The first two axes of the spherical coordinate robot, Figure 1.2.5, are revolute and orthogonal to one another, and the third axis provides prismatic radial extension. The result is a natural spherical coordinate system with a spherical work volume. The first axis of cylindrical coordinate robots, Figure 1.2.6, is a revolute base rotation. The second and third are prismatic, resulting in a natural cylindrical motion. Commercial models of Spherical and Cylindrical robots were originally very common and popular in machine tending and material handling applications. Hundreds are still in use but now there are only a few commercially available models. The decline in use of these two configurations is attributed to problems arising from use of the prismatic link for radial extension/retraction motion; a solid boom requires clearance to fully retract.

Figure 1.2.5: Hydraulic powered spherical robot (courtesy Kohol Systems, Inc.).

Parallel-Link Manipulators. For some special purpose applications, parallel-link robots are more suitable than serial link robots. These robots generally have three or six links in parallel, each link attached to a fixed base and to a moving working platform. See Figure 1.2.7. With proper design, a six-link parallel-link manipulator can have six degrees of freedom motion of the working platform. The military Link trainer is a large parallel-link robot moving a pilot's seat. These robots have greater stiffness and precision than serial-link robots, where the positioning errors of each link are compounded as one moves outwards from the base. Thus, lightweight parallel-link robots are able to precisely move large loads. These robots

1.2 Commercial Robot Configurations and Types

have been used for example in machining and surface finishing of precision industrial and aerospace components such as bulkheads and air vehicle outer skins.

Figure 1.2.6: Cylindrical arm using scissor mechanism for radial prismatic motion (courtesy of Yamaha Robotics).

The parallel-link robot is a closed-kinematic-chain system, and as such is relatively difficult to analyze [Liu and Lewis 1993]. The control system design problem is more difficult for these robots.

Drive Types of Commercial Robots

The vast majority of commercial industrial robots use electric servo-motor drives with speed reducing transmissions. Both AC and DC motors are popular. Some servo-hydraulic articulated arm robots are available now for painting applications. It is rare to find robots with servo-pneumatic drive axes. All types of mechanical transmissions are used, but the tendency is toward low- and zero-backlash type drives. Some robots use direct drive methods to eliminate the amplification of inertia and mechanical backlash associated with other drives. Joint angle position sensors, required for real-time servo-level control, are generally considered an important part of the drive train. Less often, velocity feedback sensors are provided.

Figure 1.2.7: Parallel-link robot (courtesy of ABB Robotics).

1.3 Commercial Robot Controllers

Commercial robot controllers are specialized multiprocessor computing systems that provide four basic processes allowing integration of the robot into an automation system: Motion Trajectory Generation and Following, Motion/Process Integration and Sequencing, Human User integration, and Information Integration.

Motion Trajectory Generation and Following. There are two important controller-related aspects of industrial robot motion generation. One is the extent of manipulation that can be programmed, the other is the ability to execute controlled programmed motion. A unique aspect of each robot system is its real-time servo-level motion control. The details of real-time control are typically not revealed to the user due to safety and proprietary information secrecy reasons. Each robot controller, through its operating system programs, converts digital data from higher-level coordinators into coordinated arm motion through precise computation and high-speed distribution and communication of the individual axis motion commands which are executed by individual joint servo-controllers. Most commercial robot controllers operate at a sample period of 16 msec. The real-time motion controller invariably uses classical independent-joint proportional-integral-derivative (PID) control or simple modifications of PID. This makes commercially available controllers suitable for point-to-point motion, but most are not suitable for following continuous position/velocity profiles or exerting prescribed forces without considerable programming effort, if at all.

Recently, more advanced controllers have appeared. The AdeptWindows family of automation controllers (*http://www.adept.com*) integrates robotics, motion control, machine vision, force sensing, and manufacturing

1.3 Commercial Robot Controllers

logic in a single control platform compatible with Windows 98 & Windows NT/2000. Adept motion controllers can be configured to control other robots and custom mechanisms, and are standard on a variety of systems from OEMs.

Motion/Process Integration and Sequencing. Motion/process integration involves coordinating manipulator motion with process sensors or other process controller devices. The most primitive process integration is through discrete digital input/output (I/O). For example a machine controller external to the robot controller might send a one bit signal indicating that it is ready to be loaded by the robot. The robot controller must have the ability to read the digital signal and to perform logical operations (if then, wait until, do until, etc.) using the signal. That is, some robot controllers have some programmable logic controller (PLC) functions built in. Coordination with sensors (e.g. vision) is also often provided.

Human Integration. The controller's human interfaces are critical to the expeditious setup and programming of robot systems. Most robot controllers have two types of human interface available: computer style CRT/keyboard terminals for writing and editing program code off-line, and *teach pendants*, which are portable manual input terminals used to command motion in a telerobotic fashion via touch keys or joy sticks. Teach pendants are usually the most efficient means available for positioning the robot, and a memory in the controller makes it possible to play back the taught positions to execute motion trajectories. With practice, human operators can quickly teach a series of points which are chained together in playback mode. Most robot applications currently depend on the integration of human expertise during the programming phase for the successful planning and coordination of robot motion. These interface mechanisms are effective in unobstructed workspaces where no changes occur between programming and execution. They do not allow human interface during execution or adaptation to changing environments.

More recent advanced robot interface techniques are based on *behavior-based programming*, where various specific behaviors are programmed into the robot controller at a low level (e.g. pick up piece, insert in machine chuck). The behaviors are then sequenced and their specific motion parameters specified by a higher-level machine supervisor as prescribed by the human operator. Such an approach was used in [Mireles and Lewis 2001].

Information Integration. Information integration is becoming more important as the trend toward increasing flexibility and agility impacts robotics. Many commercial robot controllers now support information integration functions by employing integrated PC interfaces through the communications ports (e.g. RS-232), or in some through direct connections to the robot controller data bus. Recent integration efforts are making it possible to interface robotic workcells to the internet to allow remote site monitoring and control. There are many techniques for this, the most convenient of which is LabVIEW 6.1, which doe not require programming in Java.

1.4 Sensors

Much of the information in this section was prepared by Kok-Meng Lee [Lewis 1998]. Sensors and actuators [Tzou and Fukuda 1992] function as *transducers*, devices through which high-level workcell Planning, Coordination, and Control systems interface with the hardware components that make up the workcell. Sensors are a vital element as they convert states of physical devices into signals appropriate for input to the workcell PC&C control system; inappropriate sensors can introduce errors that make proper operation impossible no matter how sophisticated or expensive the PC&C system, while innovative selection of sensors can make the control and coordination problem much easier.

Sensors are of many different types and have many distinct uses. Having in mind an analogy with biological systems, *proprioceptors* are sensors internal to a device that yield information about the internal state of that device (e.g. robot arm joint-angle sensors). *Exteroceptors* yield information about other hardware external to a device. Sensors yield outputs that are either analog or digital; digital sensors often provide information about the status of a machine or resource (gripper open or closed, machine loaded, job complete). Sensors produce outputs that are required at all levels of the PC&C hierarchy, including uses for:

- servo-level feedback control (usually analog proprioceptors)

- process monitoring and coordination (often digital exteroceptors or part inspection sensors such as vision)

- failure and safety monitoring (often digital— e.g. contact sensor, pneumatic pressure-loss sensor)

- quality control inspection (often vision or scanning laser).

Sensor output data must often be processed to convert it into a form meaningful for PC&C purposes. The sensor plus required signal processing is shown as a *Virtual Sensor*. It functions as a *data abstraction* — a set of data plus operations on that data (e.g. camera, plus framegrabber, plus signal processing algorithms such as image enhancement, edge detection, segmentation, etc.). Some sensors, including the proprioceptors needed for servo-level feedback control, are integral parts of their host devices, and so processing of sensor data and use of the data occurs within that device; then, the sensor data is incorporated at the servocontrol level or Machine Coordination level. Other sensors, often vision systems, rival the robot manipulator in sophistication and are coordinated by a Job Coordinator, which treats them as valuable shared resources whose use is assigned to jobs that need them by some priority assignment (e.g. dispatching) scheme. An interesting coordination problem is posed by so-called *active sensing*, where, e.g., a robot may hold a scanning camera, and the camera effectively takes charge of the motion coordination problem, directing the robot where to move to effect the maximum reduction in entropy (increase in information) with subsequent images.

Types of Sensors

This section summarizes sensors from an operational point of view. More information on functional and physical principles can be found in [Fraden 1993], [Fu et al. 1987], [Snyder 1985].

Tactile Sensors. Tactile sensors rely on physical contact with external objects. Digital sensors such as limit switches, microswitches, and vaccuum devices give binary information on whether contact occurs or not. Sensors are available to detect the onset of slippage. Analog sensors such as spring-loaded rods give more information. Tactile sensors based on rubberlike carbon- or silicon-based elastomers with embedded electrical or mechanical components can provide very detailed information about part geometry, location, and more. Elastomers can contain resistive or capacitive elements whose electrical properties change as the elastomer conmpresses. Designs based on LSI technology can produce tactile grid pads with, e.g., 64 × 64 'forcel' points on a single pad. Such sensors produce 'tactile images' that have properties akin to digital images from a camera and require similar data processing. Additional tactile sensors fall under the classification of 'force sensors' discussed subsequently.

Proximity and Distance Sensors. The noncontact proximity sensors include devices based on the Hall effect or inductive devices based on the electromagnetic effect that can detect ferrous materials within about 5 mm. Such sensors are often digital, yielding binary information about whether or not an object is near. Capacitance-based sensors detect any nearby solid or liquid with ranges of about 5mm. Optical and ultrasound sensors have longer ranges.

Distance sensors include time-of-flight rangefinder devices such as sonar and lasers. The commercially available Polaroid sonar offers accuracy of about 1 in. up to 5 feet, with angular sector accuracy of about 15 deg. For 360 deg. coverage in navigation applications for mobile robots, both scanning sonars and ring-mounted multiple sonars are available. Sonar is typically noisy with spurious readings, and requires low-pass filtering and other data processing aimed at reducing the false alarm rate. The more expensive laser rangefinders are extremely accurate in distance and have very high angular resolution.

Position, Velocity, and Acceleration Sensors. Linear position-measuring devices include linear potentiometers and the sonar and laser rangefinders just discussed. Linear velocity sensors may be laser- or sonar-based Doppler-effect devices.

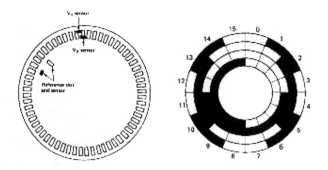

Figure 1.4.1: Optical Encoders. (a) Incremental optical encoder. (b) Absolute optical encoder with n= 4 using Grey code. (Snyder, W.E., 1985. Industrial Robots, Prentice-Hall, NJ, with permission.)

Joint-angle position and velocity proprioceptors are an important part of the robot arm servocontrol drive axis. Angular position sensors include potentiometers, which use dc voltage, and resolvers, which use ac voltage and have accuracies of 15 min. Optical encoders can provide extreme accuracy using digital techniques. Incremental optical encoders use three optical sensors and a single ring of alternating opaque/clear areas, Figure 1.4.1(a), to provide angular position relative to a reference point and angular velocity information; commercial devices may have 1200 slots per turn. More expensive absolute optical encoders, Figure 1.4.1(b), have n concentric rings of alternating opaque/clear areas and require n optical sensors. They offer increased accuracy and minimize errors associated with data reading and

1.4 Sensors

transmission, particularly if they employ the Grey code, where only one bit changes between two consecutive sectors. Accuracy is $360^0/2^n$ with commercial devices having $n = 12$ or so.

Gyros have good accuracy if repeatability problems associated with drift are compensated for. Directional gyros have accuracies of about 1.5 deg. Vertical gyros have accuracies of 0.5 deg and are available to measure multi-axis motion (e.g. pitch and roll). Rate gyros measure velocities directly with thresholds of 0.05 deg/sec or so.

Various sorts of accelerometers are available based on strain gauges (next paragraph), gyros, or crystal properties. Commercial devices are available to measure accelerations along three axes. A popular new technology involves microelectromechanical systems (MEMS), which are either surface or bulk micromachined devices. MEMS accelerometers are very small, inexpensive, robust, and accurate. MEMS sensors have especially been used in the automotive industry [Eddy 1998].

Force and Torque Sensors. Various torque sensors are available, though they are often not required; for instance, the internal torques at the joints of a robot arm can be computed from the motor armature currents. Torque sensors on a drilling tool, for instance, can indicate when tools are becoming dull. Linear force can be measured using load cells or strain gauges. A strain gauge is an elastic sensor whose resistance is a function of applied strain or deformation. The piezoelectric effect, the generation of a voltage when a force is applied, may also be used for force sensing. Other force sensing techniques are based on vacuum diodes, quartz crystals (whose resonant frequency changes with applied force), etc.

Robot arm force-torque wrist sensors are extremely useful in dexterous manipulation tasks. Commercially available devices can measure both force and torque along three perpendicular axes, providing full information about the Cartesian force vector F. Standard transformations allow computation of forces and torques in other coordinates. Six-axis force-torque sensors are quite expensive.

Photoelectric Sensors. A wide variety of photoelectric sensors are available, some based on fibreoptic principles. These have speeds of response in the neighborhood of 50 microsec with ranges up to about 45 mm, and are useful for detecting parts and labeling, scanning optical bar codes, confirming part passage in sorting tasks, etc.

Other Sensors. Various sensors are available for measuring pressure, temperature, fluid flow, etc. These are useful in closed-loop servo-control applications for some processes such as welding, and in job coordination and/or safety interrupt routines in others.

Sensor Data Processing

Before any sensor can be used in a robotic workcell, it must be calibrated. Depending on the sensor, this could involve significant effort in experimentation, computation, and tuning after installation. Manufacturers often provide calibration procedures though in some cases, including vision, such procedures may not be obvious, requiring reference to the published scientific literature. Time-consuming recalibration may be needed after any modifications to the system.

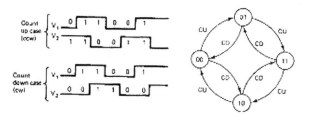

Figure 1.4.2: Signal Processing using FSM for Optical Encoders (a) Phase relations in incremental optical encoder output. (b) Finite state machine to decode encoder output into angular position. (Snyder 1985).

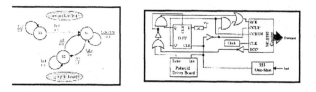

Figure 1.4.3: Hardware design from FSM. (a) FSM for sonar transducer control on a mobile robot. (b) Sonar driver control system from FSM.

Particularly for more complex sensors such as optical encoders, significant sensor signal conditioning and processing is required. This might include amplification of signals, noise rejection, conversion of data from analog to digital or from digital to analog, and so on. Hardware is usually provided for such purposes by the manufacturer and should be considered as part of the sensor package for robot workcell design. The sensor, along with its signal processing hardware and software algorithms may be considered as a data abstraction and is called the 'virtual sensor'.

If signal processing does need to be addressed, it is often very useful to use finite state machine (FSM) design. A typical signal from an incremental optical encoder is shown in Figure 1.4.2(a); a FSM for decoding this into the angular position is given in Figure 1.4.2(b). FSM are very easy to convert directly to hardware in terms of logical gates. A FSM for sequencing a sonar is given in Figure 1.4.3(a); the sonar driver hardware derived from this FSM is shown in Figure 1.4.3(b).

A particular problem is obtaining angular velocity from angular position measurements. All too often the position measurements are simply differenced using a small sample period to compute velocity. This is guaranteed to lead to problems if there is any noise in the signal. It is almost always necessary to employ a low-pass-filtered derivative where velocity samples v_k are computed from position measurement samples p_k using, e.g.,

$$v_k = \alpha v_{k-1} + (1 - \alpha) \frac{p_k - p_{k-1}}{T},$$

where T is the sample period and α is a small filtering coefficient. A similar approach is needed to compute acceleration.

Vision Systems, Cameras, and Illumination. Typical commercially available vision systems conform to the RS-170 standard of the 1950's, so that frames are acquired through a framegrabber board at a rate of 30 frames/sec. Images are scanned; in a popular US standard, each complete scan or *frame* consists of 525 lines of which 480 contain image information. This sample rate and image resolutions of this order are adequate for most applications with the exception of vision-based robot arm servoing. Robot vision system cameras are usually TV cameras— either the solid-state charge-coupled device (CCD), which is responsive to wavelengths of light from below 350nm (ultraviolet) to 1100nm (near infrared) and has peak response at approximately 800nm, or the charge injection device (CID), which offers a similar spectral response and has a peak response at approximately 650nm. Both *line-scan* CCD cameras, having resolutions ranging between 256 and 2048 elements, and *area-scan* CCD cameras are available. Medium-resolution area-scan cameras yield images of 256 × 256, though high-resolution devices of 1024 × 1024 are by now available. Line-scan cameras are suitable for applications where parts move past the camera, e.g., on conveyor belts. Framegrabbers often support multiple cameras, with a common number being four, and may support black-and-white or color images.

If left to chance, illumination of the robotic workcell will probably result in severe problems in operations. Common problems include low-contrast images, specular reflections, shadows, and extraneuos details. Such problems can be corrected by overly sophisticated image processing, but all of this can be avoided by some proper attention to details at the workcell design stage. Illumination techniques include spectral filtering, selection of suitable spectral characteristics of the illumination source, diffuse-lighting techniques, backlighting (which produces easily processed silhouettes), structured-lighting (which provides additional depth information and simplifies object detection and interpretation), and directional lighting.

REFERENCES

[Decelle 1988] Decelle, L. S., "Design of a Robotic Workstation For Component Insertions," *AT&T Technical Journal*, March/April 1988, Volume 67, Issue 2. pp 15-22.

[Eddy 1998] Eddy, D.S., and D.R. Sparks, "Application of MEMS technology in automotive sensors and actuators," Proc. IEEE, vol. 86, no. 8, pp. 1747-1755, Aug. 1998.

[Fraden 1993] Fraden, J. *AIP Handbook Of Modern Sensors, Physics, Design, and Applications*, American Institute of Physics. 1993.

[Fu et al. 1987] Fu, K.S., R.C. Gonzalez, and C.S.G. Lee, *Robotics*, McGraw-Hill, New York, 1987.

[Jamshidi et al. 1992] Jamshidi, M., Lumia, R., Mullins, J., and Shahinpoor, M., 1992. *Robotics and Manufacturing: Recent Trends in Research, Education, and Applications*, Vol. 4, ASME Press, New York.

[Lewis and Fitzgerald 1997] Lewis, F.L., M. Fitzgerald, and K. Liu "Robotics," in The Computer Science and Engineering Handbook, Allen B. Tucker, Jr. ed., Chapter 33, CRC Press, 1997.

[Lewis 1998] Lewis, F.L., "Robotics," in Handbook of Mechanical Engineering, F. Kreith ed., chapter 14, CRC Press, 1998.

[Liu and Lewis 1993] Liu, K., F.L. Lewis, G. Lebret, and D. Taylor, "The singularities and dynamics of a Stewart Platform Manipulator," J. Intelligent and Robotic Systems, vol. 8, pp. 287-308, 1993.

[Mireles and Lewis 2001] Mireles, J., and F.L. Lewis, "Intelligent Material Handling: Development and implementation of a matrix-based discrete-event controller," IEEE Trans. Industrial Electronics, vol. 48, no. 6, pp. 1087-1097, Dec. 2001.

REFERENCES

[NAM 1998] National Assoc. Manufacturers Report, "Technology on the Factory Floor III," ed. P.M. Swadimass, NAM Pub. Center, 1-800-637-3005, Aug. 1998.

[Pugh 1983] Pugh, A., ed., 1983. *Robotic Technology*, IEE Control Engineering Series 23, Pergrinus, London.

[Snyder 1985] Snyder, W. E., 1985, *Industrial Robots: Computer Interfacing and Control*, Prentice-Hall, Inc., Englewood Cliffs, New Jersey, USA.

[Tzou and Fukuda 1992] Tzou, H.S., and Fukuda, T., *Precision Sensors, Actuators, and Systems*, Kluwer Academic, 1992.

Chapter 2

Introduction to Control Theory

In this chapter we review various control theory concepts that are used in the control of robots. We first review the state-space formulation for both linear and nonlinear systems and present the stability concepts needed in the sequel. The chapter is intended to introduce modern control concepts, but even readers with a background in control theory may wish to consult it for notation and convenience.

2.1 Introduction

The control of robotic manipulators is a mature yet fruitful area for research, development, and manufacturing. Industrial robots are basically positioning and handling devices. Therefore, a useful robot is one that is able to control its movement and the interactive forces and torques between the robot and its environment. This book is concerned with the control aspect of robotic manipulators. To control usually requires the availability of a mathematical model and of some sort of intelligence to act on the model. The mathematical model of a robot is obtained from the basic physical laws governing its movement. Intelligence, on the other hand, requires sensory capabilities and means for acting and reacting to the sensed variables. These actions and reactions of the robot are the result of controller design.

In this chapter we review the concepts of control theory that are needed in this book. All proofs are omitted, but references are made to more specialized books and papers where proofs are provided. Once a satisfactory model of the robot dynamics is obtained as described in Chapter 2, the automatic control concepts presented in the current chapter may be used to modify the actions and reactions of the robot to different stimuli. Subsequent chapters will therefore deal with the application of control principles to the robot equations. The particular controller used will depend on the complexity of the mathematical model, the application at hand, the available resources, and a host of other criteria.

We begin the chapter in section 2.2 with a review of the state-space description for linear, continuous- and discrete-time systems. A similar review of nonlinear systems is presented in section 2.3. The Equilibria of nonlinear systems is reviewed in section 2.4, while concepts of vector spaces is presented in section 2.5. Stability theory is presented in section 2.6, which constitutes the bulk of the chapter. In Section 2.7, Lyapunov stability results are presented while input-output stability concepts are presented in section 2.8. Advanced stability concepts are compiled to make later developments more concise in section 2.9. In section 2.10 we review some useful theorems and lemmas. In section 2.11 we the basic linear controller designs from a state-space point of view, and the chapter is concluded in Section 2.12.

2.2 Linear State-Variable Systems

Many physical systems such as the robots considered in this book are described by *differential or difference equations*. These describing equations, which are usually obtained from fundamental physical laws, provide the starting point for the analysis and control of systems. There are, of course, some systems which are so complicated that describing differential (or difference) equations are not available. We do not consider those systems in this book.

In this section we study the state-space model of physical systems that are linear. We limit ourselves to systems described by ordinary differential equations which will lead to a finite-dimensional state space. Partial differential equations, leading to infinite-dimensional systems, are needed to study flexible robotic manipulators, but those are not considered in this textbook. We stress that the material of this chapter is intended as a quick introduction to these topics and will not be comprehensive. The readers are referred to [Kailath 1980], [Antsaklis and Michel 1997] for more rigorous presentations of linear control systems.

Continuous-Time Systems

A continuous-time system is said to be *linear* if it obeys the principle of *superposition*; that is, if the output $y_1(t)$ results from the input $u_1(t)$

2.2 Linear State-Variable Systems

and the output $y_2(t)$ results from the input $u_2(t)$, then the output resulting from $u(t) = \alpha_1 u_1(t) + \alpha_2 u_2(t)$ is given by $y(t) = \alpha_1 y_1(t) + \alpha_2 y_2(t)$, where α_1 and α_2 are scalar constants. Linear, single-input/single-output (SISO), continuous-time, time-invariant systems are described by linear, scalar, constant-coefficient ordinary differential equations such as

$$\frac{d^n y(t)}{dt^n} + a_{n-1} \frac{d^{n-1} y(t)}{dt^{n-1}} + \cdots + a_1 \frac{dy(t)}{dt} + a_0 y(t)$$
$$= b_n \frac{d^n u(t)}{dt^n} + b_{n-1} \frac{d^{n-1} u(t)}{dt^{n-1}} + \cdots + b_1 \frac{du(t)}{dt} + b_0 u(t) \quad (2.2.1)$$

where $a_i, b_i, i = 0, \cdots, n$ are scalar constants, $y(t)$ is a scalar output and $u(t)$ is a scalar input. Moreover, we are given for some time t_0 the *initial conditions*,

$$y(t_0), \frac{dy}{dt}(t_0), \cdots, \frac{d^{n-1}y}{dt^{n-1}}(t_0), u(t_0), \frac{du}{dt}(t_0), \cdots, \frac{d^{n-1}u}{dt^{n-1}}(t_0).$$

Note that the input $u(t)$ is differentiated at most as many times as the output $y(t)$. Otherwise, the system is said to be *non-dynamic*.

State-Space Realization

The *state* of the system is defined as a sufficient set of variables, which when specified at time t_0 along with the input $u(t), t \geq t_0$, is sufficient to completely determine the behavior of the system for all $t \geq t_0$ [Kailath 1980]. The *state vector* then contains all necessary variables needed to determine the future behavior of any signal in the system. By definition, such a state vector $x(t)$ is not unique, a feature that will be exploited later. In fact, if x is a state vector then so is any $\bar{x}(t) = Tx(t)$, where T is any $n \times n$ invertible matrix. For the continuous-time system described in (2.2.1), the following choice of a state vector is possible:

$$\dot{x}_1(t) = x_2(t)$$
$$\dot{x}_2(t) = x_3(t)$$
$$\vdots = \vdots$$
$$\dot{x}_{n-1}(t) = x_n(t)$$
$$\dot{x}_n(t) = -a_0 x_1(t) - a_1 x_2(t) - \cdots - a_{n-1} x_n(t) + u(t) \quad (2.2.2)$$

where $\dot{x}_i = \frac{dx_i}{dt}, i = 1, 2, \cdots, n$. The input-output equation then reduces to

$$y(t) = b_0 x_1(t) + b_1 x_2(t) + \cdots + b_{n-1} x_n(t) + b_n u(t). \quad (2.2.3)$$

A more compact formulation of 2.2.2 and 2.2.3 is given by

$$\dot{x}(t) = Ax(t) + bu(t)$$
$$y(t) = cx(t) + du(t) \qquad (2.2.4)$$

where

$$A = \begin{bmatrix} 0 & 1 & 0 & \cdots & 0 & 0 \\ 0 & 0 & 1 & \cdots & 0 & 0 \\ 0 & 0 & 0 & \cdots & 0 & 0 \\ \vdots & \vdots & \vdots & \cdots & \vdots & \vdots \\ 0 & 0 & 0 & \cdots & 1 & 0 \\ 0 & 0 & 0 & \cdots & 0 & 1 \\ -a_0 & -a_1 & -a_2 & \cdots & -a_{n-2} & -a_{n-1} \end{bmatrix},$$

$$b = \begin{bmatrix} 0 \\ 0 \\ 0 \\ \vdots \\ 0 \\ 0 \\ 1 \end{bmatrix},$$

$$c = [b_0 \; b_1 \; b_2 \cdots b_{n-2} \; b_{n-1}], \quad d = b_n. \qquad (2.2.5)$$

This particular state-space representation is known as the *controllable canonical form* [Kailath 1980], [Antsaklis and Michel 1997]. In general, a linear, time-invariant, continuous-time system will have more than one input and one output. In fact, $u(t)$ is an $m \times 1$ vector and $y(t)$ is a $p \times 1$ vector. The differential equations relating $u(t)$ to $y(t)$ will not be presented here, but the state-space representation of the multi-input/multi-output (MIMO) system becomes

$$\dot{x}(t) = Ax(t) + Bu(t)$$
$$y(t) = Cx(t) + Du(t) \qquad (2.2.6)$$

where A is $n \times n$, B is $n \times m$, C is $p \times n$, and D is $p \times m$. For the specific forms of $A, B, C,$ and D, the reader is again referred to [Kailath 1980], [Antsaklis and Michel 1997]. A block diagram of (2.2.6) is shown in Figure 2.2.1a. Note that the minimal number of states is equal to the required number of initial conditions in order to find a unique solution to the set of differential equations.

2.2 Linear State-Variable Systems

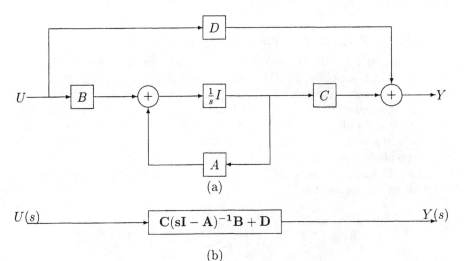

(b)

Figure 2.2.1: (a) State-space block diagram of (2.2.6); (b) Transfer function block diagram of (2.2.6)

EXAMPLE 2.2-1: Double Integrator

Consider a SISO system described by

$$\ddot{y}(t) = u(t)$$

This system is known as the double integrator and represents a wide variety of physical systems described by *Newton's Law*. In order to obtain a state-space description, let

$$\begin{aligned} x_1 &= y \\ x_2 &= \dot{x}_1 = \dot{y} \end{aligned}$$

so

$$\begin{aligned} \dot{x} &= \begin{bmatrix} 0 & 0 \\ 0 & 1 \end{bmatrix} x + \begin{bmatrix} 0 \\ 1 \end{bmatrix} u \\ y &= \begin{bmatrix} 1 & 0 \end{bmatrix} x \end{aligned}$$

∎

EXAMPLE 2.2-2: Two-Platform System

Consider the MIMO mechanical system shown in Figure 2.2.2 which represents a 2 platform system used to isolate experiments from external disturbances. There are 2 inputs to the system given by u_2 which causes the ground to move and u_1 which causes the platform m_1 to move. The system also has 2 outputs, namely the motion y_1 of platform m_1 and the motion y_2 of platform m_2. The experiments will be conducted on top of platform m_1 and therefore, one would like to minimize the size of y_1. The differential equations describing this system are obtained using Newton's second law:

$$\frac{d^2 y_1}{dt^2} = \frac{-k_1}{m_1} y_1 + \frac{k_1}{m_1} y_2 - \frac{c_1}{m_1} \dot{y}_1 + \frac{c_1}{m_1} \dot{y}_2 + \frac{1}{m_1} u_1$$

$$\frac{d^2 y_2}{dt^2} = \frac{-k_1}{m_2} y_1 + \frac{k_1 - k_2}{m_2} y_2 + \frac{c_1}{m_2} \dot{y}_1 - \frac{c_1 + c_2}{m_2} \dot{y}_2 + \frac{k_2}{m_2} y_3 + \frac{c_2}{m_2} \dot{y}_3$$

$$\frac{d^2 y_3}{dt^2} = u_2.$$

A state-space formulation of this system can be obtained by choosing

$$x_1 = y_1 \ ; \ x_2 = \dot{y}_1 \ ; \ x_3 = y_2 \ ;$$
$$x_4 = \dot{y}_2 \ ; \ x_5 = y_3 \ ; \ x_6 = \dot{y}_3$$

$$\dot{x} = \begin{bmatrix} 0 & 1 & 0 & 0 & 0 & 0 \\ \frac{-k_1}{m_1} & \frac{c_1}{m_1} & \frac{k_1}{m_1} & \frac{c_1}{m_1} & 0 & 0 \\ 0 & 0 & 0 & 1 & 0 & 0 \\ \frac{-k_1}{m_2} & \frac{c_1}{m_2} & \frac{k_1-k_2}{m_2} & \frac{-(c_1+c_2)}{m_2} & \frac{k_2}{m_2} & \frac{c_2}{m_2} \\ 0 & 0 & 0 & 0 & 0 & 1 \\ 0 & 0 & 0 & 0 & 0 & 0 \end{bmatrix} x + \begin{bmatrix} 0 & 0 \\ \frac{1}{m_1} & 0 \\ 0 & 0 \\ 0 & 0 \\ 0 & 0 \\ 0 & 1 \end{bmatrix} u$$

■

Transfer Functions

Another equivalent representation of linear, time-invariant, continuous-time systems is given by their *transfer function*, which relates the input of the system $u(t)$ to its output $y(t)$ in the Laplace variable s or in the frequency domain. It is important to note that the transfer function description has

2.2 Linear State-Variable Systems

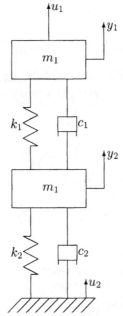

Figure 2.2.2: Two-platform system

no information about the initial conditions of the system and, as such, will not provide a unique output to a particular input unless all initial conditions are zero [Antsaklis and Michel 1997], [Kailath 1980]. The transfer function formalism, however, is important in practice, since many engineers are familiar with frequency-domain specifications. In addition, the identification of many systems may be effectively performed in the frequency domain [Ljung 1999]. It is therefore imperative that one should be able to move between the state-space (or modern) description and the transfer function (or classical) description.

Let us consider the system described by (2.2.6) and take its Laplace transform,

$$sX(s) - x(0) = AX(s) + BU(s)$$
$$Y(s) = CX(s) + DU(s) \qquad (2.2.7)$$

where $X(s), U(s)$, and $Y(s)$ are the Laplace transforms of $x(t), u(t)$, and $y(t)$ respectively. The initial state vector is $x(0)$. By eliminating $X(s)$ between the two equations in (2.2.7), we find the following relation:

$$Y(s) = \left[C(sI - A)^{-1}B + D\right]U(s) + C(sI - A)^{-1}x(0) \qquad (2.2.8)$$

As mentioned previously, the transfer function is obtained as the relationship between the input $U(s)$ and the output $Y(s)$ when $x(0) = 0$, that is,

$$Y(s) = \left[C(sI - A)^{-1}B + D\right]U(s). \qquad (2.2.9)$$

The transfer function of this particular linear, time-invariant system is given by

$$P(s) = C(sI - A)^{-1}B + D \qquad (2.2.10)$$

such that (see Fig.2.2.1)

$$Y(s) = P(s)U(s) \qquad (2.2.11)$$

EXAMPLE 2.2-3: Transfer Function of Double Integrator

Consider the system of Example 2.2.1. It is easy to see that the transfer function $\frac{Y(s)}{U(s)}$ is

$$\frac{Y(s)}{U(s)} = \frac{1}{s^2}.$$

∎

Discrete-Time Systems

In the discrete-time case, a *difference equation* is used to described the system as follows:

$$\begin{aligned}&y(k+n) + a_{n-1}y(k+n-1) + \cdots + a_1 y(k+1) + a_0 y(k) \\ &= b_n u(k+n) + b_{n-1}u(k+n-1) + \cdots + b_1 u(k+1) + b_0 u(k),\end{aligned} \qquad (2.2.12)$$

where $a_i, b_i, i = 0, \cdots, n$ are scalar constants, $y(k)$ is the output, and $u(k)$ is the input at time k. Note that the output at time $k + n$ depends on the input at time $k + n$ but not on later inputs; otherwise, the system would be *non-causal*.

2.2 Linear State-Variable Systems

State-Space Representation

In a similar fashion to the continuous-time case, the following state-vector is defined:

$$\begin{aligned}
x_1(k+1) &= x_2(k) \\
x_2(k+1) &= x_3(k) \\
&\vdots \\
x_{n-1}(k+1) &= x_n(k) \\
x_n(k+1) &= -a_0 x_1(k) - a_1 x_2(k) - \cdots - a_{n-1} x_n(k) + u(k).
\end{aligned} \tag{2.2.13}$$

The input-output equation then reduces to

$$y(k) = b_0 x_1(k) + b_1 x_2(k) + \cdots + b_{n-1} x_n(k) + u(k) \tag{2.2.14}$$

A more compact formulation of (2.2.7) and (2.2.8) is given by

$$\begin{aligned}
x(k+1) &= Ax(k) + bu(k) \\
y(k) &= cx(k) + du(k)
\end{aligned} \tag{2.2.15}$$

where

$$A = \begin{bmatrix}
0 & 1 & 0 & \cdots & 0 & 0 \\
0 & 0 & 1 & \cdots & 0 & 0 \\
0 & 0 & 0 & \cdots & 0 & 0 \\
\vdots & \vdots & \vdots & \cdots & \vdots & \vdots \\
0 & 0 & 0 & \cdots & 1 & 0 \\
0 & 0 & 0 & \cdots & 0 & 1 \\
-a_0 & -a_1 & -a_2 & \cdots & -a_{n-2} & -a_{n-1}
\end{bmatrix},$$

$$b = \begin{bmatrix} 0 \\ 0 \\ 0 \\ \vdots \\ 0 \\ 0 \\ 1 \end{bmatrix},$$

$$c = \begin{bmatrix} b_0 & b_1 & b_2 & \cdots & b_{n-2} & b_{n-1} \end{bmatrix}, \quad d = b_n. \tag{2.2.16}$$

The MIMO case is similar to the continuous-time case and is given by

$$\begin{aligned} x(k+1) &= Ax(k) + Bu(k) \\ y(k) &= Cx(k) + Du(k) \end{aligned} \quad (2.2.17)$$

where A is $n \times n$, B is $n \times m$, C is $p \times n$, and D is $p \times m$.

In many practical cases, such as in the control of robots, the system is a continuous-time system, but the controller is implemented using digital hardware. This will require the designer to translate between continuous- and discrete-time systems. There are many different approaches to "discretizing" a continuous-time system, some of which are discussed in Chapter 3. The interested reader in this very important aspect of the control problem is referred to [Åström and Wittenmark 1996], [Franklin et al. 1997].

EXAMPLE 2.2-4: Double Integrator in Discrete Time

Recall Example 2.2.1 which presented a model of the double integrator or Newton's system. One discrete-time version of the differential equation is given by the following difference equation

$$y(k+2) = Ty(k+1) + y(k) + (\frac{T^2}{2} + T)u(k)$$

where T is the sampling period in seconds. If we choose $x_1(k) = y(k)$ and $x_2(k) = x_1(k+1)$, we obtain the state-space description

$$\begin{aligned} x(k+1) &= \begin{bmatrix} 0 & 1 \\ 1 & T \end{bmatrix} x(k) + \begin{bmatrix} \frac{T^2}{2} \\ T \end{bmatrix} u(k), \\ y(k) &= \begin{bmatrix} 1 & 0 \end{bmatrix} x(k). \end{aligned}$$

∎

Transfer Function Representation

In a similar fashion to the continuous-time case, a linear, time-invariant, discrete-time system given by (2.2.17) may be described in the Z-transform domain, from input $U(z)$ to output $Y(z)$ by its transfer function $P(z)$ such that

$$Y(z) = P(z)U(z)$$

2.3 Nonlinear State-Variable Systems

where

$$P(z) = C(zI - A)^{-1}B + D$$

Note that the Z transform is used in the discrete-time case versus the Laplace transform in the continuous-time case.

EXAMPLE 2.2-5: Tranfer Function of Discrete-Tiem Double Integrator

The transfer function of the Example 2.2.4 is given by

$$\frac{Y(z)}{U(z)} = \frac{T^2}{2}\frac{z+1}{(z-1)^2}.$$

∎

2.3 Nonlinear State-Variable Systems

In many cases, the underlying physical behavior may not be described using linear state-variable equations. This is the case of robotic manipulators where the interaction between the different links is described by nonlinear differential equations, as shown in Chapter 3. The state-variable formulation is still capable of handling these systems, while the transfer function and frequency-domain methods fail. In this section we deal with the nonlinear variant of the preceding section and stress the classical approach to nonlinear systems as studied in [Khalil 2001], [Vidyasagar 1992] and in [Verhulst 1997], [LaSale and Lefschetz 1961], [Hahn 1967].

Continuous-Time Systems

A nonlinear, scalar, continuous-time, time-invariant system is described by a nonlinear, scalar, constant-coefficient differential equation such as

$$\frac{d^n y(t)}{dt^n} = h[y(t), y^{(1)}(t), \cdots, y^{(n-1)}(t), u(t), u^{(1)}(t), \cdots, u^{(n)}(t)] \quad (2.3.1)$$

where $y(t)$ is the output and $u(t)$ is the input to the system under consideration. As with the linear case, we define the state vector x by its components as follows:

$$\begin{aligned}
\dot{x}_1 &= x_2 \\
\dot{x}_2 &= x_3 \\
&\vdots \\
\dot{x}_{n-1} &= x_n \\
\dot{x}_n &= h[x_1(t), x_2(t), \cdots, x_n(t), u(t), u^{(1)}(t), \cdots, u^{(n)}(t)]
\end{aligned} \quad (2.3.2)$$

The output equation then reduces to:

$$y(t) = x_1(t) \qquad (2.3.3)$$

A more compact formulation of (2.3.2) and (2.3.3) is given by

$$\begin{aligned}
\dot{x}(t) &= f[x(t), U(t)] \\
y(t) &= cx(t)
\end{aligned} \qquad (2.3.4)$$

where

$$U(t) = [u(t) \; u^{(1)}(t) \; \cdots \; u^{(n-1)}(t)]^T$$

and

$$c = [1 \; 0 \; 0 \; \cdots \; 0]. \qquad (2.3.5)$$

EXAMPLE 2.3-1: Nonlinear Systems

We present 2 examples illustrating such concepts:

1. Consider the damped pendulum equation

$$\ddot{y} + k\dot{y} + \sin y = 0$$

A state-space description is obtained by choosing $x_1 = y$, $x_2 = \dot{y}$, leading to

$$\begin{bmatrix} \dot{x}_1 \\ \dot{x}_2 \end{bmatrix} = \begin{bmatrix} x_2 \\ \sin x_1 - x_2 \end{bmatrix}$$

The time history of $y(t)$ is shown in Figure 2.3.1.

2.3 Nonlinear State-Variable Systems

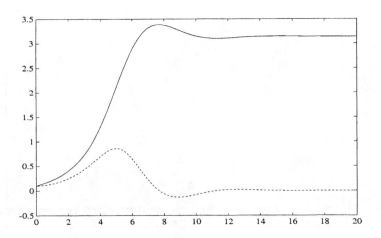

Figure 2.3.1: Damped pendulum trajectory.

2. A classical nonlinear system is the Van der Pol oscillator which is described by

$$\ddot{y} + (y^2 - 1)\dot{y} + y = 0$$

or in state-space with $x_1 = y$ and $x_2 = \dot{y}$,

$$\begin{bmatrix} \dot{x}_1 \\ \dot{x}_2 \end{bmatrix} = \begin{bmatrix} x_2 \\ (1 - x_1^2)x_2 - x_1 \end{bmatrix}$$

The time history of $y(t)$ and the phase-plane plot (i.e. x_2 versus x_1) is shown in Figure 2.3.2. ∎

EXAMPLE 2.3-2: Rigid Robot Dynamics

A rigid robot is described by the following equations

$$M(q)\ddot{q} + V(q, \dot{q}) + G(q) = \tau$$

where $M(q)$ is an $n \times n$ Inertia matrix, q and its derivatives are $n \times 1$ vectors of generalized coordinates, $V(q, \dot{q})$, $G(q)$ and τ are $n \times 1$ vectors containing velocity-dependent torques, gravity torques, and input

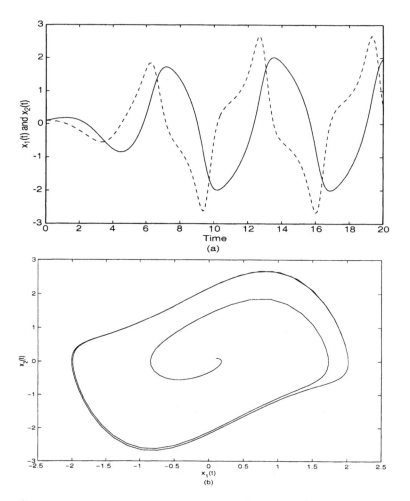

Figure 2.3.2: Van der Pol Oscillator Time Trajectories: (a) Time history, (b) phase plane.

torques respectively. In this example, we will concentrate on writing the n coupled differential equations into a state-space form. In fact, let a state vector x be

$$x = \begin{bmatrix} q \\ \dot{q} \end{bmatrix},$$

and the input vector be $u = \tau$ and suppose the output vector is $y = q$. Due to some special properties of rigid robots (see Chapter 2), the

2.3 Nonlinear State-Variable Systems

matrix $M(q)$ is known to be invertible so that

$$\begin{bmatrix} \dot{q} \\ \ddot{q} \end{bmatrix} = \begin{bmatrix} \dot{q} \\ -M^{-1}(V+G) \end{bmatrix} + \begin{bmatrix} 0 \\ M^{-1} \end{bmatrix} \tau$$

$$y = \begin{bmatrix} I & 0 \end{bmatrix} \begin{bmatrix} q \\ \dot{q} \end{bmatrix}$$

or

$$\begin{aligned} \dot{x} &= F(x) + G(x)u = f(x,u) \\ y &= [I\ 0]x \end{aligned} \qquad (1)$$

where

$$F(x) = \begin{bmatrix} \dot{q} \\ -M^{-1}(V+G) \end{bmatrix}, G(x) = \begin{bmatrix} 0 \\ M^{-1} \end{bmatrix}$$

∎

Discrete-Time Systems

A nonlinear, scalar, discrete-time, time-invariant system is described by a nonlinear, scalar, constant-coefficient difference equation such as,

$$\begin{aligned} y(k+n) = &\ h[y(k+n-1), \cdots, y(y+1), y(k), u(k), u(k-1), \cdots, \\ & u(k+n)], \end{aligned} \qquad (2.3.6)$$

where $y(.)$ and u are as defined before. A simple choice of state variables will lead to

$$\begin{aligned} x_1(k+1) &= x_2(k) \\ x_2(k+1) &= x_3(k) \\ &\vdots \\ x_{n-1}(k+1) &= x_n(k) \\ x_n(k+1) &= h[x_n(k), \cdots, x_2(k), x_1(k), u(k), u(k-1), \cdots, \\ & u(k+n)], \end{aligned} \qquad (2.3.7)$$

or, more compactly, as

$$x(k+1) = f[x(k), U(k)]$$
$$y(k) = cx(k) \qquad (2.3.8)$$

where $U(k)$ and c are defined similarly to those given in equation (2.3.4).

EXAMPLE 2.3-3: Logistics Equation

Consider the scalar system

$$y(k+1) = \lambda y(k)[1 - y(k)]$$

which leads to the state-space representation

$$x(k+1) = \lambda x(k)[1 - x(k)]$$
$$y(k) = = x(k)$$

∎

We will not emphasize the study of discrete nonlinear systems since robots are described by differential equations. However, as discussed in Chapter 4, robot controllers are usually implemented using digital controllers. It will therefore be advantageous to be able to translate between continuous- and discrete-time description of nonlinear dynamical systems as discussed in [åAström and Wittenmark 1995], [Franklin et al. 1997].

2.4 Nonlinear Systems and Equilibrium Points

In this section, we concentrate on systems described by (2.2.15) with the additional requirement that $u(t)$ is specified as a function of the state $x(t)$, i.e.

$$\dot{x}(t) = f[t, x(t)] \qquad (2.4.1)$$

This will allow us to concentrate on the *Analysis* problem. We require a few definitions which we shall now introduce.

2.4 Nonlinear Systems and Equilibrium Points

DEFINITION 2.4-1 *The system (2.4.1) is said to be* autonomous *if $f[t, x(t)]$ is not explicitly dependent on time, i.e.*

$$\dot{x}(t) = f[x(t)] \qquad (1)$$

■

EXAMPLE 2.4-1: Nonautonomous System

Both systems introduced in Example 2.3.1 are autonomous while the system described by

$$\dot{x} = tx^2$$

is not.

■

DEFINITION 2.4-2 *A vector $x_e \in \Re^n$ is a* fixed *or* equilibrium *point of (2.4.1) at time t_0 if*

$$f[t, x_e] = 0, \forall t \geq t_0 \qquad (1)$$

■

EXAMPLE 2.4-2: Equilibrium Point of Autonomous System

The system described by

$$\begin{bmatrix} \dot{x}_1 \\ \dot{x}_2 \end{bmatrix} = \begin{bmatrix} x_2 \\ \cos[x_1] - 1 \end{bmatrix}$$

is autonomous and it has an equilibrium point at the origin of $\Re^2$.

■

Note that:

- If a system is autonomous, then an equilibrium point at time t_0 is also an equilibrium point at all other times.

- If x_e is an equilibrium point at time t_0 of the non-autonomous system (2.4.1), then x_e is an equilibrium point of (2.4.1) for all $t_1 \geq t_0$.

EXAMPLE 2.4-3: Equilibrium Point of Nonautonomous System

Consider the system

$$\dot{x} = tx^2 - 1$$

which is non-autonomous. It has NO EQUILIBRIUM POINTS. Although it might seem that it has 2 equilibrium points $x_{e1} = -1$ and $x_{e2} = 1$ at time $t_0 = 1$. However, these are not equilibrium points for times since at times $t \geq 1$ the conditions of equilibrium does not hold. ∎

Some books also used the terms *stationary* or *singular* points to denote equilibrium points.

EXAMPLE 2.4-4: Damped Pendulum

Recall the pendulum in Example 2.3.1 and let us try to find its equilibrium points. Note first that the system is autonomous so that we do not need to specify the particular time t_0 and then note that the pendulum is at equilibrium if both

$$x_2 = 0 \text{ and } \sin(x_1) = 0$$

Therefore the equilibrium points are at

$$\begin{bmatrix} x_{1e} \\ x_{2e} \end{bmatrix} = \begin{bmatrix} n\pi \\ 0 \end{bmatrix}, n = 0, \pm 1, \pm 2, \cdots$$

It is obvious that the pendulum is at equilibrium when it is hanging straight up or straight down with zero velocity. ∎

DEFINITION 2.4-3 *An equilibrium point x_e at t_0 of (2.4.1) is said to be* isolated *if there is a neighborhood N of x_e which contains no other equilibrium points besides x_e.* ∎

EXAMPLE 2.4-5: Equilibrium Points of Pendulum

The equilibrium points of the pendulum are isolated. On the other hand, a system described by $\dot{x} = 0$ has for equilibrium points any point in R and therefore, none of its equilibrium points is isolated. ∎

2.5 Vector Spaces, Norms, and Inner Products

In this section, we will discuss some properties of nonlinear differential equations and their solutions. We will need many concepts such as vector spaces and *norms* which we will introduce briefly. The reader is referred to [Boyd and Barratt], [Desoer and Vidyasagar 1975], [Khalil 2001] for proofs and details.

Linear Vector Spaces

In most of our applications, we need to deal with (linear) real and complex vector spaces which are defined subsequently.

DEFINITION 2.5-1 *A real linear vector space (resp. complex linear vector space is a set V, equipped with 2 binary operations: the addition (+) and the scalar multiplication (.) such that*

1. $x + y = y + x$, $\forall x, y \in V$

2. $x + (y + z) = (x + y) + z$, $\forall x, y, z \in V$

3. *There is an element 0_V in V such that $x + 0_V = 0_V + x = x$, $\forall x \in V$*

4. *For each $x \in V$, there exists an element $-x \in V$ such that $x + (-x) = (-x) + x = 0_V$*

5. *For all scalars $r_1, r_2 \in \Re$ (resp. $c_1, c_2 \in \mathbb{C}$), and each $x \in V$, we have $r_1.(r_2.x) = (r_1 r_2).x$ (resp. $c_1.(c_2.x) = (c_1 c_2).x$)*

6. *For each $r \in \Re$ (resp. $c \in \mathbb{C}$), and each $x_1, x_2 \in V$, $r.(x_1 + x_2) = r.x_1 + r.x_2$ (resp. $c.(x_1 + x_2) = c.x_1 + c.x_2$)*

7. *For all scalars $r_1, r_2 \in \Re$ (resp. $c_1, c_2 \in \mathbb{C}$), and each $x \in V$, we have $(r_1 + r_2).x = r_1.x + r_2.x$ (resp. $(c_1 + c_2).x = c_1.x + c_2.x$)*

8. *For each $x \in V$, we have $1.x = x$ where 1 is the unity in $\Re$ (resp. in $\mathbb{C}$).*

∎

EXAMPLE 2.5-1: Vector Spaces

The following are linear vector spaces with the associated scalar fields: $\Re^n$ with $\Re$, and $\mathbb{C}^n$ with $\mathbb{C}$.

∎

DEFINITION 2.5-2 *A subset M of a vector space V is a subspace if it is a linear vector space in its own right. One necessary condition for M to be a subspace is that it contains the zero vector.* ∎

We can equip a vector space with many functions. One of which is the inner product which takes two vectors in V to a scalar either in $\Re$ or in $\mathbb{C}$, the other one is the norm of a vector which takes a vector in V to a positive value in $\Re$. The following section discusses the norms of vectors which is then followed by a section on inner products.

Norms of Signals and Systems

A *norm* is a generalization of the ideas of distance and length. As stability theory is usually concerned with the size of some vectors and matrices, we give here a brief description of some norms that will be used in this book. We will consider first the norms of vectors defined on a vector space X with the associated scalar field of real numbers $\Re$, then introduce the matrix induced norms, the function norms and finally the system-induced norms or operator gains.

Vector Norms

We start our discussion of norms by reviewing the most familiar normed spaces, that is the spaces of vectors with constant entries. In the following, $\| a \|$ denotes the absolute value of a for a real a or the magnitude of a if a is complex.

DEFINITION 2.5-3 *A norm $\| \cdot \|$ of a vector x is a real-valued function defined on the vector space X such that*

1. $\| x \| > 0$ *for all $x \in X$ with $\| x \| = 0$ if and only if $x = 0$.*

2. $\| \alpha x \| = | \alpha | \| x \|$ *for all $x \in X$ and any scalar α.*

2.5 Vector Spaces, Norms, and Inner Products

3. $\| x+y \| \leq \| x \| + \| y \|$ for all $x,y \in X$.

■

EXAMPLE 2.5-2: Vector Norms (1)

The following are common norms in $X = \Re^n$ where $\Re^n$ is the set of $n \times 1$ vectors with real components.

1. 1-norm: $\| x \|_1 = \sum_{i=1}^{n} | x_i |$.
2. 2-norm: $\| x \|_2 = \sqrt{\sum_{i=1}^{n} | x_i^2 |}$, also known as the Euclidean norm
3. p-norm: $\| x \|_p = (\sum_{i=1}^{n} | x_i^p |)^{\frac{1}{p}}$.
4. ∞ - norm : $\| x \|_\infty = \max | x_i | \ \forall i = 1, \cdots, n$.

■

EXAMPLE 2.5-3: Vector Norms (2)

Consider the vector

$$x = \begin{bmatrix} 1 \\ -2 \\ 2 \end{bmatrix}$$

Then, $\| x \|_1 = 5$, $\| x \|_2 = 2$ and $\| x \|_\infty = 2$.

■

We now present an important property of norms of vectors in $\Re^n$ which will be useful in the sequel.

LEMMA 2.5-1: Let $\| x \|_a$ and $\| x \|_b$ be any two norms of a vector $x \in \Re^n$. Then there exists finite positive constants k_1 and k_2 such that

$$k_1 \| x \|_a \leq \| x \|_b \leq k_2 \| x \|_a \ \forall x \in \Re^n$$

■

The two norms in the lemma are said to be equivalent and this particular property will hold for any two norms on $\Re^n$.

EXAMPLE 2.5-4: Equivalent Vector Norms

1. It can be shown that for $x \in \Re^n$

$$||x||_1 \leq \sqrt{n}||x||_2$$

$$||x||_\infty \leq ||x||_1 \leq n||x||_\infty$$

$$||x||_2 \leq \sqrt{n}||x||_\infty$$

2. Consider again the vector of Example 2.5.3. Then we can check that

$$||x||_1 \leq \sqrt{3}||x||_2$$

$$||x||_\infty \leq ||x||_1 \leq 3||x||_\infty$$

$$||x||_2 \leq \sqrt{3}||x||_\infty$$

∎

Matrix Norms

In systems applications, a particular vector x may be operated on by a matrix A to obtain another vector $y = Ax$. In order to relate the sizes of x and Ax we define the *induced matrix norm* as follows.

DEFINITION 2.5-4 *Let $|| x ||$ be a given norm of $x \in \Re^n$. Then each $m \times n$ matrix A, has an* induced norm *defined by*

$$||A||_i = \sup_{||x||=1} ||Ax||$$

where sup stands for the supremum. ∎

It is always imperative to check that the proposed norms verify the conditions of Definition 2.5.3. The newly defined matrix norm may also be shown to satisfy

$$||AB||_i \leq ||A||_i \, ||B||_i$$

for all $n \times m$ matrices A and all $m \times p$ matrices B.

2.5 Vector Spaces, Norms, and Inner Products

EXAMPLE 2.5-5: Induced Matrix Norms

Consider the ∞ induced matrix norm, the 1 induced matrix norm and the 2 induced matrix norm,

$$||A||_{i\infty} = \max_i \sum_j |a_{ij}|$$

$$||A||_{i1} = \max_j \sum_i |a_{ij}|$$

$$||A||_{i2} = \sqrt{\lambda_{max}(A^T A)}$$

where λ_{max} is the maximum eigenvalue. As an illustration, consider the matrix

$$A = \begin{bmatrix} 1 & -1 & 2 \\ 2 & 3 & -2 \\ -1 & 0 & 1 \end{bmatrix}$$

Then, $||A||_{i1} = max(4,4,5) = 5$, $||A||_{i2} = 4.4576$, and $||A||_{i\infty} = max(4,7,2) = 7$.

∎

Function Norms

Next, we review the norms of time-dependent functions and vectors of functions. These constitute an important class of signals which will be encountered in controlling robots.

DEFINITION 2.5-5 *Let $f(.) : [0, \infty) \longrightarrow R$ be a uniformly continuous function. A function f is uniformly continuous if for any $\epsilon > 0$, there is a $\delta(\epsilon)$ such that*

$$|t - t_0| < \delta(\epsilon) \implies |f(t) - f(t_0)| < \epsilon$$

Then, f is said to belong to L_p if for $p \in [1, \infty)$,

$$\int_0^\infty |f(t)|^p dt < \infty$$

f is said to belong to L_∞ if it is bounded i.e. if $\sup_{t \in [0,\infty)} | f(t) | \leq B < \infty$, where sup $f(t)$ denotes the supremum of $f9t)$ i.e. the smallest number that

is larger than or equal to the maximum value of $f(t)$. L_1 denotes the set of signals with finite absolute area, while L_2 denotes the set of signals with finite total energy. ∎

The following definition of the norms of vector functions is not unique. A discussion of these norms is found in [Boyd and Barratt].

DEFINITION 2.5-6 *Let L_p^n denote the set of $n \times 1$ vectors of functions f_i, each of which belonging to L_p. The norm of $f \in L_p^n$ is*

$$\|f(.)\|_p = [\int_0^\infty \sum_{i=1}^n |f_i(t)|^p dt]^{1/p}$$

for $p \in [1, \infty)$ and

$$\|f(.)\|_\infty = \max_{1 \leq i \leq n} \|f_i(t)\|_\infty$$

∎

Some common norms of scalar signals $u(t)$ that are persistent (i.e. $\lim_{t \to \infty} u(t) \neq 0$) are the following:

1. $\|u\|_{rms} = \left[\lim_{T \to \infty} \frac{1}{T} \int_0^T u^2(t) dt\right]^{1/2}$ which is valid for signals with finite steady-state power.

2. $\|u\|_\infty = \sup_{t \geq 0} |u(t)|$ which is valid for bounded signals but is dependent on outliers.

3. $\|u\|_a = \lim_{T \to \infty} \frac{1}{T} \int_0^T |u(t)| dt$ which measures the steady-state average resource consumption.

For vector signals, we obtain:

1. $\|u\|_\infty = \max_{1 \leq i \leq n} \|u_i\|_\infty$

2. $\|u\|_{rms} = \left[\lim_{T \to \infty} \frac{1}{T} \int_0^T u^T(t) u(t) dt\right]^{1/2}$

3. $\|u\|_a = \lim_{T \to \infty} \frac{1}{T} \int_0^T \sum_{i=1}^n |u(t)| dt$

Note that $\|u\|_\infty \geq \frac{1}{\sqrt{n}} \|u\|_{rms} \geq \frac{1}{n} \|u\|_a$.

On the other hand, if signals do not persist, we may find their L_2 or L_1 norms as follows:

1. $\|u\|_1 = \int_0^\infty |u(t)| dt$ which measures the total resource consumption

2.5 Vector Spaces, Norms, and Inner Products

2. $||u||_2 = \left[\int_0^\infty u^2(t)dt\right]^{1/2} = \left[\frac{1}{2\pi}\int_{-\infty}^\infty |U(jw)|^2 dw\right]^{1/2}$ which measures the total energy.

EXAMPLE 2.5-6: Function Norms

1. The function $f(t) = e^{-t}$ belongs to L_1. In fact, $|| e^{-t} ||_1 = 1$. The function $f(t) = \frac{1}{t+1}$ belongs to L_2. The sinusoid $f(t) = 2\sin t$ belongs to L_∞ since its magnitude is bounded by 2 and $|| 2\sin t ||_\infty = 2$.

2. Suppose the vector function $x(t)$ has continuous and real-valued components, i.e.

$$x : [a, b] \longrightarrow \Re^n$$

where $[a, b]$ is a closed-interval on the real line R. We denote the set of such functions x by $\mathbb{C}^n[a, b]$. Then, let us define the real-valued function

$$||x(.)|| = \sup_{t \in [a,b]} ||x(t)||$$

where $|| x(t) ||$ is any previously defined norm of $x(t)$ for a fixed t. It can be verified that $|| x(.) ||$ is a norm on the set $\mathbb{C}^n[a, b]$ and may be used to compare the size of such functions [Desoer and Vidyasagar 1975]. In fact, it is very important to distinguish between $|| x(t) ||$ and $|| x(.) ||$. The first is the norm of a fixed vector for a particular time t while the second is the norm of a time-dependent vector. It is this second norm (which was introduced in definition 2.5.6) that we shall use when studying the stability of systems.

3. The vector $f(t) = [e^{-t} \ -e^{-t} \ -(1+t)^{-2}]^T$ is a member of L_1^3. On the other hand, $f(t) = [e^{-t} \ -e^{-t} \ -(1+t)^{-1}]^T$ is a member of L_2^3 and L_∞^3.

■

In some cases, we would like to deal with signals that are bounded for finite times but may become unbounded as time goes to infinity. This leads us to define the extended L_p spaces. Thus consider the function

$$f_T(t) = \begin{cases} f(t) & \text{if } t \leq T \\ 0 & \text{if } t > T \end{cases} \quad (2.5.1)$$

then, the extended L_p space is defined by

$$L_{pe} = \{f(t) : f_T(t) \in L_p\}$$

where $T < \infty$. We also define the norm on L_pe as

$$\|f(.)_T\|_p = \|f(.)\|_{Tp}$$

Similar definitions are available for L_p^n and the interested reader is referred to [Boyd and Barratt], [Desoer and Vidyasagar 1975].

EXAMPLE 2.5-7: Extended L_p Spaces

The function $f(t) = t$ belongs to L_{pe} for any $p \in [1, \infty]$ but not to L_p. ∎

System Norms

We would like next to study the effect of a multi-input-multi-output (MIMO) system on a multidimensional signal. In other words, what happens to a time-varying vector $u(t)$ as it passes through a MIMO system H? Let H be a system with m inputs and l outputs, so that its output to the input $u(t)$ is given by

$$y(t) = (Hu)(t)$$

We say that H is L_p stable if Hu belongs to L_p^l whenever u belongs to L_p^m and there exists finite constants $\gamma > 0$ and b such that

$$\|Hu\|_p \leq \gamma \|u\|_p + b$$

If $p = \infty$, the system is said to be bounded-input-bounded-output (BIBO) stable.

DEFINITION 2.5-7 *The L_p gain of the system H is denoted by $\gamma_p(H)$ and is the smallest γ such that a finite b exists to verify the equation.*

$$\|Hu\|_p \leq \gamma \|u\|_p + b$$

∎

Therefore, the gain γ_p characterizes the amplification of the input signal as it passes through the system. The following lemma characterizes the gains of linear systems and may be found in [Boyd and Barratt].

LEMMA 2.5-2: *Given the linear system H such that an input $u(t)$ results in an output $y(t) = (Hu)(t) = \int_0^t h(t-\tau)u(\tau)d\tau$ and suppose H is BIBO stable, then*

2.5 Vector Spaces, Norms, and Inner Products

1. $\gamma_p(H)$ is $\leq \| h \|_1 \quad \forall p \in [1, \infty]$

2. $\gamma_\infty(H) = \int_0^\infty | h(t) | \, dt$

3. $\gamma_2(H) = max_{w \in R} \| H(jw) \| \leq \gamma_\infty(H)$

∎

EXAMPLE 2.5-8: System Norms

1. Consider the system

$$H(s) = \frac{1}{s+2}$$

so that the impulse response is

$$h(t) = \begin{cases} e^{-2t} & \text{if } t > 0 \\ 0 & \text{if } t < 0 \end{cases} \tag{1}$$

Note that $H(s)$ is BIBO stable. Then

$$\gamma_\infty(H) = 0.5$$

$$\gamma_2(H) = 0.5$$

2. Consider the system

$$H(s) = \begin{bmatrix} \frac{1}{s^2+k_v s+k_p} \\ \frac{s}{s^2+k_v s+k_p} \end{bmatrix}$$

where k_v and k_p are positive constants. The system is therefore BIBO stable. Then

$$\gamma_\infty(H) = max\{1/k_p, \, 4/ek_v\}$$

$$\gamma_2(H) = \frac{\sqrt{1+k_p}}{k_v}$$

where e = 2.7183 is the base of natural logarithms.

This concludes our brief review of norms as they will be used in this book.

Inner Products

An inner product is an operation between two vectors of a vector space which will allow us to define geometric concepts such as orthogonality and Fourier series, etc. The following defines an inner product.

DEFINITION 2.5-8 *An inner product defined over a vector space V is a function $<.,.>$ defined from V to F where F is either $\Re$ or $\mathbb{C}$ such that $\forall x, y, z, \in V$*

1. *$<x,y> = <y,x>^*$ where the $<.,.>^*$ denotes the complex conjugate.*
2. *$<x, y+z> = <x,y> + <x,z>$*
3. *$<x, \alpha y> = \alpha <x,y>, \forall \alpha \in F$*
4. *$<x,x> \geq 0$ where the 0 occurs only for $x = 0_V$*

■

EXAMPLE 2.5-9: Inner Products

The usual dot product in $\Re^n$ is an inner product.

■

We can define a norm for any inner product by

$$\| x \| = \sqrt{<x,x>} \quad \forall x \in V \tag{2.5.2}$$

Therefore a norm is a more general concept: A vector space may have a norm associated with it but not an inner product. The reverse however is not true. Now, with the norm defined from the inner product, a complete vector space in this norm (i.e. one in which every Cauchy sequence converges) is known as a Hilbert Space

Matrix Properties

Some matrix properties play an important role in the study of the stability of dynamical systems. The properties needed in this book are collected in this section. We will assume that the readers are familiar with elementary

2.5 Vector Spaces, Norms, and Inner Products

DEFINITION 2.5-9 *All matrices in this definition are square and real.*

- Positive Definite: *A real $n \times n$ matrix A is positive definite if $x^T A x > 0$ for all $x \in \Re^n$, $x \neq 0$.*

- Positive Semidefinite: *A real $n \times n$ matrix A is positive semidefinite if $x^T A x \geq 0$ for all $x \in \Re^n$.*

- Negative Definite: *A real $n \times n$ matrix A is negative definite if $x^T A x < 0$ for all $x \in \Re^n$, $x \neq 0$.*

- Negative Semidefinite: *A real $n \times n$ matrix A is negative semidefinite if $x^T A x \leq 0$ for all $x \in \Re^n$.*

- Indefinite: *A is indefinite if $x^T A x > 0$ for some $x \in \Re^n$ and $x^T A x < 0$ for other $x \in \Re^n$.*

∎

Note that
$$x^T A x = x^T \frac{(A + A^T)}{2} x = x^T A_s x$$
where A_s is the symmetric part of A. Therefore, the test for the definiteness of a matrix may be done by considering only the symmetric part of A.

THEOREM 2.5-1: *Let $A = [a_{ij}]$ be a symmetric $n \times n$ real matrix. As a result, all eigenvalues of A are real. We then have the following*

- Positive Definite: *A real $n \times n$ matrix A is positive definite if all its eigenvalues are positive.*

- Positive semidefinite: *A real $n \times n$ matrix A is positive definite if all its eigenvalues are nonnegative.*

- Negative Definite: *A real $n \times n$ matrix A is negative definite if all its eigenvalues are negative.*

- Negative Semidefinite: *A real $n \times n$ matrix A is negative semidefinite if all its eigenvalues are non-positive.*

- Indefinite: *A real $n \times n$ matrix A is indefinite if some of its eigenvalues are positive and some are negative.*

∎

THEOREM 2.5-2: Rayleigh-Ritz *Let A be a real, symmetric $n \times n$ positive-definite matrix. Let λ_{min} be the minimum eigenvalue and λ_{max} be the maximum eigenvalue of A. Then, for any $x \in \Re^n$,*

$$\lambda_{min}[A]||x||^2 \leq x^T A x \leq \lambda_{max}[A]||x||^2$$

■

THEOREM 2.5-3: Gerschgorin *Let $A = [a_{ij}]$ be a symmetric $n \times n$ real matrix. Suppose that*

$$|a_{ii}| > \sum_{j=1}^{n} |a_{ij}| \; ; \text{ for all } i = 1,...,n \; ; \; j \,!= i$$

If all the diagonal elements are positive, i.e. $a_{ii} > 0$, then the matrix A is positive definite. ■

EXAMPLE 2.5-10: Positive Definite Matrics

Consider the matrix

$$A = \begin{bmatrix} 4 & -4 \\ -2 & 6 \end{bmatrix} \qquad (1)$$

Its symmetric part is given by

$$A_s = \begin{bmatrix} 4 & -3 \\ -3 & 6 \end{bmatrix} \qquad (2)$$

This matrix is positive-definite since its eigenvalues are both positive (1.8377, 8.1623). Of course, Gershgorin's theorem could have been used since the diagonal elements of A_s are all positive and

$$|a_{11}| = 4 \quad |a_{12}| = 3 \; ; \; |a_{22}| = 6 \quad |a_{21}| = 3$$

On the other hand, consider a vector $x = [x_1 \; x_2]^T$ and its 2-norm, then

$$0.3944(x_1^2 + x_2^2) \leq 4x_1^2 - 6x_1x_2 + 6x_2^2 \leq 7.6056(x_1^2 + x_2^2)$$

as a result of Rayleigh-Ritz theorem.

■

2.6 Stability Theory

The first stability concept we study, concerns the behavior of free systems, or equivalently, that of forced systems with a given input. In other words, we study the stability of an equilibrium point with respect to changes in the initial conditions of the system. Before doing so however, we review some basic definitions. These definitions will be stated in terms of continuous, nonlinear systems with the understanding that discrete, nonlinear systems admit similar results and linear systems are but a special case of nonlinear systems.

Let x_e be an equilibrium (or fixed) state of the free continuous-time, possibly time-varying nonlinear system

$$\dot{x}(t) = f(x,t), \qquad (2.6.1)$$

i.e. $f(x_e, t) = 0$, where x, f are $n \times 1$ vectors.

We will first review the stability of an equilibrium point x_e with the understanding that the stability of the state $x(t)$ can always be obtained with a translation of variables as discussed later. The stability definitions we use can be found in [Khalil 2001], [Vidyasagar 1992].

DEFINITION 2.6-1 *In all parts of this definition x_e is an equilibrium point at time t_0, and $\| \cdot \|$ denote any function norm previously defined.*

1. *Stability: x_e is stable in the sense of Lyapunov (SL) at t_0, if starting close enough to x_e at t_0, the state will always stay close to x_e at later times. More precisely, x_e is SL at t_0, if for any given $\epsilon > 0$, there exists a positive $\delta(\epsilon, t_0)$ such that if*

    $$\|x_0 - x_e\| < \delta(\epsilon, t_0)$$

    then

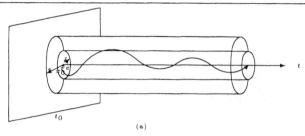

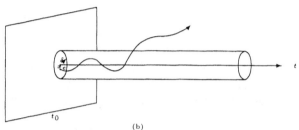

Figure 2.6.1: (a) Stability of x_e at t_0; (b) Instability of x_e at t_0.

$$||x(t) - x_e|| < \epsilon \, ; \quad \text{for all } t \geq t_0$$

x_e is stable in the sense of Lyapunov if it is stable for any given t_0. See Figure 2.6.1a.

2. **Instability:** x_e is unstable in the sense of Lyapunov (UL), if no matter how close to x_e the state starts, it will not be confined to the vicinity of x_e at some later time. In other words, x_e is unstable if it is not stable at t_0. See Figure 2.6.1b for an illustration.

3. **Convergence:** x_e is convergent (C) at t_0, if states starting close to x_e will eventually converge to x_e. In other words, x_e is convergent at t_0 if for any positive ϵ_1, there exists a positive $\delta_1(t_0)$ and a positive $T(\epsilon_1, x_0, t_0)$ such that if

$$||x_0 - x_e|| < \delta_1(t_0)$$

then

$$||x(t) - x_e|| < \epsilon_1 \, ; \quad \text{for all } t \geq t_0 + T(\epsilon_1, x_0, t_0)$$

2.6 Stability Theory

x_e is convergent, if it is convergent for any t_0. See Figure 2.6.2 for illustration.

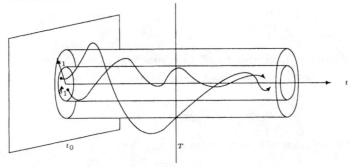

Figure 2.6.2: Convergence of x_e at t_0

4. **Asymptotic Stability:** x_e is asymptotically stable (AS) at t_0 if states starting sufficiently close to x_e will stay close and will eventually converge to it. More precisely, x_e is AS at t_0 if it is both convergent and stable at t_0. x_e is AS if it is AS for any t_0. An illustration of an AS equilibrium point is shown in Figure 2.6.3.

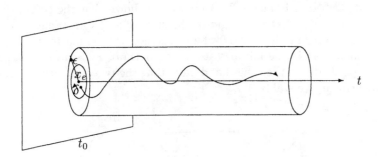

Figure 2.6.3: Asymptotic stability of x_e at t_0

5. **Global Asymptotic Stability:** x_e is globally asymptotically stable (GAS) at t_0 if any initial state will stay close to x_e and will eventually converge to it. In other words, x_e is GAS if it is stable at t_0, and if every $x(t)$ converges to x_e as time goes to infinity. x_e is GAS if it is GAS

for any t_0 and the system is said to be GAS in this case, since it can only have one equilibrium point x_e. See Figure 2.6.4.

∎

EXAMPLE 2.6-1: Stability of Various Systems

1. Consider the scalar time-varying system given by

$$\dot{y} = \frac{-y}{1+t}$$

the solution of this equation for all $t \geq t_0$ is

$$y(t) = x_0 \frac{1+t_0}{1+t}$$

The equilibrium point is located at $x_e = y_e = 0$. Let us use the 1-norm given by $|y|$ and suppose that our aim is to keep $|y(t)| < \epsilon$ for all $t \geq t_0$. It can be seen that our objective is achieved if

$$|y(t_0)| = |x_0| \leq \delta = \frac{\epsilon}{1+t_0}$$

The origin is therefore a stable equilibrium point of this system. This is further illustrated in Figure 2.6.5

2. The damped pendulum system has many equilibrium points as described in Example 2.3.1. It can be shown that the equilibrium point located at the origin of the state-space is unstable. This is illustrated in Figure 2.6.6 where it is seen that no matter how close to the origin the initial state is, the norm of $x(t)$ can not be pre-specified. On the other hand, note that the two equilibrium points at $[\pi, 0]$ are stable.

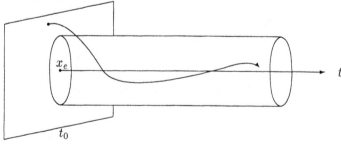

Figure 2.6.4: Global asymptotic stability of x_e at t_0

2.6 Stability Theory

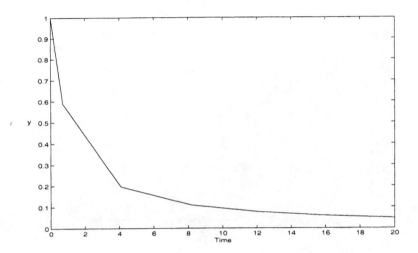

Figure 2.6.5: Time history for $\dot{y} = \frac{-y}{1+t}$.

3. The origin is an equilibrium point of the Van der Pol oscillator. However, and as shown in Figure 2.6.7, it is an unstable equilibrium point. In fact, suppose the following norm is used per definition 2.5.6, and let $\epsilon \neq 1$. Therefore, we would like

$$sup_{t \in [0,\infty)} \sqrt{x_1^2(t) + x_2^2(t)} \leq 1$$

for all $t > t_0$. As can be seen from Figure 2.6.7, no matter how close to the origin x_0 is, i.e. no matter how small δ is, the trajectory will eventually leave the ball of radius $\epsilon = 1$.

4. The origin is a stable equilibrium point of the robot described in Example 2.3.1 whenever the following choices are made

$$K_v = \text{diag}(k_{vi}) \; ; \; K_p = \text{diag}(k_{pi})$$

where $k_{vi} > 0$ and $k_{pi} > 0$ for all $i = 1, \cdots n$.   ∎

EXAMPLE 2.6-2: Stability versus Convergence

Consider the following system

$$\begin{bmatrix} \dot{x}_1 \\ \dot{x}_2 \end{bmatrix} = \begin{bmatrix} x_1(1 - x_1) \\ \sin^2(\frac{x_2}{2}) \end{bmatrix}$$

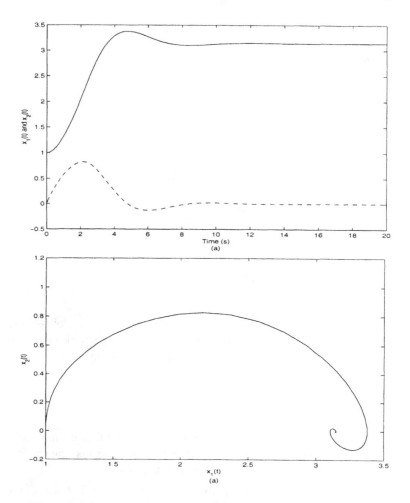

Figure 2.6.6: Damped pendulum: (a)time history; (b) phase plane

There are 2 equilibrium points located at (0,0) and (1,0), both of which are unstable (check!). On the other hand, (1,0) is convergent since all trajectories will eventually converge to it after some time T. Before T however, there is no guarantee that a trajectory will stay within some ϵ of (1,0) no matter how close the initial state is to (1,0). In fact, suppose the system starts at $x(0)$, then the state-vector is given at any $t \geq 0$ by

2.6 Stability Theory

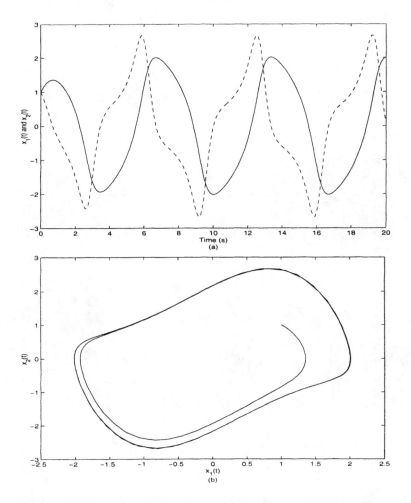

Figure 2.6.7: Van der Pol oscillator: (a)time history; (b) phase plane

$$\begin{bmatrix} x_1(t) \\ x_2(t) \end{bmatrix} = \begin{bmatrix} \frac{x_1(0)e^t}{x_1(0)[e^t-1]+1} \\ 2\mathrm{arccot}[\cot(x_2(0)/2) - t/2] \end{bmatrix}$$

See Figure 2.6.8 for illustration of the behavior of the state vector. ∎

Note that stability and asymptotic stability are local concepts in the sense that, if the initial perturbation δ is too large, the subsequent states $x(t)$ may stray arbitrarily far from x_e. There exists, therefore, a region

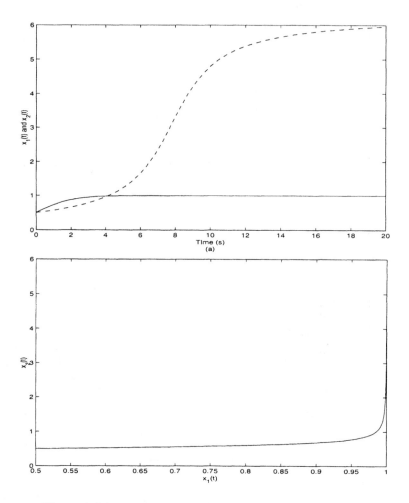

Figure 2.6.8: Example 2.6.2 (a) time history; (b) phase plane

centered at x_e and given by $R_{\delta_2(t_0)} = \{x_0 : \mid x_0 - x_e \mid < \delta_2(t_0)\}$ such that both stability and asymptotic stability will result for any state starting in $R_{\delta_2(t_0)}$, but not for states starting outside of it. This region is called the *domain of attraction* of x_e. The equilibrium state x_e is GAS if $R_{\delta_2(t_0)} = \Re^n$.

Note also that all previous stability definitions depended on the initial time t_0, so that the region of attraction may vary with varying initial times. If the system (2.6.1) were independent of time (or autonomous), then the stability concepts in definition 2.6.1 are indeed independent of t_0 and they

2.6 Stability Theory

will be equivalent to the stability concepts defined next. On the other hand, and even though the system (2.6.1) is time-dependent, we would like to have its stability properties not depend on t_0 since that would later provide us with a desired degree of robustness. This leads us to define the uniform stability concepts [Khalil 2001].

DEFINITION 2.6-2 *In all parts of this definition, x_e is an equilibrium point at time t_0.*

1. *Uniform Stability: x_e is uniformly stable (US) over $[t_0, \infty)$ if $\delta(\epsilon, t_0)$ in definition 2.6.1 is independent of t_0.*

2. *Uniform Convergence: x_e is uniformly convergent (UC) over $[t_0, \infty)$ if $\delta_1(t_0)$ and $T(\epsilon_1, x_0, t_0)$ of definition 2.6.1 can be chosen independent of t_0.*

3. *Uniform Asymptotic Stability: x_e is uniformly, asymptotically stable (UAS) over $[t_0, \infty)$, if it is both US and UC.*

4. *Global Uniform Asymptotic Stability: x_e is globally, uniformly, asymptotically stable (GUAS) if it is US, and UC.*

5. *Global Exponential Stability: x_e is globally exponentially stable (GES) if there exists $\alpha > 0$, and $\beta\ 0$ such that for all $x_0 \in \Re^n$,*

$$\|x(t) - x_e\|\ \alpha \|x_0\| e^{-\beta(t-t_0)}\ ;\ t \geq t_0$$

■

Note that GES implies GUAS, and see Figure 2.6.9 for an illustration of uniform stability concepts.

EXAMPLE 2.6-3: Uniform stability

1. Consider the damped Mathieu equation,

$$\dot{x}_1 = x_2$$

$$\dot{x}_2 = -x_2 - (2 + \sin t)x_1$$

The origin is a US equilibrium point as shown in Figure 2.6.10

2. The scalar system

$$\dot{x} = -x/t$$

has an equilibrium point at the origin which is UC.

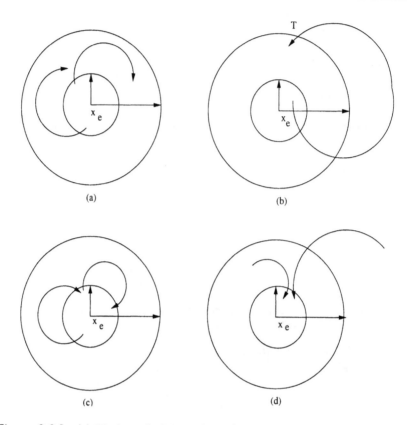

Figure 2.6.9: (a) Uniform Stability of x_e; (b) uniform convergence of x_e; (c) uniform asymptotic stability of x_e; (d) global uniform asymptotic stability of x_e

3. The origin is a UAS equilibrium point for

$$\dot{x} = -x/t$$

4. The system

$$\dot{x}(t) = -x^2(t)$$

has an equilibrium point $x_e = 0$ which is GUAS.

5. Consider the system

$$\dot{x}(t) = -(1+x^2)x$$

the origin is then a GES equilibrium point since the solution is given by

2.6 Stability Theory

$$x(t) = x_0 e^{-\int_0^t (1+x^2(\tau))d\tau}$$

so that

$$|x(t)| \leq |x_0|e^{-t}$$

See Figure 2.6.11, for an illustration of the time history of $x(t)$. ∎

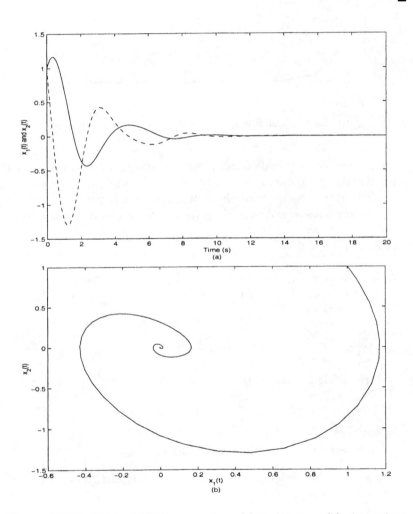

Figure 2.6.10: Damped Mathieu equation: (a)time history; (b) phase plane

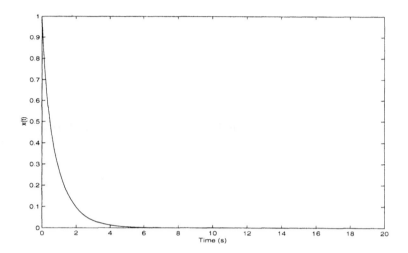

Figure 2.6.11: Example 2.6.3e: (a)time history; (b) phase plane

In many cases, a bound on the size of the state is all that is required in terms of stability. This is a less stringent requirement than Lyapunov stability. It is instructive to study the subtle difference between the definition of *Boundedness* below and that of *Lyapunov stability* in Definition 2.6.1.

DEFINITION 2.6-3

1. *Boundedness:* x_e is bounded (B) at t_0 if states starting close to x_e will never get too far. In other words, x_e is bounded at t_0 if for each $\delta > 0$ such that

$$||x_0 - x_e|| < \delta$$

there exists a positive $\epsilon(r, t_0) < \infty$ such that for all $t \geq t_0$

$$||x(t) - x_e|| \quad \epsilon(r, t_0)$$

x_e is bounded if it is bounded for any t_0.

2. *Uniform Boundedness:* x_e is uniformly bounded (UB) over $[t_0, \infty)$ if $\epsilon(r, t_0)$ can be made independent of t_0.

3. Uniform Ultimate Boundedness: x_e is said to be uniformly, ultimately bounded (UUB), if states starting close to x_e will eventually become

2.6 Stability Theory

bounded. More precisely, x_e is UUB if for any $\delta > 0$, $\epsilon > 0$, there exists a finite time $T(\epsilon, \delta)$ such that whenever $\|x_0 - x_e\| < \delta$, the following is satisfied

$$\|x(t) - x_e\| \leq \epsilon$$

for all $t \geq T(\epsilon, \delta)$.

4. Global Uniform Ultimate Boundedness: x_e is said to be globally, uniformly, ultimately bounded (GUUB) if for $\epsilon > 0$, there exists a finite time $T(\epsilon)$ such that

$$\|x(t) - x_e\| \leq \epsilon$$

for all $t \geq T(\epsilon)$. See Figure 2.6.12 for an illustration of the boundedness stability concepts. ∎

EXAMPLE 2.6-4: Boundedness

1. The second-order system given by

$$\dot{x}_1 = x_2$$

$$\dot{x}_2 = -x_2 \sin^2 t - (1 + e^{-t})x_1$$

has a uniformly bounded equilibrium point at the origin as shown in Figure 2.6.13.

2. The second-order system given by

$$\dot{x}_1 = x_2$$

$$\dot{x}_2 = -x_1^3 - x_1 x_2 + e^{-t}$$

has an UUB equilibrium point at $x_e = 0$, as shown in Figure 2.6.14. ∎

Note that in general, we are interested in the stability of the motion $x(t)$ when the system is perturbed from its trajectory. In other words, how far does $x(t)$ get from its nominal trajectory if the initial state is perturbed?

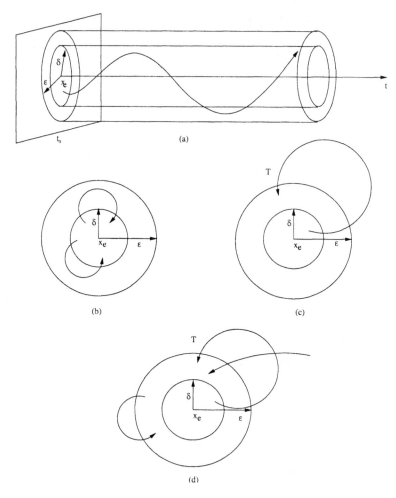

Figure 2.6.12: (a) Boundedness of x_e at t_0; (b) uniform boundedness of x_e; (c) uniform ultimate boundedness of x_e; (d) global uniform boundedness of x_e.

This problem can always be reduced to the stability of the origin of a non-autonomous system by letting

$$z = x_e - x(t)$$

and

$$\dot{z}(t) = g(z, t) = f(z + x, t) - f(x, t)$$

and studying the stability of the equilibrium point $z_e = 0$.

2.6 Stability Theory

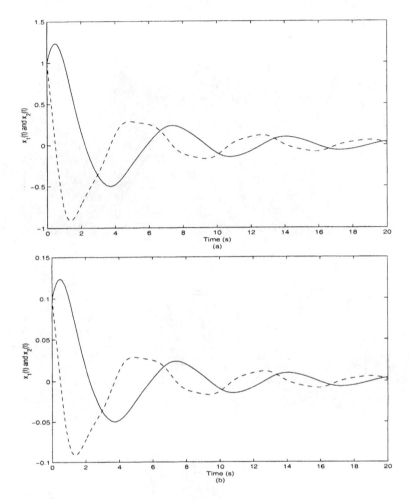

Figure 2.6.13: Example 2.6.4-a: (a)$x_1(0) = x_2(0) = 1$ (b) $x_1(0) = x_2(0) = 0.1$

EXAMPLE 2.6-5: Stability of the Origin

1. Consider the damped pendulum of Example 2.3.1a. Its equilibrium points are at $[n\pi \ 0]^T$, $n = 0, \pm 1, \cdots$. The stability of these points can be studied from the stability of the origin of the system

$$\dot{z} = \begin{bmatrix} z_2 \\ \sin(x_1) - \sin(z_1 + x_1) - kz_2 \end{bmatrix}$$

2. Consider the rigid robot equations of Example 2.3.2, and assume

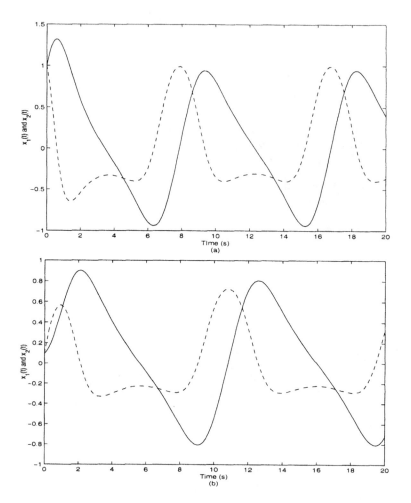

Figure 2.6.14: Example 2.6.4-b: (a)$x_1(0) = x_2(0) = 1$ (b) $x_1(0) = x_2(0) = 1$

that a desired trajectory is specified by

$$x_d(t) = \begin{bmatrix} q_d(t) \\ \dot{q}_d(t) \end{bmatrix}$$

Therefore, we can define the new system by choosing $z = x_d - x$ so that

$$\dot{z} = \begin{bmatrix} \dot{z} \\ \ddot{q}_d + M^{-1}(V + G) \end{bmatrix} + \begin{bmatrix} 0 \\ -M^{-1} \end{bmatrix} \tau$$

and verify that $z_e = 0$ is the desired equilibrium point of the modified system if $x_e = x_d$ is the desired equilibrium trajectory of the robot.

∎

2.7 Lyapunov Stability Theorems

Lyapunov stability theory deals with the behavior of unforced nonlinear systems described by the differential equations

$$\dot{x}(t) = f[x(t), t], \ t \geq 0, \ x \in \Re^n \qquad (2.7.1)$$

where without loss of generality, the origin is an equilibrium point of (2.7.1). It may seem to the reader that such a theory is not needed since all we had to do in the examples of the previous section is to solve the differential equations, and study the time evolution of a norm of the state vector. There are at least two reasons why Lyapunov theory is needed. The first is that Lyapunov theory will allow us to determine the stability of a particular equilibrium point without actually solving the differential equations. This, as is well known to any student of nonlinear differential equations, is a large saving. The second and related reason for using Lyapunov theory is that it provides us with qualitative results to the stability questions, which may be used in designing stabilizing controllers of nonlinear dynamical systems.

We shall first assume that any necessary conditions for (2.7.1) to have a unique solution are satisfied [Khalil 2001], [Vidyasagar 1992]. The unique solution corresponding to $x(t_0) = x_0$ is $x(t, t_0, x_0)$ and will be denoted simply as $x(t)$. Before we actually introduce Lyapunov's theorems, we review certain classes of functions which will simplify the statement of Lyapunov theorems.

Functions of Class K

Consider a continuous function $\alpha : \Re \longrightarrow \Re$.

DEFINITION 2.7-1 *We say that α belongs to class K, if*

1. *$\alpha(0) = 0$*

2. *$\alpha(x) > 0$, for all $x > 0$*

3. *α is nondecreasing, i.e. $\alpha(x_1) \geq \alpha(x_2)$ for all $x_1 > x_2$.*

∎

EXAMPLE 2.7-1: Class K Functions

The function $\alpha(x) = x^2$ is a class K function. The function $\alpha(x) = x^2 + 1$ is not a class K function because (1) fails. On the other hand, $\alpha(x) = -x^2$ is not a class K function because (2) and (3) fail.

■

DEFINITION 2.7-2 *In the following, $\Re^+ = [0, \infty)$.*

1. Locally Positive Definite: *A continuous function $V : \Re^+ \times \Re^n \longrightarrow R$ is locally positive definite (l.p.d) if there exists a class K function $\alpha(.)$ and a neighborhood N of the origin of $\Re^n$ such that*

$$V(t,x) \geq \alpha(\|x\|)$$

for all $t \geq 0$, and all $x \in N$.

2. Positive Definite: *The function V is said to be positive definite (p.d) if $N = \Re^n$.*

3. Negative and Local Negative Definite: *We say that V is (locally) negative definite (n.d) if $-V$ is (locally) positive definite.*

■

EXAMPLE 2.7-2: Locally Positive Definite Functions

[Vidyasagar 1992] The function $V(t,x) = x_1^2 + \cos^2(x_2)$ is l.p.d but not p.d, since $V(t,x) = 0$ at $x = (0, \pi/2)$. On the other hand, $V(t,x) = e^{-t}(x_1^2 + x_2^2)$ is not even l.p.d because $V(t,x) \longrightarrow 0$ as $t \longrightarrow \infty$ for any x. The function $V(t,x) = (1+t)(x_1^2 + x_2^2)$ is p.d.

■

DEFINITION 2.7-3

In the following, $\Re^+ = [0, \infty)$.

1. Locally Decrescent: *A continuous function $V : \Re^+ \times \Re^n \longrightarrow R$ is locally decrescent if There exists a class K function $\beta(.)$ and a neighborhood N of the origin of $\Re^n$ such that*

$$V(t,x) \leq \beta(\|x\|)$$

for all $t \geq 0$ and all $x \in N$.

2.7 Lyapunov Stability Theorems

2. Decrescent: *We say that V is decrescent if $N = \Re^n$.*

■

EXAMPLE 2.7-3: Decrescent Functions

[Vidyasagar 1992] The function $V(t,x) = e^{-t}[x_1^2 + \sin^2(x_2)]$ is locally but not globally decrescent. On the other hand, $V(t,x) = e^{-t}[x_1^2 + x_2^2]$ is globally decrescent.

■

DEFINITION 2.7-4 *Given a continuously differentiable function V : $\Re^+ \times \Re^n \longrightarrow R$, together with a system of differential equations (2.7.1), the derivative of V along (2.7.1) is defined as a function $\dot{V}$: $\Re^+ \times \Re^n \longrightarrow R$ given by*

$$\dot{V}(t,x) \equiv \frac{dV(t,x)}{dt} = \frac{\partial V(t,x)}{\partial t} + [\frac{\partial V(t,x)}{\partial x}]^T f(t,x)$$

■

EXAMPLE 2.7-4: Lyapunov Functions

Consider the function $V(t,x) = e^{-t}[x_1^2 + x_2^2]$ of Example 2.7.3 and assume given a system

$$\dot{x} = \begin{bmatrix} \frac{x_1}{2} - x_2^2 \\ \frac{x_2}{2} - x_1^2 \end{bmatrix}$$

Then, the derivative of $V(t,x)$ along this system is

$$\dot{V}(t,x) = -2e^{-t}(x_1^2 + x_2^2)$$

■

Lyapunov Theorems

We are now ready to state Lyapunov Theorems, which we group in Theorem 2.7.1. For the proof, see [Khalil 2001], [Vidyasagar 1992].

THEOREM 2.7-1:   Lyapunov

Given the nonlinear system

$$\dot{x} = f(t,x) \; ; \; x(0) = x_0$$

with an equilibrium point at the origin, i.e. $f(t,0) = 0$, and let N be a neighborhood of the origin of size ϵ i.e.

$$N = \{x \; ; \; ||x|| \leq \epsilon\}$$

Then

1. **Stability:** *The origin is stable in the sense of Lyapunov, if for $x \in N$, there exists a scalar function $V(t,x)$ with continuous partial derivative such that*

    (a) $V(t,x)$ *is positive definite*

    (b) $\dot{V}$ *is negative semi-definite*

2. **Uniform Stability:** *The origin is uniformly stable if in addition to (a) and (b) $V(t,x)$ is decrescent for $x \in N$.*

3. **Asymptotic Stability:** *The origin is asymptotically stable if $V(t,x)$ satisfies (a) and $\dot{V}(t,x)$ is negative definite for $x \in N$.*

4. **Global Asymptotic Stability:** *The origin is globally, asymptotically stable if $V(t,x)$ verifies (a), and $\dot{V}(t,x)$ is negative definite for all $x \in \Re^n$, i.e. if $N = \Re^n$.*

5. **Uniform Asymptotic Stability:** *The origin is UAS if $V(t,x)$ satisfies (a), $V(t,x)$ is decrescent, and $\dot{V}(t,x)$ is negative definite for $x \in N$.*

6. **Global Uniform Asymptotic Stability:** *The origin is GUAS if $N = \Re^n$, and if $V(t,x)$ satisfies (a), $V(t,x)$ is decrescent, $\dot{V}(t,x)$ is negative definite and $V(t,x)$ is radially unbounded, i.e. if it goes to infinity uniformly in time as $|| x || \longrightarrow \infty$.*

7. **Exponential Stability:** *The origin is exponentially stable if there exists positive constants α, β, γ such that*

    $$\alpha \, || \, x \, ||^2 \leq V(t,x) \leq \beta \, || \, x \, ||^2 \text{ and } \dot{V}(t,x) \leq -\gamma \, || \, x \, ||^2 \; \forall x \in N.$$

8. **Global Exponential Stability:** *The origin is globally exponential stable if it is exponentially stable for all $x \in \Re^n$.*

∎

The function $V(t,x)$ in the theorem is called a **Lyapunov function**. Note that the theorem provides sufficient conditions for the stability of the origin and that the inability to provide a Lyapunov function candidate has no indication on the stability of the origin for a particular system.

2.7 Lyapunov Stability Theorems

EXAMPLE 2.7-5: Stability via Lyapunov Functions

1. Consider the system described by

$$\begin{aligned} \dot{x}_1 &= x_2 \\ \dot{x}_2 &= -(t+0.5)x_2 - e^{-t}x_1 \end{aligned}$$

   and choose a Lyapunov function candidate

$$V(t,x) = x_1^2 + x_2^2$$

   Then the origin may be shown to be a stable equilibrium point.

2. Consider the Mathieu equation described in Example 2.6.3. Let the Lyapunov function candidate be given by

$$V(t,x) = x_1^2 + \frac{x_2^2}{2 + \sin t}$$

   The origin is then shown to be a US equilibrium point.

3. The system given in Example 2.6.3-5, has a GES equilibrium point at the origin. This may be shown by considering a Lyapunov function candidate

$$V(x) = x^2$$

   which leads to

$$\dot{V}(x) = -x^2(1 + x^2)$$

   Then,

$$0.5x^2 \leq V(x) \leq 2x^2$$

   and

$$\dot{V}(x) \leq -x^2$$

   The above inequalities hold for any $x \in \Re^n$.

4. Let

$$\dot{x}_1 = -x_1 - e^{-2t}x_2$$
$$\dot{x}_2 = x_1 - x_2$$

and pick a Lyapunov function candidate

$$V(t,x) = x_1^2 + (1+e^{-2t})x_2^2$$

so that

$$\dot{V} = -2[x_1^2 + x_1 x_2 + (1+e^{-2t})x_2^2] \leq -2[x_1^2 + x_1 x_2 + x_2^2]$$
$$= -(x_1+x_2)^2 - (x_1^2 + x_2^2)$$

so that the origin is SL.

■

Lyapunov Theorems may be used to design controllers that will stabilize a nonlinear system such as a robot. In fact, if one chooses a Lyapunov function candidate $V(t,x)$, then finding its total derivative $\dot{V}(t,x)$ will exhibit an explicit dependence on the control signal. By choosing the control signal to make $\dot{V}(t,x)$ negative definite, stability of the closed-loop system is guaranteed. Unfortunately, it is not always easy to guarantee the global asymptotic stability of an equilibrium point using Lyapunov Theorem. This is due to the fact, that $\dot{V}(t,x)$ may be shown to be negative but not necessarily negative-definite. If the open-loop system were **autonomous**, Lyapunov theory is greatly simplified as shown in the next section.

The Autonomous Case

Suppose the open-loop system is not autonomous, i.e. is not explicitly dependent on t, then a time-independent Lyapunov function candidate $V(x)$ may be obtained and the positive definite conditions are greatly simplified as described next.

LEMMA 2.7-1: *A time invariant continuous function $V(x)$ is positive definite if $V(0) = 0$ and $V(x) > 0$ for $x \neq 0$. It is locally positive definite if the above holds in a neighborhood of the origin* ■

Note that the condition that $V(0) = 0$ is not necessary and that as long as $V(0)$ is bounded above the Lyapunov results hold without modification.

2.7 Lyapunov Stability Theorems

With the above simplification, the Lyapunov results hold except that no distinction is made between uniform and regular stability results.

EXAMPLE 2.7-6: Uniform Stability via Lyapunov Functions

Consider again the damped pendulum described by

$$\ddot{\theta}(t) + \dot{\theta} + \sin\theta = 0$$

and obtain a state-space by choosing $x_1 = \theta$ and $x_2 = \dot{\theta}$, then a Lyapunov function candidate is

$$V(x) = (1 - \cos x_1) + \frac{x_2^2}{2}$$

so that $\dot{V} = -x_2^2 \leq 0$ and the origin is SL and actually USL. ∎

EXAMPLE 2.7-7: Uniform Stability via Lyapunov Functions

This example illustrates the local asymptotic, uniform stability of the origin for the system

$$\begin{aligned}\dot{x}_1 &= x_1(x_1^2 + x_2^2 - 2) - 4x_1 x_2^2 \\ \dot{x}_2 &= 4x_1^2 + x_2(x_1^2 + x_2^2 - 2)\end{aligned} \quad (1)$$

Choose $V(x) = x_1^2 + x_2^2$, then

$$\dot{V}(x) = 2(x_1^2 + x_2^2)((x_1^2 + x_2^2 - 2)$$

which is strictly less than zero for all $x_1^2 + x_2^2 < 2$. ∎

Sometimes, and although $\dot{V}(x)$ is only non-positive, LaSalle's theorem [LaSale and Lefschetz 1961], [Khalil 2001] may be used to guarantee the global asymptotic stability of the equilibrium point as described in the next theorem.

THEOREM 2.7-2: LaSalle

Given the autonomous nonlinear system

$$\dot{x} = f(x) \,;\, x(0) = x_0$$

and let the origin be an equilibrium point. Then,

1. **Asymptotic Stability:** *Suppose a Lyapunov function $V(x)$ has been found such that for $x \in N \subset \Re^n$, $V(x) > 0$ and $\dot{V}(x) \leq 0$. Then the origin is asymptotically stable if and only if $\dot{V}(x) = 0$ only at $x = 0$.*

2. **Global Asymptotic Stability:** *The origin is GAS if $N = \Re^n$ above and $V(x)$ is radially unbounded.*

■

Unfortunately, in many applications with time-varying trajectories, the open-loop systems are not autonomous, and more advanced results such as the ones described later will be called upon to show global asymptotic stability.

EXAMPLE 2.7-8: LaSalle Theorem

Consider the autonomous system

$$\dot{x}_1 = x_2$$

$$\dot{x}_2 = \frac{-x_1^2}{x_2} - x_2 + x_1$$

The origin is an equilibrium point. Moreover, consider a Lyapunov function candidate

$$V(x) = x_1^2 + x_2^2$$

Leading to

$$\dot{V}(x) = -2(x_1 - x_2)^2 \leq 0$$

Since $\dot{V}(x) = 0$ for all $x_1 = x_2$, we need to check whether the origin is the only point where $\dot{V}(x) = 0$. It can be seen from the state equation that $x_1 = x_2$ can only happen at the origin, therefore the origin is GAS.

■

2.7 Lyapunov Stability Theorems

Note that LaSalle's theorem is actually more general than we have described. In fact, it can be used to ascertain the stability of sets rather than just an equilibrium point. The basic idea is that since $V(x)$ is lower bounded $V(x) > c$, then the derivative $\dot{V}$ has to gradually vanish, and that the trajectory is eventually confined to the set where $\dot{V} = 0$. The following definitions is useful in explaining the more general LaSalle's theorem.

DEFINITION 2.7-5 *A set G is said to be an invariant set of a dynamical system if every trajectory which starts in G remains in G.* ∎

As examples of invariant sets, we can give the whole state-space, an equilibrium point and a limit cycle. By using this concept, LaSalle was able to describe the convergence to sets rather than to just equilibrium points. For example, we can use this result to show that the limit cycle is a stable attracting set of the Van der Pol oscillator.

The Linear Time-Invariant Case

In the case where the system under consideration is linear and time-invariant, Lyapunov theory is well developed and the choice of a Lyapunov function is simple. In fact, in this case, the various stability concepts in definitions 2.6.1 are identical. Lyapunov theory then provides necessary as well as sufficient conditions for stability as discussed in this section. For the proofs consult [Khalil 2001].

THEOREM 2.7-3: *Given a linear time-invariant system*

$$\dot{x}(t) = Ax(t)$$

The system is stable if and only if there exists a positive definite solution P to the equation

$$A^T P + PA = -Q$$

where Q is an arbitrary positive-definite matrix. ∎

Note that the stability of the whole system was obtained in the last theorem since in this case, the origin is the unique equilibrium point and its stability is equivalent to the system being stable. In addition, no reference was made to what kind of stability is implied since all stability concepts are equivalent in the very special case of linear, time-invariant systems [Khalil 2001]. Also note that this result is equivalent to testing that all eigenvalues of A have negative real parts [Kailath 1980]. In the following, we include a table summarizing Lyapunov Stability theorems.

Lyapunov Stability Theorems

1. Autonomous Systems: $\dot{x} = f(x)$, such that

   - The origin is an equilibrium point.
   - The set $B_r = \{x \in \Re^n;\ \|x\| < r\}$.
   - There exists $V(x)$ continuous, continuously differentiable, $V(0) = 0$

$V(x)$	$\dot{V}(x) = \nabla V(x)\dot{x}$	Lyapunov Stability
$> 0,\ \forall x \in B_r$	$\leq 0,\ \forall x \in B_r$	Uniformly Stable (US)
$> 0,\ \forall x \in B_r$	$< 0,\ \forall x \in B_r$	Asymptotically US (AUS)
$> 0,\ \forall x \in \Re^n$, $V(x) \to \infty$ as $\|x\| \to \infty$	$< 0,\ \forall x \in \Re^n$	Globally UAS (GUAS)

2. Non-autonomous Systems: $\dot{x} = f(t, x)$, such that

   - The origin is an equilibrium point at $t = t_0$.
   - The set $B_r = \{x \in \Re^n;\ \|x\| < r\}$.
   - There exists $V(t, x)$ continuous, continuously differentiable in t and x, $V(t, 0) = 0$
   - $\alpha(\|x\|)$, $\beta(\|x\|)$, and $\gamma(\|x\|)$ class K functions.

$V(t, x)$	$\dot{V}(x) = \frac{\partial V(t,x)}{\partial t} + \nabla V(t, x)\dot{x}$	Lyapunov Stability
$\geq \alpha(\|x\|),\ \forall x \in B_r$	$\leq 0,\ \forall x \in B_r$	S at t_0
$\beta(\|x\|) \geq V(t,x) \geq \alpha(\|x\|)$, $\forall x \in B_r$	$\leq 0,\ \forall x \in B_r$	US
$\geq \alpha(\|x\|),\ \forall x \in B_r$	$\leq -\gamma(\|x\|),\ \forall x \in B_r$	AS at t_0
$\beta(\|x\|) \geq V(t,x) \geq \alpha(\|x\|)$, $\forall x \in B_r$	$\leq -\gamma(\|x\|),\ \forall x \in B_r$	UAS
$\beta(\|x\|) \geq V(t,x) \geq \alpha(\|x\|)$, $\forall x \in \Re^n$, $V(t,x) \to \infty$ as $\|x\| \to \infty$	$\leq -\gamma(\|x\|),\ \forall x \in \Re^n$	GUAS

Lyapunov Theorems may be used to design controllers that can stabilize linear time-invariant systems as described in the next example.

2.7 Lyapunov Stability Theorems

EXAMPLE 2.7-9: Stability of PD Controllers for Rigid Robots

Consider the rigid robot example and the torque input of Example 2.3.2. The resulting linear system is given by

$$\dot{x} = \begin{bmatrix} 0 & I \\ -K_p & -K_v \end{bmatrix} x$$

The equilibrium point is $x_e = [0^T \; 0^T]^T$. It is then easy to find K_p and K_v to stabilize the equilibrium point. In fact, let $Q = I$ and consider the Lyapunov equation of theorem 2.7.3

$$A^T P + PA = -I$$

which reduces to

$$K_p^T P_2^T + P_2 K_p = I$$

$$K_v^T P_3^T + P_3 K_v = I + P_2 + P_2^T$$

$$P_1 = P_2 K_v + K_p^T P_3$$

where

$$P = P^T = \begin{bmatrix} P_1 & P_2 \\ P_2^T & P_3 \end{bmatrix}$$

The solution of these equations will provide a stabilizing controller for the robot. In particular, the choices of K_p and K_v of Example 2.6.3(part 5) will make the origin a GES equilibrium. ∎

Convergence Rate

Although Lyapunov stability theory does not directly give an indication of the transient behavior of the system, it may actually be used to estimate the convergence rate for linear systems. In order to see this, consider the following inequalities which are a result of Rayleigh-Ritz theorem 2.5.2,

$$x^T \lambda_{min}(P) x \leq x^T P x \leq x^T \lambda_{max}(P) x \qquad (2.7.2)$$

then, note

$$\dot{V} = -x^T Q x \leq -\lambda_{min}(Q) x^T x \qquad (2.7.3)$$
$$\leq -\lambda_{min}(Q) x^T x \frac{\lambda_{max}(P)}{\lambda_{max}(P)}$$
$$\leq \frac{-\lambda_{min}(Q)}{\lambda_{max}(P)} V = -\gamma V$$

Now, we can show by separation of variables and integrating that

$$x^T \lambda_{min}(P) x \leq V(t) \leq V(0) e^{-\gamma t} \qquad (2.7.4)$$

or that

$$\| x \|_2^2 \leq \frac{V(0)}{\lambda_{min}(P)} e^{-\gamma t} \qquad (2.7.5)$$

Therefore, $x(t)$ is approaching the origin at rate faster than $\gamma/2$. In fact, it can be shown that the fastest convergence rate estimate is obtained when $Q = I$.

Krasovskii Theorem

There are some cases when a Lyapunov function for autonomous nonlinear systems is easily obtained using Krasovkii's theorem stated below.

THEOREM 2.7-4: *Consider the autonomous nonlinear system $\dot{x} = f(x)$ with the origin being an equilibrium point. Let $A(x) = \partial f / \partial x$. Then, a sufficient condition for the origin to be AS is that there exists 2 symmetric positive-definite matrices, P and Q such that for all $x \neq 0$, the matrix*

$$F(x) = A(x)^T P + P A(x) + Q \qquad (1)$$

is ≥ 0 in some ball B about the origin. The function $V(x) = f(x)^T P f(x)$ is then a Lyapunov function for the system. If $B = \Re^n$ and if $V(x)$ is radially unbounded then the system is GAS. ∎

2.7 Lyapunov Stability Theorems

EXAMPLE 2.7-10: Krasovkii's Theorem

Consider the nonlinear system described by

$$\dot{x}_1 = f(x_1) + g(x_2)$$
$$\dot{x}_2 = x_1 + ax_2$$

where $f(0) = g(0) = 0$. Let us find the Jacobian matrix

$$A(x) = \begin{bmatrix} \frac{df}{dx_1} & \frac{dg}{x_2} \\ 1 & a \end{bmatrix}$$

Let $P = I$, $Q = \epsilon I$ and check the conditions on the system which would look like

$$\begin{bmatrix} 2\frac{df}{dx_1} + \epsilon & \frac{dg}{x_2} + 1 \\ \frac{dg}{x_2} + 1 & 2a + \epsilon \end{bmatrix}$$

which should be ≥ 0. ■

On the other hand, suppose we have the linear time-invariant system

$$\dot{x} = Ax + Bu \; ; \; x(0) = x_0$$

$$y = Cx$$

The transfer function is then given by

$$P(s) = C(sI - A)^{-1}B$$

Note that $P(s)$ is strictly-proper. The following stability result then holds [Desoer and Vidyasagar 1975].

THEOREM 2.7-5: *Suppose $P(s)$ is a stable transfer function then*

1. *If $u(t) \in L_\infty$ i.e. $u(t)$ is bounded, then so is $y(t)$ and $\dot{y}(t)$.*

2. *If $\lim_{t \to \infty} u(t) = 0$ then $\lim_{t \to \infty} y(t) = 0$.*

3. If $u(t) \in L_2$, then $lim_{t \to \infty} y(t) = 0$.

■

EXAMPLE 2.7-11: Input/Output Stability of Rigid Robots

Consider the closed-loop robot of Example 2.7.9. Its input/output behavior is described by a set of n decoupled differential equations

$$\ddot{q}_i + k_{vi}\dot{q} + k_{pi}q = u_i \; ; \; i = 1, 2, ,..., n$$

If $\tau = -K_v\dot{q} - K_p q + V + G + u$. Then the transfer function between each $U_i(s)$ and $Q_i(s)$ is

$$\frac{Q_i(s)}{U_i(s)} = P_i(s) = \frac{1}{s^2 + k_{vi}s + k_{pi}}$$

Note that all $P_i(s)$ are stable if k_{vi} and k_{pi} are both positive. Assume for the purposes of illustration that $k_{vi} = 3$ and $k_{pi} = 2$, and that $u_i = \sin(t)$. Note that u_i is bounded and let us find the output $y_i(t)$.

$$y_i(t) = -0.2e^{-2t} + 0.5e^{-t} - 0.32\cos(t + 0.32)$$

which is bounded above by 0.62 and below by -0.02. The derivative of $y(t)$ is also bounded. On the other hand, suppose the input is $u_i(t) = e^{-3t}$ then the output is

$$y_i(t) = 0.5e^{-t} - e^{-2t} + 0.5e^{-3t}$$

Since $lim_{t \to \infty} u_i(t) = 0$, then so is $lim_{t \to \infty} y_i(t) = 0$.

■

2.8 Input/Output Stability

When dealing with nonlinear systems, stability in the sense of Lyapunov does not necessarily imply that a **bounded input** will result in a **bounded output**. This fact is shown in the next example.

2.8 Input/Output Stability

EXAMPLE 2.8-1: Input/Output Versus Lyapunov Stability

Consider the time-varying system

$$\dot{y}(t) + \frac{y(t)}{t} = u(t)$$

The system is asymptotically stable with a single equilibrium point at $y_e = 0$. On the other hand, a unit step input (which is definitely bounded) starting at $t = 0$ will lead to the response

$$y(t) = \frac{t}{2}$$

which grows unbounded as t increases. ∎

Therefore, we need to discuss the conditions under which a bounded input will result in a bounded output [Boyd and Barratt], [Desoer and Vidyasagar 1975]. This was actually presented when discussing the system-induced norms (see Definition 2.5.7) and the current discussion should serve to contrast these concepts with Lyapunov stability. Consider the nonlinear system

$$\dot{x}(t) = f[x(t),\ t,\ u(t)]\ ;\quad x(t_0) = x_0 \qquad (2.8.1)$$
$$y(t) = g[x(t),\ t,\ u(t)] \qquad (2.8.2)$$

DEFINITION 2.8-1 *The dynamical system (2.8.1) is bounded-input-bounded-output (BIBO) stable if for any*

$$||u(t)|| \leq M < \infty$$

there exist finite $\gamma > 0$ and b such that

$$||y(t)|| \leq \gamma M + b$$

∎

Note that BIBO stability implies the uniform boundedness of all equilibrium states.

EXAMPLE 2.8-2: BIBO Stability

Consider the system

$$y(t) = u^2(t)$$

It is BIBO stable since for any input $u(t)$ such that $\mid u(t) \mid < M < \infty$, the output is bounded by M^2.

■

2.9 Advanced Stability Results

In this section we review some advanced stability concepts. These results will be used in showing the closed-loop stability of systems when robust or adaptive controllers are used. If the reader is only interested in implementing these controllers, this section may be skipped. On the other hand, anyone interested in designing new controllers should be aware of the results presented here.

Passive Systems

Given a nonlinear system shown in Figure 2.9.1, we are interested in studying the stability of such a system based on input-output measurements only. Motivated by energy concepts in network theory, such as

$\frac{d}{dt}$ [Stored Energy] = [External Power Input] +[Internal power Generation]

One can study the internal stability of all kinds of systems. In general, the external power input is the scalar product $y^T u$ of an input effort or flow u and an output flow or effort y. The last equation then takes on the form

$$\dot{V} = y^T u - g(t) \qquad (2.9.1)$$

In many cases (e.g. isolated system) $g(t) = 0$ and one can use the stored Energy

$$V(t) = \int_0^t y^T(r)u(r)dr \qquad (2.9.2)$$

2.9 Advanced Stability Results

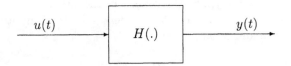

Figure 2.9.1: Input-output description of nonlinear system

as a Lyapunov function candidate.

The **passivity** of nonlinear systems was defined as follows [Narendra and Taylor 1973], [Ortega et al. 1998].

DEFINITION 2.9-1 *Consider the system shown in Figure 2.9.1 and assume that it has the same number of inputs and outputs i.e. $u(t)$ and $y(t)$ have the same dimension.*

1. **Passivity:** *The system is said to be passive if*

$$\int_0^T y^T(t)u(t)dt \geq \gamma$$

for all finite $T > 0$ and some $\gamma > -\infty$.

2. **Strict Passivity:** *The system is said to be strictly passive if there exists a $\delta > 0$ and $\gamma > -\infty$ such that*

$$\int_0^T y^T(t)u(t)dt \geq \delta \int_0^T u^T(t)u(t)dt + \gamma$$

for all finite $T > 0$. ■

EXAMPLE 2.9-1: Passivity of Rigid Robots

Consider again the robot equation

$$M(q)\ddot{q} + V(q,\dot{q}) + G(q) = \tau$$

Suppose the robot is representing a system whose input is τ and whose output is the joint velocity $\dot{q}$. Let the sum of the kinetic energy and potential energy of the robot be denote by the Hamiltonian H and recall [Ortega and Spong 1988]

$$\frac{dH}{dt} = \dot{q}^T \tau$$

then

$$\int_0^T \dot{q}^T(t)\tau(t)dt = H(t) - H(0) \geq -H(0)$$

which proves that from τ to $\dot{q}$, the rigid robot is a passive system. ∎

A passive system is in effect one that does not create energy.

Positive-Real Systems

If the system under consideration is linear and time-invariant, then passivity is equivalent to **positivity** and may be tested in the frequency domain [Narendra and Taylor 1973]. In fact, let us describe positive-real systems and discuss some of their properties. Consider the multi-input-multi-output linear time-invariant system

$$\dot{x} = Ax + Bu$$

$$y = Cx + Du$$

where x is an n vector, u is an m vector, y is a p vector, A, B, C, and D are of the appropriate dimensions. The corresponding transfer function matrix is

$$P(s) = C(sI - A)^{-1}B + D$$

We will assume that the system has an equal number of inputs and outputs, i.e. $p = m$. To simplify our notation we will denote the **Hermitian** part of a real, rational transfer matrix $T(s)$ by $He[T(s)] = \frac{1}{2}[T(s) + T^T(s^*)]$ where s^* is the complex conjugate of s. A number of definitions have been given for SPR functions and matrices [Narendra and Taylor 1973]. It appears that the most useful definition for control applications is the following.

DEFINITION 2.9-2 *An $m \times m$ matrix $T(s)$ of proper real rational functions which is not identically zero is positive-real or PR if*

2.9 Advanced Stability Results

1. All elements of $T(s)$ have no poles in the region $Re(s) > 0$,

2. Any poles of $T(s)$ on the jw axis are simple with positive-definite residues, and

3. The matrix $He[T(s)]$ is positive semidefinite for $Re(s) > 0$.

■

EXAMPLE 2.9-2: PR Matrices
Consider the matrix

$$T(s) = \begin{bmatrix} 1/(s+1) & 2/s \\ 1 & (s+1)/(s+2) \end{bmatrix}$$

This matrix is PR as can be checked.

■

DEFINITION 2.9-3 An $m \times m$ matrix $T(s)$ of proper real rational functions which is not identically zero is strictly-positive-real or SPR if

1. All elements of $T(s)$ have no poles in the region $Re(s) \geq 0$, and

2. The matrix $He[T(s)]$ is positive definite for $Re(s) > 0$.

■

EXAMPLE 2.9-3: SPR Matrices
Consider the matrix

$$T(s) = \begin{bmatrix} 1/(s+1) & 2/(s+3) \\ 1 & (s+1)/(s+2) \end{bmatrix}$$

This matrix is SPR as can be checked.

■

Lure's Problem

Consider the following feedback interconnected system

$$\begin{aligned} \dot{x}(t) &= Ax(t) + Bu(t) \\ y(t) &= Cx(t) + Du(t) \\ u(t) &= -\phi[t, y(t)] \end{aligned} \quad (2.9.3)$$

where ϕ is continuous in both arguments. Lure then stated the following **Absolute Stability** problem which became known as **Lure's Problem**: Suppose the system described by the above equations is given where:

1. All eigenvalues of A have negative real parts or that A has one eigenvalue at the origin while the rest of them is in the open left-half plane (OLHP), and

2. (A, b) is controllable, and

3. (c, A) is observable, and

4. The nonlinearity $\phi(.,.)$ satisfies

    (a) $\phi(t, 0) = 0 \; ; \; \forall t \geq 0$ and
    (b) $y\phi[t, y(t)] \geq 0 \; ; \; \forall y \in R, \; \forall t \geq 0$

Then, find conditions on the linear system (A, B, C, D) such that $x = 0$ is a GAS equilibrium point of the closed-loop system.

Note that sometimes when we know more about the nonlinearity ϕ, the last condition above is replaced by the following:

$$k_1 y^2 \leq y\phi[t, y(t)] \leq k_2 y^2 \; ; \; \forall y \in R, \; \forall t \geq 0$$

where $k_2 \geq k_1 \geq 0$ are constants. We then say that ϕ belongs to the sector $[k_1, k_2]$.

The first attempt to solve this problem was made by Aizerman in what is now known as Aizerman's conjecture [Vidyasagar 1992] followed by the efforts of Kalman [Vidyasagar 1992]. The correct solution however, was not available until the

In order to present the correct results, we need to present the KY and MKY lemmas.

The MKY Lemma

The following lemmas are versions of the Meyer-Kalman-Yakubovich (MKY) lemma which appears in [Narendra and Taylor 1973], [Khalil 2001] amongst other places, and will be useful in designing adaptive controllers for robots.

LEMMA 2.9-1: Meyer-Kalman-Yakubovitch *Let the system (2.9.3) with $D = 0$ be controllable. Then the transfer function $c(sI - A)^{-1}b$ is SPR if and only if*

2.9 Advanced Stability Results

1. For any symmetric, positive-definite Q, there exists a symmetric, positive-definite P solution of the Lyapunov equation

$$A^T P + PA = -Q$$

2. The matrices B and C satisfy

$$C = B^T P$$

∎

The MKY Lemma gives conditions under which a transfer matrix has a degree of robustness. Note that the conditions depend on both the input and output matrices and thus a particular system may be SPR for a certain choice of input/output pairs and not SPR for others. A modified version of the KY lemma which relaxes the condition of controllability is given next.

LEMMA 2.9-2: **Meyer-Kalman-Yakubovitch** *Given vector b, an asymptotically stable A, a real vector v, scalars $\gamma \geq 0$ and $\epsilon > 0$, and a positive-definite matrix Q, then, there exist a vector q and a symmetric positive definite P such that*

1.
$$A^T P + PA = -qq^T - \epsilon Q$$

2.
$$b^T P = v^T + \sqrt{\gamma} q^T$$

if and only if

1. ϵ is small enough and,

2. the transfer function $\gamma/2 + v^T(sI - A)^{-1}b$ is SPR

∎

In most of our applications, $q = 0$. These lemmas find many applications in nonlinear systems. It is usually possible to divide a nonlinear system into a linear feed-forward subsystem, and a nonlinear, passive feedback. The challenge is then to make the linear subsystem SPR so that the stability of the combined system is guaranteed.

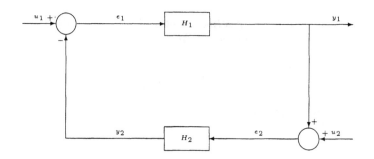

Figure 2.9.2: Feedback interconnection of two nonlinear systems.

2.10 Useful Theorems and Lemmas

Consider the block diagram shown in Figure 2.9.2. The blocks labelled H_1 and H_2 represent two systems (linear or nonlinear) which operate on the inputs e_1 and e_2 as follows

$$y_1 = H_1 e_1 = H_1(u_1 - y_2)$$

$$y_2 = H_2 e_2 = H_2(u_2 + y_1)$$

Let H_1 be an $m \times p$ matrix function, and H_2 be a $p \times m$ matrix function. Therefore, u_1 and y_2 are $p \times 11$ vectors, while u_2 and y_1 are $m \times 11$ vectors. The first theorem gives sufficient conditions to guarantee that the closed-loop system shown in Figure 2.9.2 is BIBO stable and is given in [Desoer and Vidyasagar 1975].

Small-Gain Theorem

THEOREM 2.10-1: Small gain Theorem Let $H_1 : L_{ep}^p \longrightarrow L_{ep}^m$ and $H_2 : L_{ep}^m \longrightarrow L_{ep}^p$. Therefore H_1 and H_2 satisfy the inequalities

$$\|(H_1 e_1)_T\| \leq \gamma_1 \|e_{1T}\| + \beta_1 \; ; \; \gamma_1 > 0, \; \beta_1 \in R$$

$$\|(H_2 e_2)_T\| \leq \gamma_2 \|e_{2T}\| + \beta_2 \; ; \; \gamma_2 > 0, \; \beta_2 \in R$$

for all $T \in [0, \infty)$ and suppose that $u_1 \in L_\infty^p$, $u_2 \in L_\infty^m$. If

$$\gamma_1 \gamma_2 < 1$$

2.10 Useful Theorems and Lemmas

Then e_1, y_2, $\in L_\infty^p$ and y_1, $e_2 \in L_\infty^p$. ∎

Basically, the small-gain theorem states that a feedback interconnection of two systems is BIBO stable if the loop-gain is less than unity. In other words, if a signal traverses the feedback-loop and decreases in magnitude then the closed-loop system can not go unstable.

EXAMPLE 2.10-1: Applications of Small-Gain Theorem

Let the feedback connection of Figure 2.9.2 be such that

$$Y_1(s) = \frac{0.5}{s+1} E_1(s)$$

$$y_2(t) = \sqrt{|e_2(t)|}$$

Then, we can show that

$$\gamma_1 = 0.5 \; ; \; \beta_1 = 0$$

$$\gamma_2 = 1 \; ; \; \beta_2 = 1$$

Since $\gamma_1 \gamma_2 = 0.5 < 1$, the feedback connection is BIBO stable. ∎

The next theorem provides a test for the stability of a nonlinear system based on the stability of its linear part. In fact, a certain degree of robustness is achieved if the linear system is exponentially stable as described in [Anderson et al. 1986].

Total Stability Theorem

THEOREM 2.10-2: **Total Stability** *Consider the state-space system described by*

$$\dot{x} = A(t)x + f(t,x) + g(t,x) \; ; \; x(0) = x_0$$

where

1. *The system $\dot{x} = A(t)x$ is exponentially stable i.e. there are some $a > 0$ and $K \geq 1$, such that*

$$||x(t)|| \leq K e^{-at} \; ; \; t \geq 0$$

2. $f(t,0) = 0$ *i.e. the origin is an equilibrium point of* $f(t,x)$
3. $\| f(t,x_1) - f(t,x_2) \| \leq \beta_1 \| x_1 - x_2 \|$, *for some* $\beta_1 > 0$
4. $\| g(t,x_1) \| \leq \beta_2 r$, *for some* $\beta_2 > 0$
5. $\| g(t,x_1) - g(t,x_2) \| \leq \beta_2 \| x_1 - x_2 \|$
6. $\| x_0 \| < r/K$, $\frac{(\beta_1+\beta_2)K}{a} < 1$.

Then, there exists a unique solution $x(t)$ *to 1.5.5 and*

$$\|x(t)\| \leq Ke^{(\beta_1 K-a)t}\|x_0\| + \frac{K\beta_2}{a-\beta_1 K}r[1 - e^{(\beta_1 K-a)t}] \leq r$$

∎

The total stability theorem will be used to design controllers that will make the linear part of the system exponentially stable. In effect, this theorem guarantees that if the linear part of the system is "very" stable (exponentially stable), the destabilizing effect of the bounded nonlinearities may not be sufficient to destabilize the system and the state will remain bounded.

EXAMPLE 2.10-2: Total Stability

Consider the nonlinear system

$$\dot{x}(t) = -2x(t) + \frac{x}{2+t} + 1$$

Let $|x_0| < 1$, and note the following

$$K = 1 \; ; \; a = 2 \; ; \; \beta_1 = 0.5 \; ; \; \beta_2 = 1.$$

Note first that all conditions of the theorem are satisfied, then there exists a unique solution $x(t)$ which is bounded by

$$|x(t)| \leq e^{\frac{-3}{2}t}|x_0| + \frac{2}{3}(1 - e^{\frac{-3}{2}t})$$

∎

A version of the Bellman-Gronwall lemma is proved in [Sastry and Bodson 1989] and is presented next.

2.10 Useful Theorems and Lemmas

LEMMA 2.10-1: **Bellman-Gronwall** *Let $x(.), a(.), b(.) : [0, \infty) \longrightarrow [0, \infty)$, and $T \geq 0$. Suppose that for all $t \in [0, T]$, the following inequality holds*

$$x(t) \leq \int_0^t a(\tau)x(\tau)d\tau + b(t)$$

Then for all $t \in [0, T]$

$$x(t) \leq \int_0^t [a(\tau)b(\tau)e^{\int_\tau^t a(\sigma)d\sigma}]d\tau + b(t)$$

If $b(t)$ is further a constant, the following holds

$$x(t) \leq b e^{\int_0^t a(\sigma)d\sigma}$$

■

EXAMPLE 2.10-3: Bellman-Gronwall Lemma Example

Consider a LTI system given by

$$\dot{x} = Ax + Bu$$

and assume that A is a stable matrix. Then the solution of the state equation is given by

$$x(t) = e^{At}x(0) + \int_0^t e^{A(t-\tau)}Bu(\tau)d\tau$$

or

$$e^{-At}x(t) = x(0) + \int_0^t e^{-A\tau}Bu(\tau)d\tau$$

which, when taking the norm of each side becomes

$$||e^{-At}x(t)|| \leq ||x(0)|| + \int_0^t ||e^{-A\tau}||.||B||.||u(\tau)||d\tau$$

We can then use the Bellman-Gronwall lemma by letting $a(\tau) = ||B||.||u(\tau)||$ and $b(t) = ||x(0)||$, to obtain

$$\|e^{-At}x(t)\| \leq \|x(0)\|e^{\int_0^t \|B\|\cdot\|u(\sigma)\|d\sigma}$$

or finally

$$\|x(t)\| \leq \|e^{At}\|\cdot\|x(0)\|e^{\int_0^t \|B\|\cdot\|u(\sigma)\|d\sigma}$$

∎

The next Lemma may be used in the case of non-autonomous systems and leads to results similar to LaSalle's theorem. For a proof see for example [Khalil 2001]

LEMMA 2.10-2: **Barbalat** *Let $f(t)$ be a differentiable function of t.*

- **First Version:** *If $\dot{f}(t) = df/dt$ is uniformly continuous and $\lim_{t\to\infty} f(t) = k < \infty$, then $\lim_{t\to\infty} \dot{f}(t) = 0$*

- **Second Version:** *If $f(t) \geq 0$, $\dot{f}(t) \leq 0$ and $\ddot{f}(t)$ bounded, then $\lim_{t\to\infty} \dot{f}(t) = 0$.*

∎

EXAMPLE 2.10-4: Barbalat Lemma Example
1. Consider $f(t) = e^{-t} + 1$, then $dot f(t) = -e^{-t}$ which is uniformly continuous. On the other hand $\lim_{t\to\infty}(e^{-t} + 1) = 1$ therefore $\lim_{t\to\infty} \dot{f}(t) = 0$.
2. As a second example, consider $f(t) = 1/(1+t)$, with $t > 0$. Using the second version of Barbalat's lemma, we can show that $\lim_{t\to\infty} \dot{f}(t) = 0$.

∎

LEMMA 2.10-3: *Consider the quadratic equation*

$$P(x) = ax^2 + bx + c$$

where $a, b,$ and c are positive constants. Then $P(x) < 0$ if

$$b > 2\sqrt{ac}$$

for all $x_1 < x < x_2$ where

$$x_1 = \frac{-b - \sqrt{b^2 - 4ac}}{2a}, \quad x_2 = \frac{-b + \sqrt{b^2 - 4ac}}{2a}$$

2.11 Linear Controller Design

THEOREM 2.10-3: Let $V(x)$ be a Lyapunov function of a continuous-time system that satisfies the following properties:

$$\begin{aligned} \lambda_1 ||x||^2 &< V(x) < \lambda_2 ||x||^2 \\ \dot{V}(x) &< 0 \quad x_1 < x < x_2 \\ x(0) &= 0 \end{aligned}$$

then $x(t)$ is uniformlt bounded ∎

THEOREM 2.10-4: Let $V(x)$ be a Lyapunov function of a continuous-time system that satisfies the following properties:

$$\begin{aligned} \gamma_1(||x||) &\leq V(x) \leq \gamma_2(||x||) \\ \dot{V}(x) &\leq -\gamma_3(||x||) + \gamma_3(\eta) \end{aligned}$$

where η is a positive constant, γ_1 and γ_2 are continuous, strictly increasing functions, and γ_3 is a continuous, nondecreasing function. Then, if

$$\dot{V}(x) < 0, \quad \text{for } ||x(t)|| > \eta$$

$x(t)$ is uniformly ultimately bounded. In addition, if $x(0) = 0$, $x(t)$ is uniformly bounded. ∎

2.11 Linear Controller Design

The purpose of control design is to make the robot respond in a predictable and desirable fashion to a set of input signals. In this section we review how a linear controller may be designed in order that the behavior of the robot-controller combination is acceptable. It seems obvious that the firs requirement on the robot-controller is its stability. Therefore, the first function of the controller is to stabilize the robot when it is moving in space. In fact, we can now pose the problem we wish to solve:

Linear Control Design Problem: Consider a LTI system. Design a feedback controller that operates on the output $y(t)$ without differentiation, and generates an input $u(t)$ so that the system goes from any initial state $x(0)$ to a specified desired final state x_d in some finite time. This problem has 2 parts to it:

1. Controllability: Can the problem be solved if $y = x$?

2. Observability: If so, how do I get x from y?

In the next section we introduce these concepts and find tests that will allow us to answer the questions posed. Given is a MIMO LTI system

$$\dot{x}(t) = Ax(t) + Bu(t); \ x(0) = x_0$$
$$y(t) = Cx(t) + Du(t) \qquad (2.11.1)$$

where $x \in R^n$, $u \in R^m$, $y \in R^p$ and A, B, C, D are of the appropriate dimensions.

DEFINITION 2.11-1 A state x_0 is controllable if there exists an input $u(t)$, for $0 \le t \le t_1$ such that $x(t_1) = 0$, for some finite t_1. The system itself is said to be controllable if all $x_0 \in R^n$ are controllable. ■

Note that the input $u(t)$ is not a feedback function of the state $x(t)$. Let us recall the solution of the differential equation given by

$$x(t) = exp(At)x_0 + \int_0^t e^{A(t-\tau)} Bu(\tau) d\tau$$

and evaluate it at t_1 assuming that x_0 is controllable so that $x(t_1) = 0$ to obtain

$$x_0 = -\int_0^{t_1} e^{-A\tau} Bu(\tau) d\tau \qquad (2.11.2)$$

Therefore, if x_0 is controllable, it has to satisfy the last equation. This however is not an easy test. Fortunately, we have a much simpler test given in the following theorem.

THEOREM 2.11-1: *A necessary and sufficient condition for the complete controllability of the LTI systems is that, The $n \times nm$ matrix*

2.11 Linear Controller Design

$$\mathcal{C} = [B \ \ AB \ \ A^2B \ \ \cdots \ \ A^{n-1}B] \tag{1}$$

is full rank, i.e. rank$\mathcal{C} = n$

■

Next, We show that controllability is sufficient to stabilize the system (2.11.1) by placing the eigenvalues of the A matrix

THEOREM 2.11-2: Let (A, b) be controllable. Then, there exists a constant gain matrix K such that $u = -Kx + v$ will place the eigenvalues of $A - bk$ anywhere in the s plane. ■

The state-feedback gain needed to place the eigenvalues may be found in the single-input case using Ackermann's formula (see for example [Antsaklis and Michel 1997]). Assume the desired closed-loop eigenvalues are specified as the roots of the equation

$$\phi(s) = s^n + a_{n-1}s^{n-1} + \cdots + a_1 s + a_0. \tag{2.11.3}$$

Then, the state-feedback controller is given by

$$\begin{aligned} u &= -Kx + v \\ K &= [0 \ 0 \ \cdots \ 0 \ 1]\mathcal{C}^{-1}\phi(A) \end{aligned} \tag{2.11.4}$$

where $\mathcal{C}$ is the controllability matrix of (2.11.1), and $\phi(A)$ is the matrix obtained by evaluating $\phi(s)$ at A, i.e.

$$\phi(A) = A^n + a_{n-1}A^{n-1} + \cdots + a_1 A + a_0 I. \tag{2.11.5}$$

EXAMPLE 2.11-1: Control Design for Double Integrator

Consider the double integrator system

$$\begin{aligned} \dot{x} &= \begin{bmatrix} 0 & 1 \\ 0 & 0 \end{bmatrix} x + \begin{bmatrix} 0 \\ 1 \end{bmatrix} u \\ y &= [1 \ \ 0] x \end{aligned}$$

Suppose that the desired closed-loop poles are the roots of the equation

$$s^2 + 2\zeta w_n s + w_n^2.$$

Then by Ackermann's formula

$$u = -Kx$$
$$K = [0\ 1]\begin{bmatrix} 0 & 1 \\ 1 & 0 \end{bmatrix}\begin{bmatrix} w_n^2 & 2\zeta w_n \\ 0 & w_n^2 \end{bmatrix} = [w_n^2\ 2\zeta w_n] \qquad (1)$$

∎

Next, we introduce the concept of observability and find an observability test.

DEFINITION 2.11-2 The state $x(0) = x_0$ is said to be observable if knowledge of $u(t)$, $y(t)$; $0 \le t_1$ is enough to uniquely determine x_0. Note that once $x(0)$ is known, so is $x(t)$ for $t \ge 0$. The system is said to be completely observable if every initial state is observable. ∎

This problem is a dual problem to the controllability problem as will become obvious in the following.

THEOREM 2.11-3: *A necessary and sufficient condition for the complete observability of the LTI systems is that The $np \times n$ matrix*

$$\mathcal{O} = \begin{bmatrix} C \\ CA \\ CA^2 \\ \vdots \\ CA^{n-2} \\ CA^{n-1} \end{bmatrix} \qquad (1)$$

is full rank, i.e. rank$\mathcal{O} = n$ ∎

EXAMPLE 2.11-2: Observer-Controller Design

Consider the system

$$\dot{x} = \begin{bmatrix} 0 & 1 \\ 0 & 0 \end{bmatrix} x + \begin{bmatrix} 0 \\ 1 \end{bmatrix} u$$
$$y = [0\ 1]x$$

2.11 Linear Controller Design

then

$$\mathcal{C} = \begin{bmatrix} 0 & 1 \\ 1 & 0 \end{bmatrix}$$

$$\mathcal{O} = \begin{bmatrix} 0 & 1 \\ 0 & 0 \end{bmatrix}$$

Therefore, this system is controllable but not observable. On the other hand, the system

$$\dot{x} = \begin{bmatrix} 0 & 0 \\ 1 & 0 \end{bmatrix} x + \begin{bmatrix} 0 \\ 1 \end{bmatrix} u$$
$$y = [0 \ 1] x$$

then

$$\mathcal{C} = \begin{bmatrix} 0 & 0 \\ 1 & 0 \end{bmatrix}$$

$$\mathcal{O} = \begin{bmatrix} 0 & 1 \\ 1 & 0 \end{bmatrix}$$

so this one is observable but not controllable. The trouble is that both of these realizations have the same transfer function $H(s) = c(sI-A)^{-1}b = s/s^2 = s!$.

■

Suppose we have 2 realizations of the transfer function $H(s)$ related by a state-space transformation,

$$\dot{x} = Ax + bu$$
$$y = cx + du \quad (2.11.6)$$

and

$$\dot{\bar{x}} = T^{-1}ATx + T^{-1}bu = \bar{A}\bar{x} + \bar{b}u$$
$$y = cT\bar{x} + du = \bar{c}\bar{x} + \bar{d}u$$

Then, let us find the controllability matrix $\bar{\mathcal{C}}$ of $\bar{\Sigma}$

$$\begin{aligned}\bar{\mathcal{C}} &= [T^{-1}b \ \ T^{-1}ATT^{-1}b \ \ \cdots \ \ T^{-1}A^{n-1}b] \\ &= T^{-1}\mathcal{C}\end{aligned} \qquad (2.11.7)$$

Therefore, rank$\mathcal{C}$ = rank $\bar{\mathcal{C}}$, and controllability is therefore preserved under state-space transformation.

By combining the concepts of observability and controllability, we can design compensators that solve the **Linear Control Design** problem. In fact, the following theorem summarizes linear control design.

THEOREM 2.11-4: *The* **Linear Control Design** *problem is solvable for a system 2.11.1, if and only if the state-space realization is both observable and controllable.* ∎

These compensators are known as the observer-controller compensators and are shown for example in Figure 2.11.1. In the SISO case, a transfer function admits a state-space realization which is completely observable and controllable if and only if no pole-zero cancellations occur [Kailath 1980], [Antsaklis and Michel 1997]. The next example shows an observer-controller compensator chosen to obtain desired closed-loop poles.

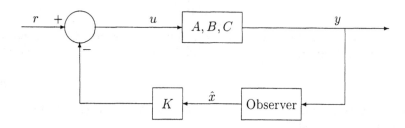

Figure 2.11.1

EXAMPLE 2.11-3: Observer-Controller Design

Consider again the double integrator system. It is easily verified that it is a controllable and observable system. Suppose that the desired closed-loop poles are the roots of the equation

2.11 Linear Controller Design

$$s^2 + 2\zeta w_n s + w_n^2$$

This can be achieved using the structure of Figure 2.11.1 using the gain $K = [w_n^2 \ \ 2\zeta w_n]$ found in Example 2.11.1

∎

The double integrator system being both controllable and observable, also lends itself to the most general controller design technique for linear systems, namely the two-degree -f freedom (2-DOF) design. This allows the controlled closed-loop system to match any desired model as descried in [Kailath 1980]. This approach goes beyond the pole or eigenvalue placement designs, as one can now move both poles **and** zeros, a goal not achievable with the observer-controller compensators. This design is illustrated in the following example.

EXAMPLE 2.11-4: Two DOF Design

Consider again the double integrator system, and assume the desired closed-loop transfer function is given by

$$T(s) = \frac{as+b}{s^2+k_v s+k_p}$$

where a, b, k_v, and k_p are given constants. Then using the model-following configuration in Figure 2.11.2, and simplifying terms, we obtain

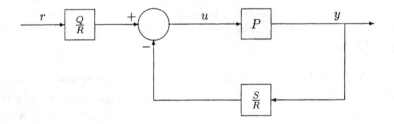

Figure 2.11.2

$$\frac{Q(s)}{s^2 R(s) + S(s)} = \frac{as+b}{s^2 + k_v s + k_p}$$

or

$$(s^2 + k_v s + k_p)Q(s) = (as + b)[s^2 R(s) + S(s)].$$

At this stage, we may choose for example $R(s) = 1$ and suppose for the purpose of illustration that $a = b = k_p = 1$ and that $k_v = 9$. Then we solve for $S(s) = 9s$ and $Q(s) = s$. Note that in this case, the feedback block $S(s)/R(s)$ is improper but this does not cause any problems in cases (such as robot control) where velocity measurements are available.

■

The question addressed next is to find conditions on (2.11.1) so that a static-output feedback controller will render the closed-loop system SPR. We will consider the SISO case only since this is the only case encountered in this book.

$$\begin{aligned} \dot{x} &= (A - \gamma BKC)x + BKr \\ y &= Cx \end{aligned} \qquad (2.11.8)$$

or in the frequency-domain

$$Y(s) = [I + \gamma P(s)K]^{-1} P(s) K R(s) = \frac{P(s)K}{1 + \gamma P(s)K} R(s) \quad (2.11.9)$$

We present a simple frequency domain result to show the existence of K and γ that will render the closed-loop system SPR. The result first appeared in [Gu 1988].

THEOREM 2.11-5: *Let system (2.2.11) have no common poles and zeros. Then there exists a nonsingular K and a positive scalar γ such that the closed-loop system (2.11.8, 2.11.9) is SPR, if and only if $P(s)$ has no zeros in the right half plane and if $P(s)$ has n poles and $n - 1$ zeros. In fact, one such K is given by*

$$K = P(CB)^{-1}$$

where P is any symmetric, positive-definite matrix.

■

EXAMPLE 2.11-5: SPR Design

Consider the double integrator system

$$\dot{x} = \begin{bmatrix} 0 & 1 \\ 0 & 0 \end{bmatrix} x + \begin{bmatrix} 0 \\ 1 \end{bmatrix} u$$

$$y = [c_1 \quad c_2] x$$

where $c_2 \neq 0$ and c_1, c_2 are of the same sign. The open-loop transfer function is then

$$P(s) = \frac{c_1 + c_2 s}{s^2}$$

which has relative degree 1 and is minimum phase. Note that $CB = c_2$ which is invertible. The gain K is chosen to be

$$K = \frac{p}{c_2}$$

where $p > 0$. The closed-loop system is then

$$T(s) = \frac{p(c_1 + c_2 s)}{c_2 s^2 + \gamma p c_2 s + \gamma p c_1}$$

which is stable for any choice of $\gamma > 0$ and is SPR if $\gamma > c_1/(pc_2$ (check!). ■

2.12 Summary and Notes

This chapter is meant to be a review of various control theory concepts that will be used in the remainder of the book. The emphasis has been to include enough material so that readers with little or no background in control theory are able to follow the design of robot controllers. Simplifications were often made to concentrate on the systems studied in this book, namely, mechanical systems and systems described by the Lagrange-Euler equations. In particular, the double integrator system was illustrated in detail, since it will result from applying a preliminary nonlinear feedback on mechanical manipulators. In Chapter 3, we introduce the dynamical description of rigid robot manipulators, and in subsequent chapters, the control concepts reviewed in this chapter will be implemented on those manipulators.

Most of the results reviewed here can be found in greater details in the books by [Kailath 1980], [Antsaklis and Michel 1997] for the linear systems case, and in [Khalil 2001], [Vidyasagar 1992] in the nonlinear systems case.

REFERENCES

[Anderson et al. 1986] B.D.O. Anderson and R. Bitmead and C.R. Johnson, Jr. and P.V. Kokotovic and R.L. Kosut and I.M.Y. Mareels and L. Praly and B.D. Riedle. "Stability of Adaptive Systems: Passivity and Averaging Analysis". *MIT Press*, Cambridge, MA, 1986.

[Antsaklis and Michel 1997] P. Antsaklis and A.N. Michel. "Linear Systems". *McGraw Hill*, New York, 1997.

[Åström and Wittenmark 1995] K.Åström and B. Wittenmark. "Adaptive Control". *Addison-Wesley*, Reading, MA, 2nd edition, 1995.

[Åström and Wittenmark 1996] K.Åström and B. Wittenmark. "Computer-Controlled Systems: Theory and Design", *Prentice-Hall*, Englewood Cliffs, NJ, 1996.

[Boyd and Barratt] S. Boyd and G. Barratt. "Linear Controller Design: Limits of Performance", *Prentice-Hall*, Englewood Cliffs, NJ, 1991.

[Desoer and Vidyasagar 1975] C. Desoer and M. Vidyasagar. "Feedack Systems: Input-Output Properties", *Academic Press*, New York, 1975.

[Franklin et al. 1997] G. Franklin and D. Powell and M. Workman. "Digital Control of Dynamic Systems", *Addison-Wesley*, Reading, MA, 1997.

[Gu 1988] G. Gu. "Stabilizability Conditions for Multivariable Uncertain systems via Output Feedback", *IEEE Trans. Autom. Control*, vol. AC-35, pp. 988–992, 1988.

[Hahn 1967] W. Hahn. "Theory and Applications of Lyapunov's Direct Method". *Prentice-Hall*, Englewood Cliffs, NJ, 1967.

[Horn and Johnson 1991] R. Horn and C. Johnson. "Topics in Matrix Analysis". *Cambridge University Press*, New York, 1991.

[Kailath 1980] T. Kailath. "Linear Systems". *Prentice-Hall*, Englewood Cliffs, NJ, 1980.

[Khalil 2001] . H. Khalil. "Nonlinear Systems". *Prentice-Hall*, Englewood Cliffs, NJ, 2001.

[LaSale and Lefschetz 1961] P. LaSale and S. Lefschetz. "Stability by Lyapunov's Direct Method". *Academic Press*, New York, 1961.

[Ljung 1999] L. Ljung. "System Identification-Theory for the User". *Prentice-Hall*, Englewood Cliffs, NJ, 1999.

[Narendra and Taylor 1973] K.S. Narendra and J.H. Taylor. "Frequency Domain Criteria for Absolute Stability". *Academic Press*, New York, 1973.

[Niculescu and Abdallah 2000] S.-I. Niculescu and C.T. Abdallah. "Delay Effects on Static Output Feedback Stabilization". *Submitted IEEE CDC 2000*, Sydney, Australia, 2000.

[Ortega and Spong 1988] R. Ortega and M. Spong. "Adaptive Motion Control of Rigid Robots: A Tutorial". *Proc. IEEE Conf. Dec. and Cont.*, pp. 1575–1584, 1988.

[Ortega et al. 1998] R.Ortega and A.Loria and P.J. Nicklasson and H. Sira-Ramirez. "Passivity-based Control of Euler-Lagrange Systems".*Springer-Verlag*, Berlin, 1998.

[Strang 1988] G. Strang. "Linear Algebra and its Applications". *Brooks/Cole*, Stamford, CT, 1988.

[Koltchinskii et al. 1999] V. Koltchinskii and C. T. Abdallah and M. Ariola and P. Dorato and D. Panchenko. "Statistical Learning Control of Uncertain Systems: It is better than it seems". *IEEE Trans. on Automatic Control*, February 1999.

[Sastry and Bodson 1989] S. Sastry and M. Bodson. "Adaptive Control: Stability, Convergence, and Robustness". *Prentice-Hall*, Englewood Cliffs, NJ, 1989.

[ATM 1996] "ATM Forum Traffic Management Specification". *ATM Forum Traffic Management Working Group AF-TM-0056.000*, version 4.0, April 1996.

[Niculescu et al. 1997] S.-I. Niculescu and E. Verriest and L. Dugard and J.-M. Dion. "Stability and robust stability of time-delay systems: A

REFERENCES

guided tour". *LNCIS, Springer-Verlag*, London, vol. 228, pp.1–71, 1997.

[CDC 1999] "CDC 1999 Special session on time-delay systems". *Proc. 1999 CDC*, Phoenix, AZ, 1999.

[Verhulst 1997] H. Verhulst. "Nonlinear Differential Equations and Dynamical Systems". *Springer-Verlag*, Berlin, 1997.

[Vidyasagar 1992] M. Vidyasagar. "Nonlinear Systems Analysis". *Prentice-Hall*, Englewood Cliffs, NJ, 1992.

Chapter 3

Robot Dynamics

This chapter provides the background required for the study of robot manipulator control. The arm dynamical equations are derived both in the second-order differential equation formulation and several state-variable formulations. Some important properties of the dynamics are introduced. We show how to include the dynamics of the arm actuators, which may be electric or hydraulic motors.

3.1 Introduction

Robotics is a complex field involving many diverse disciplines, such as physics, properties of materials, statics and dynamics, electronics, control theory, vision, signal processing, computer programming, and manufacturing. In this book our main interest is control of robot manipulators. The purpose of this chapter is to study the dynamical equations needed for the study of robot control.

For those desiring a background in control theory, Chapter 2 is provided. For those desiring a background in the basics of robot manipulators, in Appendix A we examine the geometric structure of robot manipulators, covering basic manipulator configurations, kinematics, and inverse kinematics. There we review as well the manipulator Jacobian, which is essential for control in Cartesian or workspace coordinates, where the desired trajectories of the arm are usually specified to begin with.

The robot dynamics are derived in Section 3.2. Lagrangian mechanics are used in this derivation. In Section 3.3 we review some fundamental properties of the arm dynamical equation that are essential in subsequent chapters for the derivation of robot control schemes. These are summarized in Table 3.3.1, which is referred to throughout the text.

The arm dynamics in Section 3.2 are in the form of a second-order vector differential equation. In Section 3.4 we show several ways to convert this formulation to a *state-variable description*. The state-variable description is a first-order vector differential equation that is extremely useful for developing many arm control schemes. Feedback linearization techniques and Hamiltonian mechanics are used in this section.

The robot arm dynamics in Section 3.2 are given in joint-space coordinates. In Section 3.5 we show a very general approach to obtaining the arm dynamical description in any desired coordinates, including Cartesian or workspace coordinates and the coordinates of a camera frame or reference.

In Section 3.6 we analyze the electrical or hydraulic actuators that perform the work required to move the links of a robot arm. It is shown how to incorporate dynamical models for the actuators into the arm dynamics to provide a complete dynamical description of the arm-plus-actuator system. This finally leaves us in a position to move on to the next chapters, where robot manipulator control design is discussed.

3.2 Lagrange-Euler Dynamics

For control design purposes, it is necessary to have a mathematical model that reveals the dynamical behavior of a system. Therefore, in this section we derive the dynamical equations of motion for a robot manipulator. Our approach is to derive the kinetic and potential energy of the manipulator and then use Lagrange's equations of motion.

In this section we ignore the dynamics of the electric or hydraulic motors that drive the robot arm; actuator dynamics is covered in Section 3.6.

Force, Inertia, and Energy

Let us review some basic concepts from physics that will enable us to better understand the arm dynamics [Marion 1965]. In this subsection we use boldface to denote vectors and normal type to denote their magnitudes.

The *centripetal force* of a mass m orbiting a point at a radius r and angular velocity ω is given by

$$F_{\text{cent}} = \frac{mv^2}{r} = m\omega^2 r = m\dot{\theta}^2 r. \tag{3.2.1}$$

See Figure 3.2.1. The linear velocity is given by

$$\mathbf{v} = \omega \times \mathbf{r}, \tag{3.2.2}$$

which in this case means simply that $v = \omega r$.

3.2 Lagrange-Euler Dynamics

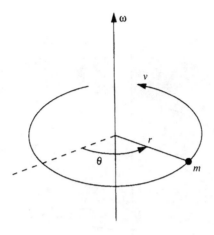

Figure 3.2.1: Centripetal force.

Imagine a sphere (i.e., the earth) rotating about its center with an angular velocity of ω_0. See Figure 3.2.2. The *Coriolis force* on a body of mass m moving with velocity $\mathbf{v}$ on the surface of the sphere is given by

$$\mathbf{F}_{\text{cor}} = -2m\omega_0 \times \mathbf{v} \qquad (3.2.3)$$

Using the right-handed screw rule (i.e., if the fingers rotate ω_0 into $\mathbf{v}$, the thumb points in the direction of $\omega_0 \times \mathbf{v}$, we see that, in the figure, the Coriolis force acts to deflect m to the right.

In a low-pressure weather system, the air mass moves toward the center of the low. The Coriolis force is responsible for deflecting the air mass to the right and so causing a counterclockwise circulation known as cyclonic flow. The result is the swirling motion in a hurricane. A brief examination of Figure 3.2.2 reveals that in the southern hemisphere $\mathbf{F}_{\text{cor}}$ deflects a moving mass to the left, so that a low-pressure system would have a clockwise wind motion.

Since $\omega_0 = \dot{\theta}$ and $v = R\dot{\varphi}$, we may write

$$F_{\text{cor}} = -2m\dot{\theta}\dot{\varphi}R\sin(90° + \varphi) = -2mR\dot{\theta}\dot{\varphi}\cos\varphi. \qquad (3.2.4)$$

It is important to note that the centripetal force involves the square of a single angular velocity, while the Coriolis force involves the product of two distinct angular velocities.

The *kinetic energy* of a mass moving with a linear velocity of v is

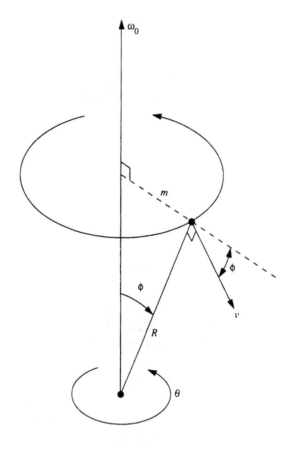

Figure 3.2.2: Coriolis force.

$$K = \tfrac{1}{2}mv^2. \tag{3.2.5}$$

The *rotational kinetic energy* of the mass in Figure 3.2.1 is given by

$$K_{\text{rot}} = \tfrac{1}{2}\mathbf{I}\omega^2, \tag{3.2.6}$$

where the *moment of inertia* is

$$I = \int_{\text{vol}} \rho(r) r^2 \, dr, \tag{3.2.7}$$

with $\rho(r)$ the mass distribution at radius r in a volume. In the simple case shown where m is a point mass, this becomes

3.2 Lagrange-Euler Dynamics

$$I = mr^2. \quad (3.2.8)$$

Therefore,

$$K_{\text{rot}} = \tfrac{1}{2} mr^2 \dot{\theta}^2. \quad (3.2.9)$$

The *potential energy* of a mass m at a height h in a gravitational field with constant g is given by

$$P = mgh. \quad (3.2.10)$$

The origin, corresponding to zero potential energy, may be selected arbitrarily since only differences in potential energy are meaningful in terms of physical forces.

The *momentum* of a mass m moving with velocity v is given by

$$\mathbf{p} = m\mathbf{v}. \quad (3.2.11)$$

The *angular momentum* of a mass m with respect to an origin from which the mass has distance r is

$$\mathbf{p}_{\text{ang}} = \mathbf{r} \times \mathbf{p}. \quad (3.2.12)$$

The *torque* or *moment* of a force $\mathbf{F}$ with respect to the same origin is defined to be

$$\mathbf{N} = \mathbf{r} \times \mathbf{F}. \quad (3.2.13)$$

Lagrange's Equations of Motion

Lagrange's equation of motion for a conservative system are given by [Marion 1965]

$$\frac{d}{dt} \frac{\partial L}{\partial \dot{q}} - \frac{\partial L}{\partial q} = \tau, \quad (3.2.14)$$

where q is an n-vector of generalized coordinates q_i, τ is an n-vector of generalized forces τ_i, and the *Lagrangian* is the difference between the kinetic and potential energies

$$L = K - P. \quad (3.2.15)$$

In our usage, q will be the joint-variable vector, consisting of joint angles θ_i; (in degrees or radians) and joint offsets d_i (in meters). Then τ is a vector that has components n_i of torque (newton-meters) corresponding to the joint angles, and f_i of force (newtons) corresponding to the joint offsets. Note that we denote the scalar components of τ by lowercase letters.

We shall use Lagrange's equation to derive the general robot arm dynamics. Let us first get a feel for what is going on by considering some examples.

EXAMPLE 3.2-1: Dynamics of a Two-Link Polar Arm

The kinematics for a two-link planar revolute/prismatic (RP) arm are given in Example A.2-3. To determine its dynamics examine Figure 3.2.3, where the joint-variable and joint-velocity vectors are

$$q = \begin{bmatrix} \theta \\ r \end{bmatrix}, \dot{q} = \begin{bmatrix} \dot{\theta} \\ \dot{r} \end{bmatrix}. \tag{1}$$

The corresponding generalized force vector is

$$\tau = \begin{bmatrix} n \\ f \end{bmatrix} \tag{2}$$

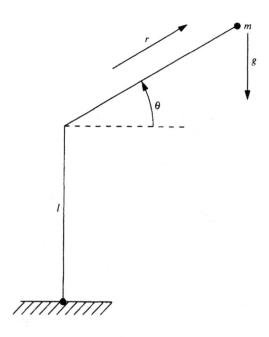

Figure 3.2.3: Two-link planar RP arm.

3.2 Lagrange-Euler Dynamics

with n a torque and f a force. The torque n and force f may be provided by either motors or hydraulic actuators. We discuss the dynamics of actuators in Section 3.6.

To determine the arm dynamics, we must now compute the quantities required for the Lagrange equation.

a. Kinetic and Potential Energy

The total kinetic energy due to the angular motion $\dot{\theta}$ and the linear motion $\dot{r}$ is

$$K = \tfrac{1}{2}mr^2\dot{\theta}^2 + \tfrac{1}{2}m\dot{r}^2 \tag{3}$$

and the potential energy is

$$P = mgr\sin\theta \tag{4}$$

b. Lagrange's Equation

The Lagrangian is

$$L = K - P = \tfrac{1}{2}mr^2\dot{\theta}^2 + \tfrac{1}{2}m\dot{r}^2 - mgr\sin\theta \tag{5}$$

Now we obtain

$$\frac{\partial L}{\partial \dot{q}} = \begin{bmatrix} \frac{\partial L}{\partial \dot{\theta}} \\ \frac{\partial L}{\partial \dot{r}} \end{bmatrix} = \begin{bmatrix} mr^2\dot{\theta} \\ m\dot{r} \end{bmatrix} \tag{6}$$

$$\frac{d}{dt}\frac{\partial L}{\partial \dot{q}} = \begin{bmatrix} mr^2\ddot{\theta} + 2mr\dot{r}\dot{\theta} \\ m\ddot{r} \end{bmatrix} \tag{7}$$

$$\frac{\partial L}{\partial q} = \begin{bmatrix} -mgr\cos\theta \\ mr\dot{\theta}^2 - mg\sin\theta \end{bmatrix} \tag{8}$$

Therefore, (3.2.14) shows that the arm dynamical equations are

$$mr^2\ddot{\theta} + 2mr\dot{r}\dot{\theta} + mgr\cos\theta = n \tag{9}$$

$$m\ddot{r} - mr\dot{\theta}^2 + mg\sin\theta = f \tag{10}$$

This is a set of *coupled nonlinear differential equations* which describe the motion $q(t) = [\theta(t) \quad r(t)]^T$ given the control input torque $n(t)$ and force $f(t)$. We shall show how to determine $q(t)$ given the control inputs $n(t)$ and $f(t)$ by *computer simulation* in Chapter 4.

Given our discussion on forces and inertias it is easy to identify the terms in the dynamical equations. The first terms in each equation are acceleration terms involving *masses and inertias*. The second term in (9) is a *Coriolis term*, while the second term in (10) is a *centripetal term*. The third terms are *gravity* terms.

c. Manipulator Dynamics

By using vectors, the arm equations may be written in a convenient form. Indeed, note that

$$\begin{bmatrix} mr^2 & 0 \\ 0 & m \end{bmatrix} \begin{bmatrix} \ddot{\theta} \\ \ddot{r} \end{bmatrix} + \begin{bmatrix} 2mr\dot{r}\dot{\theta} \\ -mr\dot{\theta}^2 \end{bmatrix} + \begin{bmatrix} mgr\cos\theta \\ mg\sin\theta \end{bmatrix}$$
$$= \begin{bmatrix} n \\ f \end{bmatrix} \tag{11}$$

We symbolize this vector equation as

$$M(q)\ddot{q} + V(q,\dot{q}) + G(q) = \tau \tag{12}$$

Note that, indeed, the inertia matrix $M(q)$ is a function of q (i.e., of θ and r), the *Coriolis/centripetal vector* $V(q,\dot{q})$ is a function of q and $\dot{q}$, and the *gravity vector* $G(q)$ is a function of q.

■

EXAMPLE 3.2-2: Dynamics of a Two-Link Planar Elbow Arm

In Example A.2-2 are given the kinematics for a two-link planar RR arm. To determine its dynamics, examine Figure 3.2.4, where we have assumed that the link masses are concentrated at the ends of the links. The joint variable is

$$q = [\theta_1 \quad \theta_2]^T \tag{1}$$

and the generalized force vector is

$$\tau = [\tau_1 \quad \tau_2]^T \tag{2}$$

3.2 Lagrange-Euler Dynamics

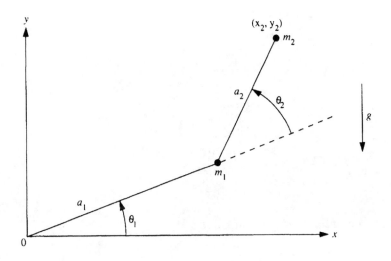

Figure 3.2.4: Two-link planar RR arm.

with τ_1, and τ_2 torques supplied by the actuators.

a. Kinetic and Potential Energy

For link 1 the kinetic and potential energies are

$$K_1 = \tfrac{1}{2} m_1 a_1^2 \dot{\theta}_1^2 \qquad (3)$$
$$P_1 = m_1 g a_1 \sin \theta_1. \qquad (4)$$

For link 2 we have

$$x_2 = a_1 \cos \theta_1 + a_2 \cos(\theta_1 + \theta_2) \qquad (5)$$
$$y_2 = a_1 \sin \theta_1 + a_2 \sin(\theta_1 + \theta_2) \qquad (6)$$
$$\dot{x}_2 = -a_1 \dot{\theta}_1 \sin \theta_1 - a_2 (\dot{\theta}_1 + \dot{\theta}_2) \sin(\theta_1 + \theta_2) \qquad (7)$$
$$\dot{y}_2 = a_1 \dot{\theta}_1 \cos \theta_1 + a_2 (\dot{\theta}_1 + \dot{\theta}_2) \cos(\theta_1 + \theta_2), \qquad (8)$$

so that the velocity squared is

$$\begin{aligned} v_2^2 &= \dot{x}_2^2 + \dot{y}_2^2 \\ &= a_1^2 \dot{\theta}_1^2 + a_2^2 (\dot{\theta}_1 + \dot{\theta}_2)^2 + 2 a_1 a_2 (\dot{\theta}_1^2 + \dot{\theta}_1 \dot{\theta}_2) \cos \theta_2. \end{aligned} \qquad (9)$$

Therefore, the kinetic energy for link 2 is

$$\begin{aligned} K_2 &= \tfrac{1}{2} m_2 v_2^2 \\ &= \tfrac{1}{2} m_2 a_1^2 \dot{\theta}_1^2 + \tfrac{1}{2} m_2 a_2^2 (\dot{\theta}_1 + \dot{\theta}_2)^2 \\ &\quad + m_2 a_1 a_2 (\dot{\theta}_1^2 + \dot{\theta}_1 \dot{\theta}_2) \cos \theta_2. \end{aligned} \qquad (10)$$

The potential energy for link 2 is

$$P_2 = m_2 g y_2 = m_2 g \left[a_1 \sin \theta_1 + a_2 \sin(\theta_1 + \theta_2) \right]. \qquad (11)$$

b. Lagrange's Equation

The Lagrangian for the entire arm is

$$\begin{aligned} L &= K - P = K_1 + K_2 - P_1 - P_2 \\ &= \tfrac{1}{2}(m_1 + m_2) a_1^2 \dot{\theta}_1^2 + \tfrac{1}{2} m_2 a_2^2 (\dot{\theta}_1 + \dot{\theta}_2)^2 \\ &\quad + m_2 a_1 a_2 (\dot{\theta}_1^2 + \dot{\theta}_1 \dot{\theta}_2) \cos \theta_2 \\ &\quad - (m_1 + m_2) g a_1 \sin \theta_1 \\ &\quad - m_2 g a_2 \sin(\theta_1 + \theta_2). \end{aligned} \qquad (12)$$

The terms needed for (3.2.14) are

3.2 Lagrange-Euler Dynamics

$$\frac{\partial L}{\partial \dot{\theta}_1} = (m_1 + m_2)a_1^2 \dot{\theta}_1 + m_2 a_2^2 (\dot{\theta}_1 + \dot{\theta}_2)$$
$$\qquad + m_2 a_1 a_2 (2\dot{\theta}_1 + \dot{\theta}_2) \cos\theta_2$$

$$\frac{d}{dt}\frac{\partial L}{\partial \dot{\theta}_1} = (m_1 + m_2)a_1^2 \ddot{\theta}_1$$
$$\qquad + m_2 a_2^2 (\ddot{\theta}_1 + \ddot{\theta}_2) + m_2 a_1 a_2 (2\ddot{\theta}_1 + \ddot{\theta}_2) \cos\theta_2$$
$$\qquad - m_2 a_1 a_2 (2\dot{\theta}_1 \dot{\theta}_2 + \dot{\theta}_2^2) \sin\theta_2$$

$$\frac{\partial L}{\partial \theta_1} = -(m_1 + m_2)g a_1 \cos\theta_1 - m_2 g a_2 \cos(\theta_1 + \theta_2)$$

$$\frac{\partial L}{\partial \dot{\theta}_2} = m_2 a_2^2 (\dot{\theta}_1 + \dot{\theta}_2) + m_2 a_1 a_2 \dot{\theta}_1 \cos\theta_2$$

$$\frac{d}{dt}\frac{\partial L}{\partial \dot{\theta}_2} = m_2 a_2^2 (\ddot{\theta}_1 + \ddot{\theta}_2) + m_2 a_1 a_2 \ddot{\theta}_1 \cos\theta_2$$
$$\qquad - m_2 a_1 a_2 \dot{\theta}_1 \dot{\theta}_2 \sin\theta_2$$

$$\frac{\partial L}{\partial \theta_2} = -m_2 a_1 a_2 (\dot{\theta}_1^2 + \dot{\theta}_1 \dot{\theta}_2) \sin\theta_2 - m_2 g a_2 \cos(\theta_1 + \theta_2).$$

Finally, according to Lagrange's equation, the arm dynamics are given by the two coupled nonlinear differential equations

$$\tau_1 = \left[(m_1 + m_2)a_1^2 + m_2 a_2^2 + 2m_2 a_1 a_2 \cos\theta_2\right] \ddot{\theta}_1$$
$$\qquad + \left[m_2 a_2^2 + m_2 a_1 a_2 \cos\theta_2\right] \ddot{\theta}_2 - m_2 a_1 a_2 (2\dot{\theta}_1 \dot{\theta}_2 + \dot{\theta}_2^2) \sin\theta_2$$
$$\qquad + (m_1 + m_2)g a_1 \cos\theta_1 + m_2 g a_2 \cos(\theta_1 + \theta_2) \qquad (13)$$
$$\tau_2 = \left[m_2 a_2^2 + m_2 a_1 a_2 \cos\theta_2\right] \ddot{\theta}_1 + m_2 a_2^2 \ddot{\theta}_2 + m_2 a_1 a_2 \dot{\theta}_1^2 \sin\theta_2$$
$$\qquad + m_2 g a_2 \cos(\theta_1 + \theta_2). \qquad (14)$$

c. Manipulator Dynamics

Writing the arm dynamics in vector form yields

$$M(q)\begin{bmatrix}\ddot\theta_1\\\ddot\theta_2\end{bmatrix}+\begin{bmatrix}-m_2a_1a_2(2\dot\theta_1\dot\theta_2+\dot\theta_2^2)\sin\theta_2\\m_2a_1a_2\dot\theta_1^2\sin\theta_2\end{bmatrix}$$
$$+\begin{bmatrix}m_1+m_2)ga_1\cos\theta_1+m_2ga_2\cos(\theta_1+\theta_2)\\m_2ga_2\cos(\theta_1+\theta_2)\end{bmatrix}$$
$$=\begin{bmatrix}\tau_1\\\tau_2\end{bmatrix}.$$
(15)

where

$$M(q)=$$
$$\begin{bmatrix}(m_1+m_2)a_1^2+m_2a_2^2+2m_2a_1a_2\cos\theta_2 & m_2a_2^2+m_2a_1a_2\cos\theta_2\\m_2a_2^2+m_2a_1a_2\cos\theta_2 & m_2a_2^2\end{bmatrix}$$
(16)

These manipulator dynamics are in the standard form

$$M(q)\ddot q+V(q,\dot q)+G(q)=\tau \qquad (17)$$

with $M(q)$ the inertia matrix, $V(q,\dot q)$ the Coriolis/centripetal vector, and $G(q)$ the gravity vector. Note that $M(q)$ is symmetric.

■

EXERCISE 3.2-3: Dynamics of a Three-Link Cylindrical Arm

We study the kinematics of a three-link cylindrical arm in Example A.2-1. In Figure 3.2.5 the joint variable vector is

$$q=[\theta \quad h \quad r]^T \qquad (1)$$

Show that the manipulator dynamics are given by

3.2 Lagrange-Euler Dynamics

$$\begin{bmatrix} J + m_2 r^2 & & \\ & m_1 + m_2 & \\ & & m_2 \end{bmatrix} \begin{bmatrix} \ddot{\theta} \\ \ddot{h} \\ \ddot{r} \end{bmatrix} + \begin{bmatrix} 2m_2 r \dot{r} \dot{\theta} \\ 0 \\ -m_2 r \dot{\theta}^2 \end{bmatrix}$$

$$+ \begin{bmatrix} 0 \\ (m_1 + m_2)g \\ 0 \end{bmatrix}$$

$$= \begin{bmatrix} n_1 \\ f_2 \\ f_3 \end{bmatrix} \tag{2}$$

with J the inertia of the base link and the force vector

$$\tau = \begin{bmatrix} n_1 & f_2 & f_3 \end{bmatrix}^T \tag{3}$$

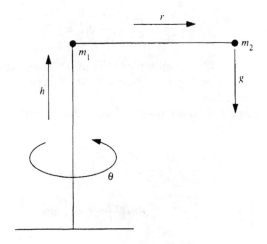

Figure 3.2.5: Three-link cylindrical arm.

■

Derivation of Manipulator Dynamics

We have shown in several examples how to apply Lagrange's equation to compute the dynamical equations of any given robot manipulator. In the examples the dynamics we found always had the special form

$$M(q)\ddot{q} + V(q, \dot{q}) + G(q) = \tau, \tag{3.2.16}$$

with q the joint-variable vector and τ the generalized force/torque vector. In this subsection we derive the dynamics for a general robot manipulator. They will be of this same form.

To obtain the general robot arm dynamical equation, we determine the arm kinetic and potential energies, then the Lagrangian, and then substitute into Lagrange's Equation (3.2.14) to obtain the final result [Paul 1981, Lee et al. 1983, Asada and Slotine 1986, Spong and Vidyasagar 1989].

Arm Kinetic Energy. Given a point on link i with coordinates of $^i r$ with respect to frame i attached to that link, the base coordinates of the point are

$$r = T_i\, {}^i r, \qquad (3.2.17)$$

where T_i is the 4×4 homogeneous transformation defined in Appendix A. Note that T_i is a function of joint variables $q_1, q_2, \ldots, q_i$. Consequently, the velocity of the point in base coordinates is

$$v = \frac{dr}{dt} = \sum_{j=1}^{i} \left[\frac{\partial T_i}{\partial q_j} \dot{q}_j \right] {}^i r \qquad (3.2.18)$$

Since $\partial T_i/\partial q_j = 0, j > i$, we may replace the upper summation limit by n, the number of links. The 4×4 matrices $\partial T_i/\partial q_j$ may be computed if the arm matrices T_i are known.

The kinetic energy of an infinitesimal mass dm at $^i r$ that has a velocity of $v = [v_x \ v_y \ v_z]^T$ is

$$dK_i = \tfrac{1}{2}(v_x^2 + v_y^2 + v_z^2)\, dm = \tfrac{1}{2} \operatorname{trace}(v^T v)\, dm = \tfrac{1}{2} \operatorname{trace} vv^T)\, dm$$

$$= \tfrac{1}{2} \operatorname{trace} \left[\sum_{j=1}^{n} \sum_{k=1}^{n} \frac{\partial T_i}{\partial q_j} ({}^i r\, {}^i r^T\, dm) \frac{\partial T_i^T}{\partial q_k} \dot{q}_j \dot{q}_k \right] dm.$$

$$(3.2.19)$$

Thus the total kinetic energy for link i is given by

$$K_i = \int_{\text{link } i} dK_i.$$

Substituting for dK_i from (3.2.19), we may move the integral inside the summations. Then, defining the 4×4 *pseudo-inertia matrix* for link i as

$$I_i \equiv \int_{\text{link } i} {}^i r\, {}^i r^T\, dm, \qquad (3.2.20)$$

3.2 Lagrange-Euler Dynamics

we may write the kinetic energy of link i as

$$K_i = \tfrac{1}{2} \text{trace} \left[\sum_{j=1}^{n} \sum_{k=1}^{n} \frac{\partial T_i}{\partial q_j} I_i \frac{\partial T_i^T}{\partial q_k} \dot{q}_j \dot{q}_k \right]. \qquad (3.2.21)$$

Let us briefly discuss the pseudo-inertia matrix before proceeding to find the arm total kinetic energy. Let

$$^i r = [x \quad y \quad z \quad 1]^T$$

be the coordinates in frame i of the infinitesimal mass dm. Then, expanding (3.2.20) yields

$$I_i = \begin{bmatrix} \int x^2 \, dm & \int yx \, dm & \int zx \, dm & \int x \, dm \\ \int xy \, dm & \int y^2 \, dm & \int zy \, dm & \int y \, dm \\ \int xz \, dm & \int yz \, dm & \int z^2 \, dm & \int z \, dm \\ \int x \, dm & \int y \, dm & \int z \, dm & \int dm \end{bmatrix}, \qquad (3.2.22)$$

where the integrals are taken over the volume of link i. This is a constant matrix that is evaluated once for each link. It depends on the geometry and mass distribution of link i. In fact, in terms of the link i *moments of inertia*

$$I_{xx} = \int (y^2 + z^2) \, dm$$

$$I_{yy} = \int (x^2 + z^2) \, dm$$

$$I_{zz} = \int (x^2 + y^2) \, dm, \qquad (3.2.23)$$

cross-products of inertia

$$I_{xy} = \int xy \, dm$$

$$I_{xz} = \int xz \, dm$$

$$I_{yz} = \int yz \, dm, \qquad (3.2.24)$$

and *first moments*

$$m\bar{x} = \int x\, dm$$

$$m\bar{y} = \int y\, dm$$

$$m\bar{z} = \int z\, dm, \qquad (3.2.25)$$

with m the total mass of link i, and

$$^i\bar{r} = [\bar{x}\ \ \bar{y}\ \ \bar{z}\ \ 1]^T \qquad (3.2.26)$$

the coordinates in frame i of the center of gravity of link i, we may write

$$I_i = \begin{bmatrix} \frac{-I_{xx}+I_{yy}+I_{zz}}{2} & I_{xy} & I_{xz} & m\bar{x} \\ I_{xy} & \frac{I_{xx}-I_{yy}+I_{zz}}{2} & I_{yz} & m\bar{y} \\ I_{xz} & I_{yz} & \frac{I_{xx}+I_{yy}-I_{zz}}{2} & m\bar{z} \\ m\bar{x} & m\bar{y} & m\bar{z} & m \end{bmatrix}. \qquad (3.2.27)$$

These quantities are either tabulated in the arm manufacturer's specifications or may be computed from quantities tabulated there.

Returning now to our development, the total arm kinetic energy may be written as

$$K = \sum_{i=1}^{n} K_i = \frac{1}{2} \sum_{i=1}^{n} \text{trace} \left[\sum_{j=1}^{n} \sum_{k=1}^{n} \frac{\partial T_i}{\partial q_j} I_i \frac{\partial T_i^T}{\partial q_k} \dot{q}_j \dot{q}_k \right]. \qquad (3.2.28)$$

Since the trace of a sum of matrices is the sum of the individual traces, we may interchange summations and the trace operator to obtain

$$K = \frac{1}{2} \sum_{j=1}^{n} \sum_{k=1}^{n} m_{jk}(q) \dot{q}_j \dot{q}_k$$

or

$$K = \frac{1}{2} \dot{q}^T M(q) \dot{q}, \qquad (3.2.29)$$

where the $n \times n$ arm inertia matrix $M(q)$ has elements defined as

$$m_{jk}(q) = \sum_{i=1}^{n} \text{trace} \left[\frac{\partial T_i}{\partial q_j} I_i \frac{\partial T_i^T}{\partial q_k} \right]. \qquad (3.2.30)$$

3.2 Lagrange-Euler Dynamics

Since $\partial T_i/\partial q_j = 0$ for $j > i$, we may write this more efficiently as

$$m_{jk}(q) = \sum_{i=\max(j,k)}^{n} \text{trace}\left[\frac{\partial T_i}{\partial q_j} I_i \frac{\partial T_i^T}{\partial q_k}\right]. \quad (3.2.31)$$

Equation (3.2.29) is what we have been seeking; it provides a convenient expression for the arm kinetic energy in terms of known quantities and the joint variables q. Since $m_{jk} = m_{kj}$, the inertia matrix $M(q)$ is *symmetric*. Since the kinetic energy is positive, vanishing only when the generalized velocity $\dot{q}$ equal zero, the inertia matrix $M(q)$ is also *positive definite*. Note that the kinetic energy depends on q and $\dot{q}$.

Arm Potential Energy. If link i has a mass m_i and a center of gravity $^i\bar{r}$ expressed in the coordinates of its frame i, the potential energy of the link is given by

$$P_i = -m_i g^T T_i {}^i\bar{r}, \quad (3.2.32)$$

where the *gravity vector* is expressed in base coordinates as

$$g = [g_x \quad g_y \quad g_z \quad 0]^T. \quad (3.2.33)$$

If the arm is level, at sea level, and the base z-axis is directed vertically upward, then

$$g = [0 \quad 0 \quad -9.8062 \quad 0]^T, \quad (3.2.34)$$

with units of m/s^2.

The total arm potential energy, therefore, is

$$P = -\sum_{i=1}^{n} m_i g^T T_i {}^i\bar{r}. \quad (3.2.35)$$

Note that P depends only on the joint variables q, not on the joint velocities $\dot{q}$.

Noting that $m_i {}^i\bar{r}$ is the last column of the link i pseudo-inertia matrix T_i, we may write

$$P(q) = -\sum_{i=1}^{n} g^T T_i(q) I_i e_4 \quad (3.2.36)$$

with e_4 the last column of the 4×4 identity matrix (i.e., $e_4 = [0 \quad 0 \quad 0 \quad 1]^T$).

Lagrange's Equation. The arm Lagrangian is

$$L(q, \dot{q}) = K(q, \dot{q}) - P(q) = \tfrac{1}{2}\dot{q}^T M(q)\dot{q} - P(q). \tag{3.2.37}$$

It is a fundamental property that the kinetic energy is a quadratic function of the joint velocity vector and the potential energy is independent of $\dot{q}$.

The terms required in Lagrange's Equation (3.2.14) are now given by

$$\frac{\partial L}{\partial \dot{q}} = \frac{\partial K}{\partial \dot{q}} = M(q)\dot{q} \tag{3.2.38}$$

$$\frac{d}{dt}\frac{\partial L}{\partial \dot{q}} = M(q)\ddot{q} + \dot{M}(q)\dot{q} \tag{3.2.39}$$

$$\frac{\partial L}{\partial q} = \frac{1}{2}\frac{\partial}{\partial q}(\dot{q}^T M(q)\dot{q}) - \frac{\partial P(q)}{\partial q}. \tag{3.2.40}$$

Therefore, the arm dynamical equation is

$$M(q)\ddot{q} + \dot{M}(q)\dot{q} - \tfrac{1}{2}\frac{\partial}{\partial q}(\dot{q}^T M(q)\dot{q}) + \frac{\partial P(q)}{\partial q} = \tau. \tag{3.2.41}$$

Defining the *Coriolis/centripetal vector*

$$V(q)\ddot{q} = \dot{M}(q)\dot{q} - \frac{1}{2}\frac{\partial}{\partial q}(\dot{q}^T M(q)\dot{q}) = \dot{M}\dot{q} - \frac{\partial K}{\partial q} \tag{3.2.42}$$

and the *gravity vector*

$$G(q) = \frac{\partial P(q)}{\partial q}, \tag{3.2.43}$$

we may write

$$M(q)\ddot{q} + V(q, \dot{q}) + G(q) = \tau, \tag{3.2.44}$$

which is the final form of the robot dynamical equation we have been seeking.

The units of elements of $M(q)$ corresponding to revolute joint variables $q_i = \theta_i$ are kg-m^2. The units of the elements of $M(q)$ corresponding to prismatic joint variables $q_i = d_i$ are kilograms. The units of elements of $V(q, \dot{q})$ and $G(q)$ corresponding to revolute joint variables are kg-m^2/s^2. The units of elements of $V(q, \dot{q})$ and $G(q)$ corresponding to prismatic joint variables are kg-m/s^2.

3.3 Structure and Properties of the Robot Equation

In this section we investigate the detailed structure and properties of the dynamical arm equations, for this structure should be reflected in the form of the control law. The controller is simpler and more effective if the known properties of the arm are incorporated in the design stage.

In reality, a robot arm is always affected by friction and disturbances. Therefore, we shall generalize the arm model we have just derived by writing the manipulator dynamics as

$$M(q)\ddot{q} + V(q,\dot{q}) + F(\dot{q}) + G(q) + \tau_d = \tau, \qquad (3.3.1)$$

with q the joint variable n-vector and τ the n-vector of generalized forces. $M(q)$ is the inertia matrix, $V(q,\dot{q})$ the Coriolis/centripetal vector, and $G(q)$ the gravity vector. We have added a *friction* term

$$F(\dot{q}) = F_v \dot{q} + F_d \qquad (3.3.2)$$

with F_v the coefficient matrix of *viscous friction* and F_d a *dynamic friction* term. Also added is a *disturbance* τ_d, which could represent, for instance, any inaccurately modeled dynamics.

Friction is not an easy term to model, and indeed, may be the most contrary term to describe in the manipulator dynamics model. Some more discussion on friction may be found in [Schilling 1990].

We shall sometimes write the arm dynamics as

$$M(q)\ddot{q} + N(q,\dot{q}) + \tau_d = \tau, \qquad (3.3.3)$$

where

$$N(q,\dot{q}) \equiv V(q,\dot{q}) + F(\dot{q}) + G(q) \qquad (3.3.4)$$

represents nonlinear terms.

Let us examine the structure and properties of each of the terms in the robot dynamics equation. This study will offer us a great deal of insight which we use in deriving robot control schemes in subsequent chapters. A summary of the properties we discover is given in Table 3.3.1, to which we refer in the remainder of the book. As we develop each property, it will be worthwhile to refer to Examples 3.2.1 to 3.2.3 in order to verify that the properties indeed hold there. At the end of this section we illustrate in Example 3.3.1 several of the properties for a two-link planar elbow arm.

Table 3.3.1: The Robot Equation and Its Properties

$$M(q)\ddot{q} + V(q,\dot{q}) + F(\dot{q}) + G(q) + \tau_d = \tau$$

or

$$M(q)\ddot{q} + N(q,\dot{q}) + \tau_d = \tau$$

where

$$N(q,\dot{q}) \equiv V(q,\dot{q}) + F(\dot{q}) + G(q)$$

Inertia Matrix:
$M(q)$ is symmetric and positive definite.
$\mu_1 I \leq M(q) \leq \mu_2 I$
$m_1 \leq \|M(q)\| \leq m_2$

Coriolis/Centripetal Vector:
$V(q,\dot{q})$ is quadratic in $\dot{q}$
$\|V(q,\dot{q})\| \leq v_b \|\dot{q}\|^2$
$V(q,\dot{q}) = V_m(q,\dot{q})\dot{q}$
$S(q,\dot{q}) \equiv \dot{M}(q) - 2V_m(q,\dot{q})$ is a skew-symmetric matrix

Friction Terms:
$F(\dot{q}) = F_v(\dot{q}) + F_d(\dot{q})$
$F_v = \text{diag}\{v_1\}$
$F_d(\dot{q}) = K_d \, \text{sgn}(\dot{q})$, with $K_d = \text{diag}\{k_i\}$
$\|F_v\dot{q} + F_d(\dot{q})\| \leq v\|\dot{q}\| + k$

Gravity Vector:
$$\|G(q)\| \leq g_b$$

Disturbance Term:
$$\|\tau_d\| \leq d$$

Linearity in the Parameters:
$M(q)\ddot{q} + V(q,\dot{q}) + F_v\dot{q} + F_d(\dot{q}) + G(q)$
$= M(q)\ddot{q} + N(q,\dot{q}) \equiv W(q,\dot{q},\ddot{q})\varphi$

Properties of the Inertia Matrix

As we have seen, $M(q)$ is symmetric and positive definite. In fact, the arm

3.3 Structure and Properties of the Robot Equation

kinetic energy is

$$K = \tfrac{1}{2}\dot{q}^T M(q)\dot{q}. \qquad (3.3.5)$$

Some expressions for M are given in the next subsection.

Another vital property of $M(q)$ is that it is bounded above and below. That is,

$$\mu_1 I \leq M(q) \leq \mu_2 I \qquad (3.3.6)$$

with μ_1, and μ_2 scalars that may be computed for any given arm (see Example 3.3.1). When we say that $\mu_1 I \leq M(q)$, for instance, we mean that $(M(q) - \mu_1 I)$ is positive semidefinite. That is,

$$x^T(M - \mu_1 I)x \geq 0$$

for all $x\varepsilon$ R^n.

Likewise, the inverse of the inertia matrix is bounded, since

$$\frac{1}{\mu_2} I \leq M^{-1}(q) \leq \frac{1}{\mu_1} I. \qquad (3.3.7)$$

If the arm is revolute, the bounds μ_1 and μ_2 are constants, since q appears only in $M(q)$ through sin and cos terms, whose magnitudes are bounded by 1 (see Examples 3.2.2 and 3.3.1). On the other hand, if the arm has prismatic joints, then μ_1 and μ_2 may be scalar functions of q. See Example 3.2.1, where $M(q)$ is bounded above by $\mu_2 = mr^2$ (if $r > 1$).

The boundedness property of the inertia matrix may also be expressed as

$$m_1 \leq \|M(q)\| \leq m_2, \qquad (3.3.8)$$

where any induced matrix norm can be used to define the positive scalars m_1 and m_2.

Properties of the Coriolis/Centripetal Term

A glance at (3.2.42) reveals a problem that, if not understood, can make the study of robot dynamics confusing. Simplification of this $V(q,\dot{q})$ term would require taking the derivative of a matrix [i.e., $M(q)$] with respect to the n-vector q. However, such derivatives are not matrices, but tensors of order three-that is, they must be represented by three indices, not two. There are several ways to get around this problem, involving several definitions of some new quantities.

Kronecker Product Analysis of $V(q, \dot{q})$. Let us first examine the term $V(q, \dot{q})$ from the point of view of the Kronecker product [Brewer 1978], defined for two matrices $A \in R^{n \times m}$, $B \in R^{p \times q}$ as

$$A \otimes B = [a_{ij} B] \in R^{np \times mq}, \qquad (3.3.9)$$

where A has elements a_{ij} and $[a_{ij} B]$ means the $np \times mq$ block matrix composed of the $p \times q$ blocks $a_{ij} B$. Thus, for $A \in R^{3 \times 3}$, we have

$$A \otimes B = \begin{bmatrix} a_{11} B & a_{12} B & a_{13} B \\ a_{21} B & a_{22} B & a_{23} B \\ a_{31} B & a_{32} B & a_{33} B \end{bmatrix}.$$

For matrices $A(q)$, $B(q)$, with $q \in R^n$, define the *matrix derivative* as

$$\frac{\partial A}{\partial q} = \begin{bmatrix} \frac{\partial A}{\partial q_1} \\ \vdots \\ \frac{\partial A}{\partial q_n} \end{bmatrix}. \qquad (3.3.10)$$

Then we may prove the product rule

$$\frac{\partial}{\partial q}[A(q) B(q)] = (I_n \otimes A) \frac{\partial B}{\partial q} + \frac{\partial A}{\partial q} B, \qquad (3.3.11)$$

with I_n the $n \times n$ identity.

Now we may examine the Coriolis/centripetal vector $V(q, \dot{q})$ [Koditschek 1984, Gu and Loh 1988]. Using (3.3.11) twice on (3.2.42), we may obtain

$$V(q, \dot{q}) = \dot{M}(q) \dot{q} - \frac{1}{2} \left[\frac{\partial}{\partial q} (\dot{q}^T M(q)) \right] \dot{q}$$

$$V(q, \dot{q}) = \dot{M}(q) \dot{q} - \frac{1}{2} (I_n \otimes \dot{q}^T) \frac{\partial M}{\partial q} \dot{q}, \qquad (3.3.12)$$

or

$$V(q, \dot{q}) = \left[\dot{M}(q) - \tfrac{1}{2} U(q, \dot{q}) \right] \dot{q}, \qquad (3.3.13)$$

where

$$U(q, \dot{q}) \equiv (I_n \otimes \dot{q}^T) \frac{\partial M}{\partial q}. \qquad (3.3.14)$$

$$V(q, \dot{q}) = V_{m1}(q, \dot{q}) \dot{q} \qquad (3.3.15)$$

3.3 Structure and Properties of the Robot Equation

with the matrix coefficient given by

$$V_{m1}(q,\dot{q}) \equiv \dot{M} - \tfrac{1}{2}U. \qquad (3.3.16)$$

To find an equivalent expression for V_{m1}, note that

$$\dot{M}(q) = \sum_{i=1}^{n} \frac{\partial M}{\partial q_i}\dot{q}_i, \qquad (3.3.17)$$

which may be written as

$$\dot{M} = \begin{bmatrix} \dot{q}_1 & & \\ & \ddots & \\ & & \dot{q}_1 \end{bmatrix} \cdots \begin{bmatrix} \dot{q}_n & & \\ & \ddots & \\ & & \dot{q}_n \end{bmatrix} \begin{bmatrix} \frac{\partial M}{\partial q_1} \\ \vdots \\ \frac{\partial M}{\partial q_n} \end{bmatrix}$$

$$= (\dot{q}^T \otimes I_n)\frac{\partial M}{\partial q} \qquad (3.3.18)$$

or as

$$\dot{M} = \begin{bmatrix} \frac{\partial M}{\partial q_1} & \cdots & \frac{\partial M}{\partial q_n} \end{bmatrix} \begin{bmatrix} \begin{bmatrix} \dot{q}_1 \\ \vdots \\ \dot{q}_n \end{bmatrix} \\ \ddots \\ \begin{bmatrix} \dot{q}_1 \\ \vdots \\ \dot{q}_n \end{bmatrix} \end{bmatrix} = \begin{bmatrix} \frac{\partial M}{\partial q} \end{bmatrix}^T (\dot{q} \otimes I_n). \qquad (3.3.19)$$

(Note that $\partial M/\partial q_i$ is symmetric.) Therefore,

$$V_{m1}(q,\dot{q}) = \left[(\dot{q} \otimes I_n) - \tfrac{1}{2}(I_n \otimes \dot{q}^T)\right]\frac{\partial M}{\partial q}, \qquad (3.3.20)$$

whence using appropriate definitions we may write

$$V(q,\dot{q}) = V_v(\dot{q})V_p(q)\dot{q}. \qquad (3.3.21)$$

It is also possible to write (see the Problems)

$$V(q,\dot{q}) = V_{p1}(q)V_{v1}(\dot{q})\dot{q}. \qquad (3.3.22)$$

Since $V_v(\dot{q})$ is linear in $\dot{q}$, it follows that $V(q,\dot{q})$ is *quadratic* in $\dot{q}$. In fact, it can be shown (see the Problems) that

$$V(q,\dot{q}) = \begin{bmatrix} \dot{q}^T V_1(q)\dot{q} \\ \dot{q}^T V_2(q)\dot{q} \\ \vdots \\ \dot{q}^T V_n(q)\dot{q} \end{bmatrix} = (I_n \otimes \dot{q}^T) \begin{bmatrix} V_1(q) \\ \vdots \\ \vdots \\ V_n(q) \end{bmatrix} \dot{q} \equiv (I_n \otimes \dot{q}^T)\overline{V}(q)\dot{q}$$

(3.3.23)

for appropriate definition of $V_i(q)$ [Craig 1988]. Indeed, the $V_i(q)$ are symmetric $n \times n$ matrices.

Since $V(q, \dot{q})$ is quadratic in $\dot{q}$, it can be bounded above by a quadratic function of $\dot{q}$. That is,

$$\|V(q,\dot{q})\| \leq v_b(q) \|\dot{q}\|^2, \qquad (3.3.24)$$

with $v_b(q)$ a known scalar function and $\|\cdot\|$ any appropriate norm. For a revolute arm, v_b is a constant independent of q. See Examples 3.2.2 and 3.3.1, where the quadratic terms in $\dot{q}$ are multiplied by $\sin\theta_2$, whose magnitude is bounded by 1. On the other hand, for an arm with prismatic joints $v_b(q)$ may be function of q; see Examples 3.2.1 and 3.2.3, where $V(q,\dot{q})$ has a term in r multiplying the quadratic terms $\dot{r}\dot{\theta}$ and $\dot{\theta}^2$.

To assist in determining $v_b(q)$ for a given robot arm, note that $\|I_n \otimes \dot{q}^T\| = \|\dot{q}\|$, so that

$$\|V(q,\dot{q})\| \leq \|\overline{V}(q)\| \cdot \|\dot{q}\|^2,$$

where $\overline{V}(q)$ is defined in (3.3.23). Therefore, for a revolute arm

$$v_b = \sup_q \|\overline{V}(q)\|. \qquad (3.3.25)$$

We may note that

$$(I_n \otimes \dot{q})\dot{q} = (\dot{q} \otimes I_n)\dot{q} = \begin{bmatrix} \dot{q}_1 \dot{q} \\ \dot{q}_2 \dot{q} \\ \vdots \\ \dot{q}_n \dot{q} \end{bmatrix} \equiv <\dot{q}\dot{q}>. \qquad (3.3.26)$$

This is an n^2-vector consisting of all possible products of the components of $\dot{q}$. This and (3.3.19) allow us to demonstrate that

$$\dot{M}\dot{q} = U^T \dot{q}. \qquad (3.3.27)$$

In this proof, we also need the identity

3.3 Structure and Properties of the Robot Equation

$$(A \otimes B)^T = A^T \otimes B^T \qquad (3.3.28)$$

for any matrices A and B. *Note:* It is not true that $\dot{M} = U^T$.

Now we may use these various identities to show that

$$V(q,\dot{q}) = \left[\frac{\partial M}{\partial q}\right]^T (\dot{q} \otimes I_n)\dot{q} - \frac{1}{2}(I_n \otimes \dot{q}^T)\frac{\partial M}{\partial q}\dot{q}$$

$$V(q,\dot{q}) = \left[U^T(q,\dot{q}) - \tfrac{1}{2}U(q,\dot{q})\right]\dot{q}. \qquad (3.3.29)$$

Thus

$$V(q,\dot{q}) = V_{m2}(q,\dot{q})\dot{q}. \qquad (3.3.30)$$

with

$$V_{m2}(q,\dot{q}) \equiv U^T - \tfrac{1}{2}U. \qquad (3.3.31)$$

Note that, in general, $V_{m1} \neq V_{m2}$.

In terms of M and U, the arm dynamics may be written as

$$M(q)\ddot{q} + \left[U^T - \tfrac{1}{2}U\right]\dot{q} + G(q) = \tau. \qquad (3.3.32)$$

At this point we may prove an identity that is extremely useful in constructing advanced control schemes. We call it the *skew-symmetric property*; it shows that the derivative of $M(q)$ and the Coriolis vector are related in a very particular way. In fact,

$$\begin{aligned}
\dot{q}^T(\dot{M} - 2V_{m2})\dot{q} &= \dot{q}^T(U^T - 2U^T + U)\dot{q} \\
&= \dot{q}^T(U - U^T)\dot{q} = 0
\end{aligned} \qquad (3.3.33)$$

since a matrix minus its transpose is always skew symmetric. This important identity holds also if V_{m1} is used in place of V_{m2}.

It is important to note that the first equality in (3.3.33) holds because $(\dot{M} - 2V_{m2})$ multiplies $\dot{q}$. That is, $\dot{M} \neq U^T$, so that it is not necessarily true that $(\dot{M} - 2V_{m2})$ itself is skew symmetric. However, it is possible to define a matrix $V_m(q,\dot{q})$ such that

$$V(q,\dot{q}) = V_m(q,\dot{q})\dot{q} \qquad (3.3.34)$$

and

$$S(q,\dot{q}) \equiv \dot{M}(q) - 2V_m(q,\dot{q}) \qquad (3.3.35)$$

is skew symmetric, so that $x^T S x = 0$ for all $x \in R^n$. Indeed, according to (3.3.13), (3.3.27) we may define

$$V_m(q,\dot{q}) = {1\!/\!2}(\dot{M} + U^T - U), \qquad (3.3.36)$$

for then the skew-symmetric matrix is nothing but

$$S(q,\dot{q}) = U - U^T. \qquad (3.3.37)$$

This V_m is the standard one used in several modern adaptive and robust control algorithms, and it is the definition we shall use in the remainder of the book. Thus we shall write the arm equation either as

$$M(q)\ddot{q} + V(q,\dot{q}) + G(q) = \tau, \qquad (3.3.38)$$

or

$$M(q)\ddot{q} + V_m(q,\dot{q})\dot{q} + G(q) = \tau. \qquad (3.3.39)$$

Note that it is possible to split $V(q,\dot{q})$ into its Coriolis and centripetal components as

$$V(q,\dot{q}) = V_{\text{cor}}(q) <\dot{q}\dot{q}>' + V_{\text{cen}}(q) <\dot{q}^2>, \qquad (3.3.40)$$

where

$$<\dot{q}^2> \equiv \begin{bmatrix} \dot{q}_1^2 & \dot{q}_2^2 & \cdots & \dot{q}_n^2 \end{bmatrix}^T \qquad (3.3.41)$$

and $<\dot{q}\dot{q}>'$ is (3.3.26) with all the square terms $\dot{q}_1^2$ removed [Craig 1988] (see the Problems).

Componentwise Analysis of $V(q,\dot{q})$. An alternative to the Kronecker product analysis of the Coriolis/centripetal vector is an analysis in terms of the scalar components of $V(q,\dot{q})$, which yields additional insight.

In terms of the components $m_{kj}(q)$ of the inertia matrix $M(q)$ we may write (3.2.38)-(3.2.40) componentwise as

3.3 Structure and Properties of the Robot Equation

$$\frac{\partial L}{\partial \dot{q}_k} = \sum_j m_{kj}(q)\dot{q}_j \qquad (3.3.42)$$

$$\frac{d}{dt}\frac{\partial L}{\partial \dot{q}_k} = \sum_j \left(m_{kj}(q)\ddot{q}_j + \left[\frac{d}{dt}m_{kj}(q)\right]\dot{q}_j \right)$$

$$= \sum_j m_{kj}\ddot{q}_j + \sum_{ij} \frac{\partial m_{kj}}{\partial q_i}\dot{q}_i\dot{q}_j \qquad (3.3.43)$$

$$\frac{\partial L}{\partial q_k} = \frac{1}{2}\sum_{ij} \frac{\partial m_{ij}}{\partial q_k}\dot{q}_i\dot{q}_j - \frac{\partial P}{\partial q_k}, \qquad (3.3.44)$$

where all sums are over the number of joints n. Now, the Lagrange equation shows that the arm dynamics are expressed componentwise as

$$\sum_j m_{kj}\ddot{q}_j + \sum_{i,j} \left[\frac{\partial m_{kj}}{\partial q_i} - \frac{1}{2}\frac{\partial m_{ij}}{\partial q_k}\right]\dot{q}_i\dot{q}_j + \frac{\partial P}{\partial q_k} = \tau_k, \quad k = 1, \ldots, n. \qquad (3.3.45)$$

with n the number of joints.

By interchanging the order of summation and taking advantage of symmetry,

$$\sum_{i,j} \frac{\partial m_{kj}}{\partial q_i}\dot{q}_i\dot{q}_j = \frac{1}{2}\sum_{i,j} \left[\frac{\partial m_{kj}}{\partial q_i} + \frac{\partial m_{ki}}{\partial q_j}\right]\dot{q}_i\dot{q}_j. \qquad (3.3.46)$$

Therefore, we may define

$$v_{ijk} \equiv \frac{1}{2}\left[\frac{\partial m_{kj}}{\partial q_i} + \frac{\partial m_{ki}}{\partial q_j} - \frac{\partial m_{ij}}{\partial q_k}\right] \qquad (3.3.47)$$

and write the arm dynamics as

$$\sum_j m_{kj}\ddot{q}_j + \sum_{i,j} v_{ijk}\dot{q}_i\dot{q}_j + \frac{\partial P}{\partial q_k} = \tau_k, \quad k = 1, \ldots, n. \qquad (3.3.48)$$

The cyclic symmetry of the v_{ijk} is what allows us to derive the important properties of the Coriolis/centripetal vector $V(q, \dot{q})$, which corresponds to the second term in this equation. The quantities v_{ijk} are known as *Christoffel symbols (of the first kind)* [Borisenko and Tarapov 1968].

The matrix $V_m(q, \dot{q})$ defined in (3.3.36) has components v_{kj} given by

$$v_{kj} \equiv \sum_i v_{ijk}(q)\dot{q}_i. \tag{3.3.49}$$

Properties of the Gravity, Friction, and Disturbance

Properties of the Gravity Term $G(q)$. According to (3.2.43) and (3.2.36),

$$G(q) = \frac{\partial P}{\partial q} = -\sum_{i-1}^{n} \frac{\partial}{\partial q}(g^T T_i(q) I_i e_4),$$

whence using (3.3.11) twice reveals

$$G(q) = -\sum_{i-1}^{n} \frac{\partial}{\partial q}(g^T T_i(q)) I_i e_4$$

$$G(q) = -\sum_{i-1}^{n} (I_n \otimes g^T) \frac{\partial T_i}{\partial q} I_i e_4. \tag{3.3.50}$$

A bound on the gravity term may be derived for any given robot arm. Thus

$$\|G(q)\| \leq g_b(q), \tag{3.3.51}$$

where $\|\cdot\|$ is any appropriate vector norm and g_b is a scalar function that may be determined for any given arm (see Example 3.3.1). For a revolute arm, g_b is a constant independent of the joint vector q, but for an arm with prismatic links, g_b may depend on q. See the examples in Section 3.2 and Example 3.3.1 to verify these claims.

Properties of the Friction Term $F(\dot{q})$. The friction in the arm Equation (3.3.1) is of the form

$$F(\dot{q}) = F_v \dot{q} + F_d(\dot{q}) \tag{3.3.52}$$

with F_v the coefficient matrix of viscous friction, and F_d a dynamic friction term. The friction coefficients are among the parameters most difficult to determine for a given arm and, in fact, (3.3.52) represents only an approximate mathematical model for their influence. For more discussion, see [Craig 1988, Schilling 1990].

Since friction is a local effect, we may assume that $F(\dot{q})$ is *uncoupled among the joints*, so that

3.3 Structure and Properties of the Robot Equation

$$F(\dot{q}) = \text{vec}\{f_i(\dot{q}_i)\} \equiv \begin{bmatrix} f_1(\dot{q}_1) \\ \vdots \\ f_n(\dot{q}_n) \end{bmatrix} \qquad (3.3.53)$$

with $f_i(\cdot)$ known scalar functions that may be determined for any given arm. We have defined the vec$\{\cdot\}$ function for future use.

The viscous friction may often be assumed to have the form

$$F_v \dot{q} = \text{vec}\{v_i \dot{q}_i\} \qquad (3.3.54)$$

with v_i known constant coefficients. Then $F_v = \text{diag}\{v_i\}$, a diagonal matrix with entries v_i. The dynamic friction may often be assumed to have the form

$$F_d(\dot{q}) = \text{vec}\{k_i \, \text{sgn}(\dot{q}_i)\}, \qquad (3.3.55)$$

with k_i known constant coefficients and the signum function defined for a scalar x by

$$\text{sgn}(x) = \begin{cases} +1, & x > 0 \\ \text{indeterminate}, & x = 0 \\ -1, & x < 0. \end{cases} \qquad (3.3.56)$$

Then

$$F_d(\dot{q}) = K_d \, \text{sgn}(\dot{q}), \qquad (3.3.57)$$

with $K_d = \text{diag}\{k_i\}$ the *coefficient matrix of dynamic friction* and the signum function defined for a vector x by

$$\text{sgn}(x) = \text{vec}\{\text{sgn}(x_i)\}. \qquad (3.3.58)$$

A bound on the friction terms may be assumed of the form

$$\|F_v \dot{q} + F_d(\dot{q})\| \leq v \|\dot{q}\| + k, \qquad (3.3.59)$$

with **v** and **k** known for a specific arm and $\|\cdot\|$ a suitable norm.

Another friction term that may be included in $F(\dot{q})$ is the *static friction*, which has components of the form

$$F_{si} = \text{sgn}(\dot{q}_i)\left[(k_{si} - k_i) \exp\left(-|\dot{q}_i|/\varepsilon\right)\right], \qquad (3.3.60)$$

where k_{si} is the coefficient of static friction for joint i and ϵ is a small positive parameter. We shall generally ignore this term.

Properties of the Disturbance Term. The arm Equation (3.3.1) has a disturbance term τ_d, which could represent inaccurately modeled dynamics, and so on. We shall assume that it is bounded so that

$$\|\tau_d\| \leq d, \tag{3.3.61}$$

where **d** is a scalar constant that may be computed for a given arm and $\|\cdot\|$ is any suitable norm.

Linearity in the Parameters

The robot dynamical equation enjoys one last property that will be of great use to us in Chapter 5. Namely, it is *linear in the parameters*, a property first exploited in [Craig 1988] in adaptive control. This is important, since some or all of the parameters may be unknown; thus the dynamics are linear in the unknown terms.

This property may be expressed as

$$\begin{aligned} & M(q)\ddot{q} + V(q,\dot{q}) + F_v\dot{q} + F_d(\dot{q}) + G(q) \\ = & M(q)\ddot{q} + N(q,\dot{q}) \equiv W(q,\dot{q},\ddot{q})\varphi \end{aligned} \tag{3.3.62}$$

with φ the parameter vector and $W(q,\dot{q},\ddot{q})$ a matrix of robot functions depending on the joint variables, joint velocities, and joint accelerations. This matrix may be computed for any given robot arm and so is known. See Example 3.3.1. Note that the disturbance τ_d is not included in this equation.

EXAMPLE 3.3-1: Structure and Bounds for Two-Link Planar Elbow Arm

The dynamics of a two-link planar arm are given in Example 3.2.2. We should now like to compute the structural matrices defined in this section, as well as the bounds needed in Table 3.3.1. The friction bounds are straightforward, so we do not mention them here. The dynamical matrices are

3.3 Structure and Properties of the Robot Equation

$$M(q) = \begin{bmatrix} (m_1 + m_2)a_1^2 + m_2 a_2^2 + 2m_2 a_1 a_2 \cos\theta_2 & m_2 a_2^2 + m_2 a_1 a_2 \cos\theta_2 \\ m_2 a_2^2 + m_2 a_1 a_2 \cos\theta_2 & m_2 a_2^2 \end{bmatrix}$$

$$V(q,\dot{q}) = \begin{bmatrix} -m_2 a_1 a_2 (2\dot{\theta}_1 \dot{\theta}_2 + \dot{\theta}_2^2) \sin\theta_2 \\ m_2 a_1 a_2 \dot{\theta}_1^2 \sin\theta_2 \end{bmatrix}$$

$$G(q) = \begin{bmatrix} (m_1 + m_2)g a_1 \cos\theta_1 + m_2 g a_2 \cos(\theta_1 + \theta_2) \\ m_2 g a_2 \cos(\theta_1 + \theta_2) \end{bmatrix}.$$

The selection of a suitable norm in Table 3.3.1 is not always straightforward. In the control algorithms to be developed in subsequent chapters, we prove suitable performance in terms of some norm, which can often be any norm desired. For implementation of the controller, a specific norm must be selected and the bounds evaluated. This choice often depends simply on which norm makes it possible to evaluate the bounds in the table. For instance, choosing the 2–norm for vectors requires the evaluation of the maximum singular value of $M(q)$, a very difficult task.

Selecting the ∞–norm for vectors means determining at each sampling time the element [of $V(q(t), \dot{q}(t))$ for instance] with the largest magnitude. This requires decision logic, and the norm may not be continuous. Therefore, let us use the 1–norm in this example. The corresponding matrix induced norm is then the maximum absolute column sum (Chapter 2).

a. Bounds on the Inertia Matrix

The evaluation of μ_1, and μ_2 amounts to the determination of the minimum and maximum eigenvalues of $M(q)$ over all q. This is not an easy affair and requires the solution of some quadratic equations, although it can be carried out without too much trouble using software such as Mathematica or Maple. Thus, let us find m_1 and m_2.

The induced 1–norm for $M(q)$ is the maximum absolute column sum. In determining bounds for this norm, it is important to consider the range of allowed motion of the joint angles. To illustrate, suppose that θ_1, and θ_2 are limited by $\pm\pi/2$. Then the 1–norm is always given in terms of column 1 as

$$\|M(q)\|_1 = |(m_1 + m_2)a_1^2 + m_2 a_2^2 + 2m_2 a_1 a_2 \cos\theta_2| + |m_2 a_2^2 + m_2 a_1 a_2 \cos\theta_2|,$$

which is bounded above for all θ_2 by

$$M_2 = (m_1 + m_2)a_1^2 + 2m_2 a_2^2 + 3m_2 a_1 a_2$$

and below by

$$M_1 = (m_1 + m_2)a_1^2 + 2m_2 a_2^2.$$

Since the arm is revolute and $\cos\theta_2$ is bounded above and below, M_2 and M_1 are constants. It is important to note that if the arm is revolute/prismatic (RP), so that the joint variables are (θ_1, a_2), the bounds are functions of q.

b. Bounds on the Coriolis and Gravity Terms

The bound v_b on the Coriolis/centripetal vector is found using

$$\begin{aligned}\|V(q,\dot{q})\|_1 &= \left|m_2 a_1 a_2 (2\dot{\theta}_1 \dot{\theta}_2 + \dot{\theta}_2^2)\sin\theta_2\right| + \left|m_2 a_1 a_2 \dot{\theta}_1^2 \sin\theta_2\right| \\ &\leq m_2 a_1 a_2 \left(\left|\dot{\theta}_1\right| + \left|\dot{\theta}_2\right|\right)^2 \equiv v_b \|\dot{q}\|^2,\end{aligned}$$

whence $v_b = m_2 a_1 a_2$.

Similarly, for the gravity bound,

$$\begin{aligned}\|G(q)\|_1 &= |(m_1 + m_2)g a_1 \cos\theta_1 + m_2 g a_2 \cos(\theta_1 + \theta_2)| \\ &\quad + |m_2 g a_2 \cos(\theta_1 + \theta_2)| \\ &\leq (m_1 + m_2)g a_1 + 2 m_2 g a_2 \equiv g_b.\end{aligned}$$

Notice that if the arm is RP, then v_b and g_b are functions of q.

c. Coriolis/Centripetal Structural Matrices

We now list the various structural matrices for $V(q,\dot{q})$ discussed in this section. Their computation is left as an exercise (see the Problems).

3.3 Structure and Properties of the Robot Equation

$$U = (I \otimes \dot{q}^T)\frac{\partial M}{\partial q}$$

$$= \begin{bmatrix} 0 & 0 \\ -(2\dot{\theta}_1 + \dot{\theta}_2)m_2 a_1 a_2 \sin \theta_2 & -\dot{\theta}_1 m_2 a_1 a_2 \sin \theta_2 \end{bmatrix}$$

The Coriolis/centripetal matrices:

$$V_{m1} = \dot{M} - \tfrac{1}{2}U$$

$$= \begin{bmatrix} -2\dot{\theta}_2 m_2 a_1 a_2 \sin \theta_2 & -\dot{\theta}_2 m_2 a_1 a_2 \sin \theta_2 \\ (\dot{\theta}_1 - \tfrac{1}{2}\dot{\theta}_2) m_2 a_1 a_2 \sin \theta_2 & \tfrac{1}{2}\dot{\theta}_1 m_2 a_1 a_2 \sin \theta_2 \end{bmatrix}$$

$$V_{m2} = U^T - \tfrac{1}{2}U$$

$$= \begin{bmatrix} 0 & -(2\dot{\theta}_1 + \dot{\theta}_2)m_2 a_1 a_2 \sin \theta_2 \\ (\dot{\theta}_1 + \tfrac{1}{2}\dot{\theta}_2) m_2 a_1 a_2 \sin \theta_2 & \tfrac{1}{2}\dot{\theta}_1 m_2 a_1 a_2 \sin \theta_2 \end{bmatrix}$$

$$V_m = \tfrac{1}{2}(M + U^T - U)$$

$$= \begin{bmatrix} -\dot{\theta}_2 m_2 a_1 a_2 \sin \theta_2 & -(\dot{\theta}_1 + \dot{\theta}_2)m_2 a_1 a_2 \sin \theta_2 \\ \dot{\theta}_1 m_2 a_1 a_2 \sin \theta_2 & 0 \end{bmatrix}$$

The skew-symmetric matrix $S(q, \dot{q})$:

$$S(q, \dot{q}) = U - U^T$$

$$= \begin{bmatrix} 0 & (2\dot{\theta}_1 + \dot{\theta}_2)m_2 a_1 a_2 \sin \theta_2 \\ -(2\dot{\theta}_1 + \dot{\theta}_2)m_2 a_1 a_2 \sin \theta_2 & 0 \end{bmatrix}$$

The symmetric matrices $V_1(q, \dot{q})$, $V_2(q, \dot{q})$:

$$V_1 = \begin{bmatrix} 0 & -m_2 a_1 a_2 \sin \theta_2 \\ -m_2 a_1 a_2 \sin \theta_2 & -m_2 a_1 a_2 \sin \theta_2 \end{bmatrix}$$

$$V_2 = \begin{bmatrix} m_2 a_1 a_2 \sin \theta_2 & 0 \\ 0 & 0 \end{bmatrix}$$

The position/velocity decomposition matrices:

$$V_p(q) = \begin{bmatrix} 0 & 0 \\ 0 & 0 \\ -2m_2 a_1 a_2 \sin\theta_2 & -m_2 a_1 a_2 \sin\theta_2 \\ -m_2 a_1 a_2 \sin\theta_2 & 0 \end{bmatrix}$$

$$V_v(\dot q) = \begin{bmatrix} \tfrac{1}{2}\dot\theta_1 & -\tfrac{1}{2}\dot\theta_2 & \dot\theta_2 & 0 \\ 0 & \dot\theta_1 & -\tfrac{1}{2}\dot\theta_1 & \tfrac{1}{2}\dot\theta_2 \end{bmatrix}$$

d. The Robot Function Parameter Matrix W

The robot dynamics are linear in the parameters. For the purposes of adaptive control, one should select the parameter vector φ in Table 3.3.1 so that it contains the unknown parameters.

The dynamics, including friction, can be written as

$$\begin{aligned}
\tau_1 &= \left[(m_1 + m_2) a_1^2 + m_2 a_2^2 + 2 m_2 a_1 a_2 \cos\theta_2\right] \ddot\theta_1 \\
&\quad + \left[m_2 a_2^2 + m_2 a_1 a_2 \cos\theta_2\right] \ddot\theta_2 - m_2 a_1 a_2 (2\dot\theta_1 \dot\theta_2 + \dot\theta_2^2) \sin\theta_2 \\
&\quad + (m_1 + m_2) g a_1 \cos\theta_1 + m_2 g a_2 \cos(\theta_1 + \theta_2) \\
&\quad + v_1 \dot\theta_1 + k_1 \operatorname{sgn}\left(\dot\theta_1\right) \\
\tau_2 &= \left[m_2 a_2^2 + m_2 a_1 a_2 \cos\theta_2\right] \ddot\theta_1 + m_2 a_2^2 \ddot\theta_2 + m_2 a_1 a_2 \dot\theta_1^2 \sin\theta_2 \\
&\quad + m_2 g a_2 \cos(\theta_1 + \theta_2) + v_2 \dot\theta_2 + k_2 \operatorname{sgn}\left(\dot\theta_2\right).
\end{aligned}$$

The second mass m_2 includes the mass of the payload. This and the friction coefficients are often unknown. Therefore, select

$$\varphi = \begin{bmatrix} m_1 & m_2 & k_1 & v_1 & k_2 & v_2 \end{bmatrix}^T.$$

Then the matrix $W(q, \dot q, \ddot q)$ of known robot functions becomes

$$W = \begin{bmatrix} w_{11} & w_{12} & w_{13} & w_{14} & 0 & 0 \\ 0 & w_{22} & 0 & 0 & w_{25} & w_{26} \end{bmatrix}$$

with

3.3 Structure and Properties of the Robot Equation

$w_{11} = a_1^2 \ddot{\theta}_1 + g a_1 \cos \theta_1$

$w_{12} = \left[a_1^2 + a_2^2 + 2a_1 a_2 \cos \theta_2\right] \ddot{\theta}_1 + \left[a_2^2 + a_1 a_2 \cos \theta_2\right] \ddot{\theta}_2$
$\quad - a_1 a_2 \left(2\dot{\theta}_1 \dot{\theta}_2 + \dot{\theta}_2^2\right) \sin \theta_2 + g a_1 \cos \theta_1 + g a_2 \cos (\theta_1 + \theta_2)$

$w_{13} = \operatorname{sgn}\left(\dot{\theta}_1\right)$

$w_{14} = \dot{\theta}_1$

$w_{22} = \left[a_2^2 + a_1 a_2 \cos \theta_2\right] \ddot{\theta}_1 + a_2^2 \ddot{\theta}_2 + a_1 a_2 \dot{\theta}_1^2 \sin \theta_2 + g a_2 \cos (\theta_1 + \theta_2)$

$w_{25} = \operatorname{sgn}\left(\dot{\theta}_2\right)$

$w_{26} = \dot{\theta}_2$.

The reader should verify that with these definitions, the dynamics may be expressed as $\tau = W\varphi$. The matrix W is computed from measured joint positions, and their velocities and accelerations.

∎

Passivity and Conservation of Energy

The "Newtonian" form of the manipulator dynamics given in Table 3.3.1 obscures some important physical properties, which we should like to explore here [Koditschek 1984], [Ortega and Spong 1988], [Johansson 1990], [Slotine and Li 1987], [Slotine 1988]. Note that the dynamics can be written in terms of the skew-symmetric matrix $S(q, \dot{q})$ as

$$M(q)\ddot{q} + \tfrac{1}{2}\left(\dot{M}(q) - S(q,\dot{q})\right)\dot{q} = \tau - G(q), \qquad (3.3.63)$$

where friction and τ_d are ignored. Now, with K the

$$\frac{dK}{dt} = \frac{1}{2}\frac{d}{dt}\dot{q}^T M(q) \dot{q} = \dot{q}^T M \ddot{q} + \tfrac{1}{2}\dot{q}^T \dot{M}\dot{q},$$

whence (3.3.63) yields

$$\dot{K} = \tfrac{1}{2}\dot{q}^T S \dot{q} + \dot{q}^T (\tau - G) \qquad (3.3.64)$$

or

$$\dot{K} = \dot{q}^T (\tau - G). \qquad (3.3.65)$$

This is a statement of the conservation of energy, with the right-hand side representing the power input from the net external forces. The skew symmetry of $S = (\dot{M} - 2V_m)$ is nothing more than a statement that the fictitious forces $S(q,\dot{q})\dot{q}$ do no work. The work done by the external forces is given by

$$K = \int \dot{q}^T (\tau - G)\, dt. \tag{3.3.66}$$

Recall at this point the passivity property of the robot arm from $\tau(t)$ to $\dot{q}(t)$ (Section 1.5), which merely states that the arm cannot create energy. From a controls point of view, a passive system cannot go unstable. A problem with some popular control schemes (e.g., standard computed torque, Section 3.4) is that they destroy the passivity property, resulting in possible instability if the system parameters are not exactly known or disturbances are present. Passivity-based designs ensure that the closed-loop system is passive (see Section 4.3, the references cited above, and [Anderson 1989]).

This analysis does not include the friction terms. A reasonable assumption regardless of the form of $f(\dot{q})$ is that friction is *dissipative*, so that $f_i(x)$ lies in the first and third quadrants only. This is equivalent to

$$\dot{q}^T F(\dot{q}) \geq 0 \tag{3.3.67}$$

Under this assumption, friction does not destroy the passivity of the manipulator. It is then simple to modify a controller designed for (3.3.63) to include the friction [Slotine 1988]. The dissipative nature of friction allows one to increase the system's bandwidth beyond classical limits.

3.4 State-Variable Representations and Feedback Linearization

The robot arm dynamical equation in Table 3.3.1 is

$$M(q)\ddot{q} + V(q,\dot{q}) + F_v \dot{q} + F_d(\dot{q}) + G(q) + \tau_d = \tau, \tag{3.4.1}$$

with $q(t) \in \mathbb{R}^n$ the joint variable vector and $\tau(t)$ the control input. $M(q)$ is the inertia matrix, $V(q,\dot{q})$ the Coriolis/centripetal vector, $F_v(\dot{q})$ the viscous friction, $F_d(\dot{q})$ the dynamic friction, $G(q)$ the gravity, and τ_d a disturbance. These terms satisfy the properties shown in Table 3.3.1. We may also write the dynamics as

$$M(q)\ddot{q} + N(q,\dot{q}) + \tau_d = \tau, \tag{3.4.2}$$

with the nonlinear terms represented by

3.4 State-Variable Representations and Feedback Linearization

$$N(q, \dot{q}) \equiv V(q, \dot{q}) + F_v \dot{q} + F_d(\dot{q}) + G(q). \tag{3.4.3}$$

In this section we intend to show some equivalent formulations of the arm dynamical equation.

The nonlinear state-variable representation discussed in Chapter 2,

$$\dot{x} = f(x, u, t) \tag{3.4.4}$$

has many properties which are useful from a controls point of view. The function $u(t)$ is the control input and $x(t)$ is the state vector, which describes how the energy is stored in a system. We show here how to place (3.4.1) into such a form. In Chapter 4 we show how to use computers to *simulate* the behavior of a robot arm using this nonlinear state-variable form. Throughout the book we shall use the state-space formulation repeatedly for controls design, either in the nonlinear form or in the linear form

$$\dot{x} = Ax + Bu. \tag{3.4.5}$$

In this section we also present a general approach to *feedback linearization* for the nonlinear robot equation, which involves redefining variables in a methodical way to yield a linear state equation in terms of a dynamical variable we are interested in. This variable could be, for instance, the joint variable $q(t)$, a Cartesian position, or the position in a camera frame of reference.

Hamiltonian Formulation

The arm equation was derived using Lagrangian mechanics. Here, let us use Hamiltonian mechanics [Marion 1965] to derive a state-variable formulation of the manipulator dynamics [Arimoto and Miyazaki 1984], [Gu and Loh 1985]. Let us neglect the friction terms $F(\dot{q}) = F_v(\dot{q}) + F_d(\dot{q})$ and the disturbance τ_d for simplicity; they may easily be added at the end of our development.

In Section 3.2 we expressed the arm Lagrangian as

$$L = K - P = \tfrac{1}{2}\dot{q}^T M(q)\dot{q} - P(q) \tag{3.4.6}$$

with $q(t) \in R^n$ the joint variable, K the kinetic energy, P the potential energy, and $M(q)$ the arm inertia matrix. Define the *generalized momentum* by

$$p \equiv \frac{\partial L}{\partial \dot{q}} = M(q)\dot{q}. \tag{3.4.7}$$

Then we have
$$\dot{q} = M^{-1}(q) p \quad (3.4.8)$$
and the kinetic energy in terms of $p(t)$ is
$$K = \tfrac{1}{2} p^T M^{-1}(q) p. \quad (3.4.9)$$
It is worth noting that
$$K = \tfrac{1}{2} p^T \dot{q}. \quad (3.4.10)$$
Defining the *manipulator Hamiltonian* by
$$H = p^T \dot{q} - L, \quad (3.4.11)$$
Hamilton's equations of motion are
$$\dot{q} = \frac{\partial H}{\partial p} \quad (3.4.12)$$
$$-\dot{p} = \frac{\partial H}{\partial q} - \tau. \quad (3.4.13)$$
Note that
$$H = \tfrac{1}{2} p^T M^{-1}(q) p + P(q) = K + P. \quad (3.4.14)$$
Evaluating (3.4.13) yields
$$\dot{p} = -\tfrac{1}{2} \frac{\partial}{\partial q} \left(p^T M^{-1}(q) p \right) - \frac{\partial P}{\partial q} + \tau,$$
which may be expressed (see the Problems) as
$$\dot{p} = -\tfrac{1}{2} (I_n \otimes p^T) \frac{\partial M^{-1}(q)}{\partial q} p - G(q) + \tau, \quad (3.4.15)$$
where $G(q)$ is the gravity vector and $\otimes$ is the Kronecker product (see Section 3.3).

Defining the state vector $x \in \mathbb{R}^{2n}$ as
$$x = \begin{bmatrix} q^T & p^T \end{bmatrix}^T, \quad (3.4.16)$$
we see that the arm dynamics may be expressed as
$$\frac{d}{dt} \begin{bmatrix} q \\ p \end{bmatrix} = \begin{bmatrix} M^{-1}(q) p \\ -\tfrac{1}{2} (I_n \otimes p^T) \frac{\partial M^{-1}(q) p}{\partial q} \end{bmatrix} + \begin{bmatrix} 0 \\ I_n \end{bmatrix} u, \quad (3.4.17)$$

3.4 State-Variable Representations and Feedback Linearization

with the control input defined by

$$u(t) = \tau - G(q). \tag{3.4.18}$$

This is a nonlinear state equation of the form (3.4.4). It is important to note that this dynamical equation is *linear* in the control input u, which excites *each component* of the generalized momentum $p(t)$.

This *Hamiltonian state-space formulation* was used to derive a PID control law using the Lyapunov approach in [Arimoto and Miyazaki 1984] and to derive a trajectory-following control in [Gu and Loh 1985].

Position/Velocity Formulations

Alternative state-space formulations of the arm dynamics may be obtained by defining the position/velocity state $x \in R^{2n}$ as

$$x = \begin{bmatrix} q^T & \dot{q}^T \end{bmatrix}^T. \tag{3.4.19}$$

For simplicity, neglect the disturbance τ_d and friction $F_v \dot{q} + F_d(\dot{q})$ and note that according to (3.4.2), we may write

$$\frac{d}{dt}\dot{q} = -M^{-1}(q) N(q, \dot{q}) + M^{-1}(q) \tau. \tag{3.4.20}$$

Now, we may directly write the position/velocity state-space representation

$$\dot{x} = \begin{bmatrix} \dot{q} \\ -M^{-1}(q) N(q, \dot{q}) \end{bmatrix} + \begin{bmatrix} 0 \\ M^{-1}(q) \end{bmatrix} \tau, \tag{3.4.21}$$

which is in the form of (3.4.4) with $u(t) = \tau(t)$.

An alternative *linear* state equation of the form (3.4.5) may be written as

$$\dot{x} = \begin{bmatrix} 0 & I \\ 0 & 0 \end{bmatrix} x + \begin{bmatrix} 0 \\ I \end{bmatrix} u, \tag{3.4.22}$$

with control input defined by

$$u(t) = -M^{-1}(q) N(q, \dot{q}) + M^{-1}(q) \tau. \tag{3.4.23}$$

Both of these position/velocity state-space formulations will prove useful in later chapters.

Feedback Linearization

Let us now develop a general approach to the determination of linear state-space representations of the arm dynamics (3.4.1)-(3.4.2). The technique involves a linearization transformation that removes the manipulator nonlinearities. It is a simplified version of the *feedback linearization* technique in [Hunt et al. 1983, Gilbert and Ha 1984]. See also [Kreutz 1989].

The robot dynamics are given by (3.4.2) with $q \in R^n$. Let us define a general sort of output by

$$y = h(q) + s(t), \qquad (3.4.24)$$

with $h(q)$ a general predetermined function of the joint variable $q \in R^n$ and $s(t)$ a general predetermined time function. The control problem, then, will be to select the joint torque and force inputs $\tau(t)$ in order to make the output $y(t)$ go to zero.

The selection of $h(q)$ and $s(t)$ is based on the control objectives we have in mind. For instance, if $h(q) = -q$ and $s(t) = q_d(t)$, the desired joint space trajectory we would like the arm to follow, then $y(t) = q_d(t) - q(t) \equiv e(t)$ the joint space tracking error. Forcing $y(t)$ to zero in this case would cause the joint variables $q(t)$ to track their desired values $q_d(t)$, resulting in arm trajectory following.

As another example, $y(t) = [e_p^T \ e_0^T]^T$ could represent the *Cartesian space* tracking error, with $e_p \in R^3$ the position error and $e_0 \in R^3$ the orientation error. Controlling $y(t)$ to zero would then result in trajectory following directly in *Cartesian space*, which is, after all, where the desired motion is usually specified.

Finally, $-h(q)$ could represent the nonlinear transformation to a *camera frame of reference* and $s(t)$ the desired trajectory in that frame. Then $y(t)$ is the camera frame tracking error. Forcing $y(t)$ to zero would then result in tracking motion in *camera space*.

Feedback Linearizing Transformation. To determine a linear state-variable model for robot controller design, let us simply differentiate the output $y(t)$ twice to obtain

$$\dot{y} = \frac{\partial h}{\partial q}\dot{q} + \dot{s} \equiv J\dot{q} + \dot{s} \qquad (3.4.25)$$

$$\ddot{y} = \dot{J}\dot{q} + J\ddot{q} + \ddot{s}, \qquad (3.4.26)$$

where we have defined the Jacobian

$$J(q) \equiv \frac{\partial h(q)}{\partial q}. \qquad (3.4.27)$$

If $y \in R^P$, the Jacobian is a $p \times n$ matrix of the form

3.4 State-Variable Representations and Feedback Linearization

$$J(q) \equiv \frac{\partial h(q)}{\partial q} = \begin{bmatrix} \frac{\partial h}{\partial q_1} & \frac{\partial h}{\partial q_2} & \cdots & \frac{\partial h}{\partial q_n} \end{bmatrix}. \quad (3.4.28)$$

Given the function $h(q)$, it is straightforward to compute the Jacobian $J(q)$ associated with $h(q)$. In the special case where $\dot{y}$ represents the Cartesian velocity, $J(q)$ is the arm Jacobian discussed in Appendix A. Then, if all joints are revolute, the units of J are those of length.

According to (3.4.2),

$$\ddot{q} = M^{-1}(-N - \tau_d + \tau), \quad (3.4.29)$$

so that (3.4.26) yields

$$\ddot{y} = \ddot{s} + \dot{J}\dot{q} + JM^{-1}(-N - \tau_d + \tau). \quad (3.4.30)$$

Define the *control input* function

$$u(t) = \ddot{s} + \dot{J}\dot{q} + JM^{-1}(-N + \tau) \quad (3.4.31)$$

and the *disturbance* function

$$v(t) = -JM^{-1}\tau_d. \quad (3.4.32)$$

Now we may define a state $x(t) \in \mathbf{R}^{2p}$ by

$$x = \begin{bmatrix} y^T & \dot{y}^T \end{bmatrix}^T \quad (3.4.33)$$

and write the robot dynamics as

$$\frac{d}{dt}\begin{bmatrix} y \\ \dot{y} \end{bmatrix} = \begin{bmatrix} 0 & I_p \\ 0 & 0 \end{bmatrix}\begin{bmatrix} y \\ \dot{y} \end{bmatrix} + \begin{bmatrix} 0 \\ I_p \end{bmatrix} u + \begin{bmatrix} 0 \\ I_p \end{bmatrix} v. \quad (3.4.34)$$

This is a linear state-space system of the form

$$\dot{x} = Ax + Bu + Dv, \quad (3.4.35)$$

driven both by the control input $u(t)$ and the disturbance $v(t)$. Due to the special form of A and B, this system is said to be in *Brunovsky canonical form* (Chapter 2). The reader should determine the controllability matrix to verify that it is always controllable from $u(t)$.

Equation (3.4.31) is said to be a *linearizing transformation* for the robot dynamical equation. We may invert this transformation to obtain

$$\tau = MJ^{+}\left[u - \ddot{s} - \dot{J}\dot{q}\right] + N, \quad (3.4.36)$$

where J^+ is the *Moore-Penrose inverse* [Rao and Mitra 1971] of the Jacobian $J(q)$. If $J(q)$ is square (i.e., $p = n$) and nonsingular, then $J^+(q) = J^{-1}(q)$ and we may write

$$\tau = MJ^{-1}\left[u - \ddot{s} - \dot{J}\dot{q}\right] + N, \qquad (3.4.37)$$

As we shall see in Chapter 4, feedback linearization provides a powerful controls design technique. In fact, if we select $u(t)$ so that (3.4.34) is stable (e.g., a possibility is the PD feedback $u = -K_v\dot{y} - K_p y$), then the control input torque $\tau(t)$ defined by (3.4.36) makes the robot arm move in such a way that $y(t)$ goes to zero.

In the special case $y(t) = q(t)$, then $J = I$ and (3.4.34) reduces to the linear position/velocity form (3.4.22).

3.5 Cartesian and Other Dynamics

In Section 3.2 we derived the robot dynamics in terms of the time behavior of $q(t)$. According to Table 3.3.1,

$$M(q)\ddot{q} + V(q,\dot{q}) + F_v\dot{q} + F_d(\dot{q}) + G(q) + \tau_d = \tau \qquad (3.5.1)$$

or

$$M(q)\ddot{q} + N(q,\dot{q}) + \tau_d = \tau, \qquad (3.5.2)$$

where the nonlinear terms are

$$N(q,\dot{q}) \equiv V(q,\dot{q}) + F_v\dot{q} + F_d(\dot{q}) + G(q). \qquad (3.5.3)$$

We call this the dynamics of the arm formulated in joint space, or simply the *joint-space dynamics*.

Cartesian Arm Dynamics

It is often useful to have a description of the dynamical development of variables other than the joint variable $q(t)$. Consequently, define

$$y = h(q) \qquad (3.5.4)$$

with $h(q)$ a generally nonlinear transformation. Although $y(t)$ could be any variable of interest, let us think of it here as the Cartesian or task space position of the end effector (i.e., position and orientation of the end effector in base coordinates).

3.5 Cartesian and Other Dynamics

The derivation of the Cartesian dynamics from the joint-space dynamics is akin to the feedback linearization in Section 3.4. Differentiating (3.5.4) twice yields

$$\dot{y} = J\dot{q} \tag{3.5.5}$$

$$\ddot{y} = J\ddot{q} + \dot{J}\dot{q}, \tag{3.5.6}$$

where the Jacobian is

$$J \equiv \frac{\partial h}{\partial q}. \tag{3.5.7}$$

The Cartesian velocity vector is $\dot{y} = [v^T \ \omega^T]^T \in R^6$, with $v \in R^3$ the linear velocity and $\omega \in R^3$ the angular velocity. Let us assume that the number of links is $n = 6$, so that J is square. Assuming also that we are away from workspace singularities so that $|J| \neq 0$, according to (3.5.6), we may write

$$\ddot{q} = J^{-1}\ddot{y} - J^{-1}\dot{J}\dot{q}, \tag{3.5.8}$$

which is the "inverse acceleration" transformation. Substituting this into (3.5.2) yields

$$MJ^{-1}\ddot{y} + \left(N - MJ^{-1}\dot{J}\dot{q}\right) + \tau_d = \tau.$$

Recalling now the force transformation $\tau = J^T F$, with F the Cartesian force vector (see Appendix A) we have

$$J^{-T}MJ^{-1}\ddot{y} + J^{-T}\left(N - MJ^{-1}\dot{J}\dot{q}\right) + J^{-T}\tau_d = F. \tag{3.5.9}$$

This may be written as

$$\bar{M}\ddot{y} + \bar{N} + f_d = F, \tag{3.5.10}$$

where we have defined the Cartesian inertia matrix, nonlinear terms, and disturbance by

$$\bar{M} \equiv J^{-T}MJ^{-1} \tag{3.5.11}$$

$$\bar{N} \equiv J^{-T}\left(N - MJ^{-1}\dot{J}\dot{q}\right) = J^{-T}\left(N - MJ^{-1}\dot{J}J^{-1}\dot{y}\right) \tag{3.5.12}$$

$$f_d \equiv J^{-T}\tau_d. \tag{3.5.13}$$

Equation (3.5.9)-(3.5.10) gives the Cartesian or workspace dynamics of the robot manipulator.

Note that $\overline{M}$, $\overline{N}$, and $\overline{f}_d$ depend on q and $\dot{q}$, so that strictly speaking, the Cartesian dynamics are not completely given in terms of y, $\dot{y}$, $\ddot{y}$. However, $\dot{q} = J^{-1}\dot{y}$, and given $y(t)$ we could use the inverse kinematics to determine $q(t)$, so that $\overline{M}$, $\overline{N}$, $\overline{f}_d$ can be computed as functions of y and $\dot{y}$ using computer subroutines.

Structure and Properties of the Cartesian Dynamics

It is important to realize that all the properties of the joint-space dynamics listed in Table 3.3.1 carry over to the Cartesian dynamics as long as J is nonsingular [Slotine and Li 1987]. Note particularly that $\overline{M}$ is symmetric and positive definite. For a revolute arm the Jacobian has units of length and is bounded. In that case, $\overline{M}$ is bounded above and below.

Defining

$$\overline{V} \equiv J^{-T}\left(V(q, \dot{q}) - MJ^{-1}\dot{J}J^{-1}\dot{y}\right), \qquad (3.5.14)$$

it follows that

$$\overline{V} = \overline{V}_m \dot{y} \qquad (3.5.15)$$

with

$$\overline{V}_m = J^{-T}\left(V_m - MJ^{-1}\dot{J}\right)J^{-1}, \qquad (3.5.16)$$

where V_m was defined in Section 3.3.

It is easy to show that

$$S \equiv \dot{\overline{M}} - 2\overline{V}_m \qquad (3.5.17)$$

is skew-symmetric. Indeed, use the identity

$$\frac{d}{dt}\left(J^{-1}\right) = -J^{-1}\dot{J}J^{-1} \qquad (3.5.18)$$

to see that

$$\begin{aligned}
\dot{\overline{M}} - 2\overline{V}_m &= \frac{d}{dt}\left(J^{-T}\right)MJ^{-1} + J^{-T}\dot{M}J^{-1} + J^{-T}M\frac{d}{dt}\left(J^{-1}\right) - 2\overline{V}_m \\
&= -J^{-T}\dot{J}^T J^{-T}MJ^{-1} - J^{-T}MJ^{-1}\dot{J}J^{-1} + 2J^{-T}MJ^{-1}\dot{J}J^{-1} \\
&\quad + J^{-T}\left(\dot{M} - 2V_m\right)J^{-1} \\
&= J^{-T}\left(\dot{M} - 2V_m + \left[MJ^{-1}\dot{J} - \left(MJ^{-1}\dot{J}\right)^T\right]\right)J^{-1},
\end{aligned}$$

3.5 Cartesian and Other Dynamics

which is skew symmetric since $\dot{M} - 2V_m$ is.

The friction terms in the Cartesian dynamics are

$$\bar{F}_v \dot{y} + \bar{F}_s \equiv J^{-T} F_v J^{-1} \dot{y} + J^{-T} F_d(\dot{q}), \qquad (3.5.19)$$

and they satisfy bounds like those in Table 3.3.1. Notice that in Cartesian coordinates the friction effects are not decoupled (e.g., $J^{-T} F_v J^{-1}$ is not diagonal). The Cartesian gravity vector

$$\bar{G} \equiv J^{-T} G(q) \qquad (3.5.20)$$

is bounded.

The property of linearity in the parameters holds and is expressed as

$$\bar{M} \ddot{y} + \bar{N} = \bar{M} \ddot{y} + \bar{V}_m \dot{y} + \bar{F}_v \dot{y} + \bar{F}_s + \bar{G} = \bar{W}(y, \dot{y}, \ddot{y}) \varphi, \qquad (3.5.21)$$

where the known Cartesian function of robot functions is

$$\bar{W}(y, \dot{y}, \ddot{y}) = J^{-T} W(q, \dot{q}, \ddot{q}) \qquad (3.5.22)$$

and φ is the vector of arm parameters.

EXAMPLE 3.5-1: Cartesian Dynamics for Three-Link Cylindrical Arm

Let us show how to convert the joint space dynamics found in Example 3.2.3 to Cartesian dynamics. From Example A.3-1, the arm Jacobian is

$$J = \begin{bmatrix} -r\cos\theta & 0 & -\sin\theta \\ -r\sin\theta & 0 & \cos\theta \\ 0 & 1 & 0 \end{bmatrix}, \qquad (1)$$

whence its inverse is

$$J^{-1} = \begin{bmatrix} -(\cos\theta)/r & -(\sin\theta)/r & 0 \\ 0 & 0 & 1 \\ -\sin\theta & -\cos\theta & 0 \end{bmatrix}. \qquad (2)$$

From Example 3.2.3 the arm inertia matrix is

$$M = \begin{bmatrix} J + m_2 r^2 & 0 & 0 \\ 0 & m_1 + m_2 & 0 \\ 0 & 0 & m_2 \end{bmatrix} \qquad (3)$$

Applying (3.5.11) yields (verify!)

$$\bar{M} = J^{-T}MJ^{-1} = \begin{bmatrix} m_2 + \bar{J}\cos^2\theta & \bar{J}\sin\theta\cos\theta & 0 \\ \bar{J}\sin\theta\cos\theta & m_2 + \bar{J}\sin^2\theta & 0 \\ 0 & 0 & m_1 + m_2 \end{bmatrix}. \quad (4)$$

where $\bar{J} \equiv J/r^2$.

In a similar fashion, one may compute $\bar{N}$.

∎

3.6 Actuator Dynamics

We have discussed the dynamics of a rigid-robot manipulator in joint space and Cartesian coordinates. However, the robot needs actuators to move it; these are generally either electric or hydraulic motors. It is now required, therefore, to add the actuator dynamics to the arm dynamics to obtain a complete dynamical description of the arm plus actuators. A good reference on actuators and sensors is provided by [de Silva 1989].

Dynamics of a Robot Arm with Actuators

We shall consider the case of electric actuators, assuming that the motors are armature controlled. Hydraulic actuators are described by similar equations. In this subsection we suppose that the armature inductance is negligible.

The equations of the n-link robot arm from Table 3.3.1 are given by

$$M(q)\ddot{q} + V_m(q,\dot{q})\dot{q} + F(\dot{q}) + G(q) = \tau, \quad (3.6.1)$$

where $q \in R^n$ is the arm joint variable. The dynamics or the armature-controlled do motors that drive the links are given by the n decoupled equations

$$J_M\ddot{q}_M + B\dot{q}_M + F_M + R\tau = K_M v, \quad (3.6.2)$$

where $q_M = \text{vec}\{q_{Mi}\} \in R^n$, with q_{Mi}, the ith rotor position angle and $\text{vec}\{a_i\}$ denoting a vector with components a_i. The control input is the motor voltage vector $v \in R^n$.

The actuator coefficient matrices are all constants given by

3.6 Actuator Dynamics

The actuator coefficient matrices are all constants given by

$$J_M = \text{diag}\{J_{Mi}\}$$
$$B = \text{diag}\{B_{Mi} + K_{bi}K_{Mi}/R_{ai}\}$$
$$R = \text{diag}\{r_i\}$$
$$K_M = \text{diag}\{K_{Mi}/R_{ai}\},$$

(3.6.3)

where the ith motor has inertia J_{Mi}, rotor damping constant B_{Mi}, back emf constant K_{bi}, torque constant K_{Mi}, and armature resistance R_{ai}.

The gear ratio of the coupling from the ith motor to the ith arm link is r_i, which we define so that

$$q_i = r_i q_{Mi} \quad \text{or} \quad q = Rq_M.$$

(3.6.4)

If the ith joint is revolute, then r_i is a dimensionless constant less than 1. If q_i is prismatic, then r_i has units of m/rad.

The actuator friction vector is given by

$$F_M = \text{vec}\{F_{Mi}\}$$

with F_{Mi} the friction of the ith rotor.

Note that capital "M" denotes motor constants and variables, while V_m is the arm Coriolis/centripetal vector defined in terms of Christoffel symbols.

Using (3.6.4) to eliminate q_M in (3.6.2), and then substituting for τ from (3.6.1) results in the dynamics in terms of joint variables

$$\left(J_M + R^2 M\right)\ddot{q} + \left(B + R^2 V_m\right)\dot{q} + \left(RF_M + R^2 F\right) + R^2 G = RK_M v \quad (3.6.5)$$

or, by appropriate definition of symbols,

$$M'(q)\ddot{q} + V'(q,\dot{q})\dot{q} + F'(\dot{q}) + G'(q) = K'v. \quad (3.6.6)$$

Properties of the Complete Arm-Plus-Actuator Dynamics. The complete dynamics (3.6.6) has the same form as the robot dynamics (3.6.1). It is very easy to verify that the complete arm-plus-actuator dynamics enjoys the same properties as the arm dynamics that are listed in Table 3.3.1 (see the Problems). In particular, V' is one-half the difference between $\dot{M}'$ and a skew-symmetric matrix, all the boundedness assumptions hold, and linearity in the parameters holds. Thus, in future work where we design controllers, we may assume that the actuators have been included in the arm equation in Table 3.3.1

Independent Joint Dynamics. In many commercial robot arms the gear ratios r_i are very small, providing a large torque advantage in the actuator/link coupling. This has important ramifications that greatly simplify the design of robot arm controllers.

To explore this, let us write the complete dynamics by components as

$$\left(J_{Mi} + r_i^2 m_{ii}\right)\ddot{q}_i + B_i \dot{q}_i + r_i F_{Mi} = \frac{r_i K_{Mi}}{R_{ai}} v_i - r_i^2 d_i, \quad i = 1, \cdots, n, \quad (3.6.7)$$

where $B \equiv \text{diag}\{B_i\}$ and d_i is a disturbance given by

$$d_i = \sum_{j \neq i} m_{ij} \ddot{q}_j + \sum_{j,k} V_{jki} \dot{q}_j \dot{q}_k + F_i + G_i \quad (3.6.8)$$

with m_{ij} the off-diagonal elements of M', V_{jki} the tensor components of $V'\dot{q}$, F_i the friction of the ith link, and G_i the ith gravity component.

This equations reveals that if r_i is small, the arm dynamics are approximately given by n *decoupled second-order equations with constant coefficients*. The dynamical effects of joint coupling and gravity appear only as disturbance terms multiplied by r_i^2. That is, robot controls design is virtually the problem of simply *controlling the actuator dynamics*.

Unfortunately, modern high-performance tasks make the Coriolis and centripetal terms large, so that d_i is not small. Moreover, modern high-performance arms have near-unity gear ratios (e.g., direct drive arms), so that the nonlinearities must be taken into account in any conscientious controls design.

Third-Order Arm-Plus-Actuator Dynamics

An alternative model of the complete robot arm is sometimes used in controls design [Tarn et al. 1991]. It is a third-order differential equation that should be used when the motor armature inductance is not negligible.

When the armature inductances L_i are not negligible, instead of (3.6.2) we must use the armature-controlled do motor equations

$$TK'_M \dot{I} + K'_M I + B' \dot{q}_M = K_M v \quad (3.6.9)$$

$$J_M \ddot{q}_M + B_M \dot{q}_M + F_M + R\tau = K'_m I \quad (3.6.10)$$

with $I \in \mathbb{R}^n$ the vector of armature currents,

3.6 Actuator Dynamics

$$T = \text{diag}\{L_i/R_{ai}\}$$
$$K'_m = \text{diag}\{K_{Mi}\}$$
$$B' = \text{diag}\{K_{bi}K_{Mi}/R_{ai}\}$$
$$B_M = \text{diag}\{B_{Mi}\}. \quad (3.6.11)$$

It is important to note that T is a matrix of motor electric time constants. In the preceding subsection, these time constants were assume negligibly small in comparison to the motor mechanical time constant.

To determine the overall dynamics of the arm plus do motor actuators, eliminate τ between (3.6.1) and (3.6.10) to obtain an expression for I. Then, differentiate to expose explicitly $\dot{I}$. Substitute these expressions into (3.6.9) (see the Problems) to obtain dynamics of the form

$$D\frac{d^3}{dt^3}q + f(q,\dot{q},\ddot{q}) = RK_M v. \quad (3.6.12)$$

The coefficient matrix D is given by

$$D(q) = TM'(q), \quad (3.6.13)$$

so that it is negligible when L_i are small.

Dynamics with Joint Flexibility

We have assumed that the coupling between between the actuators and the robot links is provided through rigid gear trains with gear ratios of r_i. In actual practice, the coupling suffers from backlash and gear train flexibility or elasticity. Here we include the flexibility of the joints in the arm dynamic model, assuming for simplicity that $r_i = 1$.

This is not difficult to do. Indeed, suppose that the coupling flexibility is modeled as a stiff spring. Then the torque mentioned in Equations (3.6.1), (3.6.2) is nothing but

$$\tau = B_s(\dot{q}_M - \dot{q}) + K_s(q_M - q), \quad (3.6.14)$$

with $B_s = \text{diag}\{b_{si}\}$, $K_s = \text{diag}\{k_{si}\}$, and b_{si} and k_{si} the damping and spring constants of the ith gear train. Thus the dynamical equations become

$$M(q)\ddot{q} + V_m(q,\dot{q})\dot{q} + F(\dot{q}) + G(q) + B_s(\dot{q} - \dot{q}_M) + K(q - q_M) = 0, \quad (3.6.15)$$

$$J_M \ddot{q}_M + B\dot{q}_M + F_M + B_s(\dot{q}_M - \dot{q}) + K(q_M - q) = K_M v. \quad (3.6.16)$$

The structure of these equations is very different from the rigid joint arm described in Table 3.3.1. We discuss the control of robot manipulators with joint flexibility in Chapter 6 (see [Spong 1987]). The next example shows the problems that can occur in controlling flexible joint robots.

EXAMPLE 3.6-1: DC Motor with Flexible Coupling Shaft

To focus on the effects of joint flexibility, let us examine a single armature-controlled dc motor coupled to a load through a shaft that has significant flexibility. The electrical and mechanical subsystems are shown in Figure 3.6.1.

The motor electrical equation is

$$L\dot{i} = -Ri - k'_m \dot{\theta}_m + u \tag{1}$$

with $i(t)$, $u(t)$ the armature current and voltage, respectively. The back emf is $v_b = k'_m \dot{\theta}_m$.

The interaction force exerted by the flexible shaft is given by $f = b(\dot{\theta}_m - \dot{\theta}_L) + k(\theta_m - \theta_L)$, where the shaft damping and spring constants are denoted by b and k. Thus the mechanical equations of motion may be written down as

$$J_m \ddot{\theta}_m + b_m \dot{\theta}_m + b\left(\dot{\theta}_m - \dot{\theta}_L\right) + k\left(\theta_m - \theta_L\right) = k_m i \tag{2}$$

$$J_L \ddot{\theta}_L + b\left(\dot{\theta}_L - \dot{\theta}_m\right) + k\left(\theta_L - \theta_m\right) = 0, \tag{3}$$

with subscripts m and L referring, respectively, to motor parameters and load parameters. The load inertia J_L is assumed constant. The definitions of the remaining symbols may be inferred from the foregoing text.

To place these equations into state-space form, define the state as

$$x = \begin{bmatrix} i & \theta_m & \omega_m & \theta_L & \omega_L \end{bmatrix}^T, \tag{4}$$

with $\omega_m = \dot{\theta}_m$ and $\omega_L = \dot{\theta}_L$ the motor and load angular velocities. Then

3.6 Actuator Dynamics

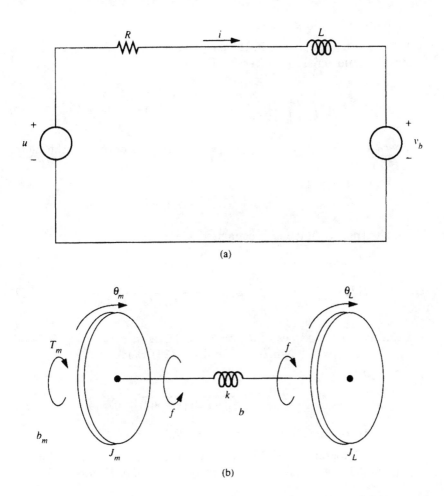

Figure 3.6.1: Dc motor with shaft compliance: (a) electrical subsystem; (b) mechanical subsystem.

$$\dot{x} = \begin{bmatrix} \frac{-R}{L} & 0 & \frac{-k'_m}{L} & 0 & 0 & 0 \\ 0 & 0 & 1 & 0 & 0 & 0 \\ \frac{k_m}{J_m} & \frac{-k}{J_m} & \frac{-(b+b_m)}{J_m} & \frac{k}{J_m} & \frac{b}{J_m} & \\ 0 & 0 & 0 & 0 & 1 & \\ 0 & \frac{k}{J_L} & \frac{b}{J_L} & \frac{-k}{J_L} & \frac{-b}{J_L} & \end{bmatrix} x + \begin{bmatrix} \frac{1}{L} \\ 0 \\ 0 \\ 0 \\ 0 \end{bmatrix} u. \quad (5)$$

a. Rigid Coupling Shaft

If there is no compliance in the coupling shaft, $\omega_m = \omega_L = \omega$ and the state equations reduce to (see the Problems)

$$\dot{x} = \begin{bmatrix} -R/L & -k'_m/L \\ k_m/J & -b_m/J \end{bmatrix} x + \begin{bmatrix} 1/L \\ 0 \end{bmatrix} u \equiv Ax + Bu, \qquad (6)$$

where $x = [i \ \ \omega]^T$, $J = J_m + J_L$. Defining the output as the motor speed gives

$$y = \begin{bmatrix} 0 & 1 \end{bmatrix} x \equiv Cx.$$

The transfer function is computed to be

$$H(s) = C(sI - A)^{-1} B = \frac{k_m}{(Ls + R)(Js + b_m) + k_m k'_m}. \qquad (7)$$

Using parameter values of $J_m = J_L = 0.1$ kg-m^2, $k_m = k'_m = 1$ V-s, $L = 0.5$ H, $b_m = 0.2$ N-m/rad/s, and $R = 5$ Ω yields

$$H(s) = \frac{10}{(s + 2.3)(s + 8.7)}, \qquad (8)$$

so that there are two real poles at $s = -2.3$, $s = -8.7$.

Using Program TRESP in Appendix B to perform a simulation (see Section 3.3) yields the step response for w shown in Figure 3.6.2.

b. Very Flexible Coupling Shaft

Coupling shaft parameters of $k = 2$ N-m/rad and $b = 0.2$ N-m/rad/s correspond to a very flexible shaft. Using these values, software like PC-MATLAB can be employed to obtain the two transfer functions

$$\frac{\omega_m}{u} = \frac{20s\left[(s+1)^2 + 4.36^2\right]}{s(s+3.05)(s+6.14)\left[(s+3.4)^2 + 5.6^2\right]} \qquad (9)$$

$$\frac{\omega_L}{u} = \frac{40s(s+10)}{s(s+3.05)(s+6.14)\left[(s+3.4)^2 + 5.6^2\right]} \qquad (10)$$

3.6 Actuator Dynamics

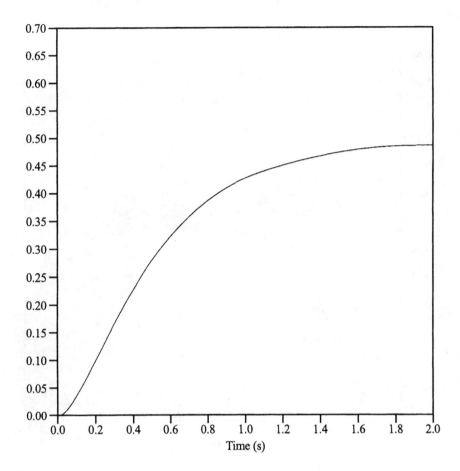

Figure 3.6.2: Step response of dc motor with no shaft flexibility. Motor speed in rad/s.

The shaft flexible mode has the poles $s = -3.4 \pm j5.6$, and so has a damping ration of $\zeta = 0.52$ and a natural frequency of $\omega = 6.55$ rad/s. Note that the system is marginally stable, with a pole at $s = 0$. It is BIBO stable due to pole-zero cancellation.

Program TRESP yielded the step response shown in Figure 3.6.3. Several points are worthy of note. Initially, the motor speed ω_m rises more quickly than in Figure 3.6.2, since the shaft flexibility means that only the rotor moment of inertia J_m initially affects the speed. Then, as the load J_L is coupled back to the motor through the shaft, the rate of increase of ω_m slows. Note also that the load speed ω_L exhibits a *delay*

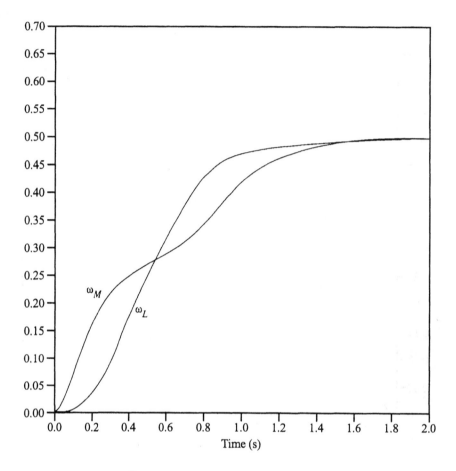

Figure 3.6.3: Step response of motor with very flexible shaft.

of approximately 0.1 s due to the flexibility in the shaft.

It is extremely interesting to note that the shaft flexibility has the effect of speeding up the slowest motor real pole [compare (8) and (9)], so that ω_L approaches its steady-state value more quickly than in the rigid-shaft case. This is due to the "whipping" action of the flexible shaft.

The shaft dynamics make the control of θ_L, which corresponds in a robot arm to the joint angle q_i, very difficult without some sort of specially designed controller.

3.7 Summary

In this chapter we have laid the foundation for a study of robot control systems. Using Lagrangian mechanics in Section 3.2, we derived the dynamics of some robot arms that will be used for demonstration designs throughout the text. We provided expressions for the general robot arm dynamics for any serial-link arm.

In Section 3.3 we studied the properties of the robot dynamics such as boundedness, linearity in the parameters, and skew symmetry that are needed in controls design. Table 3.3.1 gives a summary of our findings. We used a Kronecker product approach that yields great insight into the relations between the terms in the robot equation.

A vital form in modern control systems design is the state-variable formulation. In Section 3.4 we derived several state-space forms of the arm dynamics, setting the stage for several design techniques to be provided in subsequent chapters. The state formulation is also useful in computer simulation of robot controllers, as we see in Section 3.3.

The dynamics in Cartesian form were given in Section 3.5. The dynamics of the actuators that drive the robot manipulator links were analyzed and included in Section 3.6.

REFERENCES

[Anderson 1989] Anderson, R. J., "Passive computed torque algorithms for robots," *Proc. IEEE Conf. Decision Control*, pp. 1638-1644, Dec. 1989.

[Arimoto and Miyazaki 1984] Arimoto, S., and F. Miyazaki, "Stability and robustness of PID feedback control for robot manipulators of sensory capability," *Proc. First Int. Symp.*, pp. 783-799, MIT, Cambridge, MA, 1984.

[Asada and Slotine 1986] Asada, H., and J.-J. E. Slotine, *Robot Analysis and Control*, New York: Wiley, 1986.

[Borisenko and Tarapov 1968] Borisenko, A. I., and I. E. Tarapov, *Vector and Tensor Analysis with Applications*. Englewood Cliffs, NJ: Prentice Hall, 1968.

[Brewer 1978] Brewer, J. W., "Kronecker products and matrix calculus in system theory," *IEEE Trans. Circuits Syst.*, vol. CAS-25, no. 9, pp. 772-781, Sept. 1978.

[Craig 1988] Craig, J. J., *Adaptive Control of Mechanical Manipulators*. Reading, MA: Addison-Wesley, 1988.

[de Silva 1989] de Silva, C. W., *Control Sensors and Actuators*. Englewood Cliffs, NJ: Prentice Hall, 1989.

[Gilbert and Ha 1984] Gilbert, E. G., and I. J. Ha, "An approach to nonlinear feedback control with applications to robotics," *IEEE Trans. Syst. Man Cybern.*, vol. SMC-14, no. 6, pp. 879-884, Nov./Dec. 1984.

[Gu and Loh 1985] Gu, Y.-L., and N. K. Loh, "Dynamic model for industrial robots based on a compact Lagrangian formulation," *Proc. IEEE Conf. Decision Control*, pp. 1497-1501, 1985.

[Gu and Loh 1988] Gu, Y-L., and N. K. Loh, "Dynamic modeling and control by utilizing an imaginary robot model," *IEEE J. Robot. Autom.*, vol. 4, no. 5, pp. 532-534, Oct. 1988.

[Hunt et al. 1983] Hunt, L. R., R. Su, and G. Meyer, "Global transformations of nonlinear systems," *IEEE Trans. Autom. Control*, vol. AC-28, no. 1, pp. 24-31, Jan. 1983.

[Johansson 1990] Johansson, R., "Quadratic optimization of motion coordination and control," *IEEE Trans. Autom. Control*, vol. 35, no. 11, pp. 1197-1208, Nov. 1990.

[Koditschek 1984] Koditschek, D., "Natural motion for robot arms," *Proc. IEEE Conf. Decision Control*, pp. 733-735, Dec. 1984.

[Kreutz 1989] Kreutz, K., "On manipulator control by exact linearization," *IEEE Trans. Autom. Control*, vol. 34, no. 7, pp. 763-767, July 1989.

[Lee et al. 1983] Lee, C. S. G., R. C. Gonzalez, and K. S. Fu, *Tutorial on Robotics*. New York: IEEE Press, 1983.

[Marion 1965] Marion, J. B., *Classical Dynamics*. New York: Academic Press, 1965.

[Ortega and Spong 1988] Ortega, R., and Spong, M. W., "Adaptive motion control of rigid robots: a tutorial," *Proc. IEEE Conf. Decision Control*, pp. 1575-1584, Dec. 1988.

[Paul 1981] Paul, R. P., *Robot Manipulators*. Cambridge, MA: MIT Press, 1981.

[Rao and Mitra 1971] Rao, C. R., and S. K. Mitra, *Generalized Inverse of Matrices and Its Applications*. New York: Wiley, 1971.

[Schilling 1990] Schilling, R. J., *Fundamentals of Robotics*. Englewood Cliffs, NJ: Prentice Hall, 1990.

[Slotine 1988] Slotine, J.-J. E., "Putting physics in control: the example of robotics," *IEEE Control Syst. Mag.*, pp. 12-17, Dec. 1988.

[Slotine and Li 1987] Slotine, J.-J. E., and W. Li, "Adaptive strategies in constrained manipulation," *Proc. IEEE Conf. Robot. Autom.*, pp. 595-601, 1987.

[Spong 1987] Spong, M. W., "Modeling and control of elastic joint robots," *J. Dyn. Syst. Meas. Control*, vol. 109, pp. 310-319, Dec. 1987.

REFERENCES

[Spong and Vidyasagar 1989] Spong, M. W., and M. Vidyasagar, *Robot Dynamics and Control.* New York: Wiley, 1989.

[Tarn et al. 1991] Tarn, T.-J., A. K. Bejczy, X. Yun, and Z. Li, "Effect of motor dynamics on nonlinear feedback robot arm control," *IEEE Trans. Robot. Autom.*, vol. 7, no. 1, pp. 114-122, Feb. 1991.

PROBLEMS

Section 3.2

3.2-1 Dynamics. Find the dynamics for the spherical wrist in Example A.2-4.

3.2-2 Dynamics from Derived Equations. In Example 3.2.2 we found the dynamics of the two-link planar elbow arm from first principles. In this problem, begin with the expressions for the kinetic and potential energy in that example and:
(a) Write K in the form (3.2.29) to determine $M(q)$.
(b) Use (3.2.42) and (3.2.43) to determine $V(q,\dot{q})$ and $G(q)$.

3.2-3 Dynamics from Derived Equations. Repeat Problem 3.2-2 for the three-link arm in Example 3.2.3.

Section 3.3

3.3-1 Prove (3.3.22) by finding $V_{p1}(q)$ and $V_{v1}(q)$.

3.3-2 Prove (3.3.23) by finding the matrices $V_i(q)$.

3.3-3 Prove (3.3.27).

3.3-4 Coriolis Term. Find $V_{\text{cor}}(q)$ and $V_{\text{cen}}(q)$ in (3.3.40).

3.3-5 Coriolis Term. Demonstrate that the Coriolis/centripetal term in the dynamics equation may be expressed [Paul 1981] as $V(q,\dot{q}) = \text{vec}\{V(q,\dot{q})\}$, where

$$V_k(q,\dot{q}) = \sum_{i,j} v_{ijk} \dot{q}_i \dot{q}_j$$

with

$$v_{ijk} = \sum_{l=1}^{n} \text{trace}\left[\frac{\partial^2 T_l}{\partial q_i \partial q_j} I_l \frac{\partial T_l^T}{\partial q_k}\right]$$

and T_i defined in Appendix A. Compare this to V_{m1}, V_{m2}, V_m as defined in Section 3.2.

3.3-6 Bounds and Structure. Derive in detail the results in Example 3.3.1.

3.3-7 Bounds and Structure. Derive the bounds and structural matrices for the two-link polar arm in Example 3.2.1. Use:
(a) The 1−norm.
(b) The 2−norm .
(c) The ∞−norm.

3.3-8 Bounds and Structure. Repeat Problem 3.3-7 for the three-link cylindrical arm in Exercise 3.2.3.

3.3-9 Bounds Using 2-Norm. Derive the bounds for the two-link planar elbow arm in Example 3.3.1 using the 2−norm.

Section 3.4
3.4-1 Prove (3.4.15).

3.4-2 Hamiltonian State Formulation. Demonstrate that (3.4.15) is equivalent to

$$\dot{p} = \tfrac{1}{2}\left(\dot{M} + S\right) M^{-1} p + \tau - G,$$

with $S(q, \dot{q})$ the skew-symmetric matrix defined in Section 3.3.

3.4-3 Hamiltonian State Formulation. Use (3.4.17) to derive the Hamiltonian state-variable formulation for the two-link polar arm in Example 3.2.1.

3.4-4 Hamiltonian State Formulation. Repeat Problem 3.4-3 for the two-link planar elbow arm in Example 3.2.2.

Section 3.5
3.5-1 Cartesian Dynamics. Complete Example 3.5.1, computing the nonlinear terms $\overline{N}$ in Cartesian coordinates.

3.5-2 Cartesian Dynamics. Find the Cartesian dynamics of the two-link polar arm in Example 3.2.1.

3.5-3 Cartesian Dynamics. Find the Cartesian dynamics of the two-link planar elbow arm in Example 3.2.2.

Section 3.6
3.6-1 Actuator Dynamics. Verify that the arm-plus-actuator dynamics (3.6.6) has the properties listed in Table 3.3.1.

3.6-2 Actuator Dynamics. Derive the third-order dynamics (3.6.12), providing explicit expressions for $f(q, \dot{q}, \ddot{q})$. Verify that they reduce to (3.6.5) when L_i is negligible.

3.6-3 Flexible Coupling Shaft. Verify the state equation for the rigid-shaft case in Example 3.6.1.

Chapter 4

Computed-Torque Control

In this chapter we examine some straightforward control schemes for robot manipulators that fall under the class known as "computed-torque controllers." These generally perform well when the robot arm parameters are known fairly accurately. Some connections are given with classical robot control, and modern design techniques are provided as well. The effects of digital implementation of robot controllers are shown. Trajectory generation is outlined.

4.1 Introduction

A basic problem in controlling robots is to make the manipulator follow a preplanned desired trajectory. Before the robot can do any useful work, we must position it in the right place at the right instances. In this chapter we discuss *computed-torque control*, which yields a family of easy-to-understand control schemes that often work well in practice. These schemes involve the decomposition of the controls design problem into an *inner-loop design* and an *outer-loop design*.

In Section 4.4 we provide connections with classical manipulator control schemes based on independent joint design using PID control. In Section 4.6 we show how to use some modern design techniques in conjunction with computed-torque control. Thus this chapter could be considered as a bridge between classical design techniques of the sort used several years ago in robot control, and the modern design techniques in the remainder of the book which are needed to obtain high performance in uncertain environments.

We assume here the robot is moving in free space, having no contact with its environment. Contact results in the generation of forces. The

169

force control problem is dealt with in Chapter 7. We will also assume in this chapter that the robot is a well-known rigid system, thus designing controllers based on a fairly well-known model. Control in the presence of uncertainties or unknown parameters (e.g., friction, payload mass) requires refined approaches. This problem is dealt with using *robust control* in Chapter 4 and *adaptive control* in Chapter 5.

An actual robot manipulator may have flexibility in its links, or compliance in its gearing (joint flexibility). In Chapter 6 we cover some aspects of control with joint flexibility.

Before we can control a robot arm, it is necessary to know the *desired path* for performing a task. There are many issues associated with the path planning problem, such as avoiding obstacles and making sure that the planned path does not require exceeding the voltage and torque limitations of the actuators. To reduce the control problem to its basic components, in this chapter we assume that the ultimate control objective is to move the robot along a prescribed desired trajectory. We do not concern ourselves with the actual trajectory-planning problem; we do, however, show how to reconstruct a continuous desired path from a given table of desired points the end effector should pass through. This *continuous-path generation* problem is covered in Section 4.2.

In most practical situations robot controllers are implemented on microprocessors, particularly in view of the complex nature of modern control schemes. Therefore, in Section 4.5 we illustrate some notions of the *digital implementation of robot controllers*.

Throughout, we demonstrate how to simulate robot controllers on a computer. This should be done to verify the effectiveness of any proposed control scheme prior to actual implementation on a real robot manipulator.

4.2 Path Generation

Throughout the book we assume that there is given a prescribed path $q_d(t)$ the robot arm should follow. We design control schemes that make the manipulator follow this desired path or trajectory. *Trajectory planning* involves finding the prescribed path and is usually considered a separate design problem involving collision avoidance, concerns about *actuator saturation*, and so on. See [Lee et al. 1983].

We do not cover trajectory planning. However, we do cover two aspects of trajectory generation. First, we show how to convert a given prescribed path from Cartesian space to joint space. Then, given a table of desired points the end effector should pass through, we show how to reconstruct a continuous desired trajectory.

Converting Cartesian Trajectories to Joint Space

In robotic applications, a desired task is usually specified in the workspace or Cartesian space, as this is where the motion of the manipulator is easily described in relation to the external environment and workpiece. However, trajectory-following control is easily performed in the joint space, as this is where the arm dynamics are more easily formulated.

Therefore, it is important to be able to find the desired joint space trajectory $q_d(t)$ given the desired Cartesian trajectory. This is accomplished using the *inverse kinematics*, as shown in the next example. The example illustrates that the mapping of Cartesian to joint space trajectories may not be unique-that is, several joint space trajectories may yield the same Cartesian trajectory for the end-effector.

EXAMPLE 4.2-1: Mapping a Prescribed Cartesian Trajectory to Joint Space

In Example A.3-5 are derived the inverse kinematics for the two-link planar robot arm shown in Figure 4.2.1. Let us use them to convert a path from Cartesian space to joint space.

Suppose that we want the two-link arm to follow a given workspace or Cartesian trajectory

$$p(t) = (x(t), y(t)) \tag{1}$$

in the (x, y) plane which is a function of time t. Since the arm is moved by actuators that control its angles θ_1, θ_2, it is convenient to convert the specified Cartesian trajectory $(x(t), y(t))$ into a *joint space trajectory* $(\theta_1(t), \theta_2(t))$ for control purposes.

This may be achieved by using the inverse kinematics transformations

$$r^2 = x^2 + y^2 \tag{2}$$

$$C = \cos\theta_2 = \frac{r^2 - a_1^2 + a_2^2}{2a_1 a_2} \tag{3}$$

$$D = \pm\sqrt{1 - \cos^2\theta_2} = \pm\sqrt{1 - C^2} \tag{4}$$

$$\theta_2 = \text{ATAN2}(D, C) \tag{5}$$

$$\theta_1 = \text{ATAN2}(y, x) - \text{ATAN2}(a_2 \sin \theta_2, a_1 + a_2 \cos \theta_2) \qquad (6)$$

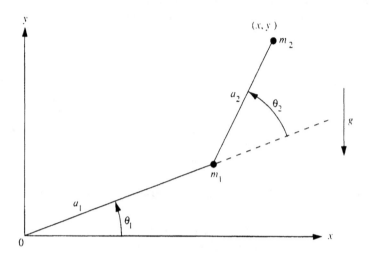

Figure 4.2.1: Two-link planar elbow arm.

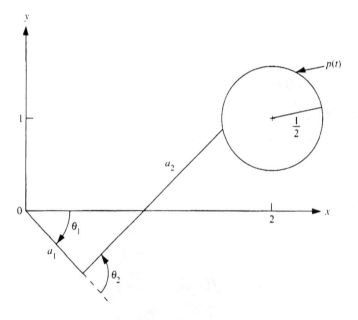

Figure 4.2.2: Desired Cartesian trajectory.

4.2 Path Generation

Suppose that the end of the arm should repeatedly trace out the circular workspace path $p(t)$ shown in Figure 4.2.2, which is described by

$$x(t) = 2 + 1/2 \cos t$$
$$y(t) = 1 + 1/2 \sin t.$$

(7)

By using these expressions for each time t in the inverse kinematics equations, we obtain the required joint-space trajectories $q(t) = (\theta_1(t), \theta_2(t))$ given in Figure 4.2.3 that yield the circular Cartesian motion of the end effector (using $a_1 = 2$, $a_2 = 2$).

We have computed the joint variables for the "elbow down" configuration. Selecting the opposite sign in (4) gives the "elbow up" joint space trajectory yielding the same Cartesian trajectory. ∎

Polynomial Path Interpolation

Suppose that a desired trajectory for the manipulator motion has been determined, either in Cartesian space or, using the inverse kinematics, in joint space. For convenience, we use the joint space variable $q(t)$ for notation. It is not possible to store the entire trajectory in computer memory, and few practically useful trajectories have a simple closed-form expression. Therefore, it is usual to store in computer memory a sequence of points $q_i(t_k)$ for each joint variable i that represent the desired values of that variable at the discrete times t_k. Thus $q(t_k)$ is a point in $\mathbb{R}^n$ that the joint variables should pass through at time t_k. We call these *via points*.

Most robot control schemes require a continuous desired trajectory. To convert the table of via points $q_i(t_k)$ to a continuous desired trajectory $q_d(t)$, we may use many options. Let us discuss here *polynomial interpolation*.

Suppose that the via points are uniformly spaced in time and define the *sampling period* as

$$T = t_{k+1} - t_k.$$ (4.2.1)

For smooth motion, on each time interval $[t_k, t_{k+1}]$ we require the desired position $q_d(t)$ and velocity $\dot{q}_d(t)$ to match the tabulated via points. This yields boundary conditions of

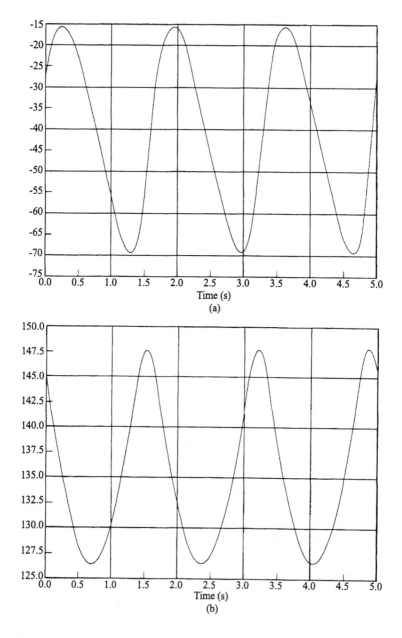

Figure 4.2.3: Required joint-space trajectories: (a) θ_1 (deg); (b) θ_2 (deg).

4.2 Path Generation

$$q_{d_i}(t_k) = q_i(t_k)$$
$$\dot{q}_{d_i}(t_k) = \dot{q}_i(t_k)$$
$$q_{d_i}(t_{k+1}) = q_i(t_{k+1})$$
$$\dot{q}_{d_i}(t_{k+1}) = \dot{q}_i(t_{k+1}). \quad (4.2.2)$$

To match these boundary conditions, it is necessary to use on $[t_k, t_{k+1}]$ the *cubic interpolating polynomial*

$$q_{d_i}(t) = a_i + (t - t_k) b_i + (t - t_k)^2 c_i + (t - t_k)^3 d_i, \quad (4.2.3)$$

which has four free variables. Then

$$\dot{q}_{d_i}(t) = b_i + 2(t - t_k) c_i + 3(t - t_k)^2 d_i \quad (4.2.4)$$

$$\ddot{q}_{d_i}(t) = 2c_i + 6(t - t_k) d_i \quad (4.2.5)$$

so that the acceleration is *linear* on each sample period.

It is easy to solve for the coefficients that guarantee matching of the boundary conditions. In fact, we see that

$$\begin{bmatrix} 1 & 0 & 0 & 0 \\ 0 & 1 & 0 & 0 \\ 1 & T & T^2 & T^3 \\ 0 & 1 & 2T & 3T^2 \end{bmatrix} \begin{bmatrix} a_i \\ b_i \\ c_i \\ d_i \end{bmatrix} = \begin{bmatrix} q_i(t_k) \\ \dot{q}_i(t_k) \\ q_i(t_{k+1}) \\ \dot{q}_i(t_{k+1}) \end{bmatrix} \quad (4.2.6)$$

This is solved to obtain the required interpolating coefficients on each interval $[t_k, t_{k+1}]$.

$$a_i = q_i(t_k)$$
$$b_i = \dot{q}_i(t_k)$$
$$c_i = \frac{3[q_i(t_{k+1}) - q_i(t_k)] - T[2\dot{q}_i(t_k) + \dot{q}_i(t_{k+1})]}{T^2}$$
$$d_i = \frac{2[q_i(t_k) - q_i(t_{k+1})] - T[\dot{q}_i(t_k) + \dot{q}_i(t_{k+1})]}{T^3}. \quad (4.2.7)$$

Note that this technique requires storing the desired position *and velocity* at each sampling point in tabular form. A variant uses a higher-order polynomial to ensure continuous position, velocity, *and acceleration* at each sample time t_k.

Although we have used the joint variable notation $q(t)$, it should be emphasized that trajectory interpolation can also be performed in Cartesian space.

Linear Function with Parabolic Blends

Using cubic interpolating polynomials, the acceleration on each sample period is linear. However, in many practical applications there are good reasons for insisting on *constant* accelerations within each sample period. For instance, any real robot has upper limits on the torques that can be supplied by its actuators. For linear systems (think of Newton's law) this translates into constant accelerations. Therefore, constant accelerations are less likely to saturate the actuators. Besides that, most industrial robot controllers are programmed to use constant accelerations on each sample period.

A constant acceleration profile is shown in Figure 4.2.4(a). The associated velocity and position profiles are shown in Figure 4.2.4(b) and 4.2.4(c). The position trajectory has three parts: a quadratic or parabolic initial portion, a linear midsection, and a parabolic final portion. Therefore, let us discuss interpolation of via points using *linear functions with parabolic blends* (LFPB).

The time at which the position trajectory switches from parabolic to linear is known as the *blend time* t_b. A position $q_{d_i}(t)$ *should be specified for each joint variable* i. The trajectory in Figure 4.2.4(c) can be written for joint i as

$$q_{d_i}(t) = \begin{cases} a_i + (t - t_k) b_i + (t - t_k)^2 c_i, & t_k \leq t < t_k + t_b \\ d_i + v_i t, & t_k + t_b \leq t < t_{k+1} - t_b \\ e_i + (t - t_{k+1}) f_i + (t - t_{k+1})^2 g_i, & t_{k+1} - t_b \leq t \leq t_{k+1}. \end{cases} \quad (4.2.8)$$

The coefficient v_i may be interpreted as the maximum velocity allowed for joint variable i. The design parameters are v_i and t_b.

It is straightforward to solve for the coefficients on each time interval $[t_k, t_{k+1}]$ that ensure satisfaction of the boundary conditions (4.2.2). The result is

4.2 Path Generation

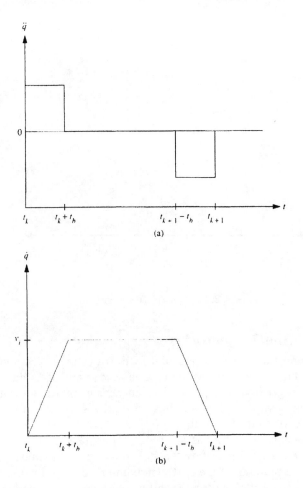

Figure 4.2.4: LFPB trajectory: (a) acceleration; (b) velocity.

$$a_i = q_i\left(t_k\right), \quad b_i = \dot{q}_i\left(t_k\right), \quad c_i = \frac{v_i - \dot{q}_i\left(t_k\right)}{2t_b},$$
$$d_i = \frac{q_i\left(t_k\right) + q_i\left(t_{k+1}\right) - v_i t_{k+1}}{2}$$
$$e_i = q_i\left(t_{k+1}\right), \quad f_i = \dot{q}_i\left(t_{k+1}\right)$$
$$g_i = \frac{v_i t_{k+1} + q_i\left(t_k\right) - q_i\left(t_{k+1}\right) + 2t_b\left[\dot{q}_i\left(t_{k+1}\right) - v_i\right]}{2t_b^2}.$$

$$(4.2.9)$$

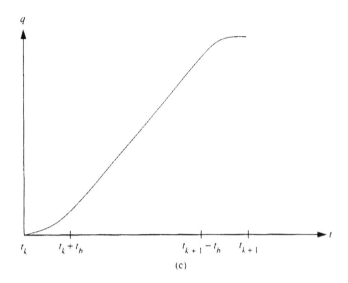

Figure 4.2.4: *(Cont.)* (c) position.

Minimum-Time Trajectories

There is an important special class of LFPB trajectories. Suppose the acceleration is limited by a maximum value of a_M and it is desired for the robot arm to get from one position to another in *minimum time*. For simplicity assume that the initial and final velocities are equal to zero. The general case is covered in [Lewis 1986a] (see the Problems).

A minimum-time trajectory is shown in Figure 4.2.5. To drive joint variable i from a rest position of $q_0 \equiv q_i(t_0)$ to a desired final rest position of $q_f \equiv q_i(t_f)$ in a minimum time t_f, the maximum acceleration a_M should be applied until the *switching time* t_s, after which time the maximum deceleration $-a_M$ should be applied until t_f. Note that both t_s and t_f depend on q_0 and q_f. We may write

$$q_i(t_s) = q_0 + \tfrac{1}{2} a_M (t_s - t_0)^2$$
$$\dot{q}_i(t_s) = a_M (t_s - t_0)$$
$$q_i(t_f) = q_i(t_s) + \dot{q}_i(t_s)(t_f - t_s) - \tfrac{1}{2} a_M (t_f - t_s)^2$$
$$\dot{q}_i(t_f) = \dot{q}_i(t_s) - a_M (t_f - t_s).$$

Then the velocity equations yield

$$\dot{q}_i(t_f) = a_M (t_s - t_0) - a_M (t_f - t_s) = 0$$

4.2 Path Generation

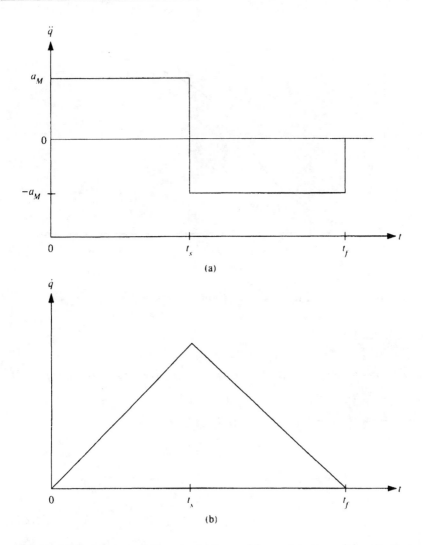

Figure 4.2.5: Minimum-time trajectory: (a) acceleration; (b) velocity.

or

$$t_s = (t_f + t_0)/2. \qquad (4.2.10)$$

That is, the switching from maximum acceleration to maximum deceleration occurs at the *half-time point*. Now simple manipulations on the position equations yield

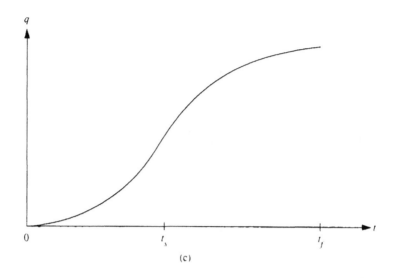

Figure 4.2.5: *(Cont.)* (c) position.

$$q_i(t_f) = q_0 + \tfrac{1}{2}a_M(t_s - t_0)^2 + a_M(t_s - t_0)(t_f - t_s) - \tfrac{1}{2}a_M(t_f - t_s)^2$$
$$= q_f$$
$$\frac{q_f - q_0}{a_M} = \tfrac{1}{2}(t_s - t_0)^2 + (t_s - t_0)(t_f - t_s) - \tfrac{1}{2}(t_f - t_s)^2$$

whence (4.2.10) yields

$$t_s = t_0 + \sqrt{(q_f - q_0)/a_M}. \qquad (4.2.11)$$

A closed-loop formulation of minimum-time control appears in [Lewis 1986a].

Unfortunately, minimum-time trajectories computed using a constant maximum acceleration are not directly relevant in robotics. This is due to the fact that an actual manipulator has a *torque limit* of τ_M. Since the robot equation (see Table 3.3.1) is nonlinear, this does not correlate to a constant upper bound on the acceleration. For instance, a robot arm has different maximum accelerations in its fully extended and fully retracted positions. For more discussion see [Kahn and Roth 1971], [Chen 1989], [Geering 1986], [Gourdeau and Schwartz 1989], [Jayasuriya and Suh 1985], [Kahn and Roth 1971], [Kim and Shin 1985], [Shin and McKay 1985].

4.3 Computer Simulation of Robotic Systems

It is very important to *simulate* on a digital computer a proposed manipulator control scheme before actually implementing it on an arm. We show here how to perform such computer simulations for robotic systems. Since most robot controllers are actually implemented in a digital fashion (Section 4.5), we also show how to simulate digital robot arm controllers.

Simulation of Robot Dynamics

There is a variety of software packages for the simulation of nonlinear dynamical systems, including SIMNON [Åström and Wittenmark 1984], MATLAB, and others. For convenience, we include in Appendix B some simulation programs that are quite useful for continuous and digital control.

All simulation programs require the user to write similar subroutines. Time response simulators that use integration routines such as Runge-Kutta all require the computation of the state derivative given the current state. In Section 3.4 we saw how to represent the robot arm equation

$$M(q)\ddot{q} + N(q,\dot{q}) + \tau_d = \tau \qquad (4.3.1)$$

in the nonlinear state-space form

$$\dot{x} = f(x,u,t), \qquad (4.3.2)$$

with $x(t)$ the state and $u(t)$ the input.

Defining a state as $x = [q^T \quad \dot{q}^T]^T$, we may write the *implicit form*

$$\begin{bmatrix} I & 0 \\ 0 & M(q) \end{bmatrix} \begin{bmatrix} \dot{q} \\ \ddot{q} \end{bmatrix} = \begin{bmatrix} \dot{q} \\ -N(q,\dot{q}) \end{bmatrix} + \begin{bmatrix} 0 \\ I \end{bmatrix} \tau + \begin{bmatrix} 0 \\ -I \end{bmatrix} \tau_d, \qquad (4.3.3)$$

with τ the arm control torque that is provided by the controller and τ_d the disturbance torque. We say that this is an implicit form since the coefficient matrix of the left-hand side means that $\dot{x} = d[q^T \quad \dot{q}^T]^T/dt$ is not given explicitly in terms of the right-hand side.

Given $x(t)$, it is necessary to provide a subroutine for the integration program that computes $\dot{x}(t)$. One approach to solving for $\dot{x}$ is to invert $M(q)$. However, due to potential numerical problems this is not recommended. Let us represent (4.3.3) as

$$E(x)\dot{x} = f(x,u,t). \qquad (4.3.4)$$

Note that in this case $u(t)$ is the vector composed of the controls $\tau(t)$ and the disturbances $\tau_d(t)$.

A simple time response program, TRESP, is given in Appendix B. Given a subroutine $F(time, x, \dot{x})$ that computes $\dot{x}$ given $x(t)$ and $u(t)$ using (4.3.4); it uses a Runge-Kutta integrator to compute the state trajectory $x(t)$. To solve for $\dot{x}$ within subroutine $F(t, x, \dot{x})$, we recommend computing $M(q)$ and $N(q, \dot{q})$, and then solving

$$M(q) \frac{d\dot{q}}{dt} = -N(q, \dot{q}) + \tau - \tau_d \qquad (4.3.5)$$

[i.e., the bottom portion of (4.3.3)] by *least-squares techniques*, which are more stable numerically than the inversion of $M(q)$. Least-squares equation solvers are readily available commercially in, for instance [IMSL], [LINPACK], and elsewhere. For simpler arms, $M(q)$ may be inverted analytically.

Throughout the book we illustrate the simulation of the arm dynamics using various control schemes.

Simulation of Digital Robot Controllers

While most robot controllers are designed in continuous time, they are implemented on actual robots digitally. That is, the control signals are only updated at discrete instants of time using a microprocessor. We discuss the implementation of digital robot arm controllers in Section 4.5. To verify that a proposed controller will operate as expected, therefore, it is highly desirable to simulate it in its digitized or discretized form prior to actual implementation.

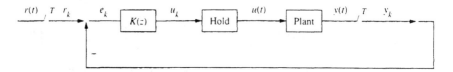

Figure 4.3.1: Digital controller.

A digital control scheme is shown in Figure 4.3.1. The *plant* or system to be controlled is a continuous-time system, and $K(z)$ is the dynamic digital controller, where z is the Z-transform variable (i.e., z^{-1} represents a unit time delay). The digital controller $K(z)$ is implemented using software code in a digital signal processor (DSP). The *reference input* $r(t)$ is the desired trajectory that $y(t)$ should follow, and e_k is the (discrete) tracking error.

The sampler with sample period T is an analog-to-digital (A/D) converter that takes the samples $y_k = y(kT)$ of the output $y(t)$ that are required by the software controller $K(z)$. The definition of $y(t)$ can vary de-

4.3 Computer Simulation of Robotic Systems

pending on the control scheme. For instance, in robot control, $y(t)$ might represent the $2n$-vector composed of $q(t)$ and $\dot{q}(t)$.

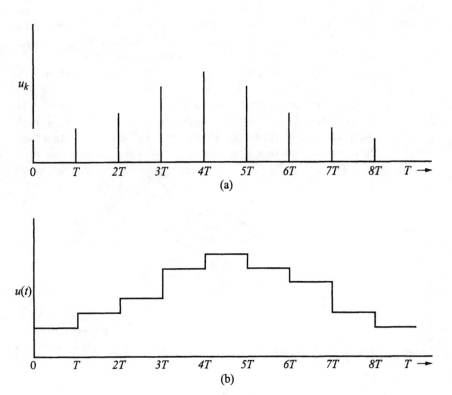

Figure 4.3.2: Data reconstruction using a ZOH: (a) discrete control sequence u_k; (b) reconstructed continuous signal $u(t)$.

The hold device in the figure is a digital-to-analog (D/A) converter that converts the discrete control samples u_k computed by the software controller $K(z)$ into the continuous-time control $u(t)$ required by the plant. It is a *data reconstruction* device. The input u_k and output $u(t)$ for a zero-order hold (ZOH) are shown in Figure 4.3.2. Note that $u(kT) = u_k$, with T the sample period, so that $u(t)$ is continuous from the right. That is, $u(t)$ is updated at times kT. The ZOH is generally used for controls purposes, as opposed to other higher-order devices such as the first-order hold, since most commercially available DSPs have a built-in ZOH.

Once a controller has been designed, it is important to *simulate* it using a digital computer before it is implemented to determine if the closed-loop response is suitable. This is especially true in robotics, since the digital controller is generally found by designing a continuous-time controller, which is then digitized using approximation techniques such as Euler's method. That is, for nonlinear systems, the controller discretization schemes are generally not exact. This results in degraded performance. To verify that the controller performance will be suitable, the simulation should provide the response at all times, including times *between* the samples.

To simulate a digital controller we may use the scheme shown in Figure 4.3.3. There, the continuous plant dynamics are contained in the subroutine $F(t, x, \dot{x})$; they are integrated using a Runge-Kutta integrator. The figure assumes a ZOH; thus the control input $u(t)$ is updated to u_k at each time kT, and then held constant until time $(k+1)T$ Note that two time intervals are involved; the sampling period T and the *Runge-Kutta integration period* $T_R \ll T$. T_R should be selected as an integral divisor of T.

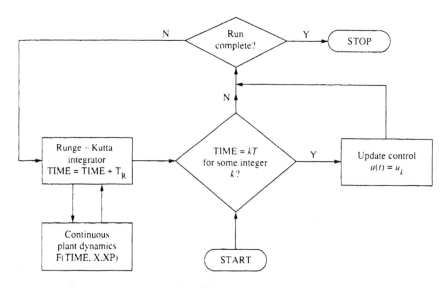

Figure 4.3.3: Digital control simulation scheme.

This simulation technique provides the plant state $x(t)$ as a continuous function of time, even at values *between* the sampling instants [in fact, it provides $x(t)$ at multiples of T_R]. This is essential in verifying acceptable *intersample behavior* of the closed-loop system prior to implementing the digital controller on the actual plant.

Program TRESP in Appendix B can be used to implement Figure 4.3.3. It is written in a modular fashion to apply to a wide variety of situations. We shall illustrate its use for the purpose of digital control in several subsequent examples. The use of such simulation software as SIMNON is quite similar.

We discuss the implementation of digital robot arm controllers in Section 4.5. Some detailed discussion on digital control, simulation, and DSP implementation of controllers is given in [Lewis 1992].

4.4 Computed-Torque Control

Through the years there have been proposed many sorts of robot control schemes. As it happens, most of them can be considered as special cases of the class of *computed-torque controllers*. Computed torque, at the same time, is a special application of *feedback linearization* of nonlinear systems, which has gained popularity in modern systems theory [Hunt et al. 1983], [Gilbert and Ha 1984]. In fact, one way to classify robot control schemes is to divide them as "computed-torque-like" or "noncomputed-torque-like." Computed-torque-like controls appear in robust control, adaptive control, learning control, and so on.

In the remainder of this chapter we explore this class of robot controllers, which includes such a broad range of designs. Computed-torque control allows us to conveniently derive very effective robot controllers, while providing a framework to bring together classical independent joint control and some modern design techniques, as well as set the stage for the rest of the book. A summary of the different computed-torque-like controllers is given at the end of the section in Table 4.4.1. We shall see that many digital robot controllers are also computed-torque-like controllers (Section 4.5).

Derivation of Inner Feedforward Loop

The robot arm dynamics are

$$M(q)\ddot{q} + V(q, \dot{q}) + F_v \dot{q} + F_d(\dot{q}) + G(q) + \tau_d = \tau \quad (4.4.1)$$

or

$$M(q)\ddot{q} + N(q, \dot{q}) + \tau_d = \tau, \quad (4.4.2)$$

with the joint variable $q(t) \in R^n$, $\tau(t)$ the control torque, and $\tau_d(t)$ a disturbance. If this equation includes motor actuator dynamics (Section 3.6), then $\tau(t)$ is an input voltage.

Suppose that a desired trajectory $q_d(t)$ has been selected for the arm motion, according to the discussion in Section 4.2. To ensure trajectory tracking by the joint variable, define an output or *tracking error* as

$$e(t) = q_d(t) - q(t). \qquad (4.4.3)$$

To demonstrate the influence of the input $\tau(t)$ on the tracking error, differentiate twice to obtain

$$\dot{e} = \dot{q}_d - \dot{q}$$
$$\ddot{e} = \ddot{q}_d - \ddot{q}.$$

Solving now for $\ddot{q}$ in (4.4.2) and substituting into the last equation yields

$$\ddot{e} = \ddot{q}_d + M^{-1}(N + \tau_d - \tau). \qquad (4.4.4)$$

Defining the control input function

$$u = \ddot{q}_d + M^{-1}(N - \tau) \qquad (4.4.5)$$

and the disturbance function

$$w = M^{-1}\tau_d \qquad (4.4.6)$$

we may define a state $x(t) \in \mathbf{R}^{2n}$ by

$$x = \begin{bmatrix} e \\ \dot{e} \end{bmatrix} \qquad (4.4.7)$$

and write the *tracking error dynamics* as

$$\frac{d}{dt}\begin{bmatrix} e \\ \dot{e} \end{bmatrix} = \begin{bmatrix} 0 & I \\ 0 & 0 \end{bmatrix}\begin{bmatrix} e \\ \dot{e} \end{bmatrix} + \begin{bmatrix} 0 \\ I \end{bmatrix}u + \begin{bmatrix} 0 \\ I \end{bmatrix}w. \qquad (4.4.8)$$

This is a linear error system in Brunovsky canonical form consisting of n pairs of double integrators $1/s^2$, one per joint. It is driven by the control input $u(t)$ and the disturbance $w(t)$. Note that this derivation is a special case of the general feedback linearization procedure in Section 3.4.

The feedback linearizing transformation (4.4.5) may be inverted to yield

$$\tau = M(\ddot{q}_d - u) + N. \qquad (4.4.9)$$

We call this the *computed-torque control law*. The importance of these manipulations is as follows. There has been no state-space transformation in going from (4.4.1) to (4.4.8). Therefore, if we select a control $u(t)$ that stabilizes (4.4.8) so that $e(t)$ goes to zero, then the nonlinear control input $\tau(t)$ given by (4.4.9) will cause trajectory following in the robot arm (4.4.1). In fact, substituting (4.4.9) into (4.4.2) yields

4.4 Computed-Torque Control

or

$$M\ddot{q} + N + \tau_d = M(\ddot{q}_d - u) + N$$

$$\ddot{e} = u + M^{-1}\tau_d, \tag{4.4.10}$$

which is exactly (4.4.8).

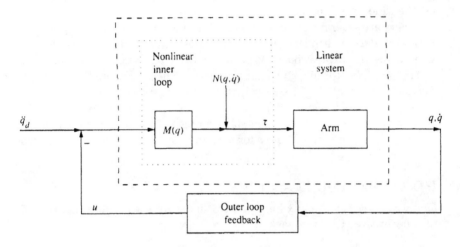

Figure 4.4.1: Computed-torque control scheme, showing inner and outer loops.

The stabilization of (4.4.8) is not difficult. In fact, the nonlinear transformation (4.4.5) has converted a complicated nonlinear controls design problem into a simple design problem for a linear system consisting of n decoupled subsystems, each obeying Newton's laws.

The resulting control scheme appears in Figure 4.4.1. It is important to note that it consists of an *inner nonlinear loop* plus an *outer control signal* $u(t)$. We shall see several ways for selecting $u(t)$. Since $u(t)$ will depend on $q(t)$ and $\dot{q}(t)$, the outer loop will be a feedback loop. In general, we may select a dynamic compensator $H(s)$ so that

$$U(s) = H(s)E(s). \tag{4.4.11}$$

$H(s)$ can be selected for good closed-loop behavior. According to (4.4.10), the closed-loop error system then has transfer function

$$T(s) = s^2 I - H(s). \tag{4.4.12}$$

It is important to realize that computed-torque depends on the inversion of the robot dynamics, and indeed is sometimes called *inverse dynamics control*. In fact, (4.4.9) shows that $\tau(t)$ is computed by substituting $\ddot{q}_d - u$ for $\ddot{q}$ in (4.4.2); that is, by solving the robot *inverse dynamics problem*. The caveats associated with system inversion, including the problems resulting when the system has non-minimum-phase zeros, all apply here. (Note that in the linear case, the system zeros are the poles of the inverse. Such non-minimum-phase notions generalize to nonlinear systems.) Fortunately for us, the rigid arm dynamics are minimum phase.

There are several ways to compute (4.4.9) for implementation purposes. Formal matrix multiplication at each sample time should be avoided. In some cases the expression may be worked out analytically. A good way to compute the torque $\tau(t)$ is to use the efficient Newton-Euler inverse dynamics formulation [Craig 1985] with $\ddot{q}_d - u$ in place of $\ddot{q}(t)$.

The outer-loop signal $u(t)$ can be chosen using many approaches, including robust and adaptive control techniques. In the remainder of this chapter we explore some choices for $u(t)$ and some variations on computed-torque control.

PD Outer-Loop Design

One way to select the auxiliary control signal $u(t)$ is as the proportional-plus-derivative (PD) feedback,

$$u = -K_v \dot{e} - K_p e. \qquad (4.4.13)$$

Then the overall robot arm input becomes

$$\tau = M(q)(\ddot{q}_d + K_v \dot{e} + K_p e) + N(q, \dot{q}) \qquad (4.4.14)$$

This controller is shown in Figure 4.4.6 with $K_i = 0$.

The closed-loop error dynamics are

$$\ddot{e} + K_v \dot{e} + K_p e = w, \qquad (4.4.15)$$

or in state-space form,

$$\frac{d}{dt}\begin{bmatrix} e \\ \dot{e} \end{bmatrix} = \begin{bmatrix} 0 & I \\ -K_p & -K_v \end{bmatrix} \begin{bmatrix} e \\ \dot{e} \end{bmatrix} + \begin{bmatrix} 0 \\ I \end{bmatrix} w. \qquad (4.4.16)$$

The closed-loop characteristic polynomial is

$$\Delta_c(s) = \left| s^2 I + K_v s + K_p \right|. \qquad (4.4.17)$$

Choice of PD Gains. It is usual to take the $n \times n$ gain matrices diagonal so that

4.4 Computed-Torque Control

$$K_v = diag\{k_{v_i}\}, \quad K_p = diag\{k_{p_i}\}. \tag{4.4.18}$$

Then

$$\Delta_c(s) = \prod_{i=1}^{n} (s^2 + k_{v_i}s + k_{p_i}), \tag{4.4.19}$$

and the error system is asymptotically stable as long as the k_{v_i} and k_{p_i} are all positive. Therefore, as long as the disturbance $w(t)$ is bounded, so is the error $e(t)$. In connection with this, examine (4.4.6) and recall from Table 3.3.1 that M^{-1} is upper bounded. Thus boundedness of $w(t)$ is equivalent to boundedness of $\tau_d(t)$.

It is important to note that although selecting the PD gain matrices diagonal results in decoupled control at the outer-loop level, it does not result in a decoupled joint-control strategy. This is because multiplication by $M(q)$ and addition of the nonlinear feedforward terms $N(q, \dot{q})$ in the inner loop scrambles the signal $u(t)$ among all the joints. Thus, information on all joint positions $q(t)$ and velocities $\dot{q}(t)$ is generally needed to compute the control $\tau(t)$ for any one given joint.

The standard form for the second-order characteristic polynomial is

$$p(s) = s^2 + 2\xi\omega_n s + \omega_n^2, \tag{4.4.20}$$

with ζ the damping ratio and ω_n the natural frequency. Therefore, desired performance in each component of the error $e(t)$ may be achieved by selecting the PD gains as

$$k_{p_i} = \omega_n^2, \quad k_{v_i} = 2\xi\omega_n, \tag{4.4.21}$$

with ζ, ω_n the desired damping ratio and natural frequency for joint error i. It may be useful to select the desired responses at the end of the arm faster than near the base, where the masses that must be moved are heavier.

It is undesirable for the robot to exhibit overshoot, since this could cause impact if, for instance, a desired trajectory terminates at the surface of a workpiece. Therefore, the PD gains are usually selected for *critical damping* $\zeta = 1$. In this case

$$k_{v_i} = 2\sqrt{k_{p_i}}, \quad k_{p_i} = k_{v_i}^2/4. \tag{4.4.22}$$

Selection of the Natural Frequency. The natural frequency ω_n governs the speed of response in each error component. It should be large for fast responses and is selected depending on the performance objectives. Thus the desired trajectories should be taken into account in selecting ω_n. We discuss now some additional factors in this choice.

There are some *upper limits* on the choice for ω_n [Paul 1981]. Although the links of most industrial robots are massive, they may have some flexibility. Suppose that the frequency of the first flexible or resonant mode of link i is

$$\omega_r = \sqrt{k_r/J} \qquad (4.4.23)$$

with J the link inertia and k_r the link stiffness. Then, to avoid exciting the resonant mode, we should select $\omega_n < \omega_r/2$. Of course, the link inertia J changes with the arm configuration, so that its maximum value might be used in computing ω_r.

Another upper bound on ω_n is provided by considerations on actuator saturation. If the PD gains are too large, the torque $\tau(t)$ may reach its upper limits.

Some more feeling for the choice of the PD gains is provided from error-boundedness considerations as follows. The transfer function of the closed-loop error system in (4.4.15) is

$$e(s) = (s^2 I + k_v s + k_p)^{-1} w(s), \qquad (4.4.24)$$

or if K_v and K_p are diagonal,

$$e_i(s) = \frac{1}{s^2 + k_{v_i} s + k_{p_i}} w(s) \equiv H(s)w(s) \qquad (4.4.25)$$

$$\dot{e}_i(s) = \frac{s}{s^2 + k_{v_i} s + k_{p_i}} w(s) \equiv sH(s)w(s). \qquad (4.4.26)$$

We assume that the disturbance and M^{-1} are bounded (Table 3.3.1), so that

$$\|w\| \leq \|M^{-1}\| \|\tau_d\| \leq \overline{m}\overline{d}, \qquad (4.4.27)$$

with $\overline{m}$ and $\overline{d}$ known for a given robot arm. Therefore,

$$\|e_i(t)\| \leq \|H(s)\| \|w\| \leq \|H(s)\| \overline{m}\overline{d} \qquad (4.4.28)$$

$$\|\dot{e}_i(t)\| \leq \|sH(s)\| \|w\| \leq \|sH(s)\| \overline{m}\overline{d}. \qquad (4.4.29)$$

Now selecting the L_2-norm, the operator gain $\|H(s)\|_2$ is the maximum value of the Bode magnitude plot of $H(s)$. For a critically damped system,

$$\sup_{\omega} \|H(j\omega)\|_2 = 1/k_{p_i}. \qquad (4.4.30)$$

Therefore,

4.4 Computed-Torque Control

$$\|e_i(t)\|_2 \leq \overline{m}\overline{d}/k_{p_i}. \qquad (4.4.31)$$

Moreover (see the Problems),

$$\sup_\omega \|j\omega H(j\omega)\|_2 = 1/k_{v_i}, \qquad (4.4.32)$$

so that

$$\|\dot{e}_i(t)\|_2 \leq \overline{m}\overline{d}/k_{v_i}. \qquad (4.4.33)$$

Thus, in the case of critical damping, the position error decreases with k_{p_i} and the velocity error decreases with k_{v_i}.

EXAMPLE 4.4-1: Simulation of PD Computed-Torque Control

In this example we intend to show the detailed mechanics of simulating a PD computed-torque controller on a digital computer.

a. Computed-Torque Control Law

In Example 3.2.2 we found the dynamics of the two-link planar elbow arm shown in Figure 4.2.1 to be

$$\begin{bmatrix} (m_1 + m_2)\,a_1^2 + m_2 a_2^2 + 2m_2 a_1 a_2 \cos\theta_2 & m2a_2^2 + m_2 a_1 a_2 \cos\theta_2 \\ m_2 a_2^2 + m_2 a_1 a_2 \cos\theta_2 & m2a_2^2 \end{bmatrix} \begin{bmatrix} \ddot{\theta}_1 \\ \ddot{\theta}_2 \end{bmatrix}$$
$$+ \begin{bmatrix} -m_2 a_1 a_2 \left(2\dot{\theta}_1\dot{\theta}_2 + \dot{\theta}_2^2\right)\sin\dot{\theta}_2 \\ m_2 a_1 a_2 \dot{\theta}_1^2 \sin\theta_2 \end{bmatrix}$$
$$+ \begin{bmatrix} (m_1 + m_2)\,ga_1 \cos\theta_1 + m_2 g a_2 \cos(\theta_1 + \theta_2) \\ m_2 g a_2 \cos(\theta_1 + \theta_2) \end{bmatrix}$$
$$= \begin{bmatrix} \tau_1 \\ \tau_2 \end{bmatrix}. \qquad (1)$$

These are in the standard form

$$M(q)\ddot{q} + V(q,\dot{q}) + G(q) = \tau. \qquad (2)$$

Take the link masses as 1 kg and their lengths as 1 m.

The PD computed-torque control law is given as

$$\tau = M(q)(\ddot{q}_d + K_v \dot{e} + K_p e) + V(q, \dot{q}) + G(q), \qquad (3)$$

with the tracking error defined as

$$e = q_d - q. \qquad (4)$$

b. Desired Trajectory

Let the desired trajectory $q_d(t)$ have the components

$$\begin{aligned} \theta_{1d} &= g_1 \sin(2\pi t/T) \\ \theta_{2d} &= g_2 \cos(2\pi t/T) \end{aligned} \qquad (5)$$

with period $T = 2$ s and amplitudes $g_i = 0.1$ rad ≈ 6 deg. For good tracking select the time constant of the closed-loop system as 0.1 s. For critical damping, this means that $K_v = \text{diag}\{k_v\}$, $K_P = \text{diag}\{k_p\}$, where

$$\begin{aligned} \omega_n &= 1/0.1 = 10 \\ k_p &= \omega_n^2 = 100, \quad k_v = 2\omega_n = 20. \end{aligned} \qquad (6)$$

It is important to realize that the selection of controller parameters such as the PD gains depends on the performance objectives-in this case, the period of the desired trajectory.

c. Computer Simulation

Let us simulate the computed-torque controller using program TRESP in Appendix B. Simulation using commercial packages such as MATLAB and SIMNON is quite similar.

The subroutines needed for TRESP are shown in Figure 4.4.2. They are worth examining closely. Subroutine SYSINP (ITx,t) is called once per Runge-Kutta integration period and generates the reference trajectory $q_d(t)$, as well as $\dot{q}_d(t)$ and $\ddot{q}_d(t)$. Note that the reference signal should be held constant during each integration period.

4.4 Computed-Torque Control

```
C     FILE 21nkct.FOR
C     IMPLEMENTATION OF COMPUTED-TORQUE CONTROLLER ON 2-LINK PLANAR ARM
C     SUBROUTINES FOR USE WITH TRESP
C
C     SUBROUTINE TO COMPUTE DESIRED TRAJECTORY
C        The trajectory value must be constant within each Runge-Kutta
C           integration interval
C
      SUBROUTINE SYSINP(IT,x,t)
      REAL x(*)
      COMMON/TRAJ/qd(6), qdp(6), qdpp(6)
      DATA g1, g2, per, twopi/0.1, 0.1, 2., 6.283/
C     COMPUTE DESIRED TRAJECTORY qd(t), qdp(t), qdpp(t)
      fact= twopi/per
      qd(1)=     g1*sin(fact*t)
      qd(2)=     g2*cos(fact*t)
      qdp(1)=    g1*fact*cos(fact*t)
      qdp(2)=   -g2*fact*sin(fact*t)
      qdpp(1)=  -g1*fact**2*sin(fact*t)
      qdpp(2)=  -g2*fact**2*cos(fact*t)
C
      RETURN
      END
C
C
C ************************************************************
C     MAIN SUBROUTINE CALLED BY RUNGE-KUTTA INTEGRATOR
      SUBROUTINE F(t,x,xp)
      REAL x(*), xp(*)
C
C     COMPUTED-TORQUE CONTROLLER
      CALL CTL(x)
C     ROBOT ARM DYNAMICS
      CALL ARM(x,xp)
C
      RETURN
      END
C
C
C ************************************************************
C     COMPUTED-TORQUE CONTROLLER SUBROUTINE
      SUBROUTINE CTL(x)
      REAL x(*),m1,m2,M11,M12,M22,N1,N2,kp,kv
      COMMON/CONTROL/t1, t2
C     The next line is to plot the errors and torques
      COMMON/OUTPUT/ e(2), ep(2), tp1, tp2
      COMMON/TRAJ/qd(6), qdp(6), qdpp(6)
      DATA m1,m2,a1,a2, g/1.,1.,1.,1., 9.8/
      DATA kp, kv/ 100,20/
C
C     COMPUTE TRACKING ERRORS
      e(1) = qd(1) - x(1)
      e(2) = qd(2) - x(2)
```

```
      ep(1)= qdp(1) - x(3)
      ep(2)= qdp(2) - x(4)

C    COMPUTATION OF M(q), N(q,qp)
      M11= (m1+m2)*a1**2 + m2*a2**2 + 2*m2*a1*a2*cos(x(2))

      M12= m2*a2**2 + m2*a1*a2*cos(x(2))
      M22= m2*a2**2
      N1= -m2*a1*a2*(2*x(3)*x(4) + x(4)**2) * sin(x(2))
      N1= N1 + (m1+m2)*g*a1*cos(x(1)) + m2*g*a2*cos(x(1)+x(2))
      N2= m2*a1*a2*x(3)**2*sin(x(2))    +m2*g*a2*cos(x(1)+x(2))

C    COMPUTATION OF CONTROL TORQUES
      s1= qdpp(1) + kv*ep(1) + kp*e(1)
      s2= qdpp(2) + kv*ep(2) + kp*e(2)
      t1= M11*s1 + M12*s2 + N1
      t2= M12*s1 + M22*s2 + N2
C    The next lines are to plot the torque
      tp1= t1
      tp2= t2

      RETURN
      END
C
C
C ***********************************************************
C    ROBOT ARM DYNAMICS SUBROUTINE
      SUBROUTINE ARM(x,xp)
      REAL x(*),xp(*),m1,m2,M11,M12,M22,MI11,MI12,MI22,N1,N2
      COMMON/CONTROL/t1, t2
C    The next line is to plot the Cartesian position
      COMMON/OUTPUT/ dum(6), y1, y2
      DATA m1,m2,a1,a2, g/1.,1.,1.,1., 9.8/
C    COMPUTATION OF M(q)
      M11= (m1+m2)*a1**2 + m2*a2**2 + 2*m2*a1*a2*cos(x(2))
      M12= m2*a2**2 + m2*a1*a2*cos(x(2))
      M22= m2*a2**2
C    INVERSION OF M(q) (For large n, use least-squares to find qpp)
      det= M11*M22 - M12**2
      MI11= M22/det
      MI12= -M12/det
      MI22= M11/det
C    NONLINEAR TERMS
      N1= -m2*a1*a2*(2*x(3)*x(4) + x(4)**2) * sin(x(2))
      N1= N1 + (m1+m2)*g*a1*cos(x(1))   + m2*g*a2*cos(x(1)+x(2))
      N2= m2*a1*a2*x(3)**2*sin(x(2))    + m2*g*a2*cos(x(1)+x(2))

C    STATE EQUATIONS
      xp(1)= x(3)
      xp(2)= x(4)
      xp(3)= MI11*(-N1 + t1) + MI12*(-N2 + t2)
      xp(4)= MI12*(-N1 + t1) + MI22*(-N2 + t2)
```

4.4 Computed-Torque Control

```
C  OUTPUT EQUATION - CARTESIAN POSITION

   y1= al*cos(x(1)) + a2*cos(x(1)+x(2))
   y2= al*sin(x(1)) + a2*sin(x(1)+x(2))

   RETURN
   END
```

Figure 4.4-2: Subroutines for simulation using TRESP.

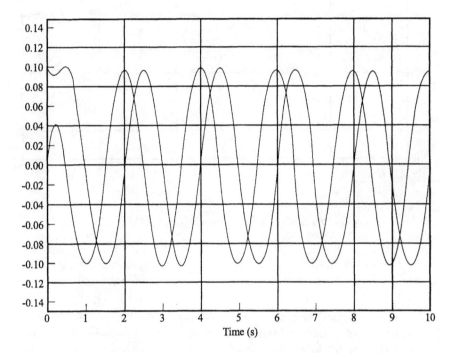

Figure 4.4.3: Joint angles $\theta_1(t)$ and $\theta_2(t)$ (red).

Subroutine F(time,x,xp) is called by Runge-Kutta and contains the continuous dynamics. This includes both the controller and the arm dynamics. The state to be integrated is $x = [\theta_1 \ \ \theta_2 \ \ \dot{\theta}_1 \ \ \dot{\theta}_2]^T$, and the subroutine should compute the state derivative $\dot{x}$ (i.e., xp, which signifies x−prime).

Subroutine CTL(x) contains the controller (3). Note the structure of this subroutine. First, the tracking error $e(t)$ and its derivative are computed. Then $M(q)$ and $N(q,\dot{q}) = V(q,\dot{q}) + G(q)$ are computed. Finally, (3) is manufactured.

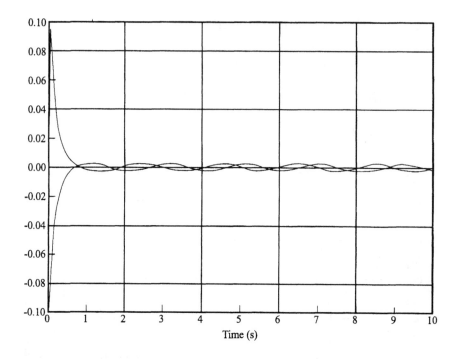

Figure 4.4.4: Tracking error $e_1(t)$, $e_2(t)$ (red).

Subroutine ARM(x,xp) contains the robot dynamics. First, $M(q)$ and $M^{-1}(q)$ are computed, and then $N(q,\dot{q})$. The state derivatives are then determined.

The results of the simulation are shown in the figures. Figure 4.4.3 shows the joint angles. Figure 4.4.4 shows the joint errors. The initial conditions result in a large initial error that vanishes within 0.6 s. Figure 4.4.5 shows the control torques; the larger torque corresponds to the inner motor, which must move two links.

It is interesting to note the ripples in $e(t)$ that appear in Figure 4.4.4. These are artifacts of the integrator. The Runge-Kutta integration period was $T_R = 0.01$ s. When the simulation was repeated using $T_R = 0.001$ s, the tracking error was exactly zero after 0.6 s. It should be zero, since computed-torque, or inverse dynamics control, is a scheme for canceling the nonlinearities in the dynamics to yield a second-order linear error system. If all the arm parameters are exactly known, this cancellation is exact. It is a good exercise to repeat this simulation using various values for the PD gains (see the Problems).

4.4 Computed-Torque Control

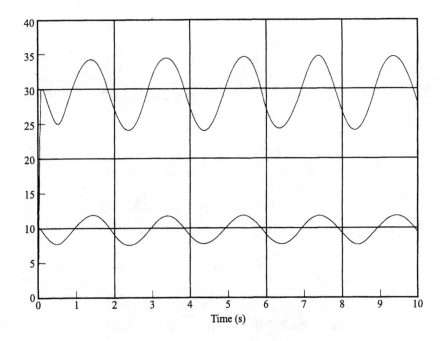

Figure 4.4.5: Torque inputs (N-m).

PID Outer-Loop Design

We have just seen that PD computed-torque control is very effective if all the arm parameters are known and there is no disturbance τ_d. However, from classical control theory we know that in the presence of constant disturbances, PD control gives a nonzero steady-state error. Consequently, we are motivated to make the system type I by including an integrator in the feedforward loop-this can be achieved using the *PID computed-torque controller*

$$\dot{\varepsilon} = e \qquad (4.4.34)$$

$$u = -K_v \dot{e} - K_p e - K_i \varepsilon, \qquad (4.4.35)$$

which yields the arm control input

$$\tau = M(q)(\ddot{q}_d + K_v \dot{e} + K_p e + K_i \varepsilon) + N(q, \dot{q}), \qquad (4.4.36)$$

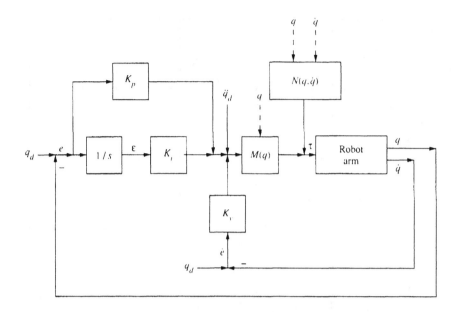

Figure 4.4.6: PID computed-torque controller.

with $\varepsilon(t)$ the integral of the tracking error $e(t)$. Thus additional dynamics have been added to the linear outer-loop compensator.

This control law is conveniently described by defining the state as $x = [\varepsilon^T \quad e^T \quad \dot{e}^T]^T \in \mathbb{R}^{3n}$ and augmenting the error dynamics (4.4.8) with an integrator, so that

$$\frac{d}{dt}\begin{bmatrix} \varepsilon \\ e \\ \dot{e} \end{bmatrix} = \begin{bmatrix} 0 & I & 0 \\ 0 & 0 & I \\ 0 & 0 & 0 \end{bmatrix}\begin{bmatrix} \varepsilon \\ e \\ \dot{e} \end{bmatrix} + \begin{bmatrix} 0 \\ 0 \\ I \end{bmatrix}u + \begin{bmatrix} 0 \\ 0 \\ I \end{bmatrix}w. \qquad (4.4.37)$$

A block diagram of the PID computed-torque controller appears in Figure 4.4.6. Then the closed-loop system is

$$\frac{d}{dt}\begin{bmatrix} \varepsilon \\ e \\ \dot{e} \end{bmatrix} = \begin{bmatrix} 0 & I & 0 \\ 0 & 0 & I \\ -K_i & -K_p & -K_v \end{bmatrix}\begin{bmatrix} \varepsilon \\ e \\ \dot{e} \end{bmatrix} + \begin{bmatrix} 0 \\ 0 \\ I \end{bmatrix}w. \qquad (4.4.38)$$

The closed-loop characteristic polynomial is

$$\Delta c(s) = \left|s^3 I + K_v s^2 + K_p s + K_i\right|. \qquad (4.4.39)$$

4.4 Computed-Torque Control

Selecting diagonal control gains

$$K_v = \text{diag}\{k_{v_i}\}, \quad K_p = \text{diag}\{k_{p_i}\}, \quad K_i = \text{diag}\{k_{i_i}\} \qquad (4.4.40)$$

gives

$$\Delta_c(s) = \prod_{i=1}^{n} \left(s^3 + k_{v_i} s^2 + k_{p_i} s + k_{i_i}\right). \qquad (4.4.41)$$

By using the Routh test it can be found that for closed-loop stability we require that

$$k_{i_i} < k_{v_i} k_{p_i}; \qquad (4.4.42)$$

that is, the integral gain should not be too large.

Actuator Saturation and Integrator Windup. It is important to be aware of an effect in implementing PID control on any actual robot manipulator that can cause serious problems if not accounted for. Any real robot arm will have limits on the voltages and torques of its actuators. These limits may or may not cause a problem with PD control, but are virtually guaranteed to cause problems with integral control due to a phenomenon known as *integrator windup* [Lewis 1992].

Consider the simple case where $\tau = k_i \varepsilon$ with $\varepsilon(t)$ the integrator output. The torque input $\tau(t)$ is limited by its maximum and minimum values $\tau_{\max}$ and $\tau_{\min}$. If $k_i \varepsilon(t)$ hits $\tau_{\max}$, there may or not may not be a problem. The problem arises if the integrator input remains positive, for then the integrator continues to integrate upwards and $k_i \varepsilon(t)$ may increase well beyond $\tau_{\max}$. Then, when the integrator input becomes negative, it may take considerable time for $k_i \varepsilon(t)$ to decrease below $\tau_{\max}$. In the meantime τ is held at $\tau_{\max}$, giving an incorrect control input to the plant.

Integrator windup is easy to correct using *antiwindup protection* in a digital controller. This is discussed in Section 4.5. The effects of uncorrected windup are demonstrated in Example 4.4.4.

The next example shows the usefulness of an integral term when there are unknown disturbances present.

EXAMPLE 4.4-2: Simulation of PID Computed-Torque Control

In Example 4.4.1 we simulated the PD computed-torque controller for a two-link planar arm. In this example we add a constant unknown

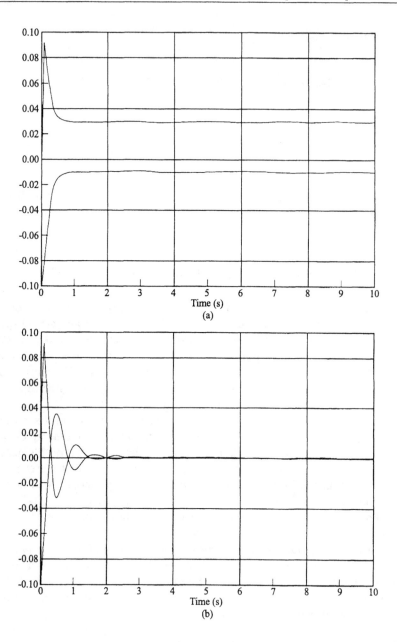

Figure 4.4.7: Computed-torque controller tracking errors $e_1(t)$, $e_2(t)$ (red): (a) PD control; (b) PID control.

4.4 Computed-Torque Control

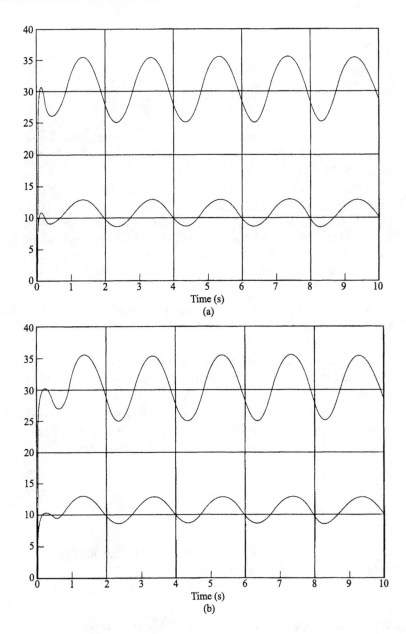

Figure 4.4.8: Computed-torque controller torque inputs (N-m): (a) PD control; (b) PID control.

disturbance to the arm dynamics and compare PD to PID computed torque.

Thus let the arm dynamics be

$$M(q)\ddot{q} + N(q,\dot{q}) + \tau_d = \tau, \tag{1}$$

with τ_d a constant disturbance with 1 N-m in each component. This could model unknown dynamics such as friction, and so on. The value of 1 N-m represents quite a large bias.

Adding 1 to the computation of the nonlinear terms N1 and N2 in subroutine arm(x,xp) in Figure 4.4.2 and using the PD computed-torque controller with $k_p = 100$, $k_v = 20$ yields the error plot in Figure 4.4.7(a). There is a unacceptable residual bias in the tracking error due to the unmodeled constant disturbance, which is not accounted for in the computed-torque law. The largest error is 0.033 rad, somewhat less than 2 deg.

Adding now an integral-error weighting term with $k_i = 500$ yields the results in Figure 4.4.7(b), which show a zero steady-state error and are quite good.

The associated control torques are shown in Figure 4.4.8, which shows that the torque magnitudes are not appreciably increased by using the integral term.

To simulate the PID control law, it is necessary to add two additional states to x in subroutine arm(x,xp). It is convenient to call them x(5) and x(6), so that the lines added to the subroutine are

$$\text{xp}(5) = \text{e}(1).$$
$$\text{xp}(6) = \text{e}(2). \tag{2}$$

Thus it is now necessary to compute e(1) and e(2) in subroutine arm(x,xp) for integration purposes.

∎

Class of Computed-Torque-Like Controllers

An entire class of *computed-torque-like* controllers can be obtained by modifying the computed-torque control law to read

$$\tau_c = \hat{M}(\ddot{q}_d - u) + \hat{N}. \tag{4.4.43}$$

4.4 Computed-Torque Control

The carets denote design choices for the weighting and offset matrices. One choice is $\hat{M} = M(q)$, $\hat{N} = N(q, \dot{q})$. The calculated control input into the robot arm is $\tau_c(t)$.

In some cases $M(q)$ is not known exactly (e.g., unknown payload mass), or $N(q, \dot{q})$ is not known exactly (e.g., unknown friction terms). Then $\hat{M}$ and $\hat{N}$ could be the best estimate we have for these terms. On the other hand, we might simply wish to avoid computing $M(q)$ and $N(q, \dot{q})$ at each sample time, or the sample period might be too short to allow this with the available hardware. From such considerations, we call (4.4.43) an "approximate computed-torque" controller.

In Table 4.4.1 are given some useful computed-torque-like controllers. As it turns out, computed torque is quite a good scheme since it has some important *robustness properties*. In fact, even if $\hat{M} \neq M$ and $\hat{N} \neq N$ the performance of controllers based on (4.4.43) can be quite good if the outer-loop gains are selected large enough. We study robustness formally in Chapter 4.

In the remainder of this chapter we consider various special choices of $\hat{M}$ and $\hat{N}$ that give some special sorts of controllers. We shall present some theorems and simulation examples that illustrate the robustness properties of computed-torque control.

Error Dynamics with Approximate Control Law. Let us now derive the error dynamics if the approximate computed-torque controller (4.4.43) is applied to the robot arm (4.4.2). Substituting $\tau_c(t)$ into the arm equation for $\tau(t)$ yields

$$M\ddot{q} + N + \tau_d = \hat{M}(\ddot{q}_d - u) + \hat{N}$$
$$\hat{M}\ddot{q}_d - M\ddot{q} = \hat{M}u + \tau_d + (N - \hat{N}).$$

Adding $M\ddot{q}_d - M\ddot{q}_d$ to the left-hand side and $Mu - Mu$ to the right gives

$$M\ddot{e} = Mu - (M - \hat{M})u + \tau_d + (M - \hat{M})\ddot{q}_d + (N - \hat{N})$$

or

$$\ddot{e} = u - \Delta u + d, \qquad (4.4.44)$$

where the inertia and nonlinear-term model mismatch terms are

$$\Delta = M^{-1}(M - \hat{M}) = I - M^{-1}\hat{M} \qquad (4.4.45)$$

$$\delta = M^{-1}(N - \hat{N}) \qquad (4.4.46)$$

Table 4.4.1: Computed-Torque-Like Robot Controllers.

Robot Dynamics:
$$M(q)\ddot{q} + V(q,\dot{q}) + F_v\dot{q} + F_d(\dot{q}) + G(q) + \tau_d = \tau$$
or
$$M(q)\ddot{q} + N(q,\dot{q}) + \tau_d = \tau$$
where
$$N(q,\dot{q}) \equiv V(q,\dot{q}) + F_v\dot{q} + F_d(\dot{q}) + G(q)$$

Tracking Error:
$$e(t) = q_d(t) - q(t)$$

PD Computed-Torque:
$$\tau = M(q)(\ddot{q}_d + K_v\dot{e} + K_p e) + N(q,\dot{q})$$

PID Computed-Torque:
$$\dot{\varepsilon} = e$$
$$\tau = M(q)(\ddot{q}_d + K_v\dot{e} + K_p e + K_i\varepsilon) + N(q,\dot{q})$$

Approximate Computed-Torque Control:
$$\tau_c = \hat{M}(\ddot{q}_d - u) + \hat{N}$$

Error System:
$$\ddot{e} = (I - \Delta)u + d$$
$$\text{where } \Delta = I - M^{-1}\hat{M}, \; \delta = M^{-1}\left(N - \hat{N}\right)$$
$$d(t) = M^{-1}\tau_d + \Delta\ddot{q}_d(t) + \delta(t)$$

PD-Gravity Control:
$$\tau_c = K_v\dot{e} + K_p e + G(q)$$

Classical PD Control:
$$\tau_c = K_v\dot{e} + K_p e$$

Classical PID Control:
$$\dot{\varepsilon} = e$$
$$\tau_c = K_v\dot{e} + K_p e + K_i\varepsilon$$

Digital Control (see Section 4.5):
$$\tau_k = M(q_k)\left(\ddot{q}_k^d + K_v\dot{e}_k + K_p e_k\right) + N(q_k,\dot{q}_k)$$

4.4 Computed-Torque Control

and the disturbance is

$$d(t) = M^{-1}\tau_d + \Delta \ddot{q}_d(t) + \delta(t). \quad (4.4.47)$$

This reduces to the error system (4.4.10) if exact computed-torque control is used so that $\Delta = 0$, $\delta = 0$. Otherwise, *the error system is driven by the desired acceleration and the nonlinear term mismatch* $\delta(t)$. Thus the tracking error will never go exactly to zero. Moreover, the auxiliary control $u(t)$ is multiplied by $(I - \Delta)$, which can make for a very difficult control problem.

Using outer-loop PD feedback so that $u(t) = -K_v\dot{e} - K_p e$ yields the error system

$$\ddot{e} + K_v\dot{e} + K_p e = \Delta(K_v\dot{e} + K_p e) + d(t). \quad (4.4.48)$$

The behavior of such systems is not obvious, even if K_v and K_p are selected for good stability of the left-hand side. There are two sorts of problems: first, the disturbance term $d(t)$, and second the function $\Delta(K_v\dot{e} + K_p e)$ of the error and its derivative.

PD-Plus-Gravity Controller

A useful controller in the computed-torque family is the *PD-plus-gravity controller* that results when $\hat{M} = I$, $\hat{N} = G(q) - \ddot{q}_d$, with $G(q)$ the gravity term of the manipulator dynamics. Then, selecting PD feedback for $u(t)$ yields

$$\tau_c = K_v\dot{e} + K_p e + G(q). \quad (4.4.49)$$

This control law was treated in [Arimoto and Miyazaki 1984], [Schilling 1990]. It is much simpler to implement than the exact computed-torque controller.

When the arm is at rest, the only nonzero terms in the dynamics (4.4.1) are the gravity $G(q)$, the disturbance τ_d, and possibly the control torque τ. The PD-gravity controller τ_c, includes $G(q)$, so that we should expect good performance for set-point tracking, that is, when a constant q_d is given so that $\dot{q}_d = 0$. The next result formalizes this. It relies on a Lyapunov proof (Chapter 1) of the sort that will be of consistent usefulness throughout the book, drawing especially on the skew-symmetry property in Table 3.3.1. Thus it is very important to understand the steps in this proof.

THEOREM 4.4-1: *Suppose that PD-gravity control is used in the arm dynamics (4.4.1) and $\tau_d = 0$, $\dot{q}_d = 0$. Then the steady-state tracking error $e = q_d - q$ is zero.*

Proof:
1. Closed-Loop System
  Ignoring friction, the robot dynamics are given by

$$M(q)\ddot{q} + V_m(q,\dot{q})\dot{q} + G(q) = \tau. \tag{1}$$

When $\dot{q}_d = 0$, the proposed control law (4.4.49) yields the closed-loop dynamics

$$M\ddot{q} + V_m\dot{q} + K_v\dot{q} - K_p e = 0. \tag{2}$$

2. Lyapunov Function
  Select now the Lyapunov function

$$V = \tfrac{1}{2}\left(\dot{q}^T M \dot{q} + e^T K_p e\right) \tag{3}$$

and differentiate to obtain

$$\dot{V} = \dot{q}^T \left(M\ddot{q} + \tfrac{1}{2}\dot{M}\dot{q} - K_p e\right). \tag{4}$$

Substituting the closed-loop dynamics (2) yields

$$\dot{V} = \dot{q}^T \left(\tfrac{1}{2}\dot{M} - V_m\right)\dot{q} = \dot{q}^T K_v \dot{q} \tag{5}$$

Now, the skew symmetry of the first term gives

$$\dot{V} = -\dot{q}^T K_v \dot{q}. \tag{6}$$

  The state is $x = [e^T \quad \dot{q}^T]^T$, so that V is positive definite but $\dot{V}$ only negative semidefinite. Therefore, we have demonstrated stability in the sense of Lyapunov, that is, that the error and joint velocity are both bounded.

3. Asymptotic Stability by LaSalle's Extension
  The asymptotic stability of the system may be demonstrated using Barbalat's lemma and a variant of LaSalle's extension [Slotine and Li 1991] (Chapter 2). Thus it is necessary to demonstrate that the only invariant set contained in the set $\dot{V} = 0$ is the origin.
  Since V is lower bounded by zero and $\dot{V}$ nonpositive, it follows that V approaches a finite limit, which can be written

4.4 Computed-Torque Control

$$-\int_0^\infty \dot{V}\,dt = const. \tag{7}$$

We now invoke Barbalat's lemma to show that $\dot{V}$ goes to zero. For this, it is necessary to demonstrate the uniform continuity of $\dot{V}$. We see that

$$\ddot{V} = -2\dot{q}^T K_v \ddot{q}. \tag{8}$$

The demonstrated stability shows that a and q are bounded, whence (2) and the boundedness of $M^{-1}(q)$ (see Table 3.3.1) reveal that $\ddot{q}$ is bounded. Therefore, $\ddot{V}$ is bounded, whence $\dot{V}$ is uniformly continuous. This guarantees by Barbalat's lemma that $\dot{V}$ goes to zero.

It is now clear that $\dot{q}$ goes to zero. It remains to show that the tracking error $e(t)$ vanishes. Note that when $\dot{q} = 0$, (2) reveals that

$$\ddot{q} = M^{-1} K_p e. \tag{9}$$

Therefore, a nonzero $e(t)$ results in nonzero $\ddot{q}$, and hence in $\dot{q} \neq 0$, a contradiction. Therefore, the only invariant set contained in $\{x(t) | \dot{V} = 0\}$ is $x(t) = 0$. This finally demonstrates that both $e(t)$ and $\dot{q}(t)$ vanish and concludes the proof. ∎

Some notes on this proof are warranted. First, note that the Lyapunov function chosen is a natural one, as it contains the kinetic energy and the "artificial potential energy" associated with the virtual spring in the control law [Slotine and Li 1991]. K_p appears in V while K_v appears in $\dot{V}$.

The proof of Lyapunov stability is quick and easy. The effort comes about in showing asymptotic stability. For this, it is required to return to the system dynamics and study it more carefully, discussing issues such as boundedness of signals, invariant sets, and so on. The fundamental property of skew symmetry is used in computing $\dot{V}$. The boundedness of $M^{-1}(q)$ is needed in Barbalat's lemma.

The next example illustrates the performance of the PD-gravity controller for trajectory tracking. In general, if $\dot{q}^d \neq 0$, then the PD-gravity controller guarantees *bounded tracking errors*. The error bound decreases, that is, the tracking performance improves, as the PD gains become larger. This will be made rigorous in Chapter 4.

EXAMPLE 4.4-3: Simulation of PD-Gravity Controller

In Example 4.4.1 we simulated the exact computed-torque control law on a two link planar manipulator. Here, let us simulate the PD-gravity controller. We shall take the same arm parameters and desired trajectory.

Assuming identical PD gains for each link, the PD-gravity control law for the two-link arm is

$$\tau_{c1} = k_v \dot{e}_1 + k_p e_1 + (m_1 + m_2) g a_1 \cos\theta_1 + m_2 g a_2 \cos(\theta_1 + \theta_2)$$
$$\tau_{c2} = k_v \dot{e}_2 + k_p e_2 + m_2 g a_2 \cos(\theta_1 + \theta_2). \tag{1}$$

This is easily simulated by commenting out several lines of code and making a few other modifications in subroutine CTL(x) of Figure 4.4.2.

For critical damping, the PD gains are selected as

$$k_p = \omega_n^2, \quad k_v = 2\omega_n. \tag{2}$$

The results of the PD-gravity controller simulation for several values of ω_n are shown in Figure 4.4.9. As ω_n, and hence the PD gains, increases, the tracking performance improves. However, no matter how large the PD gains, the tracking error never goes exactly to zero, but is bounded about zero by a ball whose radius decreases as the gains become larger. The performance for $\omega_n = 50$, corresponding to $k_p = 2500$, $k_v = 100$, would be quite suitable for many applications.

It quite important to note that the dc value of the errors is equal to zero. One can consider the gravity terms as the "dc portion" of the computed-torque control law. If they are included in computing the torques, there will be no error offset.

The associated control input torques are shown in Figure 4.4.10. It is extremely interesting to note that the torques are *smaller for the higher gains*. This is contrary to popular belief, which assumes that the control torques always increase with increasing PD gains. It is due to the fact that larger gains give "tighter" performance, and hence smaller tracking errors.

In view of the fact that the errors in Figure 4.4.9(a) have different magnitudes, it would probably be more reasonable to take the PD gains larger for the inner link than the outer link.

∎

Classical Joint Control

A simple controller that often gives good results in practice is obtained by selecting in (4.4.43)

$$\hat{M} = I, \quad \hat{N} = -\ddot{q}_d. \tag{4.4.50}$$

4.4 Computed-Torque Control

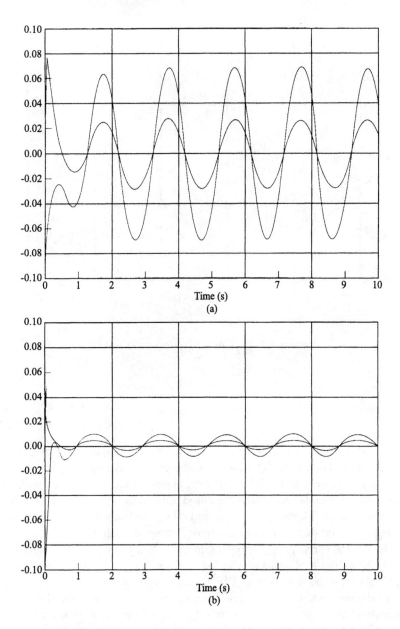

Figure 4.4.9: PD-gravity tracking error $e_1(t)$, $e_2(t)$ (red): (a) $\omega_n = 10$ rad/s; (b) $\omega_n = 25$ rad/s.

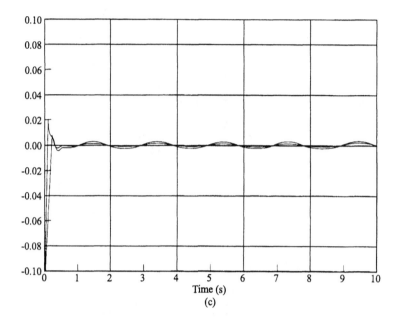

Figure 4.4.9: *(Cont.)* (c) $\omega_n = 50$ rad/s.

Then there results

$$\tau_c = -u. \tag{4.4.51}$$

By selecting u_i to depend only on joint variable i, this describes n decoupled individual joint controllers and is called *independent joint control*. Implementation on an actual robot arm is easy since the control input for joint i only depends on locally measured variables, not on the variables of the other joints. Moreover, the computations are easy and do not involve solving the complicated nonlinear robot inverse dynamics.

During the early days of robotics, independent joint control was popular [Paul 1981] since it allows a decoupled analysis of the closed-loop system using single-input/single-output (SISO) classical techniques. It is also called *classical joint control*. There are strong arguments even today by several researchers that this control scheme is always suitable in practical implementations, and that modern nonlinear control schemes are too complicated for industrial robotic applications.

A traditional analysis of independent joint control control now follows. It provides a connection with classical control notions and is important for the robot controls designer to understand. See [Franklin et al. 1986] for a

4.4 Computed-Torque Control

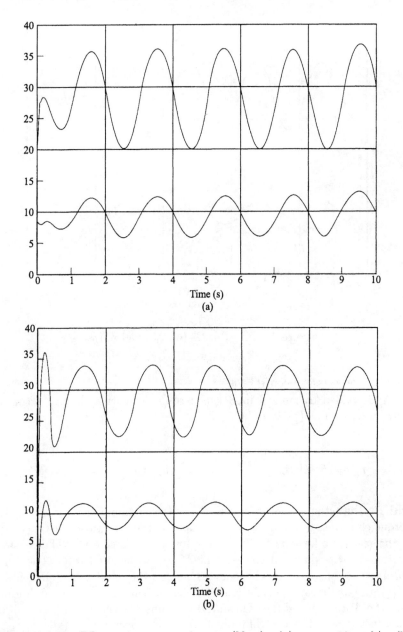

Figure 4.4.10: PD-gravity torque inputs (N-m): (a) $\omega_n = 10$ rad/s; (b) $\omega_n = 25$ rad/s.

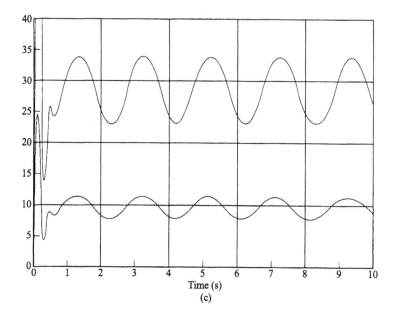

Figure 4.4.10: *(Cont.)* (c) $\omega_n = 50$ rad/s.

reference on classical control theory.

A simplified dynamical model of a robot arm with electric actuators may be written as (Section 3.6)

$$\left(J_m + r_i^2 m_{ii}\right) \ddot{\theta}_i + \left(B_m + \frac{k_b k_m}{R}\right) \dot{\theta}_i = \frac{k_m}{R} v_i - r_i d_i, \quad i = 1, \cdots, n \quad (4.4.52)$$

with J_m the actuator motor inertia, B_m the rotor damping constant, k_m the torque constant, k_b the back emf constant, R the armature resistance, and r_i the gear ratio for joint i. The motor angle is denoted $\theta_i(t)$. The constant portions of the diagonal elements of $M(q)$ are denoted m_{ii}. The time-varying portions of these elements, as well as the off-diagonal elements of $M(q)$, the nonlinear terms $N(q, \dot{q})$, and any disturbances τ_d are all lumped into the disturbance $d_i(t)$. Thus $d_i(t)$ contains the effects on joint i of all the other joints. The control input is the motor armature voltage $v_i(t)$.

Note that predominantly motor parameters appear in this equation. In fact, if the gear ratio is small, even m_{ii} may be neglected. For this reason, if the gear ratio is small, the robot arm control problem virtually reduces to the problem of controlling the actuator motors.

4.4 Computed-Torque Control

Let us denote this simplified linear time-invariant model of manipulator joint i as

$$J\ddot{\theta} + B\dot{\theta} = u - rd. \tag{4.4.53}$$

where the constant k_m/R has been incorporated into the definitions of J and B. According to the properties in Table 3.3.1, the disturbance $d(t)$ is bounded, although not generally by a constant. The bound may be a function of $q(t)$, $\dot{q}(t)$, and even $\ddot{q}(t)$. This is generally ignored in a classical analysis. The effects of $d(t)$ are, however, somewhat ameliorated by multiplication by the gear ratio r. The effects of joint compliance (Chapter 6) are also ignored in classical joint control design. [For consistency with classical notation, there is a sign change in this section in the definition of $u(t)$ compared to our previous usage.]

Now, let us consider a few selections for the control input $u(t)$.

PD Control. Selecting the PD control law

$$u = k_v \dot{e} + k_p e, \tag{4.4.54}$$

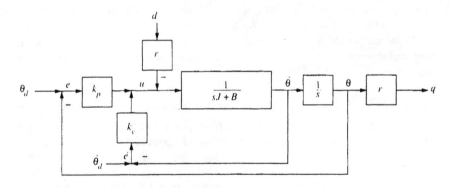

Figure 4.4.11: PD independent joint control.

with $e(t) = \theta_{d_i}(t) - \theta_i(t)$ the tracking error for motor angle i, yields the closed-loop system shown in Figure 4.4.11. Recall that the motor angle is related to the joint variable by $q_i = r_i \theta_i$. Thus $\theta_d(t)$ may be computed from $q_{d_i}(t)$. This PD controller is easily implemented on an actual arm with very little computing power.

Using Mason's theorem from classical control, the closed-loop transfer function for set-point tracking (e.g., with $\dot{\theta}_d = 0$) is found to be (see the problems)

$$\theta = \frac{k_p}{\Delta(s)}\theta_d - \frac{r}{\Delta(s)}d \qquad (4.4.55)$$

with the closed-loop characteristic polynomial

$$\Delta(s) = s^2 J + s(B + k_v) + k_p. \qquad (4.4.56)$$

The PD gains can be selected to obtain a suitable natural frequency and damping ratio, as we have seen.

At steady state, the only nonzero contribution to the disturbance $d(t)$ is the neglected gravity $G(q)$. Table 3.3.1 shows that the gravity vector is bounded by a known value g_b for a given robot arm. Therefore, at steady state,

$$|d| < g_b.$$

Using the final value theorem, the steady-state tracking error for joint i using PD control is bounded by

$$e_{ss} < rg_b/k_p. \qquad (4.4.57)$$

Therefore, for set-point tracking where a final value for q_d is specified and $\dot{q}_d = 0$, PD control with a large k_p might be suitable.

As a matter of fact, the next results show that PD control is often very suitable even for following a desired trajectory, not only for set-point control. It is proven in [Dawson 1990].

THEOREM 4.4-2: *If the PD control law (4.4.54) is applied to each joint and $e(0) = 0$, $\dot{e}(0) = 0$, the position and velocity tracking errors are bounded within a ball whose radius decreases approximately (for large k_v) as $1/\sqrt{k_v}$.* ∎

This result gives credence to those who maintain that PD control is often good enough for practical applications.

Of course, the point is that k_v cannot be increased without limit without hitting the actuator torque limits. Other schemes to be discussed in the book allow good trajectory following without such large torques.

In [Paul 1981] are discussed several methods for modifying the PD control law to obtain better performance. These include gravity compensation [which yields exactly controller (4.4.49)], and acceleration feedforward, which amounts to using

$$u = k_v \dot{e} + k_p e + J\ddot{\theta}_{d_i} \qquad (4.4.58)$$

4.4 Computed-Torque Control

for each joint. Also mentioned is joint-coupling control, which amounts to adding back into $u(t)$ some of the neglected terms in $M(q)$ and $N(q, \dot{q})$ that describe the interactions between the joints. Thus such corrections involve using better estimates for $\hat{M}$ and $\hat{N}$ in the approximate computed-torque control (4.4.43).

PID Control. It has been seen that PD independent joint control is often suitable for tracking control. However, at steady state there is a residual error (4.4.57) due to gravity. This can be removed using the *PID independent joint control law*

$$\dot{\varepsilon} = e$$
$$u = k_v \dot{e} + k_p e + k_i \varepsilon \qquad (4.4.59)$$

for each joint. See Figure 4.4.12.

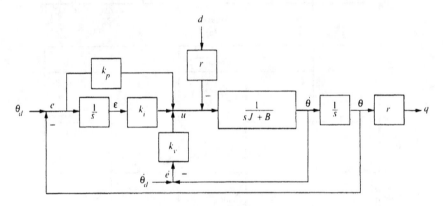

Figure 4.4.12: PID independent joint control.

Now the transfer function for set-point tracking ($\dot{\theta}_d = 0$) is

$$\theta = \frac{k_p s + k_i}{\Delta_1(s)} \theta_d - \frac{rs}{\Delta_1(s)} d, \qquad (4.4.60)$$

with closed-loop characteristic polynomial

$$\Delta_1(s) = s^3 J + s^2 (B + k_v) + s k_p + k_i. \qquad (4.4.61)$$

Now the final value theorem shows that the steady-state error for set-point control is zero. The Routh test shows that for stability it is required that

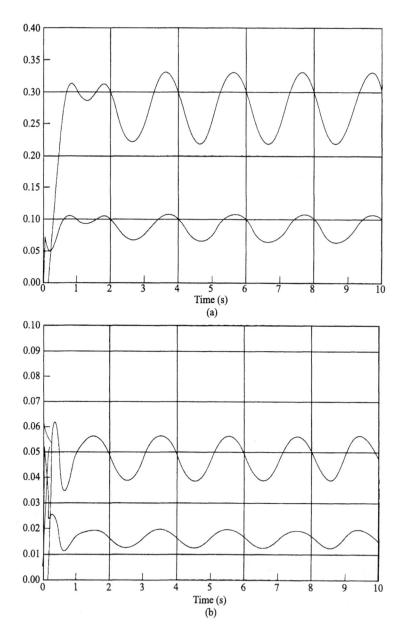

Figure 4.4.13: PD classical joint control tracking error $e_1(t)$, $e_2(t)$ (red): (a) $\omega_n = 10$ rad/s; (b) $\omega_n = 25$ rad/s.

4.4 Computed-Torque Control

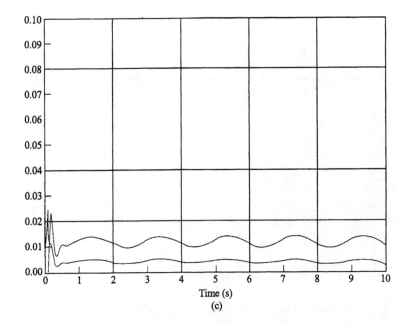

Figure 4.4.13: *(Cont.)* (c) $\omega_n = 50$ rad/s.

$$k_i < (B + k_v) k_p / J. \qquad (4.4.62)$$

In using an integral term in the control law, it is important to be aware of the possibility of *integrator windup* due to actuator saturation limits, which was discussed earlier in the section "PID Outer-Loop Design."

EXAMPLE 4.4-4: Classical Joint Control and Torque Saturation Limits

In Example 4.4.1 we simulated the exact computed-torque control law for a two-link planar elbow arm. Here we want to show the results of using PD and PID classical joint control on the same arm (with the same desired trajectory). We are also interested in demonstrating the effects of torque saturation limits. To highlight the effects without extraneous details, we shall use only the arm dynamics and no actuator dynamics or gear ratio. These may easily be included by making some slight modifications to the subroutine arm(x,xp) in Figure 4.4.2.

a. PD Independent Joint Control

For the two-link arm, PD independent joint control is simply

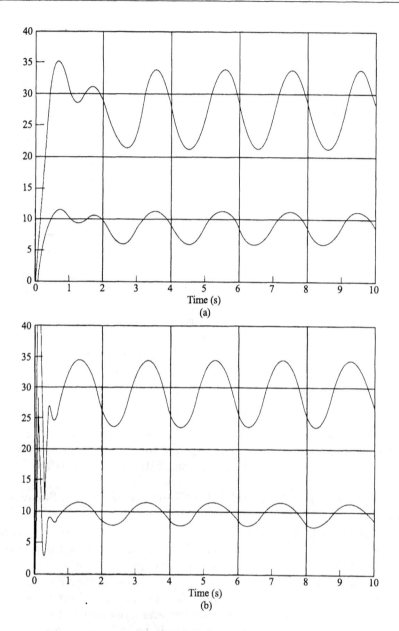

Figure 4.4.14: PD classical joint control torque inputs (N-m): (a) $\omega_n = 10$ rad/s; (b) $\omega_n = 25$ rad/s.

4.4 Computed-Torque Control

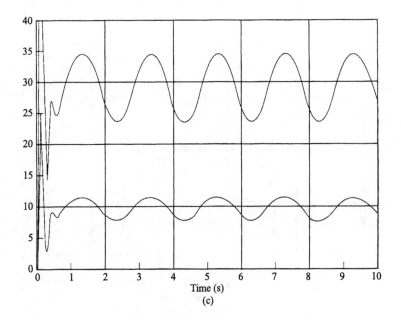

Figure 4.4.14: *(Cont.)* (c) $\omega_n = 50$ rad/s.

$$\tau_j = k_v \dot{e}_j + k_p e_j, \quad j = 1, 2, \tag{1}$$

with the tracking error $e(t) = q_d(t) - q(t)$. This control law is very simple and direct to implement on an actual arm without even using a DSP. It is much simpler than the control in Example 4.4.1.

The results of using the PD controller on the two-link arm are shown in Figure 4.4.13 for several values of a closed-loop natural frequency ω_n. Recall that the PD gains for critical damping are

$$k_p = \omega_n^2, \quad k_v = 2\omega_n. \tag{2}$$

The low-gain results ($k_p = 100$, $k_v = 20$) in part (a) of the figure are very bad (note the scale). However, the high-gain results in part (c) are much better. Even higher gains would cause a further tracking error decrease.

It is important to notice that there is in all cases a do component of the tracking error. This is due to the fact that the gravity terms are neglected and should be contrasted with the results of Example 4.4.3.

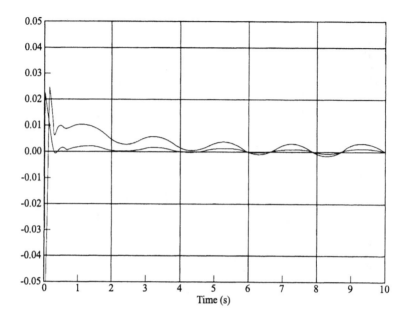

Figure 4.4.15: PD classical joint control: $k_v = 100$, $k_p = 2500$, $k_i = 1000$.

Adding the gravity terms in the control law, therefore, significantly increases the tracking performance.

The associated control torques are shown in Figure 4.4.14.

b. PID Independent Joint Control

PID independent joint control for the two-link arm is

$$\dot{\varepsilon}_j = e_j.$$
$$\tau_j = k_v \dot{e}_j + k_p e_j + k_i \varepsilon_j, \quad j = 1, 2, \tag{3}$$

This simple law is easily implemented by adding two states, x(5) and x(6), to subroutine arm(x,xp) in Figure 4.4.2 to account for the integrators (see Example 4.4.2).

The simulation in Figure 4.4.13(c) was repeated, but now adding an integral gain of $k_i = 1000$. The result is shown in Figure 4.4.15. After several iterations of the sinusoidal motion, the do value of the tracking errors goes to zero, so that the performance is much improved. That is, adding an integral term can compensate for neglecting the gravity terms

4.4 Computed-Torque Control

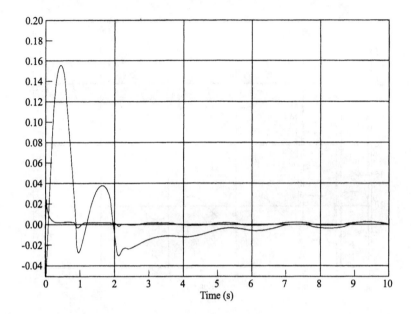

Figure 4.4.16: PID classical joint control with torque limits of ±35 N-m. Tracking error in rads.

in the control law. However, it takes awhile for the error to converge to a zero do value, and the results are still not as good as the PD-gravity controller in Example 4.4.3 for the same PD gains. On the other hand, the integral term can reject other terms besides gravity (e.g., friction of some sorts). The torque control input was virtually the same in form as Figure 4.4.14(c).

c. Actuator Saturation Limits

We discussed integrator windup due to actuator saturation limits earlier in the section "PID Outer-Loop Design." Here we should like to demonstrate its deleterious effects.

The control torque in part (b) of this example is similar to the torque in Figure 4.4.14(c); it is fairly well behaved, except for some frantic action near time zero, where the maximum positive excursion is 78 N-m.

Suppose now that there are torque limits of $t_{max} = 35$ N-m, $t_{min} = -35$ N-m. This is easily simulated by modifying subroutine CTL(x) in Figure 4.4.2 to include a saturation function by adding the lines

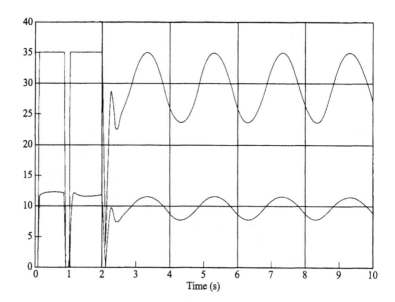

Figure 4.4.17: PID classical joint control with torque limits of ±35 N-m. Torque inputs in N-m.

$$\text{IF (abs (t1) .gt.tmax) t1} = \text{sign (tmax, t1)}$$
$$\text{IF (abs (t2) .gt.tmax) t2} = \text{sign (tmax, t2)} \qquad (4)$$

The results of the simulation with these limits are shown in Figure 4.4.16 and are terrible. The torque is shown in Figure 4.4.17.

In Section 4.5 we show how to ameliorate the effects of actuator saturation by using *antiwindup protection* in a digital controller. The simulation results for PD control with actuator limits are not as bad as the ones just shown for PID control, since saturation limits have less effect if no integrator is present. ∎

4.5 Digital Robot Control

Many robot control schemes are complicated and involve a great deal of computation for the evaluation of nonlinear terms. Therefore, they are implemented as digital control laws on digital signal processors (DSPs).

4.5 Digital Robot Control

Certain sorts of digital controllers for robot arms can be considered as members of the computer-torque-like class (see Table 4.4.1).

One approach to digital robot control is shown in Figure 4.5.1. There, a sampler is placed on $q(t)$ and $\dot{q}(t)$ to define

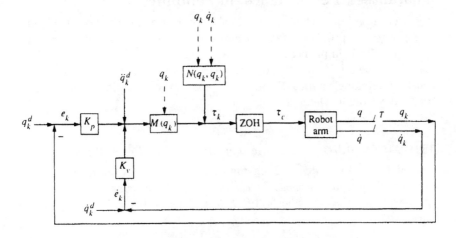

Figure 4.5.1: Digital robot control scheme.

$$q_k \equiv q(kT) \in \mathrm{R}^n$$
$$\dot{q}_k \equiv \dot{q}(kT) \in \mathrm{R}^n, \quad (4.5.1)$$

with T the sample period. Sample periods in robotic applications vary from about 1 to 20 ms. A zero-order hold is used to reconstruct the continuous-time control input $\tau(t)$ needed for the actuators from the samples τ_k. In this section we use superscripts "d" for the desired trajectory for notational ease.

This digital control law amounts to selecting

$$\hat{M} = M(q(kT)), \quad \hat{N} = N(q(kT), \dot{q}(kT)) \quad (4.5.2)$$

in the approximate computed-torque controller in Table 4.4.1 and a digital outer loop control signal u_k. Then, with PD outer-loop control, the arm control input is

$$\tau_k = M(q_k)\left(\ddot{q}_k^d + k_v \dot{e}_k + k_p e_k\right) + N(q_k, \dot{q}_k), \quad (4.5.3)$$

where the tracking error is $e(t) = q_d(t) - q(t)$. There are many variations of this control scheme. For instance, it is very reasonable to use multirate sampling and update $\hat{M}$ and $\hat{N}$ less frequently than the sampling frequency. That is, *the inner nonlinear loop can be sampled more slowly than the outer linear feedback loop.* In view of the robustness properties of computed-torque control, this works quite well in practice.

Guaranteed Performance on Sampling

It is usual in robot controls to design the controller in continuous time, providing rigorous proofs of stability and error boundedness. However, when the controller is implemented, a "small" sample period is selected and the stability is left to chance and verified by simulation studies. That this "wishful thinking" approach may not be so unreasonable is suggested by the next theorem. Define $\underline{e} = [\underline{e}^T \quad \dot{e}^T]^T$.

THEOREM 4.5-1: *Suppose that the PD digital control law (4.5.3) is applied to the robot arm. Then for every $L > 0$ there exists a T_M such that for all sampling periods T less than T_M, each error trajectory which at some time t_0 satisfies $\|\underline{e}(t_0)\| \leq L$ has $\|\underline{e}(t)\| \leq L + r$ for all $t \geq t_0$ and any $r > 0$.*

Proof:

Using the digital control law yields the error system (Table 4.4.1)

$$\ddot{e} = u - \Delta u + d, \tag{1}$$

where $d = M^{-1}\tau_d + \Delta \ddot{q}_d + \delta$ and

$$\Delta = M^{-1}(q)[M(q) - M(q_k)], \quad \delta = M^{-1}(q)[N(q,\dot{q}) - N(q_k,\dot{q}_k)]. \tag{2}$$

Defining $\underline{e} = [e^T \quad \dot{e}^T]^T$ this may be written in state-space form as

$$\dot{\underline{e}} = \begin{bmatrix} 0 & I \\ 0 & 0 \end{bmatrix} \underline{e} + \begin{bmatrix} 0 \\ I \end{bmatrix} (u - \Delta u + d) \equiv A\underline{e} + B(u - \Delta u + d). \tag{3}$$

Applying now the outer-loop digital control

$$u = -K\underline{e}_k = -K\underline{e} + K(\underline{e} - \underline{e}_k) \tag{4}$$

yields the closed-loop system

$$\dot{\underline{e}} = (A - BK)\underline{e} + BK(\underline{e} - \underline{e}_k) + B(\Delta K \underline{e}_k + d). \tag{5}$$

The feedback K is selected so that $(A - BK)$ is stable; hence

4.5 Digital Robot Control

$$(A - BK)^T P + P(A - BK) = -Q \qquad (6)$$

for some positive definite P and Q.
Selecting now the Lyapunov function

$$V(\underline{e}) = \tfrac{1}{2}\underline{e}^T P \underline{e}, \qquad (7)$$

(5) and (6) reveal

$$\dot{V}(\underline{e}) = -\tfrac{1}{2}\underline{e}^T Q \underline{e} + \underline{e}^T PB\left[K(\underline{e} - \underline{e}_k) + \mathrm{d} + \Delta K \underline{e}_k\right]. \qquad (8)$$

Assuming for simplicity that the disturbance τ_d is zero, note that

$$\dot{V}(\underline{e}_k) = -\tfrac{1}{2}\underline{e}_k^T Q \underline{e}_k \qquad (9)$$

is negative definite at each sample time. [If τ_d is nonzero but bounded, then $V(e_k)$ is negative outside some ball, and the discussion may be modified.]

The remainder of the proof follows from a theorem of [LaSalle and Lefschetz 1961] by showing that:

1. $\dot{V}(\underline{e}) \le 0$ when $\|\underline{e}\| > L$.

2. $V(\underline{e}(t_1)) < V(\underline{e}(t_2))$ for all $t_2 \ge t_1 \ge 0$ when $\|\underline{e}(t_1)\| \le L$ and $\|\underline{e}(t_2)\| > L + r$.

At issue is the fact (9) and the continuity of $\dot{V}(\underline{e})$. ∎

This result shows that good tracking is obtained for all sample periods less than some *maximum sample period* T_M which depends on the specific robot arm. A more refined result can show that the errors increase as the maximum desired acceleration $\ddot{q}_d$ increases, or equivalently, that smaller sample periods are required for larger desired accelerations.

We now discuss some issues in discretizing the inner nonlinear loop and then the outer linear loop.

Discretization of Inner Nonlinear Loop

There is no convenient exact way to discretize nonlinear dynamics. Given a nonlinear state-space system

$$\dot{x} = f(x, u), \qquad (4.5.4)$$

Euler's approximation yields

$$x_{k+1} = x_k + Tf(x_k, u_k). \qquad (4.5.5)$$

There do exist "exact" techniques for deriving discrete nonlinear robot dynamics. They rely on discretizing the robot arm dynamics in such a way that energy and momentum are conserved at each sampling instant [Neuman and Tourassis 1985]. See also [Elliott 1990]. Unfortunately, these schemes result in extremely complicated discrete dynamical equations, even for simple robot arms. It is very difficult to derive guaranteed digital control laws for them.

In this section we simply take the discretized inner nonlinear loop as given by the approximation (4.5.2).

Joint Velocity Estimates from Position Measurements

Throughout this chapter in the examples we have simulated continuous-time robot controllers assuming that both the joint positions and velocities are measured exactly. In point of fact, it is usual to measure the joint velocities using optical encoders, and then estimate the joint velocities from these position measurements. Simply computing the joint velocities using the Euler approximation

$$\dot{q}_k = (q_k - q_{k-1})/T \qquad (4.5.6)$$

is virtually doomed to failure, since this high-pass filter amplifies the encoder measurement noise.

Denote the joint velocity estimates by v_k. Then a filtered derivative can be used to compute v_k from q_k using

$$v_k = \nu v_{k-1} + (q_k - q_{k-1})/T, \qquad (4.5.7)$$

where ν is a design parameter. If ν is small, it corresponds to a fast pole near $z = 0$, which provides some high-pass filtering to reject unwanted sensor noise. Example 4.5.1 will illustrate the use of this joint velocity estimation filter.

The velocity estimation filter design can be optimized for the given encoder noise statistics by using an alpha-beta tracker to reconstruct v_k [Lewis 1986a], [Lewis 1986b], [Lowe and Lewis 1991]. This is a specialized form of Kalman filter. It should be noted that the velocity estimates are not only used in the outer linear loop for computing $\dot{e}_k$; they must be used to compute the inner nonlinear terms in (4.5.3) as well.

Discretization of Outer PD/PID Control Loop

We have seen that a useful computed-torque outer feedback loop is the PID controller. Given a continuous-time PID controller, a digital PID controller for the outer loop may be designed as follows [Lewis 1992].

4.5 Digital Robot Control

A continuous PID controller that only uses joint position measurements $q(t)$ is given by

$$u = -K^c(s)\,e$$

$$K^c(s) = k\left[1 + \frac{1}{T_I s} + \frac{T_D s}{1 + T_D s/N}\right], \qquad (4.5.8)$$

where k is the proportional gain, T_1 is the integration time constant or "reset" time, T_D is the derivative time constant. Rather than use pure differentiation, a "filtered derivative" is used which has a pole far left in the s-plane at $s = -N/T_D$. The value for N is often in the range 3 to 10; it is usually fixed by the manufacturer of the controller [Åström and Wittenmark 1984]. A special case of the PID controller, of course, is the PD controller, which is therefore also covered by this discussion.

A common approximate discretization technique for converting continuous-time controllers $K^c(s)$ to digital controllers $K(z)$ is the bilinear transform (BLT), where

$$K(z) = K^c(s') \qquad (4.5.9)$$

$$s' = \frac{2}{T}\frac{z-1}{z+1}. \qquad (4.5.10)$$

This corresponds to approximating integration by the trapezoidal rule. Under this mapping, stable continuous systems with poles at s are mapped into stable discrete systems with poles at

$$z = \frac{1 + sT/2}{1 - sT/2}. \qquad (4.5.11)$$

The finite zeros also map according to this transformation. However, the zeros at infinity in the s-plane map into zeros at $z = -1$.

Using the BLT to discretize (4.5.8) yields

$$K(z) = k\left[1 + \frac{T}{T_{Id}}\frac{z+1}{z-1} + \frac{T_{Dd}}{T}\frac{z-1}{z-\nu}\right] \qquad (4.5.12)$$

with the discrete integral and derivative time constants

$$T_{Id} = 2T_I \qquad (4.5.13)$$

$$T_{Dd} = \frac{NT}{1 + NT/2T_D} \qquad (4.5.14)$$

and the derivative-filtering pole at

$$\nu = \frac{1 - NT/2T_D}{1 + NT/2T_D}. \quad (4.5.15)$$

It is easy to implement this digital outer-loop filter in terms of difference equations on a DSP. First, write $K(z)$ in terms of z^{-1}, which is the unit delay in the time domain (i.e., a delay of T seconds), as

$$K\left(z^{-1}\right) = k\left[1 + \frac{T}{T_{Id}}\frac{1 + z^{-1}}{1 - z^{-1}} + \frac{T_{Dd}}{T}\frac{1 - z^{-1}}{1 - \nu z^{-1}}\right]. \quad (4.5.16)$$

[*Note:* There is some abuse in notation in denoting (4.5.16) as $K(z^{-1})$; this we shall accept.]

Now suppose that the control input u_k is related to the tracking error as

$$u_k = -K\left(z^{-1}\right)e_k. \quad (4.5.17)$$

Then u_k may be computed from past and present values of e_k using auxiliary variables as follows:

$$v_k^I = v_{k-1}^I + (T/T_{Id})(e_k + e_{k-1}) \quad (4.5.18)$$
$$v_k^D = \nu v_{k-1}^D + (T_{Dd}/T)(e_k - e_{k-1}) \quad (4.5.19)$$
$$-u_k = k\left(e_k + v_k^I + v_k^D\right). \quad (4.5.20)$$

The variables v_k^I and v_k^D represent the integral and derivative portions of the digital PID controller, respectively. These difference equations are easily implemented in software.

Actuator Saturation and Integrator Antiwindup Compensation

Actuator saturation leading to *integrator windup* is a problem that can occur in the outer PID loop of a robot controller. In Example 4.4.4 we saw the deleterious effects of integrator windup. It is easy to implement antiwindup protection digitally [Åström and Wittenmark 1984], [Lewis 1992]. The antiwindup protection circuit would be placed into the outer linear feedback loop of a robot control system, where the integrator (controller memory) is located.

Suppose that the controller is given in transfer function form

$$R\left(z^{-1}\right)v_k = T\left(z^{-1}\right)r_k - S\left(z^{-1}\right)w_k, \quad (4.5.21)$$

4.5 Digital Robot Control

where r_k is the reference command and w_k is the controller input, composed generally of the tracking error and the plant measured output. Note that z^{-1} is interpreted in the time domain as a unit delay of T seconds. The controller output v_k is passed through a hold device to generate the continuous plant control input $u(t)$.

We have assumed thus far that the desired plant control input $v_k \in \mathrm{R}^m$ computed by the controller can actually be applied to the plant. However, in practical systems the plant inputs (such as motor voltages, etc.) are limited by *maximum* and *minimum* allowable values. Thus the relation between the *desired plant input* v_k and the *actual plant input* u_k is given by the sort of behavior shown in Figure 4.5.2, where u_H and u_L represent, respectively, the maximum and minimum control effort allowed by the mechanical actuator. If there are no control limits, we may set $u_k = v_k$.

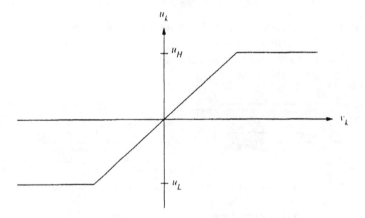

Figure 4.5.2: Actuator saturation function.

Thus, to describe the actual case in a practical control system with actuator saturation, we are forced to include *nonlinear saturation functions* in the control channels as shown in Figure 4.5.3. Consider the simple case where the controller is an integrator with input w_k and output v_k. Then all is well as long as v_k is between u_L and u_H, for in this region the plant input u_k equals v_k. However, if v_k exceeds u_H, then u_k is limited to its maximum value u_H. This in itself may not be a problem. The problem arises if w_k remains positive, for then the integrator continues to integrate and v_k may increase well beyond u_H. Then, when w_k becomes negative, it may take considerable time for v_k to decrease below u_H. In the meantime, u_k is held at u_H, giving an incorrect control input to the plant. This effect of integrator saturation is called windup. It arises because the controllers we design are generally dynamical in nature, which means that they store information or energy. Windup occurs when $R(z)$ is not a stable polynomial.

To correct integrator windup, it is necessary to *limit the state of the controller* so that it is consistent with the saturation effects being experienced by the plant input u_k. This may be accomplished as follows. Select a desired stable observer polynomial $A_0(z)$ and add $A_0(z^{-1})u_k$ to both sides to obtain

$$A_0 u_k = Tr_k - Sw_k + (A_0 - R) u_k. \qquad (4.5.22)$$

A regulator with antiwindup compensation is then given by

$$A_0 v_k = Tr_k - Sw_k + (A_0 - R) u_k. \qquad (4.5.23)$$

$$u_k = \text{sat}(v_k). \qquad (4.5.24)$$

A special case occurs when $A_0(z)$ has all n of its poles at the origin. Then the regulator displays *deadbeat behavior*; after n time steps its state remains limited to an easily computed value dependent on the values of w_k and u_k (see the Problems).

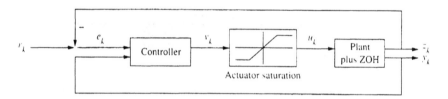

Figure 4.5.3: Control system including actuator saturation.

The next example demonstrates some aspects of digital control in robotics.

EXAMPLE 4.5-1: Simulation of Digital Robot Computed-Torque Controller

In Example 4.4.1 we simulated a continuous-time PD computed-torque (CT) controller for the two-link planar elbow arm. In this example some issues in

```
C   DIGITAL COMPUTED-TORQUE CONTROLLER FOR 2-LINK PLANAR ARM
C   SUBROUTINES FOR USE WITH TRESP

C   DIGITAL COMPUTED-TORQUE CONTROLLER SUBROUTINE
        SUBROUTINE DIG(IK,TD,x)
        REAL x(*),m1,m2,M11,M12,M22,N1,N2,kp,kv
        COMMON/CONTROL/t1, t2
C   The next line is to plot the errors and torques
```

4.5 Digital Robot Control

```
      COMMON/OUTPUT/ e(2), ep(2), tp1, tp2
      COMMON/TRAJ/qd(6), qdp(6), qdpp(6)
      DATA m1,m2,a1,a2, g/1.,1.,1.,1., 9.8/
      DATA kp, kv/ 100,20/

C   COMPUTE TRACKING ERRORS
      e(1) = qd(1)  - x(1)
      e(2) = qd(2)  - x(2)
      ep(1)= qdp(1) - x(3)
      ep(2)= qdp(2) - x(4)

C   COMPUTATION OF M(q), N(q,qp)
      M11= (m1+m2)*a1**2 + m2*a2**2 + 2*m2*a1*a2*cos(x(2))
      M12= m2*a2**2 + m2*a1*a2*cos(x(2))
      M22= m2*a2**2
      N1= -m2*a1*a2*(2*x(3)*x(4) + x(4)**2) * sin(x(2))
      N1= N1 + (m1+m2)*g*a1*cos(x(1)) + m2*g*a2*cos(x(1)+x(2))
      N2= m2*a1*a2*x(3)**2*sin(x(2))   + m2*g*a2*cos(x(1)+x(2))

C   COMPUTATION OF CONTROL TORQUES

      s1= qdpp(1) + kv*ep(1) + kp*e(1)
      s2= qdpp(2) + kv*ep(2) + kp*e(2)
      t1= M11*s1 + M12*s2 + N1
      t2= M12*s1 + M22*s2 + N2
C   The next lines are to plot the torque
      tp1= t1
      tp2= t2

      RETURN
      END
C
C
C   CONTINUOUS SUBROUTINE CALLED BY RUNGE-KUTTA INTEGRATOR
      SUBROUTINE F(t,x,xp)
      REAL x(*), xp(*)
C
C   ROBOT ARM DYNAMICS
      CALL ARM(x,xp)
C
      RETURN
      END
```

Figure 4.5-4: Subroutines for use with TRESP for digital CT control.

discretizing that controller are demonstrated. There are two objectives. First, we show the effects of sampling on the CT controller. Then a practical digital controller is designed that uses only joint position measurements from an encoder, reconstructing the velocities needed for computing the control law.

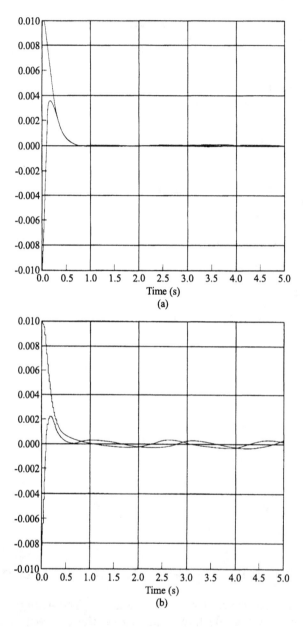

Figure 4.5.5: Joint tracking error on sampling the CT controller: (a) $T = 5$ ms; (b) $T = 20$ ms.

4.5 Digital Robot Control

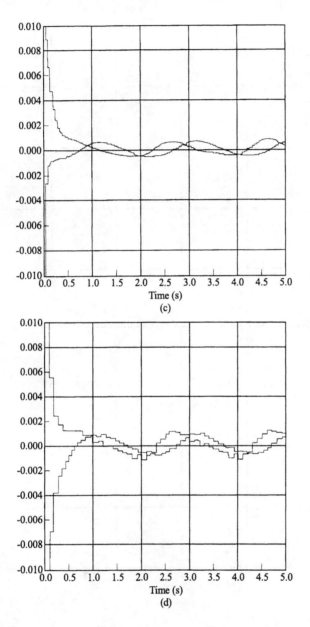

Figure 4.5.5: *(Cont.)* (c) $T = 50$ ms; (d) $T = 100$ ms.

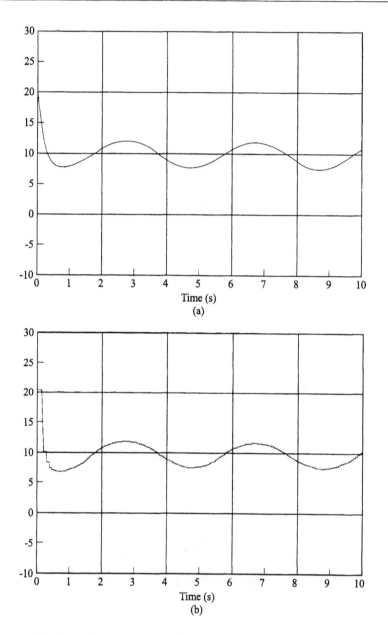

Figure 4.5.6: Link 2 torque input in the digital CT controller: (a) $T = 5$ ms; (b) $T = 50$ ms.

4.5 Digital Robot Control

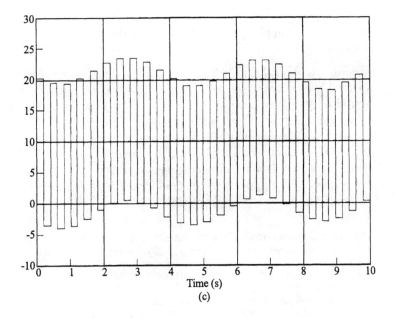

Figure 4.5.6: *(Cont.)* (c) $T = 100$ ms.

a. Effect of Sample Period on Digital Computed-Torque Controller

In Example 4.4.1 were given the subroutines needed for use with TRESP (Appendix B) to simulate the continuous-time PD CT controller. To simulate a digitized version of PD CT control, it is necessary to remove the controller subroutine [called CTL(x) in that example] from the continuous dynamics and place it into the subroutine DIG (IK,T,x) that is called by TRESP. This is in keeping with Figure 4.5.1 and the technique given in Section 4.3 for digital controller simulation.

In Figure 4.5.4 are shown the modified subroutines DIG(IK,TD,x) and F(t,x,xp) needed for digital CT controller simulation with TRESP. Subroutines SYSINP(IT,x,t) for trajectory generation and ARM(x,xp) containing the arm dynamics are the same as in Example 4.4.1.

Running now program TRESP with different sample periods T yields the tracking error plots shown in Figure 4.5.5. The Runge-Kutta integration period was 5 ms for all plots. The graphs show that for $T = 5$ ms the error was very small, in fact very similar to that in Example 4.4.1. However, as the sampling period T increased the tracking performance deteriorated.

The torque input for joint 2 is depicted in Figure 4.5.6. For small T is is virtually identical to the continuous CT controller in Example 4.4.1. A very interesting phenomenon occurs when $T = 0.1$ s. According to the plot in Figure 4.5.6(c), there is a *limit cycle*, a form of nonlinear oscillation. It is well known that sampling can induce such nonlinear effects in the closed-loop system [Lewis 1992]. According to Figure 4.5.5(d), the limit cycle is reflected in the tracking error as a periodic oscillation about $e(t) = 0$. The appearance of limit cycles is closely tied to a *loss of observability* in the sampled system.

```
C     DIGITAL COMPUTED-TORQUE CONTROLLER FOR 2-LINK PLANAR ARM
C        Uses measurements of e(kT) only, no velocity measurements

C     DIGITAL COMPUTED-TORQUE CONTROLLER SUBROUTINE
      SUBROUTINE DIG(IK,T,x)
      REAL x(*),m1,m2,M11,M12,M22,N1,N2,kp,kv
      REAL xm1(2), em1(2), v(2), vm1(2), ev(2), evm1(2), nu
      COMMON/CONTROL/t1, t2
C     The next line is to plot the errors and torques
      COMMON/OUTPUT/ e(2), ep(2), tp1, tp2
      COMMON/TRAJ/qd(6), qdp(6), qdpp(6)
      COMMON/PARAM/ nu
      DATA m1,m2,a1,a2, g/1.,1.,1.,1., 9.8/
      DATA kp, kv/ 100,20/

C     COMPUTE JOINT VELOCITIES v(I), TRACKING ERRORS, VELOCITY ERROR
      DO 10 I= 1,2
      v(I)=nu*vm1(I) + (x(I) - xm1(I)) / T
      IF (IK.EQ.0) v(I)= 0
      e(I) = qd(I) - x(I)
      ev(I)= nu*evm1(I) + (e(I) - em1(I))/T
10    IF(IK.EQ.0) ev(I)= 0

C     COMPUTATION OF M(q), N(q,qp)
      M11= (m1+m2)*a1**2 + m2*a2**2 + 2*m2*a1*a2*cos(x(2))
      M12= m2*a2**2 + m2*a1*a2*cos(x(2))
      M22= m2*a2**2
      N1= -m2*a1*a2*(2*v(1)*v(2) + v(2)**2) * sin(x(2))
      N1= N1 + (m1+m2)*g*a1*cos(x(1)) + m2*g*a2*cos(x(1)+x(2))
      N2=m2*a1*a2*v(1)**2*sin(x(2)) + m2*g*a2*cos(x(1)+x(2))

C     COMPUTATION OF CONTROL TORQUES
      s1= qdpp(1) + kp*e(1) + kv*ev(1)
      s2= qdpp(2) + kp*e(2) + kv*ev(2)
      t1= M11*s1 + M12*s2 + N1
```

4.5 Digital Robot Control

```
              t2= M12*s1 + M22*s2 + N2
      C       Store signals for next sample
              DO 20 I= 1,2
              xm1(I)= x(I)
              vm1(I)= v(I)
              em1(I)= e(I)
      20      evm1(I)= ev(I)

      C       The next lines are to plot the torque
              tp1= t1
              tp2= t2

              RETURN
              END
```

Figure 4.5-7: Digital control subroutine that uses only joint position measurements.

b. Digital Controller with Only Position Measurements

In part a we assumed that joint positions and velocities are both available as measurements. In practical situations, only the joint positions are available, often from optical encoder measurements. Therefore, here we should like to design a realistic digital CT controller that reconstructs the velocities.

The subroutines in Figure 4.5.7 uses only position measurements, estimating the joint velocities using the derivative filter (4.5.7). The resultant tracking error is shown in Figure 4.5.8. The performance is comparable to the controller in part a that used velocity measurements, with the exception of a larger initial error transient. The value of the filter pole ν was taken as 0.1.

Some implementation details are worthy of note. First, it is very important how the digital controller is initialized. Note the code lines that zero the velocity estimates in the first iteration (IK = 0). It is a good exercise to repeat this simulation deleting these lines (see the Problems).

Second, it might be thought that a reasonable procedure for finding the velocity error $\dot{e}_k = \dot{e}(kT)$ is to find an estimate v_k of the joint velocity $\dot{q}_k$ and then use

$$\dot{e}_k = \dot{q}_k^d - v_k. \qquad (1)$$

There are two disadvantages to this. First, it requires the storage in memory of the desired velocity $\dot{q}_k^d$ as well as the desired trajectory q_k^d. Second, it does not work.

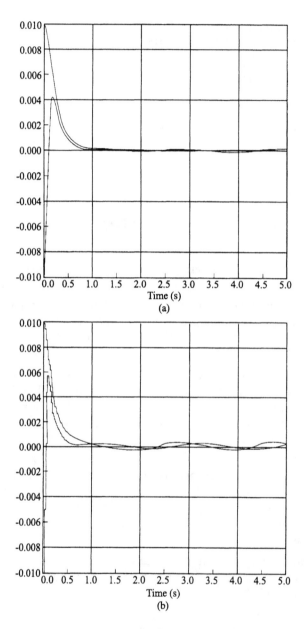

Figure 4.5.8: Tracking error for digital controller using no velocity measurements: (a) $T = 5$ ms; (b) $T = 20$ ms.

4.5 Digital Robot Control

Instead, it is necessary to use *two derivative filters*, one for estimating joint velocities $\dot{q}_k$ and one for providing an estimate e_k^v for the velocity error $\dot{e}_k$. A simulation that attempts to use 1 makes the point quite well.

Note that the velocity estimates are also used instead of $\dot{q}_k$ in computing the nonlinear terms $N(q_k, \dot{q}_k)$.

■

EXAMPLE 4.5-2: Digital PI Controller with Anti-windup Compensation

A general digital PI controller with sampling period of T seconds is given by

$$u_k = k\left[1 + \frac{T}{T_I}\frac{1}{z-1}\right]e_k. \tag{1}$$

We have sampled the continuous PI controller by the modified matched-polezero (MPZ) method [Lewis 1992] to obtain a delay of T seconds in the integrator to allow for computation time. The proportional gain is k and the reset time is T_I; both are fixed in the design stage. The tracking error is e_k.

Multiply by $(1 - z^{-1})$ to write

$$\left(1 - z^{-1}\right)u_k = k\left[\left(1 - z^{-1}\right) + Tz^{-1}/T_I\right]e_k, \tag{2}$$

which is in the transfer function form (4.5.21). The corresponding difference equation form for implementation is

$$u_k = u_{k-1} + ke_k + k\left(-1 + T/T_I\right)e_{k-1}. \tag{3}$$

This controller will experience windup problems since the autoregressive polynomial $R = 1 - z^{-1}$ has a root at $z = 1$, making it marginally stable. Thus, when u_k is limited, the integrator will continue to integrate, "winding up" beyond the saturation level.

a. Antiwindup Compensation

To correct this problem, select an observer polynomial of

$$A_0\left(z^{-1}\right) = 1 - \alpha z^{-1}, \tag{4}$$

which has a pole at some desirable location $|\alpha| < 1$. The design parameter α may be selected by simulation studies. Then the controller with antiwindup protection (4.5.23), (4.5.24) is given by

$$\left(1 - \alpha z^{-1}\right) v_k = k \left[1 + \left(-1 + \frac{T}{T_I}\right) z^{-1}\right] e_k + (1 - \alpha) z^{-1} u_k \quad (5)$$

$$u_k = \text{sat}(v_k). \quad (6)$$

The corresponding difference equations for implementation are

$$v_k = k e_k + \alpha v_{k-1} + k\left(-1 + T/T_I\right) e_{k+1} + (1 - \alpha) u_{k+1} \quad (7)$$

$$u_k = \text{sat}(v_k). \quad (8)$$

A few lines of FORTRAN code implementing this digital controller are given in Figure 4.5.9. This subroutine may be used as the control update routine DIG for the digital simulation driver program TRESP in Appendix B.

```
DIGITAL PI CONTROLLER WITH ANTIWINDUP COMPENSATION

    SUBROUTINE DIG(IK,T,x)
    REAL x(*), k
    COMMON/CONTROL/ u
    COMMON/OUTPUT/ z
    DATA k,AL,TI,ULOW,UHIGH/ 3.318. 0.9, 1., -1.5 , 1.5/
    DATA r/1./

    v= AL*v + k*(-1 + T/TI)*e + (1-AL)*u
    e= r - z
    v= k*e + v
    u= AMAX1(ULOW,v)
    u= AMIN1(UHIGH,u)

    RETURN
    END
```

Figure 4.5-9: FORTRAN code implementing PI controller with antiwindup compensation.

If $\alpha = 1$ we obtain the special case (2), which is called the *position form* and has no antiwindup compensation. If $\alpha = 0$, we obtain the *deadbeat antiwindup compensation*

4.5 Digital Robot Control

$$v_k = k \left[1 + \left(-1 + \frac{T}{T_I} \right) z^{-1} \right] e_k + u_{k-1}, \qquad (9)$$

with corresponding difference equation implementation

$$v_k = u_{k-1} + k e_k + k \left(-1 + T/T_I \right) e_{k-1}. \qquad (10)$$

If u_k is not in saturation, this amounts to updating the plant control by adding the second and third terms on the right-hand side of (10) to u_{k-1}. These terms are, therefore, nothing but $u_k - u_{k-1}$. The compensator with $\alpha = 0$ is thus called the *velocity form* of the PI controller.

b. Digital Control of DC Motor

Consider the simplified model for a dc motor given by

$$\dot{\omega} = -a\omega + bu, \qquad (11)$$

with ω the angular velocity. A motor speed controller has the form

$$u = -\left[k_1 + k_2/s \right] e, \qquad (12)$$

where $e = r - \omega$ is the tracking error, with r the desired command angular velocity. Taking $a = 1$ and $b = 1$, with $k_1 = k_2 = -3.318$, we obtain poles at $s = -1, -3.318$. The slower pole is canceled by a zero, so that the step response is fast, having only a mode like $e^{-3.318t}$.

Writing the PI controller as

$$u = k \left[1 + 1/T_I s \right] e, \qquad (13)$$

we see that

$$k = -k_1 = 3.318$$
$$T_I = k_1/k_2 = 1\,\text{s}. \qquad (14)$$

The digital controller obtained using the modified MPZ is given by (1).

The time constant of the closed-loop system is $\tau = 1/3.318 = 0.3$ s, so that a sampling period of $T = 0.05$ s is reasonable. The sampling period should be about one-tenth the time constant.

Program TRESP in Appendix B was used to obtain the response shown in Figure 4.5.10(a). No saturation limits were imposed on u_k.

Next, a saturation limit of $u_H = 1.5$ V was imposed on the control u_k. No antiwindup compensation was used (i.e., $\alpha = 1$ in Figure 4.5.9). The resulting behavior shown in Figure 4.5.10(b) displays an unacceptable overshoot.

Figure 4.5.10(c) shows that the overshoot problem is easily corrected using $\alpha = 0.9$ in the digital PI controller with antiwindup protection in Figure 4.5.9. In this example, as a decreases the step response slows down. The value of $\alpha = 0.9$ was selected after several simulation runs with different values of α.

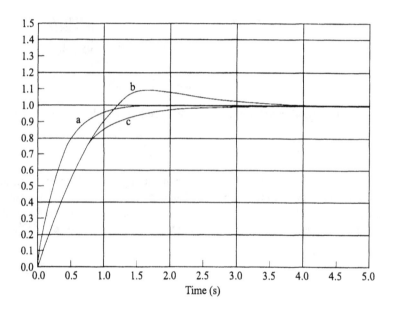

Figure 4.5.10: Angular velocity step responses using digital controller: (a) digital PI controller with no saturation limits; (b) digital PI controller with saturation limit $u_H = 1.5$V and no antiwindup compensation; (c) digital PI controller with saturation limit $u_H = 1.5$V and antiwindup compensation with $\alpha = 0.9$.

c. Antiwindup Compensation in Robotics

To implement the antiwindup compensation, it is necessary to know the maximum and minimum values u_H and u_L of the integrator output. Unfortunately, in a robot arm the motor torques are limited. These torque limits must be mapped using the feedback linearization input transformation to determine the limits on the integrator outputs see [Bobrow et al. 1983]. Thus the saturation limits needed by the antiwindup compensator are functions of the joint position q, the desired acceleration $\ddot{q}_d$, and so on. This issue is explored in the problems.

■

4.6 Optimal Outer-Loop Design

In Section 4.4 we discussed computed-torque control, showing how to select the inner control loop using exact techniques involving the inverse manipulator dynamics, as well as by a variety of approximate means. We also discussed several schemes for designing the outer linear feedback (tracking) loop. The results of our discussions are summarized in Table 4.4.1. In this section we intend to present a modern control optimal technique for selecting the outer feedback loop. Modern optimal design yields improved robustness in the presence of disturbances and unmodeled dynamics.

Several papers have dealt with "optimal" or "suboptimal" control of robot manipulators [Vukobratovic and Stokic 1983], [Lee et al. 1983], [Luo and Saridis 1985], [Johansson 1990]. Although they are not all based on a computed-torque-like approach, we would like here to present the flavor of this work by using optimal techniques to design the computed-torque outer feedback loop.

Linear Quadratic Optimal Control

First, it is necessary to review modern linear-quadratic (LQ) design. Suppose that we are given the linear time-invariant system in state-space form

$$\dot{x} = Ax + Bu. \qquad (4.6.1)$$

with $x(t) \in R^n$, $u(t) \in R^m$. It is desired to compute the state-feedback gain in

$$u = -Kx, \qquad (4.6.2)$$

so that the closed-loop system

$$\dot{x} = (A - BK)x \qquad (4.6.3)$$

is asymptotically stable. Moreover, we do not want to use too much control energy to stabilize the system, since in many modern systems (e.g., automobile, spacecraft), fuel or energy is limited.

This is a complex multivariable design problem, for the feedback gain matrix K is of dimension $m \times n$. A classical controls approach might involve, for instance, performing mn root locus designs to close the feedback loops one at a time. On the other hand, a solution that *guarantees stability* can be found using modern controls techniques simply by solving some *standard matrix design equations*. This modern approach *closes all the feedback loops simultaneously and guarantees a good gain and phase margin*.

The feedback matrix is found using modern control theory as follows. First, define a *quadratic performance index* (PI) of the form

$$J = 1/2 \int_0^\infty \left(x^T Q x + u^T R u \right) dt, \qquad (4.6.4)$$

where Q is a symmetric positive semidefinite $n \times n$ matrix (denoted $Q \geq 0$) and R is a symmetric positive definite $m \times m$ matrix ($R > 0$). That is, all eigenvalues of R are greater than zero and those of Q are greater than or equal to zero. Q is called the *state-weighting matrix* and R the *control-weighting matrix*. These matrices are *design parameters* that are selected by the engineer depending, for instance, on the desired form of the closed-loop time responses.

The *optimal LQ feedback gain* K is the one that minimizes the PI J. The motivation follows. The quadratic terms $x^T Q x$ and $u^T R u$ in the PI are generalized energy functions (e.g., the energy in a capacitor is $1/2 C v^2$, the kinetic energy of motion is $1/2 m v^2$). Suppose, then, that J is minimized in the closed-loop system (4.6.3). This means that the infinite integral of $[x^T(t) Q x(t) + u^T(t) R u(t)]$ is finite, so that this function of time goes to zero as t becomes large. However,

$$x^T(t) Q x(t) = \left\| \sqrt{Q} x(t) \right\|^2$$

$$u^T(t) R u(t) = \left\| \sqrt{R} u(t) \right\|^2$$

with the square root of a matrix defined as $M = \sqrt{M^T}\sqrt{M}$. Since these norms vanish with t and $|R| \neq 0$, the functions $y(t) = \sqrt{Q} x(t)$ and $u(t)$ both go to zero. Under the assumption that $(A, \sqrt{Q})$ is observable [Kailath 1980], $x(t)$ goes to zero if $y(t)$ does.

Therefore, *the optimal gain K guarantees that all signals go to zero with time in the closed-loop system (4.6.3)*. That is, K stabilizes $(A - BK)$.

The determination of the optimal K is easy and is a standard result in modern control theory (see, e.g., [Lewis 1986a], [Lewis 1986b]). The optimal feedback gain is simply found by solving the *matrix design equations*

4.6 Optimal Outer-Loop Design

$$K = R^{-1}B^T P \qquad (4.6.5)$$

$$A^T P + PA + Q - PBR^{-1}B^T P = 0, \qquad (4.6.6)$$

where P is a symmetric $n \times n$ auxiliary design matrix on which the optimal gain depends. The second of these is a nonlinear matrix quadratic equation known as the *Riccati equation*; it is easy to solve this equation for the auxiliary matrix P using standard routines in, for instance, MATLAB, [MATRIX$_X$ 1989], and other software design packages.

The next result is of prime importance in modern control theory and formalizes the stability discussion just given.

THEOREM 4.6-1: Let $(A, \sqrt{Q})$ be observable and (A, B) be controllable. Then:
(a) There is a unique positive definite solution P to the Riccati equation.
(b) The closed-loop system $(A - BK)$ is asymptotically stable.
(c) The closed-loop system has an infinite gain margin and 60° of phase margin. ■

Controllability was discussed in Chapter 1. Observability means roughly speaking that all the system modes have an independent effect in the PI, so that if J is bounded, all the modes independently go to zero with t. To verify these properties is easy. The system is controllable if the controllability matrix

$$U = \begin{bmatrix} B & AB & A^2B & \cdots & A^{n-1}B \end{bmatrix} \qquad (4.6.7)$$

has full rank n. The system (A, C) is observable if the observability matrix

$$V = \begin{bmatrix} C \\ CA \\ CA^2 \\ \vdots \\ CA^{n-1} \end{bmatrix} \qquad (4.6.8)$$

has full rank n. MATLAB, for instance, provides routines for these tests. Therefore, the state-weighting matrix Q may be chosen to satisfy the observability requirement.

The theorem makes this modern design approach very powerful. No matter how many inputs and states, a stabilizing feedback gain can always be found under the hypotheses that stabilizes the system. The gain is found by closing all the nm feedback loops simultaneously by solving the matrix design equations.

Linear Quadratic Computed-Torque Design

Now we apply these results to the control of the robot manipulator dynamics

$$M(q)\ddot{q} + N(q,\dot{q}) = \tau. \qquad (4.6.9)$$

According to Section 4.4, the computed-torque control law

$$\tau = M(\ddot{q}_d - u) + N \qquad (4.6.10)$$

yields the error system

$$\frac{d}{dt}\begin{bmatrix} e \\ \dot{e} \end{bmatrix} = \begin{bmatrix} 0 & I \\ 0 & 0 \end{bmatrix}\begin{bmatrix} e \\ \dot{e} \end{bmatrix} + \begin{bmatrix} 0 \\ I \end{bmatrix}u, \qquad (4.6.11)$$

which we may write as

$$\dot{x} = Ax + Bu, \qquad (4.6.12)$$

with the state defined as

$$x = \begin{bmatrix} e \\ \dot{e} \end{bmatrix}. \qquad (4.6.13)$$

Now, select the outer-loop PD feedback

$$u = -Kx = -\begin{bmatrix} K_p & K_v \end{bmatrix}x = -K_p e - K_v \dot{e}. \qquad (4.6.14)$$

To find a stabilizing gain K, select the design parameter Q in the PI as

$$Q = \text{diag}\{Q_p, Q_v\} \quad \text{with } Q_p, Q_v \in \mathbb{R}^{n \times n}$$

so that the position and velocity errors are independently weighted. Then, due to the simple form of the A and B matrices, which represent n decoupled Newton's law (i.e., double integrator) systems, the solution of the Riccati equation is easily found (see the Problems). Using this solution in (4.6.5) yields the formula for the optimal stabilizing gains

$$K_p = \sqrt{Q_p R^{-1}}, \quad K_v = \sqrt{2K_p + Q_v R^{-1}}. \qquad (4.6.15)$$

This LQ approach reveals the relation between the PD gains and some design parameters Q and R that determine the total energy in the closed-loop system. Note particularly that the relative magnitudes of $x(t)$ and $u(t)$

4.6 Optimal Outer-Loop Design

in the closed-loop system can be traded off. Indeed, if R is relatively larger than Q_p and Q_v, the control effort in the PI (4.6.4) is weighted more heavily that the state. Then the optimal control will attempt to keep $u(t)$ smaller by selecting smaller control gains; thus the response time will increase. On the other hand, selecting a smaller R will increase the PD gains and make the error vanish more quickly.

If Q_p, Q_v, and R are diagonal, so then are the PD gains K_p, K_v. The LQ approach with nondiagonal Q_p, Q_v, and R affords the possibility of outer feedback loops that are coupled between the joints, which can sometimes improve performance. Another important feature of LQ design is the guaranteed robustness mentioned in the theorem. This can be very useful in approximate computed-torque design where

$$\tau_c = \hat{M}(\ddot{q}_d - u) + \hat{N} \qquad (4.6.16)$$

and $\hat{M}$ and $\hat{N}$ can be simplified versions of $M(q)$ and $N(q, \dot{q})$. The performance of such a controller with an LQ-design outer loop can be expected to surpass that of a controller designed using arbitrary choices for K_p and K_v. This robust aspect of LQ design is explored in the problems.

It is important to note that this LQ design results in minimum closed-loop energy in terms of $e(t)$, $\dot{e}(t)$, and $u(t)$. However, the actual control input into the robot arm is

$$\tau = M(\ddot{q}_d + K_v \dot{e} + K_p e) + N. \qquad (4.6.17)$$

Although the energy in $\tau(t)$ is not minimized using this approach, we can use some norm inequalities to write

$$\|\tau\| \leq \|M(q)\| \cdot \|\ddot{q}_d\| + \|M(q)\| \cdot \|u(t)\| + \|N(q, \dot{q})\|, \qquad (4.6.18)$$

so that keeping small $\|u(t)\|$ might be expected to make $\|\tau(t)\|$ smaller. A more formal statement can be made taking into account the bounds on $\|M(q)\|$ and $\|N(q, \dot{q})\|$ given in Table 3.3.1.

Since the energy in $\tau(t)$ is not formally minimized in this approach, it is considered as a *suboptimal approach* with respect to the actual arm dynamics, although with respect to the error system and $u(t)$ it is optimal. An optimal control approach that weights $e(t)$ and $\tau(t)$ in the PI is given in [Johansson 1990].

We have derived an LQ controller using a computed-torque (i.e., feedback linearization) approach. An alternative approach that yields the same Riccati-equation-based design is to employ the full nonlinear arm dynamics and find an approximate (i.e., time-invariant) solution to the Hamilton-Jacobi-Bellman equation [Luo and Saridis 1985]; [Luo et al. 1986].

4.7 Cartesian Control

We have seen how to make a robot manipulator track a desired joint space trajectory $q_d(t)$. However, in any practical application the desired trajectories of a robot arm are given in the workspace or Cartesian coordinates. An important series of papers dealt with *resolved motion manipulator control* [Whitney 1969]; [Luh et al. 1980], [Wu and Paul 1982]. There the joint motions were resolved into the Cartesian coordinates, where the control objectives are specified. The result is that an operator can use a joystick to specify Cartesian motion (e.g., for a prosthetic device), with the arm following the specified motion. Older teleoperator devices used joysticks that directly controlled the motion of the actuators, resulting in long training times and very awkward manipulability.

There are several approaches to Cartesian robot control. For instance, one might:

1. Use the Cartesian dynamics in Section 3.5 for controls design (see the Problems).

2. Convert the desired Cartesian trajectory $y_d(t)$ to a joint-space trajectory $q_d(t)$ using the inverse kinematics. Then use the joint-space computed-torque control schemes in Table 4.4.1.

3. Use Cartesian computed-torque control.

Let us discuss the last of these.

Cartesian Computed-Torque Control

This approach begins with the joint space dynamics

$$M(q)\ddot{q} + N(q,\dot{q}) + \tau_d = \tau. \qquad (4.7.1)$$

In Section 3.4 we discussed a general feedback-linearization approach for linearizing the arm dynamics with respect to a general output. In this section the output we are interested in is the *Cartesian error*

$$e_y(t) = y_d(t) - y(t), \qquad (4.7.2)$$

with $y_d(t)$ the desired Cartesian trajectory and $y(t)$ the end-effector Cartesian position.

The problems associated with specifying the Cartesian position of the end effector are covered in Appendix A. There we see that $y(t)$ is not necessarily a 6-vector, but could in fact be the 4×4 arm T matrix. Then $y_d(t)$ is a 4×4 matrix given by

4.7 Cartesian Control

$$y_d(t) = T_d(t) = \begin{bmatrix} n_d(t) & o_d(t) & a_d(t) & p_d(t) \\ 0 & 0 & 0 & 1 \end{bmatrix}, \quad (4.7.3)$$

containing the desired orientation $(n_d(t), o_d(t), a_d(t))$ and position $p_d(t)$ of the end effector with respect to base coordinates. On the other hand, $y(t)$ could be specified (nonuniquely) using Euler angles as a 6-vector, or using quaternions as a 7-vector, or using the encoded tool configuration vector which gives $y(t) \in R^6$.

Although there are problems with specifying $y(t)$ as a 6-vector, the Cartesian error is easily specified (see the next subsection) as the 6-vector

$$e_y = \begin{bmatrix} e_p \\ e_o \end{bmatrix} \quad (4.7.4)$$

with $e_p(t)$ the position error and $e_o(t)$ the orientation error. Thus equation (4.7.2) is generally valid only as a loose notational convenience.

Let us assume that

$$y = h(q) \quad (4.7.5)$$

with $h(q)$ the transformation from $q(t)$ to $y(t)$, which is a modification of the kinematics transformation, depending on the form decided on for $y(t)$. Then the associated Jacobian is $J = \partial h / \partial q$ and

$$\dot{y} = J\dot{q}. \quad (4.7.6)$$

Now, the approach of Section 3.4, or a small modification of the derivation in Section 4.4, shows that the computed-torque control relative to $e_y(t)$ is given by

$$\tau = MJ^{-1}\left(\ddot{y}_d - \dot{J}\dot{q} - u\right) + N, \quad (4.7.7)$$

which results in the error system

$$\ddot{e}_y = u + w \quad (4.7.8)$$

with the disturbance

$$w = JM^{-1}\tau_d. \quad (4.7.9)$$

We call (4.7.7) the *Cartesian computed-torque control law*.

The outer-loop control $u(t)$ may be selected using any of the techniques already mentioned for joint-space computed-torque control (see Table 4.4.1). For PD control, for instance, the complete control law is

$$\tau = MJ^{-1}\left(\ddot{y}_d - \dot{J}\dot{q} + K_v\dot{e}_y + K_p e_y\right) + N. \quad (4.7.10)$$

A disadvantage with Cartesian computed-torque control is the necessity to compute the inverse Jacobian. To avoid inverting the Jacobian at each sample period, we might propose the *approximate Cartesian computed-torque controller*

$$\tau_c = \hat{C}\left(\ddot{y}_d - \dot{J}\dot{q} - u\right) + \hat{N}, \quad (4.7.11)$$

where $\hat{C}$ and $\hat{N}$ are approximations to MJ^{-1} and N, respectively. The error system for this control law is not difficult to compute (cf. the joint space approximation in Table 4.4.1 and see the Problems).

A PD outer feedback loop yields

$$\tau_c = \hat{C}\left(\ddot{y}_d - \dot{J}\dot{q} + K_v\dot{e}_y + K_p e_y\right) + \hat{N}. \quad (4.7.12)$$

A special case of this control law is obtained by setting $\hat{C} = I$, $\hat{N} = -\ddot{y}_d + \dot{J}\dot{q} + G(q)$, which yields the *Cartesian PD-gravity controller*

$$\tau_c = K_v\dot{e}_y + K_p e_y + G(q). \quad (4.7.13)$$

The robustness properties of computed-torque control make this a successful control law for many applications.

Simulations like those presented in this section could be carried out for Cartesian computed-torque control. The basic principles would be the same as for joint space computed-torque control (see the Problems).

Cartesian Error Computation

The actual Cartesian position may be computed from the measured joint variables using the arm kinematics in terms of the arm T matrix

$$y(t) = T(t) = \begin{bmatrix} n(t) & o(t) & a(t) & p(t) \\ 0 & 0 & 0 & 1 \end{bmatrix}, \quad (4.7.14)$$

and the desired Cartesian position may likewise be expressed as (4.7.3). Then a Cartesian position error and velocity error suitable for computed-torque control may be computed as follows [Luh et al. 1980], [Wu and Paul 1982]. Define

$$\dot{y} = \begin{bmatrix} v \\ \omega \end{bmatrix}, \quad \dot{y}_d = \begin{bmatrix} v_d \\ \omega_d \end{bmatrix} \quad (4.7.15)$$

with $v(t)$, $v_d(t)$ the actual and desired linear velocity, and $\omega(t)$, $\omega_d(t)$ the actual and desired angular velocities. Then

$$\dot{e}_y = \dot{y}_d - \dot{y} \qquad (4.7.16)$$

is easy to compute, since $\dot{y}(t)$ may be determined from the measured joint variables using (4.7.6).

The linear position error is simply given by

$$e_p = p_d - p. \qquad (4.7.17)$$

An orientation error $e_o(t)$ suitable for feedback purposes is more difficult to obtain, but may be defined as follows.

Denote the rotation transformation portions of (4.7.14) and (4.7.3), respectively, as $R(t)$, $R_d(t)$. The orientation error can be expressed in terms of a rotation of $\varphi(t)$ rads about an Euler axis of $k(t)$ that takes $R(t)$ into $R_d(t)$ (Appendix A). In fact, one may define the 3-vector

$$e_o(t) = k(t)\sin\varphi(t), \quad -\pi/2 \le \varphi(t) \le \pi/2, \qquad (4.7.18)$$

where $e_o(t)$ may be assumed small. With this definition, it can be shown that $e_o(t)$ is found from $T(t)$ and $T_d(t)$ using

$$e_o = \tfrac{1}{2}(n \times n_d + o \times o_d + a \times a_d). \qquad (4.7.19)$$

The overall Cartesian error is now given by (4.7.4). Unfortunately, with this definition of e_o, it happens that $\dot{e}_y$ is not the derivative of e_y; however, the control law (4.7.10) still yields suitable results. Alternative definitions of $e_y(t)$ and $\dot{e}_y(t)$ are given in [Wu and Paul 1982]; they are closely tied to the cross-product matrix Ω, in Appendix A and require the selection of a sampling period T.

4.8 Summary

In this chapter we showed how to generate smooth trajectories defining robot end-effector motion that passes through a set of specified points. Then we covered the important class of computed-torque controllers, which subsumes many types of robot control algorithms. Both classical and modern control algorithms are described by this class, so that computed torque provides a bridge between older and more modern algorithms for motion control.

As special types of computed-torque algorithms, we mentioned PD control, PID control, PD-plus-gravity, classical joint control, and digital control. Most robot control algorithms are implemented digitally, and computed-torque provides a rigorous framework for analyzing the effects of digitization and the size of the sampling period. This is approached by considering digital control as an approximate computed-torque law and studying the error system (cf. subsequent chapters).

We showed some aspects of modern linear quadratic outer-loop design, and concluded with a discussion of Cartesian control.

REFERENCES

[Arimoto and Miyazaki 1984] Arimoto, S., and F Miyazaki, "Stability and robustness of PID feedback control for robot manipulators of sensory capability," *Proc. First Int. Symp.*, pp. 783-799, MIT, 1984.

[Åström and Wittenmark 1984] Åström, K. J., and B. Wittenmark, *Computer Controlled Systems*. Englewood Cliffs, NJ: Prentice Hall, 1984.

[Bobrow et al. 1983] Bobrow, J. E., S. Dubowsky, and J. S. Gibson, "On the optimal control of robotic manipulators with actuator constraints," *Proc. Am. Control Conf*, pp. 782-787, June 1983.

[Chen 1989] Chen, Y.-C., "On the structure of the time-optimal controls for robotic manipulators," *IEEE Trans. Autom. Control*, vol. 34, no. 1, pp. 115-116, Jan. 1989.

[Dawson 1990] Dawson, D. M., "Uncertainties in the control of robot manipulators," Ph.D. thesis, School of Electrical Engineering, Georgia Institute of Technology, Mar. 1990.

[Elliott 1990] Elliott, D. L., "Discrete-time systems on manifolds," *Proc. IEEE Conf Decision Control*, pp. 1908-1909, Dec. 1990.

[Franklin et al. 1986] Franklin, G. F:, J. D. Powell, and A. Emami-Naeini, *Feedback Control of Dynamic Systems*. Reading, MA: Addison-Wesley, 1986.

[Geering 1986] Geering, H. P., L. Guzzella, S. A. R. Hepner, and C. H. Onder, "Time-optimal motions of robots in assembly tasks," *IEEE Trans. Autom. Control*, vol. AC-31, no. 6, pp. 512-518, June 1986.

[Gilbert and Ha 1984] Gilbert, E. G., and I. J. Ha, "An approach to nonlinear feedback control with applications to robotics," *IEEE Trans. Syst. Man Cybern.*, vol. SMC-14, no. 6, pp. s 879-884, Nov./Dec. 1984.

[Gourdeau and Schwartz 1989] Gourdeau, R., and H. M. Schwartz, "Optimal control of a robot manipulator using a weighted time-energy cost function," *Proc. IEEE Conf. Decision Control*, pp. 1628-1631, Dec. 1989.

[Hunt et al. 1983] Hunt, L. R., R. Su, and G. Meyer, "Global transformations of nonlinear systems," *IEEE Trans. Autom. Control*, vol. AC-28, no. 1, pp. 24-31, Jan. 1983.

[IMSL] IMSL, *Library Contents Document*, 8th ed. Houston, TX: International Mathematical and Statistical Libraries.

[Jayasuriya and Suh 1985] Jayasuriya, S., and M.-S. Suh, "Sub-optimal control strategies for manipulators with actuator constraints: the near minimum-time problem," *Proc. Am. Control Conf.*, pp. 61-62, June 1985.

[Johansson 1990] Johansson, R., "Quadratic optimization of motion coordination and control," *IEEE Trans. Autom. Control*, vol. 35, no. 11, pp. 1197-1208, Nov. 1990.

[Kahn and Roth 1971] Kahn, M. E., and B. Roth, "The near-minimum-time control of open-loop articulated kinematic chains," *Trans. ASME J. Dyn. Syst. Meas. Control*, pp. 164-172, Sept. 1971.

[Kailath 1980] Kailath, T. *Linear Systems*. Englewood Cliffs, NJ: Prentice Hall, 1980.

[Kim and Shin 1985] Kim, B. K., and K. G. Shin, "Suboptimal control of industrial manipulators with a weighted minimum time-fuel criterion," *IEEE Trans. Autom. Control*, vol. AC30, no. 1, pp. 1-10, Jan. 1985.

[LaSalle and Lefschetz 1961] LaSalle, J., and S. Lefschetz, *Stability by Liapunov's Direct Method*. New York: Academic Press, 1961.

[Lee et al. 1983] Lee, C. S. G., and M. H. Chen, "A suboptimal control design for mechanical manipulators," *Proc. Am. Control Conf.*, pp. 1056-1061, June 1983.

[Lewis 1986a] Lewis, F. L., *Optimal Control*. New York: Wiley, 1986 (a).

[Lewis 1986b] Lewis, F. L., *Optimal Estimation*. New York: Wiley, 1986 (b).

[Lewis 1992] Lewis, F. L., *Applied Optimal Control and Estimation*. Englewood Cliffs, NJ: Prentice Hall, 1992.

REFERENCES

[LINPACK] LINPACK, *User's Guide*, J. J. Dongarra, C. B. Moler, J. R. Bunch, and G. W. Stewart. Philadelphia: SIAM Press, 1979.

[Lowe and Lewis 1991] Lowe, J. A., and F. L. Lewis, "Digital signal processor implementation of a Kalman filter for disk drive head-positioning mechanism," in *Microprocessors in Robotic and Manufacturing Systems*, S. Tzafestas ed., pp. 369-383, 1991.

[Luh et al. 1980] Luh, J. Y. S., M. W. Walker, and R. P. C. Paul, "Resolved-acceleration control of mechanical manipulators," *IEEE Trans. Autom. Control*, vol. AC-25, no. 3, pp. 195-200, June 1980.

[Luo and Saridis 1985] Luo, G. L., and G. N. Saridis, "L-Q design of PID controllers for robot arms," *IEEE J. Robot. Autom.*, vol. RA-1, no. 3, pp. 152-159, Sept. 1985.

[Luo et al. 1986] Luo, G. L., G. N. Saridis, and C. Z. Wang, "A dual-mode control for robotic manipulators," *Proc. IEEE Conf. Decision Control*, pp. 409-414, Dec. 1986.

[MATRIX$_X$ 1989] MATRIX$_X$, Santa Clara, CA: Integrated Systems, Inc., 1989.

[Moler 1987] Moler, C., J. Little, and S. Bangert, *PC-Matlab*. Sherborn, MA: The Mathworks, 1987.

[Neuman and Tourassis 1985] Neuman, C. P., and V. D. Tourassis, "Discrete dynamic robot models," *IEEE Trans. Syst. Man and Cybern.*, vol. SMC-15, no. 2, pp. 193-204, Mar./Apr. 1985.

[Paul 1981] Paul, R. P., *Robot Manipulators*. Cambridge, MA: MIT Press, 1981.

[Schilling 1990] Schilling, R. J., Fundamentals of Robotics. Englewood Cliffs, NJ: Prentice Hall, 1990.

[Shin and McKay 1985] Shin, K. G., and N. D. McKay, "Minimum-time control of robotic manipulators with geometric path constraints," *IEEE Trans. Autom. Control*, vol. AC-30, no. 6, pp. 531-541, June 1985.

[Slotine and Li 1991] Slotine, J.-J. E, and W. Li, *Applied Nonlinear Control*. Englewood Cliffs, NJ: Prentice Hall, 1991.

[Vukobratovic and Stokic 1983] Vukobratovic and Stokic, "Contribution to the suboptimal control of manipulation robots," *IEEE Trans. Autom. Control*, vol. AC-28, no. 10, pp. 981-985, Oct. 1983.

[Whitney 1969] Whitney, D. E., "Resolved motion rate control of manipulators and human prostheses," *IEEE Trans. Man Machine Syst.*, vol. MMS-10, no. 2, pp. 47-53, June 1969.

[Wu and Paul 1982] Wu, C.-H., and R. P. Paul, "Resolved motion force control of robot manipulator," *IEEE Trans. Syst. Man Cybern.*, vol. SMC-12, no. 3, pp. 266-275, June 1982.

PROBLEMS

Section 4.2

4.2-1 Minimum-Time Control. Derive the minimum-time control switching time t_s [cf. (4.2.11)] when the initial and final velocities are not zero.

4.2-2 Polynomial Path Interpolation. It is desired to move a single joint from $q(0) = 0$, $\dot{q}(0) = 0$ through the point $q(l) = 5$, $\dot{q}(1) = 40$, to a final position/velocity of $q(2) = 10$, $\dot{q}(2) = 0$. Determine the cubic interpolating polynomials required in this two-interval path. Plot the path generated and verify that it meets the specified requirements on $q(t)$ and $\dot{q}(t)$. Plot $\dot{q}(t)$ versus $q(t)$.

4.2-3 LFPB. Repeat Problem 4.2-2 using LFPB.

4.2-4 Polynomial Path for Acceleration Matching. Derive the interpolating polynomial required to match positions, velocities, and accelerations at the via points.

Section 4.3

4.3-1 Simulation of Flexible Coupling System. Use computer simulation to reproduce the results for the motor with flexible coupling shaft in Example 3.6.1.

4.3-2 Simulation of Nonlinear System. The Van der Pol oscillator is a nonlinear system with some interesting properties. The state equation is

$$\dot{x}_1 = x_2$$

$$\dot{x}_2 = -\alpha(x_1^2 - 1)x_2 - x_1.$$

Simulate the dynamics for initial conditions of $x_1(0) = 0.1$, $x_2(0) = 0.1$. Use values for the parameter of $\alpha = 0.1$ and then $\alpha = 0.8$. Plot $x_1(t)$ and $x_2(t)$, as well as $x_2(t)$ vs. $x_1(t)$ in the phase plane. For each simulation you should clearly see the limit cycle that is characteristic of the Van der Pol oscillator.

Section 4.4

4.4-1 Prove (4.4.32).

4.4-2 PD Computed-Torque Simulation. Repeat Example 4.4.1 using various values for the PD gains. Try both critical damping and underdamping to examine the effects of overshoot on the joint trajectories.

4.4-3 Classical Joint Control. Prove (4.4.55), (4.4.57), (4.4.60), and (4.4.62). See [Franklin et al. 1986].

4.4-4 PD Computed Torque with Payload Uncertainty. The CT controller is inherently robust. In Example 4.4.1, suppose that m_2 changes from 1 kg to 2 kg at $t = 5$ s, corresponding to a payload mass being picked up. The CT controller, however, still uses a value of $m_2 = 1$. Use simulation to plot the error time history. Does the performance improve with larger PD gains?

4.4-5 PID Computed Torque with Payload Uncertainty. Repeat Problem 4.4-4 using a PID outer loop. Does the integral term help in rejecting the mass uncertainty?

4.4-6 PD Computed Torque with Friction Uncertainty. Repeat Problem 4.4-4 assuming now that $m_2 = 1$ kg stays constant and is known to the controller. However, add friction of the form $F(q, \dot{q}) = F_v \dot{q} + K_d \, \text{sgn}(\dot{q})$ (see Table 3.3.1) to the arm dynamics, but not to the CT controller. Use $v_i = 0.1$, $k_i = 0.1$. Simulate the performance for different PD gains.

4.4-7 PID Computed Torque with Friction Uncertainty. Repeat Problem 4.4-6 using a PID outer loop.

4.4-8 PD Computed Torque with Actuator Dynamics

(a) Design a CT control law for the two-link planar elbow arm with actuator dynamics (Section 3.6) of the form

$$J_M \ddot{q}_M + B \dot{q}_M + F_M + R\tau = K_M v.$$

Take the link masses and lengths as 1 kg, 1 m. Take motor parameters of $J_m = 0.1$ kg $-$ m^2, $k_m = k'_m = 1$ V-s, $b_m = 0.2$ N-m/rad/s, and $R = 5\Omega$, $F_m = 0.05 \dot{q}_m$. Set the gear ratio

REFERENCES

$r = 0.1$.

(b) Simulate the controller for various values of PD gains.

4.4-9 PD Computed Torque with Neglected Actuator Dynamics. Consider the arm-plus-dynamics in Problem 4.4-8. Suppose, however, that the CT controller was designed using only the arm dynamics and neglecting the actuator dynamics. Use the PD gains in Example 4.4.1. Simulate the control law on the arm-plus-dynamics for various values of gear ratio r. As r decreases, the performance should deteriorate.

4.4-10 PD Computed Torque Using Only Actuator Dynamics. Consider the arm-plus-dynamics in Problem 4.4-8. Design a CT controller using only the actuator dynamics and no arm dynamics. Simulate the control law on the arm-plus-dynamics for various values of gear ratio r. As r decreases, the performance should improve. For fixed r, it is also instructive to try different PD gains.

4.4-11 Classical Joint Control with Actuator Dynamics. Repeat Example 4.4.4 including actuator dynamics like those in Problem 4.4-8. Try different values of gear ratio r. Compare to Problem 4.4-10.

4.4-12 PD Computed Torque with Flexible Coupling. Combine the flexible shaft in Example 3.6.1 with the two-link arm in Example 4.4.1 to study the effects of using CT control on a robot with compliant motor coupling. Try different values of the coupling shaft parameters.

4.4-13 Error Dynamics with Approximate CT Control. Consider the two-link polar arm in Example 3.2.1 with friction of the form $F(q, \dot{q}) = F_v \dot{q} + K_d \operatorname{sgn}(\dot{q})$. Find the error dynamics (4.4.44) for the cases:
(a) Friction is not included in the CT control law.
(b) Payload mass m_2 is not exactly known in the CT control law.
(c) PD-gravity CT is used.
(d) PD classical joint control is used with no nonlinear terms.

Section 4.5
4.5-1 Digital Control Simulation
(a) Repeat Example 4.5.1 using a desired trajectory with period of 1

s instead of 2 s. Plot as well the Cartesian position $(x_2(t), y_2(t))$ of the end effector in base coordinates.

(b) Redo the simulation deleting the lines in Figure 4.5.7 that zero the initial velocity estimates.

(c) Try to simulate the digital CT controller using the alternative technique to compute $\dot{e}_k$ from $\dot{q}_k^d$ and v_k, as given in equation (1) in the example.

4.5-2 Digital Control Simulation. Convert the PD-gravity CT controller in Example 4.4.3 to a digital controller. Try several sample periods.

4.5-3 Error Dynamics for Digital Control. Find the error system in Table 4.4.1 using digital control of the form (4.5.3).

4.5-4 Antiwindup Protection. In Example 4.4.4 we saw the deleterious effects in robot control of integrator windup due to actuator saturation. In Example 4.5.2 we showed how to implement antiwindup protection on a simple PI controller. Implement antiwindup protection on the robot controller in Example 4.4.4. The issue is determining the limits on the integrator outputs given the motor torque limits. Successful and thorough completion of this problem might lead to a nice conference paper.

Section 4.6

4.6-1 Optimal LQ Outer-Loop PD Gains. Verify (4.6.15). To do this, select the Riccati solution matrix as

$$P = \begin{bmatrix} P_1 & P_2 \\ P_2^T & P_4 \end{bmatrix}$$

with $P_i \in R^n$. Substitute P, A, B, Q, R into the Riccati equation (4.6.6). You will obtain three $n \times n$ equations that can be solved for P_i. Now use (4.6.5).

4.6-2 Robust Control Using LQ Outer-Loop Design. Redo the PD-gravity simulation in Example 4.4.3 using PD gains found from LQ design. Does the LQ robustness property improve the responses found in Example 4.4.3 using nonoptimal gains? Try various choices for Qp, Q_v, and R, both diagonal and nondiagonal.

Section 4.7

4.7-1 Direct Cartesian Computed-Torque Design. Begin with the Cartesian dynamics in Section 3.5 and design a computed-torque controller. Compare it to (4.7.10).

4.7-2 Approximate Cartesian Computed-Torque. Derive the error system dynamics associated with the approximate control law (4.7.11).

4.7-3 Cartesian PD-Plus-Gravity Control. Repeat Example 4.4.3 using Cartesian computed-torque control, where the trajectory is given in workspace coordinates.

Chapter 5

Robust Control of Robotic Manipulators

In this chapter we discuss the control of robots when their dynamical model is uncertain. This may arise because the robot is carrying an unknown load or because the exact evaluation of the robot's dynamics is too costly. The robust controllers in this chapter are obtained from modifications to the controllers designed in Chapter 4.

5.1 Introduction

The control of uncertain systems is usually accomplished using either an adaptive control or a robust control philosophy. In the adaptive approach, one designs a controller that attempts to "learn" the uncertain parameters of the system and, if properly designed, will eventually be a "best" controller for the system in question. In the robust approach, the controller has a fixed structure that yields "acceptable" performance for a class of plants which include the plant in question. In general, the adaptive approach is applicable to a wider range of uncertainties, but robust controller are simpler to implement and no time is required to "tune" the controller to the particular plant. In this chapter we review different robust control designs used in controlling the motion of robots. The adaptive control approach is discussed in Chapter 6. The robust control methods presented in this chapter may be used to analyze the performance of the simple controllers used by robot manufacturers which were discussed in Chapter 4, and to suggest improvements and modifications. In fact, we are able to determine the range of applicability of the simple PID controllers of Chapter 4, as a function of the inherent lack of knowledge of the robot's dynamics.

The controllers designed in this chapter may be analyzed using input-output stability tools or state-space tools. In the input-output approach, the stability of the controlled robot is shown using the small-gain theorem or the passivity theorem. In the state-space approach, most designs are shown to be stable using Lyapunov-based arguments. See Chapter 2 for an overview of both approaches.

Consider the robot dynamics given in Chapter 3:

$$M(q)\ddot{q} + V_m(q,\dot{q})\dot{q} + F(\dot{q}) + G(q) = M(q)\ddot{q} + N(q,\dot{q}) = \tau - \tau_d \quad (5.1.1)$$

and assume that a desired trajectory in joint space is specified by the time function $x_d^T(t) = [q_d^T(t) \quad \dot{q}_d^T(t)]$. We will suppress the time dependence if no ambiguity results. Let q_d, $\dot{q}_d$, $\ddot{q}_d$, and τ_d be bounded functions of time. In a fashion similar to Chapter 4, we assume the trajectory error e to have two components:

$$e = q_d - q, \qquad \dot{e} = \dot{q}_d - \dot{q} \quad (5.1.2)$$

The controllers of this chapter assume that measurements of q and $\dot{q}$ are available. As described in Section 3.5, variables other than q and $\dot{q}$ may be measured. The Cartesian computed-torque controllers of Section 4.7 provide a setting where Cartesian trajectory is to be followed directly. We limit our discussion to the case of joint measurements with the understanding that a desired trajectory in another coordinate system may be followed by first obtaining the corresponding joint trajectory then applying the methods of this chapter.

We may, however, assume that the measurements p and $\dot{p}$ of q and $\dot{q}$ are corrupted by a bounded noise, that is,

$$p = q + w, \qquad \dot{p} = \dot{q} + w + 2 \quad (5.1.3)$$

where $||w_i|| \leq c_i$.

In Section 5.2 we discuss the computed-torque-like controllers of Section 4.4 and study their robustness properties. The section is divided into controllers whose robustness is deduced using Lyapunov stability and others whose robustness relies on input-output stability. Nonlinear controllers which are not necessarily derived from the computed-torque controllers are presented in Section 5.3. These include controllers that exploit the passivity of the robot dynamics and others which are variable-structure methods and saturation controllers without particular emphasis on the special properties of the robot. Finally, in Section 5.4 we review approaches that attempt to

5.2 Feedback-Linearization Controllers

The controllers designed in this section may be obtained as modification of the feedback-linearization (or computed-torque) controllers of Chapter 3. They are basically the computed-torque-like controllers of Section 4.4. We study both static and dynamic feedback designs and compare different controllers found in the literature. Note that such a study was started in Section 4.4 and some of the controllers introduced there will reappear in this chapter. The emphasis will be here on relating many of the controllers scattered through the literature and to give them a common theoretical justification.

We assume for simplicity that $\tau_d = 0$ in (5.1.1) and that $w_i = 0$ in (5.1.3), although the effects of bounded τ_d and w_i can be easily accounted for and will be considered in most examples. In a fashion similar to Chapter 4, the dynamics of the robot are transformed into the linear system

$$\begin{bmatrix} \dot{e} \\ \ddot{e} \end{bmatrix} = \begin{bmatrix} 0 & I \\ 0 & 0 \end{bmatrix} \begin{bmatrix} e \\ \dot{e} \end{bmatrix} + \begin{bmatrix} 0 \\ I \end{bmatrix} u$$

and

$$u = M(q)^{-1}[N(q,\dot{q}) - \tau] + \ddot{q}_d \qquad (5.2.1)$$

leading to the nonlinear computed-torque controller

$$\tau_d = M(q)[\ddot{q}_d - u] + N(q,\dot{q}) \qquad (5.2.2)$$

which, due to the invertibility of $M(q)$, gives the following closed-loop system:

$$\ddot{e} = u \qquad (5.2.3)$$

which is described by the transfer function

$$E(s) = \frac{I}{s^2} U(s). \qquad (5.2.4)$$

The problem is then reduced to finding a linear control u that will achieve a desired closed-loop performance; that is, find F, G, H, and J in

$$\dot{z} = Fz + Ge$$
$$u = Hz + Je,$$

or

$$u(t) = [H(sI - F)^{-1}G + J]e(t) \equiv C(s)e(t) \tag{5.2.5}$$

Not that the notion above indicates that $u(t)$ is the output of a system $C(s)$ when an input $e(t)$ is applied. Note also from (5.2.4) that the different joints or the robot are decoupled so that at this level, n SISO separate controllers may be designed to control the n joints of the robot. Unfortunately, the control law (5.2.2) cannot usually be implemented due to its complexity or to uncertainties present in $M(q)$ and $N(q, \dot{q})$ and to the presence of τ_d and w_i. Instead, one applies τ in (5.2.6) below where $\hat{M}$ and $\hat{N}$ are estimates of M and N,

$$\tau = \hat{M}[\ddot{q}_d - u] + \hat{N} \tag{5.2.6}$$

This in turn will reintroduce some coupling in the linear model and leads to (Fig. 5.5.1)

$$\begin{aligned} \dot{\mathbf{e}} &= A\mathbf{e} + B(u + \eta) \\ \eta &= \Delta(u - \ddot{q}_d) + M^{-1}\delta \\ \Delta &= M^{-1}\hat{M} - I_n, \qquad \delta = N - \hat{N}. \end{aligned} \tag{5.2.7}$$

Note first that Δ, δ, and therefore η are zero if $\hat{M} = M$ and $\hat{N} = N$. In general, however, the vector η is a nonlinear function of both e and u and cannot be treated as an external disturbance. It represents an internal disturbance of the globally linearized error dynamics caused by modelling uncertainties, parameter variations, external disturbances, friction terms, and maybe even noise measurements [Spong and Vidyasagar 1987]. Most commercial robots are in fact controlled with the controller given in (5.2.6) with choices of $\hat{M} = I$ and $\hat{N} = 0$. See, for example, [Luh 1983] and Section 4.4. The choice of $\hat{M}$ is validated by the powerful motors used to drive the robot links, and the gearing mechanisms used to torque the motor output to an acceptable level, while showing its speed down. The choice of $\hat{N}$ is validated by keeping the different motors from driving their links too fast, thus limiting the Coriolis and centripetal torques. Such commercial controllers are known as "non-model-based controllers" and have been used since the early days of robotics. The quest for more performance is, however, leading

5.2 Feedback-Linearization Controllers

researchers and manufacturers to use direct-drive robots and to attempt moving them at higher speeds with less powerful but more efficient motors [Asada and Youcef-Toumi 1987]. This new direction is increasing the need for more robust controllers such as the ones described next.

The approaches of this section revolve around the design of linear controllers $C(s)$ such that the complete closed-loop system in Figure 5.5.1 is stable in some suitable sense (e.g., uniformly ultimately bounded, globally asymptotically stable, $\mathcal{L}_p$ stable etc.) for a given class of nonlinear perturbation η. In other words choose $C(s)$ in (5.2.5) such that the error $e(t)$ in (5.2.8) is stable in some desired sense.

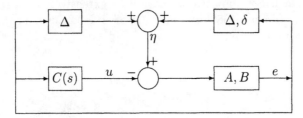

Figure 5.2.1: Feedback-linearization; uncertain structure.

The reasonable assumptions (5.2.9)-(5.2.11) below are often made for revolute-joint robots when using this approach [Spong and Vidyasagar 1987]. In the following, μ_1, μ_2, α, β_0, β_1, and β_2 are nonnegative finite constants which depend on the size of the uncertainties.

$$\frac{1}{\mu_2} \leq \ ||M^{-1}|| \leq \frac{1}{\mu_1}, \quad \mu_1 \neq 0 \qquad (5.2.8)$$

$$||\Delta|| \ \leq \ \alpha \leq 1 \qquad (5.2.9)$$

$$||\delta|| \ \leq \ \beta_0 + \beta_1 ||e|| + \beta_2 ||e||^2 \qquad (5.2.10)$$

$$||\ddot{q}_d|| \ \leq \ c. \qquad (5.2.11)$$

Recall that inequality (5.2.8) was introduced in Section 3.3, and note that the norms used in the inequalities above can be, depending on the application, either $\mathcal{L}_\infty$ or $\mathcal{L}_2$ norms. Also note that the bounds μ_i and β_i are scalar functions of q for robots with prismatic joints, and that (5.2.10) is satisfied by $\hat{M} = 0.5(\mu_1 + \mu_2)I$, so that $\alpha = (\mu_2 - \mu_1)/(\mu_2 + \mu_1)$) [Spong and Vidyasagar 1987]. Finally, (5.2.10) is a result of the properties of the Coriolis and centripetal terms discussed in Section 3.3.

We will give different representative designs of the feedback-linearization approach, starting with controllers whose behavior is studied using Lyapunov stability theory.

Lyapunov Designs

Static feedback compensators have been extensively used starting with the works of [Freund 1982] and [Tarn et al. 1984]. Consider the controller introduced in (4.4.13):

$$u = C(s)\mathbf{e} = -K\mathbf{e} \tag{5.2.12}$$

such that

$$\dot{\mathbf{e}} = A\mathbf{e} + B(u + \eta) = (A - Bk)\mathbf{e} + B\eta = A_c\mathbf{e} + B\eta. \tag{5.2.13}$$

It can be seen that by placing the poles of A_c sufficiently far in the left half-plane, the robust stability of the closed-loop system in the presence of η is guaranteed. This was shown true in [Arimoto and Miyazaki 1985] for the case where $\dot{q}_d = 0$ and $\tau_d = 0$ as described in Theorem 4.4.1 and Example 4.4.3. It was also shown true for the trajectory-following problem assuming that $e(0) = \dot{e}(0) = 0$ in [Dawson et al. 1990] as described in Theorem 4.4.2. There are as many robust controllers designed using Lyapunov stability concepts as there are ways of choosing Lyapunov function candidates, and of designing the gain K to guarantee that the Lyapunov function candidate is decreasing along the trajectories of (5.2.13). To decrease the asymptotic trajectory error, however, excessively large gains may be required (see Example 4.4.3). We therefore choose to use the passivity theorem and a choice of the gain matrix K that renders the linear part of the closed-loop system SPR. As described in Section 2.11, an output may be chosen to make the closed-loop system SPR; therefore allowing large passive uncertainties in the knowledge of $M(q)$. In fact, the state-feedback controller may be used to define an appropriate output $K\mathbf{e}$ such that the input-output closed-loop linear systems $K(sI - A + BK)^{-1}B$ is strictly positive real (SPR). Consider the following closed-loop linear system:

5.2 Feedback-Linearization Controllers

$$\dot{\mathbf{e}} = \begin{bmatrix} 0 & I \\ -K_p & -K_v \end{bmatrix} \mathbf{e} + \begin{bmatrix} 0 \\ I \end{bmatrix} u$$

$$y = [K_p \quad K_v] \begin{bmatrix} e \\ \dot{e} \end{bmatrix} = K\mathbf{e}. \tag{5.2.14}$$

It may then be shown using Theorem 2.11.5 that this system is SPR if

$$K_v^2 > K_p \tag{5.2.15}$$

with the choice of

$$Q = \begin{bmatrix} 2K_p^2 & 0 \\ 0 & 2K_v^2 - 2K_p \end{bmatrix} \tag{5.2.16}$$

such that

$$P = \begin{bmatrix} 2K_p K_v & K_p \\ K_p & K_v \end{bmatrix} \tag{5.2.17}$$

is the positive-definite solution to the Lyapunov equation

$$A_c^T P + P A_c = -Q \tag{5.2.18}$$

and

$$K = B^T P = [K_v^2/a \quad K_v], \quad a > 1. \tag{5.2.19}$$

The next theorem presents sufficient conditions for the uniform boundedness of the trajectory error.

THEOREM 5.2-1: *The closed-loop system given by (5.2.13) will be uniformly bounded if*

$$e(0) = \dot{e}(0)$$

and

$$a > 1 + \frac{1}{\mu_1}[\beta_1 + 2(\beta_2\beta_0 + \beta_2(\mu_1 + \mu_2)c)^{\frac{1}{2}}],$$

where $K_v = 2aI$ and $K_p = 4aI$.

Proof:

Consider the closed-loop system given by (5.2.8), with the controller (5.2.12), and choose the following Lyapunov function candidate:

$$V = \frac{1}{2}e^T P e + \int_0^t y^T(\sigma) \Delta y(\sigma) d\sigma \tag{1}$$

where $\frac{1}{2}e^T P e$ is the Lyapunov function corresponding to the SPR system (5.2.14). Then if $\Delta \geq 0$, we have that $V > 0$. This condition is satisfied for $\hat{M} \geq \mu_2 I$. Then differentiate to find

$$\dot{V} = -\frac{1}{2}e^T P e + eK(M^{-1}\delta - \Delta \ddot{q}_d). \tag{2}$$

To guarantee that $\dot{V} < 0$, recall the bounds (5.2.8)-(5.2.11), and write

$$\dot{V} \leq \frac{-q_1}{2}||e||^2 + \frac{k}{\mu_1}[\beta_2||e||^3 + \beta_1||e||^2 + (\beta_0 + (\mu_2 + \mu_1)c)||e||], \tag{3}$$

where $q_1 = \lambda_{min}(Q) = 2(a-1)k$, $(a-1)I < K_p = kI$. Let $K_v = 2aI < K_p = 4aI < K_v^2$. Note that $||e||$ may be factored out of (3) without affecting the sign definiteness of the equation. The uniform boundedness of the error is then guaranteed using Lemma 2.10.3 and Theorem 2.10.3 if

$$k\beta_2||e||^2 + \left(k\beta_1 - \mu_1 \frac{q_1}{2}\right)||e|| + k[\beta_0 + (\mu_1 + \mu_2)c] < 0, \tag{4}$$

which is guaranteed if

$$q_1 > 2\frac{k}{\mu_1}[\beta_1 + 2(\beta_2\beta_0 + \beta_2(\mu_1 + \mu_2)c)^{1/2}] \tag{5}$$

or

$$a > 1 + \frac{1}{\mu_1}[\beta_1 + 2(\beta_2\beta_0 + \beta_2(\mu_1 + \mu_2)c)^{1/2}] \tag{6}$$

∎

The error will be bounded by a term that goes to zero as a increases (see Theorem 2.10.3 and its proof in [Dawson et al. 1990] for details). This analysis then allows Δ to be arbitrarily large as long as $\hat{M} \geq \mu_2 I$, as shown in the next example. In fact, if N were known, global asymptotic stability is assured from the passivity theorem since in that case $\delta = 0$. The controller is summarized in Table 5.2.1.

It is instructive to study (6) and try to understand the contribution of each of its terms. The following choices will help satisfy (6).

1. Large gains K_p and K_v which correspond to a large a.

5.2 Feedback-Linearization Controllers

2. A good knowledge of N which translates into small β_i's.
3. A large μ_1 or a large inertia matrix $M(q)$.
4. A trajectory with a small c, this a small desired acceleration $\ddot{q}_d$.

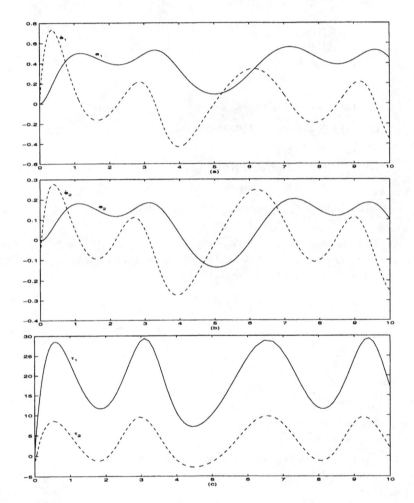

Figure 5.2.2: $K_p = 50$, $K_v = 25$ (a) errors of joint 1; (b) errors of joint 2; (c) torques of joints 1 and 2.

The following example illustrates the sufficiency of condition (6) and of the effects of larger gains K_p and K_v.

EXAMPLE 5.2-1: Static Controller (Lyapunov Design)

In all our examples in this chapter we use the two-link revolute-joint robot first described in Chapter 3, Example 3.2.2, whose dynamics are repeated here:

$$\begin{bmatrix} M_{11}(q) & M_{12}(q) \\ M_{12}(q) & M_{22}(q) \end{bmatrix} \begin{bmatrix} \ddot{q}_1 \\ \ddot{q}_2 \end{bmatrix} + \begin{bmatrix} m_1(q,\dot{q}) \\ m_2(q,\dot{q}) \end{bmatrix} = \begin{bmatrix} \tau_1 \\ \tau_2 \end{bmatrix} \quad (1)$$

where

$$\begin{aligned} M_{11}(q) &= (m_1 + m_2)a_1^2 + m_2 a_2^2 + 2 m_2 a_1 a_2 \cos(q_2) \\ M_{12}(q) &= m_2 a_2^2 + m_2 a_1 a_2 \cos(q_2) \\ M_{22}(q) &= m_2 a_2^2 \\ m_1(q,\dot{q}) &= -m_2 a_1 a_2 (2\dot{q}_1 \dot{q}_2 + \dot{q}_2^2) \sin(q_2) + (m_1 + m_2) g a_1 \cos(q_1) \\ &\quad + m_2 g a_2 \cos(q_1 + q_2) \\ m_2(q,\dot{q}) &= m_2 a_1 a_2 \dot{q}_1^2 \sin(q_2) + m_2 g q_2 \cos(q_1 + q_2) \end{aligned} \quad (2)$$

The parameters $m_1 = 1$kg, $m_2 = 1$kg, $a_1 = 1$m, $a_2 = 1$m, and $g = 9.8 m/s^2$ are given. Let the desired trajectory used in all examples throughout this chapter be described by

$$q_d = \begin{bmatrix} \sin t \\ \sin t \end{bmatrix}, \quad \dot{q}_d = \begin{bmatrix} \cos t \\ \cos t \end{bmatrix}$$

Then $||q_d||_\infty = 1$ rad, $||\dot{q}_d||_\infty = 1 rad/s^2$. It may then be shown that

$$\mu_1 = 0.1, \quad \mu_2 = 6$$

Let $\hat{N} = 0$, then

$$||\delta|| \leq \sqrt{10}||e||^2 + 2\sqrt{10}||e|| + \sqrt{10}$$

or that

$$\beta_0 = \sqrt{10}, \quad \beta_1 = 2\sqrt{10}, \quad \beta_2 = \sqrt{10}$$

Then use $\hat{M} = 6I$ and $a = 172$ to satisfy (6). In fact, these values will lead to a larger controller gains than are actually needed. Suppose instead that we let $6I$, $\hat{N} = 0$, and that

5.2 Feedback-Linearization Controllers

$$u = -\frac{50}{6}e - \frac{25}{6}\dot{e}$$
$$\tau = 6(\ddot{q}_d - u). \tag{3}$$

Note that this is basically a computed-torque-like PD controller. A simulation of the robot's trajectory is shown in Figure 5.2.2. We also start our simulation at $e(0) = \dot{e}(0) = 0$. The effect of increasing the gains is shown in Figure 5.2.3, which corresponds to the controller

$$u = -\frac{225}{6}e - \frac{30}{6}\dot{e}$$
$$\tau = 6(\ddot{q}_d - u). \tag{4}$$

Note that at least initially, more torque is required for the higher-gains case (compare Figs. 5.2.2c and 5.2.3c) but that the errors magnitude is greatly reduced by expanding more effort. ■

There are other proofs of the uniform boundedness of these static controllers. In particular, the results in [Dawson et al. 1990] provide an explicit expression for the bound on **e** in terms of the controller gains. In the interest of brevity and to present different designs, we choose to limit our development to one controller in this section.

As discussed in Section 4.4, a residual stead-state error may be present even when using an exact computed-torque controller if disturbances are present. A common cure for this problem (and one that will eliminate constant disturbances) is to introduce integral feedback as done in Section 4.4. Such a controller may again be used within a robust controller framework and will lead to similar improvements if the integrator windup problem is avoided (see Section 4.4).

In the next section we show the stability of static controllers similar to the ones designed here and use input-output stability methods to design more general dynamic compensators.

Input-Output Designs

In this section we group designs that show the stability of the trajectory error using input-output methods. In particular, we present controllers that show $\mathcal{L}_\infty$ and $\mathcal{L}_2$ stability of the error. We divide this section into a subsection that deals with static controllers such as the ones described previously,

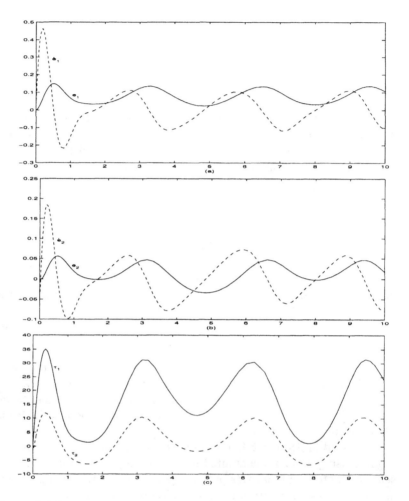

Figure 5.2.3: $K_p = 225$, $K_v = 30$ (a) errors of joint 1; (b) errors of joint 2; (c) torques of joints 1 and 2.

and a subsection dealing with the more general dynamic controllers, which have similar gains in general.

Static Controllers

These controllers have the same structure as the ones described in Section 4.4 and in the preceding section. The difference is that here we show the stability of the error using input-output concepts rather that the state-space methods implied by Lyapunov theory. In [Craig 1988] the boundedness of

5.2 Feedback-Linearization Controllers

Table 5.2.1: Static Controller, Lyapunov Design

Assumptions:

$$\frac{1}{\mu_2} \leq \|M^{-1}\| \leq \frac{1}{\mu_1}, \quad \mu_i \neq 0$$
$$\|\Delta\| \leq a \leq 1$$
$$\|\delta\| \leq \beta_0 + \beta_1\|e\| + \beta_2\|e\|^2$$
$$\|\ddot{q}_d\| \leq c$$
$$e(0) = \dot{e}(0) = 0$$

Controller:

$$\tau = \hat{M}(\ddot{q}_d + 2a\dot{e} + 4ae) + \hat{N}$$
$$\hat{M} > \mu_2 I$$
$$a > 1 + \frac{1}{\mu_1}[\beta_1 + (2\beta_2\beta_0 + \beta_2(\mu_1 + \mu_2)c)^{\frac{1}{2}}]$$

Performance:
Guaranteed Lyapunov stability of e and $\dot{e}$. The errors go to zero as a is increased.

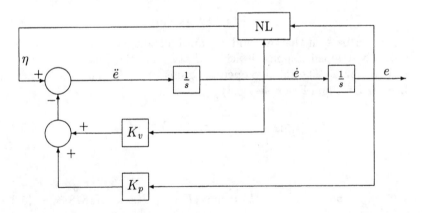

Figure 5.2.4: Block Diagram for second-order differential equation.

the error signals was shown using a static controller. The norms used in (5.2.8)-(5.2.10) are then $\mathcal{L}_\infty$ norms. The development of this controller starts with assumptions (5.2.8), (5.2.9), and a modification of (5.2.10) to

$$||\delta|| \leq \beta_0 + \beta_1||\dot{e}|| + \beta_2||\dot{e}||^2 \qquad (5.2.20)$$

This assumption is justified by the fact that N is composed of gravity and velocity-dependent terms which may be bounded independent from the position error e [see (5.1.1)]. We shall also assume that $e(0) = \dot{e}(0) = 0$. Let us then choose the state-feedback controller (5.2.12) repeated here for convenience:

$$u = -K_p e - K_v \dot{e} \qquad (5.2.21)$$

The corresponding input-output differential equation

$$\ddot{e} + K_v \dot{e} + K_p e = \eta \qquad (5.2.22)$$

A block diagram description of this equation is given in Figure 5.2.4. Consider now the transfer function from η (taken as an independent input) to e:

$$\begin{bmatrix} E(s) \\ sE(s) \end{bmatrix} = \begin{bmatrix} (s^2 + sK_v + K_p)^{-1} \\ s(s^2 + sK_v + K_p)^{-1} \end{bmatrix} \eta(s) = \begin{bmatrix} P_{11}(s) \\ P_{12}(s) \end{bmatrix} \qquad (5.2.23)$$

or

$$\mathbf{E}(s) = P_1(s)\eta(s) \qquad (5.2.24)$$

It can be seen that K_v and K_p are both diagonal, with $K_v^2 = 4K_p$, a critically damped response it achieved at every joint [see (4.4.22), (4.4.30), and (3.3.32)]. The infinity operator gains of $P_{11}(s)$ and $P_{12}(s)$ are (see Lemma 2.5.2 and Example 2.5.8)

$$\gamma_\infty(P_{11}) \equiv \gamma_{11} = \frac{1}{k_p}, \qquad \gamma_\infty(P_{12}) \equiv \gamma_{12} = \frac{4}{ek_v} \qquad (5.2.25)$$

where

$$k_p = \max\{K_{pi}\}, \qquad k_v = \max\{K_{vi}\}, \qquad e = 2.71828 \qquad (5.2.26)$$

Consider then the following inequalities:

$$||e||_{T\infty} \leq \gamma_{11}||\eta||_{T\infty}$$
$$||\dot{e}||_{T\infty} \leq \gamma_{12}||\eta||_{T\infty}$$

5.2 Feedback-Linearization Controllers

and using (5.2.8)-(5.2.11), we have that

$$||\eta||_{T\infty} \leq ac + ak_p||e||_{T\infty} + ak_v||\dot{e}||_{T\infty} + \frac{\beta_0}{\mu_1} + \frac{\beta_1}{\mu_1}||\dot{e}||_{T\infty} + \frac{\beta_2}{\mu_1}||\dot{e}||^2_{T\infty} \quad (5.2.27)$$

The following theorem presents sufficient conditions for the boundedness of the error that parallel those of Theorem 5.2.1.

THEOREM 5.2-2: *Suppose that*

$$e(0) = \dot{e}(0) = 0$$

and

$$\gamma_{12}\left(ak_v + \frac{\beta_1}{\mu_1}\right) = \frac{4a}{e} + \frac{4\beta_1}{e\mu_1 k_v} < 1 \quad (1)$$

Then the $\mathcal{L}_\infty$ stability of the error is guaranteed if

$$1 - \gamma_{11}ak_p - \gamma_{12}\left(ak_v + \frac{\beta_1}{\mu_1}\right) - 2\gamma_{12}\left[\left(a||\ddot{q}_d|| + \frac{\beta_0}{\mu_1}\right)\frac{\beta_2}{\mu_1}\right]^{\frac{1}{2}} > 0 \quad (2)$$

Proof:
The condition above results from applying the small-gain theorem to the closed-loop system, under the assumption that $\mathbf{e}(0) = 0$ so that the quadratic term $||\mathbf{e}||^2$ is small. See [Craig 1988] for details. ∎

Note that (2) reduces to

$$1 - a\left(1 + \frac{4}{e}\right) - \frac{4\beta_1}{\mu_1 e k_v} - 2\frac{4}{ek_v}\left[\left(a||\ddot{q}_d|| + \frac{\beta_0}{\mu_1}\right)\frac{\beta_2}{\mu_1}\right]^{\frac{1}{2}} > 0$$

and further to

$$k_v = \frac{4\beta_1 - 8[(a\mu_1 c + \beta_0)\beta_2]^{\frac{1}{2}}}{\mu_1 e[1 - a(1 + \frac{4}{e})]}. \quad (5.2.28)$$

Let us study the inequality above to determine the effect of each term. The following observations are made to help satisfy (5.2.28).

1. A large k_v will help satisfy the stability condition. Note: That will also imply a large k_p.

2. A good knowledge of N, which will translate into small β_i's.

3. A large μ_1 or a large inertia matrix $M(q)$.

4. A trajectory with a small $\ddot{q}_d$.

5. Robots whose inertia matrix $M(q)$ does not vary greatly throughout its workspace (i.e. $\mu_1 \approx \mu_2$)), so that a is small. Note that a small a is needed to guarantee that at least $a(1 + \frac{4}{e}) < 1$ in (5.2.28). This will translate into the severe requirement that the matrix $\hat{M}$ should be close to the inertia matrix $M(q)$ in all configurations of the robot.

The controller is summarized in Table 5.2.2.

These observations are similar to those made after inequality (6) and are illustrated in the next example.

EXAMPLE 5.2-2: Static Controller (Input-Output Design)

Consider the nonlinear controller (5.2.6), where

$$\hat{M} = \frac{M(q)}{4}, \qquad \hat{N} = 0 \tag{1}$$

Therefore,

$$a = \frac{1}{4} \tag{2}$$

Condition (1) is then satisfied if $k_v > 720$. This of course is a large bound that can be improved by choosing a better $\hat{N}$. A simulation of the closed-loop behavior for $k_p = 225$ and $k_v = 30$ is shown in Figure 5.2.5. The errors magnitudes are much smaller than those achieved with the PD controllers of Example 5.2.1 with a comparable control effort. This improvement came with the expense of knowing the inertia matrix $M(q)$ as seen in (1).

■

Dynamic Controllers

The controllers discussed so far are static controllers in that they do not have a mechanism of storing previous state information. In Chapter 4 and in this chapter, these controllers could operate only on the current position and velocity errors. In this section we present three approaches to show the robustness of dynamic controllers based on the feedback-linearization

5.2 Feedback-Linearization Controllers

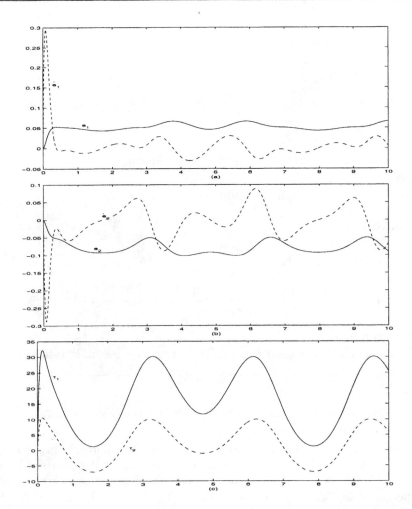

Figure 5.2.5: $K_p = 225$, $K_v = 30$ (a) errors of joint 1; (b) errors of joint 2; (c) torques of joints 1 and 2.

method. The first two approaches are one-degree-of-freedom (DOF) feedback compensators, while the last one is a two-DOF compensator.

One-Degree-of-Freedom Designs.

The first class of dynamic controllers are called one-degree-of-freedom controllers because they can only operate on the measured output of the robot. In other words, these are controllers that will take the measured signals and filter them through a dynamical system before feeding the signal back to the input. They should be contrasted with the static controllers,

Table 5.2.2: Static Controller, Input-Output Design

Assumptions:

$$\frac{1}{\mu_2} \leq ||M^{-1}|| \leq \frac{1}{\mu_1}, \quad \mu_i \neq 0$$
$$||\Delta|| \leq a \leq 1$$
$$||\delta|| \leq \beta_0 + \beta_1 ||\mathbf{e}|| + \beta_2 ||\mathbf{e}||^2$$
$$||\ddot{q}_d|| \leq c$$
$$e(0) = \dot{e}(0) = 0$$

Controller:

$$\tau = \hat{M}(\ddot{q}_d + k_v \dot{e} + \frac{k_v^2}{4}e) + \hat{N}$$

$$1 - a\left(1 + \frac{4}{e}\right) - \frac{4\beta_1}{\mu_1 e k_v} - 2\frac{4}{ek_v}\left[\left(a||\ddot{q}_d|| + \frac{\beta_0}{\mu_1}\right)\frac{\beta_2}{\mu_1}\right]^{\frac{1}{2}} > 0 \quad (5.2.29)$$

Performance:
Guaranteed boundedness of e and $\dot{e}$. The errors go to zero as k_v is increased.

---

which did not used dynamical feedback, and with the two-degree-of-freedom controllers considered next.

In [Spong and Vidyasagar 1987] the factorization approach was used to design a class of dynamic linear compensators $C(s)$, parameterized by a stable transfer matrix $Q(s)$, which guarantee that the solution $e(t)$ to the linear system (5.2.9) is bounded. The actual factorization approach design is beyond the scope of this book, but fairly general representative of the methodology as it applies to robotics is given in Example 5.2.3. In [Spong and Vidyasagar 1987] it was actually assumed that the bound on δ is linear [i.e., $\beta_2 = 0$ in 5.2.10)] before the family of all $\mathcal{L}_\infty$ stabilizing compensators of the nominal plant was found. Although the case of noisy measurements was treated in [Spong and Vidyasagar 1987], we limit ourselves to the noiseless case for simplicity. Let us first recall the system (5.2.16), while suppressing the s dependence,

$$\mathbf{e} = G(I - CG)^{-1}\eta \equiv P_1\eta \quad (5.2.30)$$

and define

5.2 Feedback-Linearization Controllers

$$u = CG(I - CG)^{-1}\eta \equiv P_2\eta \tag{5.2.31}$$

and let the operator gains of P_i, $i = 1, 2$, be given by

$$\gamma_\infty(P_i) = \gamma_i, \quad i = 1, 2. \tag{5.2.32}$$

Note that $\gamma_1 = \max \gamma_{11}, \gamma_{12}$. See Lemma 2.5.2 and Example 2.5.8. Consider then

$$\begin{aligned} \|e\|_{T\infty} &\leq \gamma_1 \|\eta\|_{T\infty} \\ \|u\|_{T\infty} &\leq \gamma_2 \|\eta\|_{T\infty} \\ \|\eta\|_{T\infty} &\leq a\|\ddot{q}_d\|_{T\infty} + a\|u\|_{T\infty} + \frac{\beta_0}{\mu_1} + \frac{\beta_1}{\mu_1}\|e\|_{T\infty}. \end{aligned} \tag{5.2.33}$$

The next theorem gives sufficient conditions for the BIBO stability of the trajectory error.

THEOREM 5.2-3: *The BIBO stability of the closed-loop system (5.2.30) will be guaranteed if*

$$1 - \frac{\beta_1 \gamma_1}{\mu_1} - a\gamma_2 > 0 \tag{1}$$

In fact, the trajectory error is bounded by

$$\begin{aligned} \|e\|_\infty &\leq \frac{\gamma_1 b}{1 - \beta_1 \gamma_1/\mu_1 - a\gamma_2} \\ b &= a\|\ddot{q}_d\|_{T\infty} + \frac{\beta_0}{\mu_1} \end{aligned} \tag{2}$$

Proof:
By the small-gain theorem. See [Spong and Vidyasagar 1987] for details. ∎

If we study (1) carefully we note that it will be satisfied if

1. μ_1 is large or $M(q)$ is large.
2. Good knowledge of N, resulting in a small β_1.
3. Small γ_1 due to a large gain of the compensator C.

4. γ_2 close to 1, which may also be obtained with a large-gain compensator C.

Note that in the limit, and as the gain of $C(s)$ becomes infinitely large, γ_1 goes to zero. This will then transform condition (1) to

$$a < \frac{1}{\gamma_2} \qquad (5.2.34)$$

It is also seen from (5.2.33)-(5.2.34) that increasing the gain k of $C(s)$ will decrease γ_1, therefore decreasing $\|\mathbf{e}\|_\infty$. A particular compensator may now be obtained by choosing the parameter $Q(s)$ to satisfy other design criteria, such as suppressing the effects of η. One can, for example, recover Graig's compensator, by choosing $C(s) = -K$ so that the control effort is given by

$$u = K\mathbf{e}. \qquad (5.2.35)$$

Then note that conditions (5.2.28) and (1) are identical if $\beta_2 = 0$ and $\gamma_2 = \gamma_{11} k_p + \gamma_{12} k_v$. Also note from (2) that a smaller $\ddot{q}_d$ results in a smaller tracking error. In fact, if $\ddot{q}_e = 0$ and $\beta_0 = 0$, the asymptotic stability of the error may be shown. Finally, note that the presence of bounded disturbance will make the bound on the error $\mathbf{e}$ larger but will not affect the stability condition (1). This controller is summarized in Table 5.2.3.

The factorization approach gives the family of all one-degree-of-freedom stabilizing compensators $C(s)$. The design methodology is illustrated for the two-link robot in the next example.

EXAMPLE 5.2-3: Dynamic Controller (Input-Output Design)

Let $G_v(s)$ of Example 4.2.1 be factored as

$$G_v(s) = [\tilde{D}(s)]^{-1}\tilde{N}(s) = N(s)D(s) \qquad (1)$$

where $N(s)$, $D(s)$, $\tilde{N}(s)$, and $\tilde{D}(s)$ are matrices of stable rational functions. We can then find

$$\begin{aligned} N(s) = \tilde{N}(s) &= \frac{1}{(s+1)^2}\begin{bmatrix} I \\ sI \end{bmatrix} \\ D(s) &= \frac{s^2}{(s+1)^2}I \\ \tilde{D}(s) &= \frac{1}{(s+1)^2}\begin{bmatrix} (s^2+2s)I & -2I \\ -sI & (s^2+1)I \end{bmatrix}. \end{aligned} \qquad (2)$$

5.2 Feedback-Linearization Controllers

Table 5.2.3: Dynamic One-DOF Controller: Design 1

Assumptions:

$$\frac{1}{\mu_2} \leq \|M^{-1}\| \leq \frac{1}{\mu_1}, \quad \mu_i \neq 0$$
$$\|\Delta\| \leq a \leq 1$$
$$\|\delta\| \leq \beta_0 + \beta_1\|\mathbf{e}\| + \beta_2\|\mathbf{e}\|^2$$
$$\|\ddot{q}_d\| \leq c$$
$$e(0) = \dot{e}(0) = 0$$

Controller:

$$\tau = \hat{M}(\ddot{q}_d + C(s)\mathbf{e}) + \hat{N}$$

with

$$\|\mathbf{e}\|_{T\infty} \leq \gamma_1 \|\eta\|_{T\infty}$$
$$\|u\|_{T\infty} \leq \gamma_2 \|\eta\|_{T\infty}$$
$$\|\eta\|_{T\infty} \leq a\|\ddot{q}_d\|_{T\infty} + a\|u\|_{T\infty} + \frac{\beta_0}{\mu_1} + \frac{\beta_1}{\mu_1}\|\mathbf{e}\|_{T\infty}$$
$$1 - \frac{\beta_1 \gamma_1}{\mu_1} - a\gamma_2 > 0$$

Performance:

$$\|\mathbf{e}\|_\infty \leq \frac{\gamma_1 b}{1 - \beta_1 \gamma_1/\mu_1 - a\gamma_2}$$
$$b = a\|\ddot{q}_d\|_{T\infty} + \frac{\beta_0}{\mu_1}$$

Next we solve the Bezout indentity [Vidyasagar 1985] for $X(s)$ and $Y(s)$, which are also stable rational functions,

$$Y(s)D(s) + X(s)N(s) = I$$

to get

$$X(s) = \frac{1}{(s+1)^2}[(1+2s)I \quad (2+4s)I], \qquad Y(s) = \frac{s^2+4s+2}{(s+1)^2}I. \quad (3)$$

Then all the stabilizing controllers are given by

$$C(s) = -[Y(s) - Q(s)\tilde{N}(s)]^{-1}[X(s) + Q(s)\tilde{D}(s)], \quad (4)$$

where $Q(s)$ is a stable rational function which is otherwise arbitrary. One choice, of course, is to let $Q(s) = 0$, which leads to the "central solution" [Vidyasagar 1985]

$$C(s) = \frac{2s+1}{s^2+4s+2}[I \quad 2I]. \quad (5)$$

One can also choose $Q(s)$ to satisfy the required performance. In particular, the following choice of $Q(s)$ is presented in [Spong and Vidyasagar 1987]:

$$Q(s) = \left[2I \quad \frac{4k + (k+2)s}{s+k}I\right], \qquad k = 1, 2, 3, \ldots, \quad (6)$$

which leads to the following controller:

$$C(s) = [C_1(s) \quad C_2(s)],$$

where

$$C_1(s) = \frac{-[2s^3 + (k+4)s^2 + (2k+1)s + k]}{s^2(s+2)}I$$

$$C_2(s) = \frac{-[(k+2)s^3 + 4(k+1)s^2 + 5ks + 2k]}{s^2(s+2)}I$$

and

$$u = C_1(s)e + C_2(s)\dot{e}. \quad (7)$$

As k increases, the disturbance rejection property of the controller is enhanced at the expense of higher gains as seen from the expression

5.2 Feedback-Linearization Controllers

of $C(s)$. A simulation of this controller for $k = 225$ is shown in Figure 5.2.6. The following observations are in order: The trajectory errors are smaller than any of the previous controllers while the torque efforts are comparable. In addition, the complexity of the controller is acceptable since the dynamics of the robot are not used in implementing the control. ∎

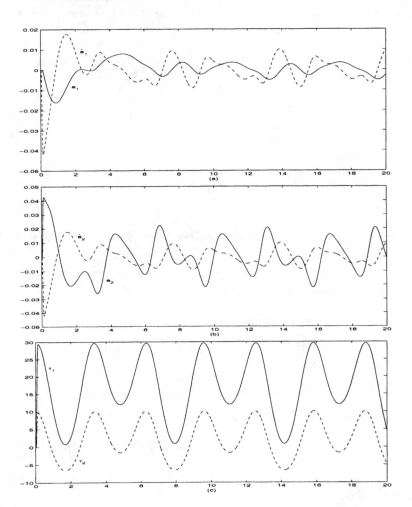

Figure 5.2.6: (a) errors of joint 1; (b) errors of joint 2; (c) torques of joints 1 and 2.

As it was discussed in [Craig 1988] and presented in Theorem 5.2.2, including the more reasonable quadratic bound will not destroy the $\mathcal{L}_\infty$ stability result of [Spong and Vidyasagar 1987]. It was shown in [Becker and Grimm 1988], however, that the $\mathcal{L}_2$ stability of the error cannot be guaranteed unless the problem is reformulated and more assumptions are made. It effect, the error will be bounded, but it may or may not have a finite energy. In particular, noisy measurements are no longer tolerated for $\mathcal{L}_2$ stability to hold. We next present an extension of the $\mathcal{L}_\infty$ stability result that applies to dynamical compensators similar to the one described in Theorem 5.2.3 but without the requirement that $\beta_2 = 0$.

THEOREM 5.2-4: *The error system of (5.2.30) is $\mathcal{L}_\infty$ bounded if*

$$e(0) = \dot{e}(0) = 0$$

and

$$1 - \gamma_2 ak - \gamma_1 \frac{\beta_1}{\mu_1} - 2\gamma_1 \left[\left(ac + \frac{\beta_0}{\mu_1}\right) \frac{\beta_2}{\mu_1}\right]^{\frac{1}{2}} > 0 \qquad (1)$$

Proof:

An extension of the small-gain theorem. See [Becker and Grimm 1988] for details. ∎

A study of (1) reveals that the following desired characteristics will help satisfy the inequality:

1. A large μ_1 due to a large $M(q)$.

2. A small γ_1 and a γ_2 close to 1, which will result from a large-gain compensator C.

3. Small β_i's, which will result from a good knowledge of N.

4. A small c due to a small $\ddot{q}_d$.

Note that Craig's conditions in Theorem 5.2.2 are recovered if $\gamma_1 = \max \gamma_{11}, \gamma_{12}$ and $\gamma_2 k = \gamma_{11} k_p + \gamma_{12} k_v$.

On the other hand, assuming that $\ddot{q}_d = 0$ and $\beta_2 = 0$, the $\mathcal{L}_2$ stability of e was shown in [Becker and Grimm 1988] if

$$1 - \gamma_2 ak - \gamma_1 \frac{\beta_1}{\mu_1} > 0 \qquad (5.2.36)$$

where $\gamma_i = \gamma_2(P_i) = \max_{\omega \in \Re} ||P_i(j\omega)||$ as given in Lemma 2.5.2. This controller is summarized in Table 5.2.4.

5.2 Feedback-Linearization Controllers

Table 5.2.4: Dynamic One-DOF Controller: Design 2

Assumptions:

$$\frac{1}{\mu_2} \leq \|M^{-1}\| \leq \frac{1}{\mu_1}, \quad \mu_i \neq 0$$
$$\|\Delta\| \leq a \leq 1$$
$$\|\delta\| \leq \beta_0 + \beta_1\|\mathbf{e}\| + \beta_2\|\mathbf{e}\|^2$$
$$\|\ddot{q}_d\| \leq c$$

Controller:

$$\tau = \hat{M}(\ddot{q}_d + C(s)\mathbf{e}) + \hat{N}$$

For $\mathcal{L}_\infty$ stability

$$1 - \gamma_2 ak - \gamma_1 \frac{\beta_1}{\mu_1} - 2\gamma_1 \left[\left(ac + \frac{\beta_0}{\mu_1}\right)\frac{\beta_2}{\mu_1}\right]^{\frac{1}{2}} > 0$$

For $\mathcal{L}_2$ stability

$$\ddot{q}_d = 0; \quad \beta_2 = 0 \quad [\text{i.e., } \hat{N} = V_m(q,\dot{q})\dot{q}]$$

$$1 - \gamma_2 ak - \gamma_1 \frac{\beta_1}{\mu_1} > 0$$

Performance:
Guaranteed boundedness of e and $\dot{e}$. The errors go to zero as the gain k of $C(s)$ is increased.

EXAMPLE 5.2-4: Dynamic One DOF Design

Use the same linear controller as that of Example 5.2.3, but assume that the quadratic velocity terms in N are known so that $\beta_2 = 0$; that is

$$\tau = \hat{M}(\ddot{q}_d - u) + V_m(q,\dot{q})\dot{q}.$$

The results are shown in Figure 5.2.7 for the parameter $k = 150$. They do not look remarkably different from those of Example 5.2.3 in

Figure 5.2.6. This is due to the fact that the velocity terms are truly negligent in this particular application. Such terms will, however, make a more vital contribution in faster trajectories. ∎

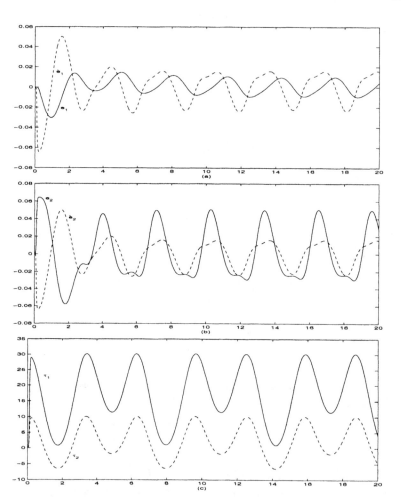

Figure 5.2.7: (a) errors of joint 1; (b) errors of joint 2; (c) torques of joints 1 and 2.

Two-Degree-of-Freedom Design.

It is well known that the two-DOF structure is the most general linear controller structure. The two-DOF design allows us simultaneously to specify the desired response to a command input and guarantee the ro-

5.2 Feedback-Linearization Controllers

bustness of the closed-loop system. This design was briefly discussed in Chapter 2, Example 2.11.4. It is in a different spirit from the other design of this chapter, because it relies on classical frequency-domain SISO concepts. The general structure is shown in Figure 5.2.8. A two-DOF robust controller was designed and simulated in [Sugie et al. 1988] and will be presented next. Let the plant be given by (5.2.5) and consider the following factorization:

$$G(s) = N(s)D^{-1}(s),$$

where

$$D(s) = s^2, \quad N(s) = I. \tag{5.2.37}$$

The following result presents a two-DOF compensator which will robustly stabilize (in the $\mathcal{L}_\infty$ sense) the error system.

THEOREM 5.2-5: *Consider the two-DOF structure of Figure 5.2.8. Let $K_1(s)$ be a stable system and $K_2(s)$ be a compensator to stabilize $G(s)$. Then the controller*

$$u = s^2 K_1 v + K_2(K_1 v - q) \tag{1}$$

will lead to the closed-loop system

$$q = K_1 v \tag{2}$$

and the closed-loop error system (5.2.13) will be $\mathcal{L}_\infty$ stable.

Proof:
With simple block-diagram manipulations, it may be shown that the closed-loop system is

$$q = K_1 v.$$

The actual robustness analysis is involved and will be omitted, but a particular design and its robustness are discussed in the next example. ∎

Note from (2) that $K_1(s)$ is used to obtain the desired closed-loop transfer function. It should then be stable, and to guarantee a zero steady-state error, we choose $v = q_d$ and make sure that the dc gain $K_1(0) = 1$. Finally, we would like $K_1(s)$ to be exactly proper (i.e., zero relative degree). $K_2(s)$, on the other hand, will assure the robustness of the closed-loop system. Therefore, $K_2(s)$ should stabilize $G(s)$ and provide suitable stability

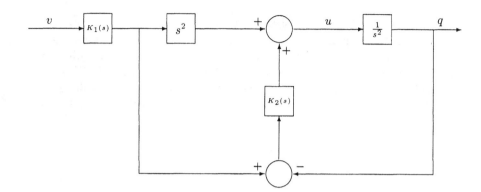

Figure 5.2.8: Two-degree-of-freedom controller structure.

margins. It should contain an integral term to achieve static accuracy. Its relative degree may be -1 since both q and $\dot{q}$ are available.

EXAMPLE 5.2-5: Dynamic Two DOF Design

Consider the two-DOF structure of Figure 5.2.8, where

$$G(s) = \frac{1}{s^2}I$$
$$v = q_d. \qquad (1)$$

Then a two-DOF regulator is described by

$$K_1(s) = \frac{w_1^2}{s^2 + a_1 w_1 s + w_1^2}I$$
$$K_2(s) = \left[b_2 w_2 s + b_1 w_2^2 + \frac{w_2^3}{s}\right]I, \qquad (2)$$

where $s^3 + b_2 s^2 + b_1 s + 1$ is a stable polynomial, and w_1, w_2, a_1, b_1, and b_2 are design parameters. Note that $K_2(s)$ is a PID controller and that the closed-loop system given by $K_1(s)$ is a second-order system with natural frequency w_1 and damping ration $\zeta = \frac{a_1}{2}$. Note also that the input to the robot becomes

$$\tau - K_2(K_1 q_d - q)s^2 K_1 q_d$$

5.2 Feedback-Linearization Controllers

or

$$\tau = K_2 q + (K_2 + s^2) K_1 q_d. \tag{3}$$

We can immediately see that the joint position vector q is filtered through the PID controller K_2. Therefore, the differentiation of q is required unless the measurement of $\dot{q}$ is available. The behavior of the nonlinear closed-loop system is shown in Figure 5.2.9, when $\hat{M} = I$, and $\hat{N} = 0$, $a_1 = 2$, $b_1 = b_2 = 10$, $w_1 = 8$, and $w_2 = 12$. It is seen that initially, the torque effort and the trajectory errors are too large. To understand the behavior of this controller, consider the controller τ in the limit (i.e., as time goes to infinity). The output of $K_1 q_d$ has settled down to its final value q_d and therefore the controller (3) becomes equivalent to a PID compensator [see Chapter 4, equation (4.4.35)]. It seems that a different structure for k_1 and K_2 is warranted because in the meantime, the two-DOF controller preforms rather poorly. This is a characteristic of the example rather than an inherent flow in the two-DOF methodology. As a matter of fact, this structure has shown better performance than the one-DOF PID compensator in [Sugie et al. 1988] for a set-tracking case. The reader is encouraged to work the problems at the end of the chapter related to this design in order to compare the performance of one- and two-DOF designs. ∎

We have this presented a large sample of controllers that are more or less computed-torque based. We have shown using different stability arguments that the computed-torque structure is inherently robust and that by increasing the gains on the outer-loop linear compensator, the position and velocity errors tend to decrease in the norm. This class of compensators constitutes by far the most common structure used by robotics manufacturers and is the simplest to implement and study. There are more compensators that would fit into this structure while appealing to some classical control applications. The PD and PID compensators may be replaced with the lead-lag compensators. These are especially appealing when only position measurements are available. Such designs are discussed in [Chen 1989] in the discrete-time case. There is also some work being done in the nonlinear observer area which is directly relevant to this problem [Canudas de Wit and Fixot 1991]. We refer the reader to the observability discussion in Section 2.11. We also suggest some of the problems at the end of this chapter, which discuss further modification of the feedback-linearization designs.

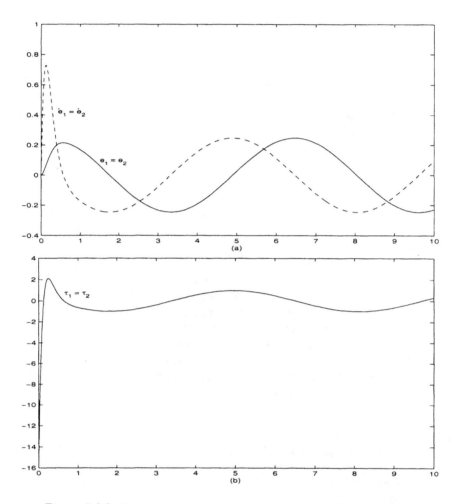

Figure 5.2.9: (a) errors of joints 1 and 2; (b) torques of joints 1 and 2.

On the other hand, there exists other types of controllers, which although less prevalent, still constitute a very important class of robot compensators. These are nonlinear controllers which do not rely directly on the feedback linearizability of the robot. Instead, the may be obtained from the passivity of its dynamics or may even bypass any special structural properties of the robot. These controllers are discussed next.

5.3 Nonlinear Controllers

There is a class of robot controllers that are not computed-torque-like controllers. These controllers are obtained directly from the robot equations without using the feedback-linearization procedure. Instead, these controller may rely on other properties of the robot (such as the passivity of its Lagrange-Euler description) or may be obtained without even considering the physics of the robot. In general, these controllers may be written as a computed-torque controller with an auxiliary, nonlinear controller added to it. The nonlinear control term introduces coupling between the different joints independently from the computed-torque term. In other words, even if the computed-torque controller is a simple PID, the nonlinear term couples all joints together as will be seen in Theorems 5.3.4 and 5.3.5, for example.

Direct Passive Controllers

First, we present controllers that rely directly on the passive structure of rigid robots as described in equations (5.1.1), where $\dot{M}(q) - 2V_m(q,\dot{q})$ is skew symmetric by an appropriate choice of $V_m(q,\dot{q})$ as described in Section 3.3.

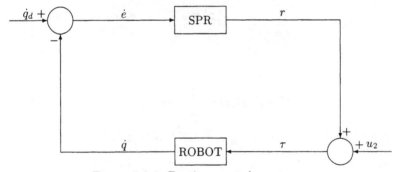

Figure 5.3.1: Passive-control structure.

Based on the passivity property, if one can close the loop from $\dot{q}$ to τ with a passive system (along with $\mathcal{L}_2$ bounded inputs) as in Figure 5.3.1, the closed-loop system will be asymptotically stable using the passivity theorem. Note that the input u_2 gives an extra degree of freedom to satisfy some performance criteria. In other words, by choosing different $\mathcal{L}_2$ bounded u_2 we may be able to obtain better trajectory tracking or noise immunity. This structure will show the asymptotic stability of $\dot{e}$ but only the Lyapunov stability of e. On the other hand, if one can show the passivity of the system, which maps τ to a new vector r which is a filtered version of e, a controller that closes the loop between $-r$ and τ will guarantee the asymptotic stability of both e and $\dot{e}$. This indirect use of the passivity property was illustrated in [Ortega and Spong 1988] and will be discussed first.

Let the controller be given by (5.3.1), where $F(s)$ is a strictly proper, stable, rational function, and K_r is a positive-definite matrix,

$$\tau = M(q)[\ddot{q}_d + C(s)e] + V_m(q,\dot{q})(\dot{q}+r) + G(q) + K_r r$$

$$r = \left[sI + \frac{C(s)}{s}\right] e = F(s)^{-1} e. \qquad (5.3.1)$$

Substituting (5.3.1) into (5.1.1) and assuming no friction [i.e., $F(\dot{q}) = 0$], we obtain

$$M(q)\dot{r} + V_m(q,\dot{q})r + K_r r = 0. \qquad (5.3.2)$$

Then it may be shown that both e and $\dot{e}$ are asymptotically stable. In fact, choose the following Lyapunov function:

$$V = \frac{1}{2} r^T M(q) r.$$

Then

$$\dot{V} = r^T M(q)\dot{r} + \frac{1}{2} r^T \dot{M}(q) r.$$

Substituting for $M(q)\dot{r}$ from (5.3.2), we obtain

$$\dot{V} = -r^T K_r r < 0.$$

Therefore, r is asymptotically stable, which can be used to show that both e and $\dot{e}$ are asymptotically stable [Slotine 1988]. This approach was used in the adaptive control literature to design passive controllers [Ortega and Spong 1988], but its modification in the design of robust controllers when M, V_m and G are not exactly known is not immediately obvious. Such modifications will be given in the variable-structure designs, but first, we present a simple controller to illustrate the robustness of passive compensators.

5.3 Nonlinear Controllers

THEOREM 5.3-1: *Consider the control law (1)*

$$\tau = \Lambda(s)\dot{e} + u_2, \tag{1}$$

where $\Lambda(s)$ is an SPR transfer function, to be chosen by the designer, and the external input u_2 is bounded in the $\mathcal{L}_2$ norm. Then $\dot{e}$ is asymptotically stable, and e is Lyapunov stable.

Proof:
Using the control law above, one gets from Figure 5.3.1,

$$r = \Lambda(s)\dot{e}. \tag{2}$$

By an appropriate choice of $\Lambda(s)$ and u_2, one can apply the passivity theorem and deduce that $\dot{e}$ and r are bounded in the $\mathcal{L}_2$ norm, and since $\Lambda(s)^{-1}$ is SPR (being inverse of an SPR function), one deduces that $\dot{e}$ is asymptotically stable because

$$\dot{e} = \Lambda(s)^{-1} r. \tag{3}$$

This will imply that the position error e is bounded but not its asymptotic stability in the case of time-varying trajectories $[q_d^T \quad \dot{q}_d^T]^T$. In the set-point tracking case, however (i.e., $\dot{q}_d = 0$), and with gravity precompensation, the asymptotic stability of e may be deduced using LaSalle's theorem. The robustness of the closed-loop system is guaranteed as long as $\Lambda(s)$ is SPR and that u_2 is $\mathcal{L}_2$ bounded, regardless of the exact values of the robot's parameters. ∎

The controller is summarized in Table 5.3.1.

Table 5.3.1: Passive Controller

Assumptions:
$\Lambda(s)$ is an SPR transfer function. u_2 is a finite energy signal.
Controller:

$$\tau = \Lambda(s)\dot{e} + u_2$$

Performance:
Guaranteed Lyapunov stability of e, asymptotic stability of $\dot{e}$. In the case where $\dot{q}_d = 0$, asymptotic stability of e is also guaranteed.

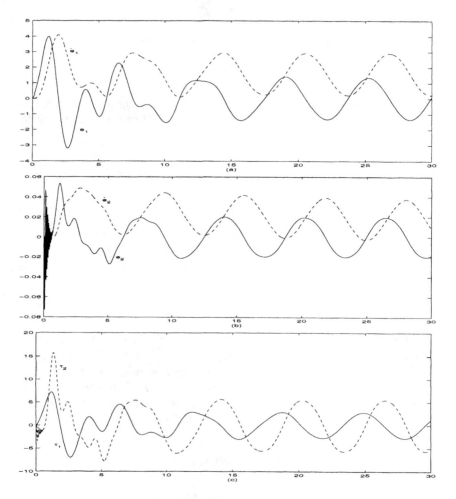

Figure 5.3.2: (a) errors of joints 1; (b) errors of joint 2; (c) torques of joints 1 and 2.

EXAMPLE 5.3-1: Passive Controllers

We choose an SPR transfer function

$$\Lambda(s) = \begin{bmatrix} \frac{s+450}{s+225} & 0 \\ 0 & \frac{s+3300}{s+10} \end{bmatrix} \quad (1)$$

and we let $u_2 = \ddot{q}_d$. Note that the desired trajectory used so far violates the assumption that u_2 should be $\mathcal{L}_2$ stable. Nevertheless, the trajectories of Figure 5.3.2 show that his controller, when started at $e(0) = \dot{e}(0) = 0$, performs rather well for the sinusoidal trajectory. Of course, since the passivity theorem provides sufficient conditions for stability, this example is not contradicting the previous theorem but merely pointing out its conservatism.

∎

Note that the compensator of Theorem 5.3.1 is a generalization of the PD compensators of Chapter 4 and of the preceding section. In [Anderson 1989] it was demonstrated, using network-theoretic concepts, that even in the absence of the contact forces, a computed-torque-like controller is not passive and may therefore cause instabilities in the presence of uncertainties. His solution to the problem consisted of using proportional-derivative (PD) controllers with variable gains $K_1(q)$ and $K_2(q)$ which depend on the inertia matrix $M(q)$, that is,

$$\tau = K_p(q)e + K_v(q)\dot{e}_2 + G(q) \qquad (5.3.3)$$

Even though $M(q)$ is not exactly known, the stability of the closed-loop error is guaranteed by the passivity of the robot and the feedback law. The advantage of this approach is that contact forces and larger uncertainties may now be accommodated. Its main disadvantage is that although robust stability is guaranteed, the closed-loop performance depends on the knowledge of $M(q)$, whose singular values are needed in order to find K_p and K_v. We will not discuss this particular design and refer the interested reader to [Anderson 1989].

Variable-Structure Controllers

In this section we group designs that use variable-structure (VSS) controllers. The VSS theory has been applied to the control of many nonlinear processes [DeCarlo et al. 1988]. One of the main features of this approach is that one only needs to drive the error to a "switching surface," after which the system is in "sliding mode" and will not be affected by any modelling uncertainties and/or disturbances.

There are two main criticisms of these controllers as they apply to robots. The first is that by ignoring the physics or the robot, these controllers will necessarily perform no better (if not worse) that controllers which exploit the structure of the Lagrange-Euler equations. The other criticism relates to the "chattering" problem commonly associated with variable-structure controllers. We shall address the second issue later in this section, and answer the first by admitting that although initial applications of variable-structure theory did indeed gloss over the physics of robots, later designs (such as the ones discussed here) by [Slotine 1985] and [Chen et al. 1990] remedied the problem.

The first application of this theory to robot control seems to be in [Young 1978], where the set-point regulation problem ($\dot{q}_d = 0$) was solved using the following controller:

$$\tau_i = \begin{cases} \tau_i^+ & \text{if } r_i(e_{1i}, \dot{q}_i) > 0 \\ \tau_i^- & \text{if } r_i(e_{1i}, \dot{q}_i) < 0 \end{cases} \quad (5.3.4)$$

where $i = 1, \ldots, n$ for an n-link robot, and r_i are the switching planes,

$$r_i(e_{1i}, \dot{q}_i) = \lambda_i e_i + \dot{q}_i, \quad \lambda_i > 0. \quad (5.3.5)$$

It is then shown, using the hierarchy of the sliding surfaces $r_1, r_2, \ldots, r_n$ and given bounds on the uncertainties in the manipulators model, that one can find τ^+ and τ^- in order to drive the error signal to the intersection of the sliding surfaces, after which the error will "slide" to zero. This controller eliminates the nonlinear coupling of the joints by forcing the system into the sliding mode. Other VSS robot controllers have since been designed. Unfortunately, for most of these schemes, the control effort as seen from (5.3.4) is discontinuous along $r_i = 0$ and will therefore create chattering, which may excite unmodelled high-frequency dynamics. In addition, these controllers do not exploit the physics of the robot and are therefore less effective than controllers that do.

To address this problem, the original VSS controllers were modified in [Slotine 1985] as described in the next theorem. Let us first define a few variables to simplify the statement of the theorem. Let

$$\begin{aligned} r &= \Lambda e + \dot{e}, \quad \Lambda = \text{diag}[\lambda_1, \lambda_2, \ldots, \lambda_n], \quad \lambda_i > 0 \\ \dot{q}_r &= \Lambda e + \dot{q}_d \\ \text{sgn}(r_i) &= [\text{sgn}(r_1), \text{sgn}(r_2), \ldots, \text{sgn}(r_n)]^T \end{aligned} \quad (5.3.6)$$

where

$$\text{sgn}(r_i) = +1 \quad \text{if } r_i > 0, \quad \text{sgn}(r_i) = -1 \quad \text{if } r_i < 0.$$

THEOREM 5.3-2: *Consider the controller*

$$\tau = \hat{M}\ddot{q}_r + \hat{V}_m \dot{q}_r + \hat{G} + K\,\text{sgn}(r),$$

5.3 Nonlinear Controllers

where

$$K = diag[k_1, k_2, \ldots, k_n], \quad k_i > 0.$$

Then the error reaches the surface

$$r = \Lambda e + \dot{e}$$

in a finite time. In addition, once on the surface, $q(t)$ will go to $q_d(t)$ exponentially fast.

Proof:
Consider the Lyapunov function candidate

$$V(r) = \frac{1}{2} r^T M(q) r.$$

Then

$$\dot{V}(r) = r^T M \dot{r} + \frac{1}{2} \dot{M}(q) r.$$

Substituting for $\dot{r}$ and using the skew-symmetric property of $\dot{M}(q) - 2V_m(q, \dot{q})$, we obtain

$$\dot{V}(r) = r^T [M \ddot{q}_r + V_m \dot{q}_r + G - \tau].$$

Then, if we use the controller given by

$$\tau = \hat{M} \ddot{q}_r + \hat{V}_m \dot{q}_r + \hat{G} + K \, sgn(r),$$

we obtain

$$\dot{V}(r) = r^T [\tilde{M} \ddot{q}_r + \tilde{V}_m \dot{q}_r + \tilde{G}] - \sum_{i=1}^{n} k_i |r_i|,$$

where

$$\tilde{M} = M - \hat{M}, \quad \tilde{V}_m = V_m - \hat{V}_m, \quad \tilde{G} = G - \hat{G}.$$

Then it is sufficient to choose

$$k_i \geq |[\tilde{M}\ddot{q}_r + \tilde{V}_m\dot{q}_r + \tilde{G}]_i| + \eta_i,$$

where $\eta_i > 0$. Therefore,

$$\dot{V}(r) \leq -\sum_{i=1}^{n} \eta_i |r_i|,$$

which implies that $r = 0$ is reached in a finite time. In addition, once in sliding mode, e converges exponentially fast to zero. ∎

The controller is summarized in Table 5.3.2.

EXAMPLE 5.3-2: VSS Controller 1

Consider the two-link robot and choose $K = 10I$ and $\Lambda = 5I$. The trajectory errors for joint 1 are given in Figure 5.3.3. Note that the chattering behavior due to the infinitely fast switching of the controller is becoming apparent. A common remedy of the problem is to sacrifice asymptotic stability by using $\text{sat}(s/\epsilon)$, $\epsilon > 0$ instead of $\text{sgn}(s)$ in the torque calculation. The saturation function is depicted in Figure 5.3.4a and will result in the errors being uniformly ultimately bounded. The behavior of such a controller is shown in Figure 5.3.5, where $\Lambda = 5I$, $K = 75I$, and $\epsilon = 0.001$. Note that we were able to greatly increase the gains in this case, which results in a smaller error trajectories. ∎

More recently, in [Chen et al. 1990], another VSS controller was introduced. We describe this controller in detail to give a flavor of a different VSS approach, which will exploit the dynamics of the robot. The assumptions required by this controller are listed below.

$$\begin{aligned} |(M - \hat{M})_{ij}| &< \bar{M}_{ij}(q) \\ |\dot{M}_{ij}| &< \bar{m}_{ij}(q,\dot{q}) \\ |[(M - \hat{M})\ddot{q}_d]_i| &< \bar{c}_i(t) \\ |(N - \hat{N})_i| &< \bar{n}_i(q,\dot{q}) \end{aligned} \quad (5.3.7)$$

Note that these bounds are different in spirit that those given in (5.2.8)-(5.2.11). The current bounds are often more useful since they depend on the uncertainty of each element of the inertia matrix M and the velocity-dependent torques N.

5.3 Nonlinear Controllers

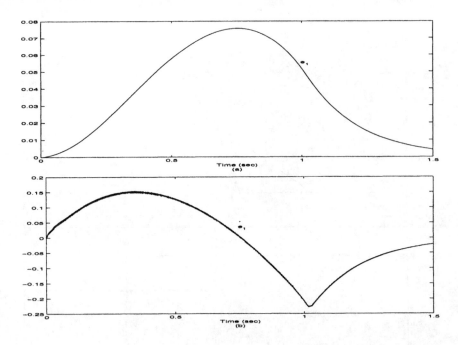

Figure 5.3.3: errors of joint 1: (a) e_1; (b) $\dot{e}_1$

THEOREM 5.3-3: *The errors e and $\dot{e}$ are asymptotically stable if the input torque is given by*

$$\tau = \hat{M}\ddot{q}_r + \hat{N} + P(t)r(t) + Q(t)sgn(r), \qquad (1)$$

where

$$\begin{aligned} P(t) &= diag[p_1(t), p_2(t), \ldots, p_n(t)], \\ Q(t) &= diag[q_1(t), q_2(t), \ldots, q_n(t)] \end{aligned} \qquad (2)$$

$$p_i(t) = \frac{1}{2}\sum_{i=1}^{n}\bar{m}_{ij} + k_i, \quad k_i > 0 \qquad (3)$$

$$q_i(t) = \sum_{i=1}^{n}\bar{M}\Lambda_{ij}|\dot{e}_j| + \bar{n}_i + \bar{c}_i. \qquad (4)$$

Proof:
    Choose the same Lyapunov function as in Theorem 5.3.2:

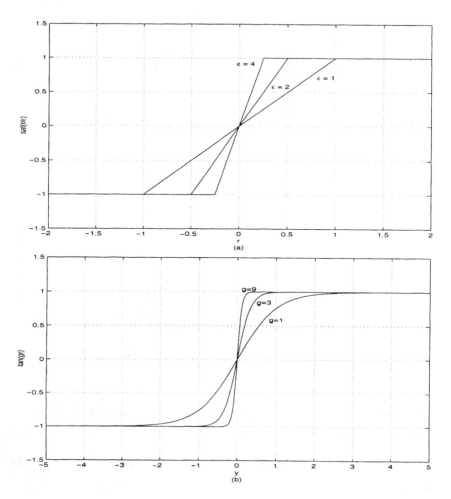

Figure 5.3.4: (a) plot of $sat(r/\epsilon)$; (b) plot of $tanh(gr)$

$$V(r) = \frac{1}{2}r^T M(q)r.$$

Then, differentiating and substituting for τ, we obtain

$$\dot{V} = -t^T \left[P - \frac{\dot{M}}{2} \right] r + r^T[-Q\,sgn(r) + \tilde{M}\Lambda\dot{e} + \tilde{N} + \tilde{M}\ddot{q}_d],$$

5.3 Nonlinear Controllers

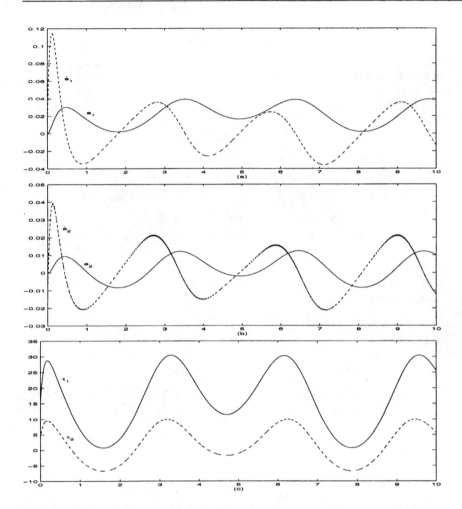

Figure 5.3.5: (a) errors of joint 1; (b) errors of joint 2; (c) torques of joints 1 and 2

which using (4)-(5.3.8) may be shown to be

$$\dot{V} \leq -r^T K r \leq 0.$$

Therefore, the surface $r(t) = 0$ is reached in a finite time, after which the exponential stability of the error results as discussed in Theorem 5.3.2. ∎

This controller is summarized in Table 5.3.3.

Table 5.3.2: Variable Structure Controller 1

Assumptions:
None

Controller:

$$r = \Lambda e + \dot{e}, \quad \Lambda = \text{diag}[\lambda_1, \lambda_2, \ldots, \lambda_n], \quad \lambda_i > 0$$
$$\dot{q}_r = \Lambda e + \dot{q}_d$$
$$\text{sgn}(r_i) = [\text{sgn}(r_1), \text{sgn}(r_2), \ldots, \text{sgn}(r_n)]^T$$
$$\text{sgn}(r_i) = +1 \quad \text{if } r_i > 0, \quad \text{sgn}(r_i) = -1 \quad \text{if } r_i < 0.$$
$$\tau = \hat{M}\ddot{q}_r + \hat{V}_m \dot{q}_r + \hat{G} + K\text{sgn}(r),$$
$$K = \text{diag}[k_1, k_2, \ldots, k_n], \quad k_i > 0.$$

Performance:
Guaranteed uniform ultimate boundedness of e and $\dot{e}$.
Guaranteed uniform boundedness if $e(0) = \dot{e}(0) = 0$.

EXAMPLE 5.3-3: VSS Controller 2 (Saturation)

The two-link robot was controlled using this controller with the following design parameters:

$$\hat{M} = \begin{bmatrix} 3 & 1 \\ 1 & 1 \end{bmatrix}, \quad \bar{M} = \begin{bmatrix} 2 & 1 \\ 1 & 0 \end{bmatrix}$$

$$\hat{N} = 0, \quad \bar{N} = \begin{bmatrix} 2\dot{q}_1\dot{q}_2 + \dot{q}_2^2 + 3(9.8) \\ \dot{q}_1^2 + 9.8 \end{bmatrix}$$

$$\bar{c} = \begin{bmatrix} 3 \\ 1 \end{bmatrix}, \quad \Lambda = \begin{bmatrix} 3 & 0 \\ 0 & 1 \end{bmatrix}, \quad K = \begin{bmatrix} 10 \\ 10 \end{bmatrix}$$

The resulting errors for joint 1 are shown in Figure 5.3.5. Due to the same chattering problems encountered in Example 5.3.2, we choose instead a controller with a saturation function and parameters identical to those in Example 5.3.2. The resulting behavior is plotted in Figure 5.3.7. ∎

As seen in Figures 5.3.3 and 5.3.5, the algorithms of Theorems 5.3.2 and 5.3.3, although using the physics of the robot, suffer from the chatter-

5.3 Nonlinear Controllers

Table 5.3.3: Variable Structure Controller 2

Assumptions:

$$\begin{aligned}
|(M - \hat{M})_{ij}| &< \bar{M}_{ij}(q) \\
|\dot{M}_{ij}| &< \bar{m}_{ij}(q, \dot{q}) \\
|[(M - \hat{M})\ddot{q}_d]_i| &< \bar{c}_i(t) \\
|(N - \hat{N})_i| &< \bar{n}_i(q, \dot{q})
\end{aligned}$$

Controller:

$$\tau = \hat{M}\ddot{q}_r + \hat{N} + P(t)r(t) + Q(t)\text{sgn}(r)$$

where

$$\begin{aligned}
P(t) &= \text{diag}[p_1(t), p_2(t), \ldots, p_n(t)], \\
Q(t) &= \text{diag}[q_1(t), q_2(t), \ldots, q_n(t)] \\
p_i(t) &= \frac{1}{2}\sum_{i=1}^{n} \bar{m}_{ij} + k_i, \quad k_i > 0 \\
q_i(t) &= \sum_{i=1}^{n} \bar{M}\Lambda_{ij}|\dot{e}_j| + \bar{n}_i + \bar{c}_i
\end{aligned}$$

Performance:
Guaranteed uniform ultimate boundedness of e and $\dot{e}$.
Guaranteed uniform boundedness if $e(0) = \dot{e}(0) = 0$.

---

ing commonly encountered in VSS control because of the presence of the sgn(r) term. A common remedy of the problem is to sacrifice asymptotic stability by using sat(r/ϵ), $\epsilon > 0$, instead of sgn(r) in the torque calculation as done in both Examples 5.3.2 and 5.3.3. We propose instead to use a term tanh(gr), where tanh is the hyperbolic tangent and g is a gain parameter that adjusts the slope of tanh around the origin as shown in Figure 5.3.7b. This term is continuously differentiable, and a good approximation of sgn(r) for large s, will result in uniformly ultimately bounded errors, and by adjusting g we are able to get similar performance to the sgn(r) controller without the chattering behavior. The next example illustrates this controller and compares it to the usual saturation controller.

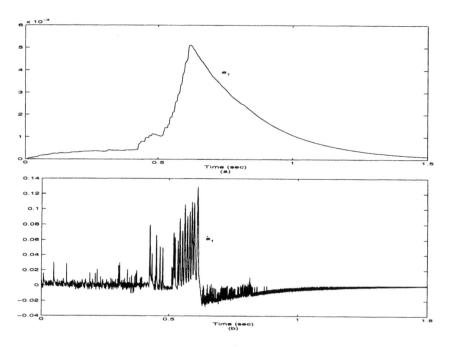

Figure 5.3.6: errors of joint 1(a) e_1; (b) $\dot{e}_1$

EXAMPLE 5.3-4: VSS Controller 2 (Hyperbolic Tangent)

Contrast the saturation controller behavior of Example 5.3.2 with the hyperbolic tangent modification with $g = 3$ in Figure 5.3.8. Similar comparisons should be made between Example 5.3.3 and the behavior shown in Figure 5.3.9, where $g = 3$ is also used.

∎

Saturation-Type Controllers

In this section we present controllers that utilize an auxiliary saturating signal to compensate for the uncertainty present in the robot dynamics as given by (5.2.1), where $V_m(q, \dot{q})$ is defined in Chapter 3.

$$M(q)\ddot{q} + V_m(q, \dot{q})\dot{q} + F(\dot{q}) + G(q) - \tau_d = \tau \qquad (5.3.8)$$

or

$$M(q)\ddot{q} + V_m(q, \dot{q})\dot{q} + Z(q, \dot{q}) = M(q)\ddot{q} + N(q, \dot{q}) = \tau \qquad (5.3.9)$$

5.3 Nonlinear Controllers

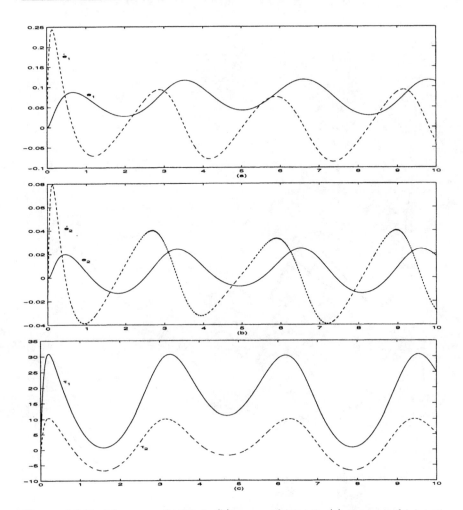

Figure 5.3.7: (a) errors of joint 1; (b) errors of joint 1; (c) torques of joints 1 and 2

Therefore, $Z(q,\dot{q})$ is an n-vector representing friction, gravity and bounded torque disturbances. The controllers introduced in this section are robust due to the fact that they are designed based on uncertainty bounds rather than on the actual values of the parameters. The following bounds are needed and may be physically justified. The ζ_i's in (5.3.11) are positive scalar constants and the trajectory error **e** is defined before.

$$\mu_1 I \leq M(q) \leq \mu_2 I \qquad (5.3.10)$$

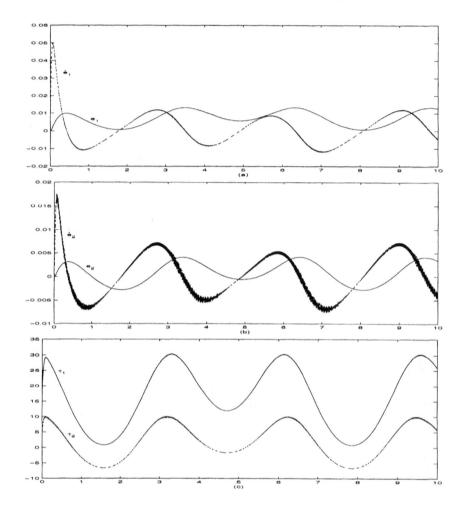

Figure 5.3.8: (a) errors of joint 1; (b) errors of joint 1; (c) torques of joints 1 and 2

$$\|N(q,\dot{q})\| = \|V_m(q,\dot{q})\dot{q} + Z(q,\dot{q})\| \leq \zeta_0 + \zeta_1\|e\| + \zeta_2\|\mathbf{e}\|^2 \quad (5.3.11)$$

A representative of this class was developed in [Spong et al. 1987] and is given as follows.

THEOREM 5.3-4: *The trajectory error* **e** *is uniformly ultimately bounded (UUB) with the controller*

5.3 Nonlinear Controllers

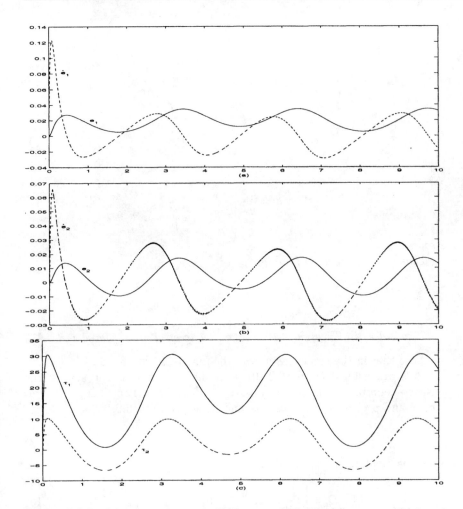

Figure 5.3.9: (a) errors of joint 1; (b) errors of joint 1; (c) torques of joints 1 and 2

$$\tau = \frac{2\mu_1\mu_2}{\mu_1 + \mu_2}[\ddot{q}_d + K_v\dot{e} + K_p e + v_r] + \hat{N}(q, \dot{q}), \quad (1)$$

where

$$K_v = k_v I, \quad K_p = k_p I$$

$$v_r = \begin{cases} (B^T P e) p (\|B^T P e\|)^{-1} & \text{if } \|B^T P e\| > \epsilon \\ (B^T P e) p / \epsilon & \text{if } \|B^T P e\| \leq \epsilon \end{cases} \quad (2)$$

and

$$p = \frac{1}{1-a}\left[a||\ddot{q}_d|| + k_p||e|| + k_v||\dot{e}|| + \frac{1}{\mu_1}\phi\right] \tag{3}$$

$$\phi = \beta_0 + \beta_1||e|| + \beta_2||e||^2 \tag{4}$$

$$a = \frac{\mu_2 - \mu_1}{\mu_1 + \mu_2} \tag{5}$$

Proof:
   Choose the Lyapunov function

$$V(\mathbf{e}) = \mathbf{e}^T P \mathbf{e}$$

and proceed as in Theorem 2.10.4. See [Spong et al. 1987] for details. ∎

Note that in the equations above, the matrix B is defined as in (5.2.3), the β_i's are defined as in (5.2.10), and the matrix P is the symmetric, positive-definite solution of the lyapunov equation (5.3.12), where Q is symmetric and positive-definite matrix and A_c is given in (5.2.14).

$$A_c^T P + P A_c = -Q \tag{5.3.12}$$

In particular, the choice of Q

$$Q = \begin{bmatrix} K_p & 0 \\ 0 & 2K_v - I \end{bmatrix}, \quad K_v > I \tag{5.3.13}$$

leads to the following P:

$$P = \begin{bmatrix} K_p + 0.5K_v & 0.5I \\ 0.5I & I \end{bmatrix}. \tag{5.3.14}$$

The expression of P in (5.3.14) may therefore be used in the expression of v_r in (2). This design is summarized in Table 5.3.4.

5.3 Nonlinear Controllers

Table 5.3.4: Saturation Controller 1

Assumptions:

$$\mu_1 I_n \leq M(q) \leq \mu_2 I_n$$
$$\|N(q,\dot{q})\| \leq \zeta_0 + \zeta_1\|e\| + \zeta_2\|e\|^2$$

Controller:

$$\tau = \frac{2\mu_1\mu_2}{\mu_1+\mu_2}[\ddot{q}_d + K_v\dot{e} + K_p e + v_r] + \hat{N}(q,\dot{q})$$

where

$$v_r = \begin{cases} (B^T Pe)p(\|B^T Pe\|)^{-1} & \text{if } \|B^T Pe\| > \epsilon \\ (B^T Pe)p/\epsilon & \text{if } \|B^T Pe\| \leq \epsilon \end{cases}$$

$$p = \frac{1}{1-a}\left[a\|\ddot{q}_d\| + k_p\|e\| + k_v\|\dot{e}\| + \frac{1}{\mu_1}\phi\right]$$

$$\phi = \beta_0 + \beta_1\|e\| + \beta_2\|e\|^2$$

$$a = \frac{\mu_2 - \mu_1}{\mu_1 + \mu_2}$$

Performance:
Guaranteed uniform ultimate boundedness of e and $\dot{e}$.
Guaranteed uniform boundedness if $e(0) = \dot{e}(0) = 0$.

EXAMPLE 5.3-5: Saturation Controller 1

The following design parameters were chosen in simulating this controller:

$$K_p = 15I, \quad K_v = 30I, \quad \mu_1 = 0.15, \quad \mu_2 = 6.5$$
$$\beta_0 = 50, \quad \beta_1 = 10, \quad \beta_2 = 10, \quad \epsilon = 5$$
$$\hat{C} = 0, \quad \hat{Z} = 0,$$

and

$$e(0) = \dot{e}(0) = 0$$

The same trajectory is followed by the two-link robot as shown in Figure 5.3.10. Note that although the trajectory errors seem to be diverging, they are indeed ultimately bounded and may be shown to be so by running the simulation for a long time. ∎

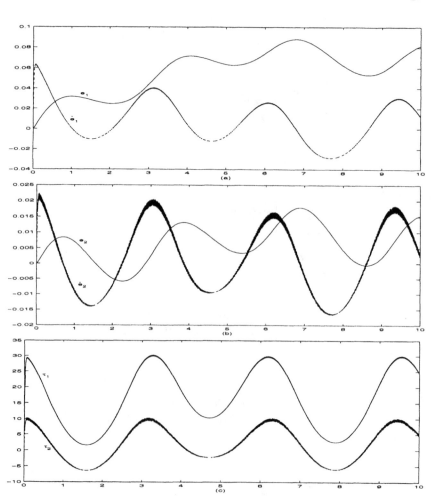

Figure 5.3.10: (a) errors of joint 1; (b) errors of joint 1; (c) torques of joints 1 and 2

Upon closer examination of Spong's controller in Theorem 5.3.4, it becomes clear that v_r depends on the servo gains K_p and K_v through p. This might obscure the effect of adjusting the servo gains and may be avoided

5.3 Nonlinear Controllers

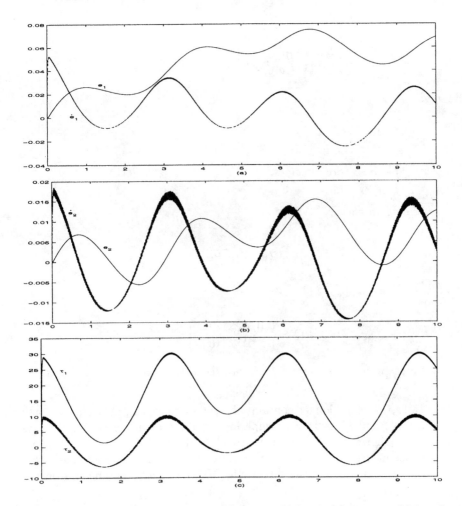

Figure 5.3.11: (a) errors of joint 1; (b) errors of joint 1; (c) torques of joints 1 and 2

as described in [Dawson et al. 1990].

THEOREM 5.3-5: *The trajectory error* **e** *is uniformly ultimately bounded (UUB) with the controller*

$$\tau = K_v \dot{e} + v_r(p, e, \dot{e}, \epsilon), \tag{1}$$

where

$$v_r = \begin{cases} (\frac{e}{2}+\dot{e})p(\|\frac{e}{2}+\dot{e}\|)^{-1} & \text{if } \|\frac{e}{2}+\dot{e}\| > \epsilon \\ (\frac{e}{2}+\dot{e})p/\epsilon & \text{if } \|\frac{e}{2}+\dot{e}\| \leq \epsilon \end{cases}$$

and

$$p = \delta_0 + \delta_1\|e\| + \delta_2\|\dot{e}\|^2 \qquad (2)$$

where γ_i's are positive scalars.

Proof:
We again choose

$$V(\mathbf{e}) = \mathbf{e}^T P \mathbf{e}$$

and

$$Q = \begin{bmatrix} 0.5K_p & 0 \\ 0 & K_v \end{bmatrix}$$

and proceed as in Theorem 2.10.4. See [Dawson et al. 1990] for details. ∎

Note that p no longer contains the servo gains and, as such, one may adjust K_p and K_v without tampering with the auxiliary control v_r. As was also shown in [Dawson et al. 1990], if the initial error $e(0) = 0$ and by choosing $K_v = 2K_p = k_v I$, the tracking error may be bounded by the following, which shows the direct effect of the control parameters on the tracking error:

$$\|e\| \leq \left[\frac{(8k_v + 6\mu_2)\epsilon}{k_v}\right]^{\frac{1}{2}}. \qquad (5.3.15)$$

Finally, note that if $e(0) = \dot{e}(0) = 0$, the uniform boundedness of $\mathbf{e}(t)$ may be deduced. This controller is given in Table 5.3.5.

EXAMPLE 5.3-6: Saturation Controller 2
In this example, let

$$K_p = 15I, \quad K_v = 30I,$$
$$\delta_0 = 50, \quad \delta_1 = 10, \quad \delta_2 = 10, \quad \epsilon = 0.1.$$

5.3 Nonlinear Controllers

Table 5.3.5: Saturation Controller 2

Assumptions:

$$\mu_1 I_n \leq M(q) \leq \mu_2 I_n$$

Controller:

$$\tau = K_v \dot{e} + v_r(p, e, \dot{e}, \epsilon)$$

where

$$v_r = \begin{cases} (\frac{e}{2} + \dot{e}) p (||\frac{e}{2} + \dot{e}||)^{-1} & \text{if } ||\frac{e}{2} + \dot{e}|| > \epsilon \\ (\frac{e}{2} + \dot{e}) p / \epsilon & \text{if } ||\frac{e}{2} + \dot{e}|| \leq \epsilon \end{cases}$$

$$p = \delta_0 + \delta_1 ||e|| + \delta_2 ||\dot{e}||^2$$

Performance:
Guaranteed uniform ultimate boundedness of e and $\dot{e}$.
Guaranteed uniform boundedness if $e(0) = \dot{e}(0) = 0$.

The results of this simulation are presented in Figure 5.3.11. The errors are ultimately bounded as may be seen after a long simulation run.

■

Note that the last two controllers, although using a continuous term v_r, may be modified using the hyperbolic tangent term introduced in Example 5.3.4. In fact, there are many extensions to these types of controllers. A particularly simple one that uses linearity of the dynamics with respect to the unknown physical parameters of the robot 3.3.59 was given in (3.3.62), where $p = ||W|| \Phi_m ax$. In this formulation, W is a matrix containing time-varying but known terms, while Φ is a vector of unknown but constant terms in the dynamic formulation (3.3.59) repeated here:

$$\tau = W(q, \dot{q}, \ddot{q}) \Phi. \tag{5.3.16}$$

More on this formulation and its usage is presented in Chapter 6.

5.4 Dynamics Redesign

In this section we present two other approaches to design robust controllers. The first starts with the mechanical design of the robot and proposes to design robots such that their dynamics are simple and decoupled. It then solves the robust controller problem by eliminating its causes. The second approach may be recast into one of the approaches discussed previously, but it presents such a novel way to looking at the problem that we decided to include it separately.

Decoupled Designs

It was shown throughout the previous chapters that the controller complexity is directly dependent on that of the robot dynamics. Thus it would make sense to design robots such that they have simple dynamics making their control much easier. This approach is advocated in [Asada and Youcef-Toumi 1987]. In fact, it is shown that certain robotic structures will have a decoupled dynamical structures resulting in a decoupled set of n SISO nonlinear systems which are easier controlled than the one MIMO nonlinear system. The decoupling is achieved by modifying the dimensions and mass properties of the arm to cancel out the velocity-dependent terms and decouple the inertia matrix. An illustrative example of such robots is given in the next example.

EXAMPLE 5.4-1: Decoupled Design

Consider the robot described in Figure 5.4.1. This mechanism is known as the five-bar linkage and its dynamics are described when the following condition holds:

$$\frac{m_4 g_4}{m_3 g_3} = \frac{l_2}{l_1}$$

by

$$\tau = \begin{bmatrix} I_1 + I_3 + m_1 g_1^2 + m_4 l_1^2 \left(1 + \frac{g_3 g_4}{l_1 l_2}\right) & 0 \\ 0 & I_2 + I_4 + m_3 l_2^2 + m_4 g_4^2 \left(1 + \frac{l_1 l_2}{g_3 g_4}\right) \end{bmatrix} \ddot{q}.$$

Note that the inertia matrix is decoupled and position independent. The controller given by

$$\tau = M(\ddot{q}_d + K_v \dot{e} + K_p e)$$

5.4 Dynamics Redesign

with positive-definite K_p and K_v is enough to achieve any desired level of performance. In particular, consider the following case:

$$I_1 = I_2 = I_3 = I_4 = 1$$

$$\begin{aligned} m_1 &= 5, & m_2 &= \tfrac{1}{2}, & m_3 &= 2, & m_4 &= 1, \\ l_1 &= 1, & l_2 &= \tfrac{1}{2}, & l_3 &= \tfrac{3}{4}, & l_4 &= 1, \\ g_1 &= \tfrac{1}{2}, & g_2 &= \tfrac{1}{4}, & g_3 &= \tfrac{1}{2}, & g_4 &= \tfrac{1}{2}. \end{aligned}$$

The closed-loop error system is a linear second-order system which will be stable with positive-definite gains. ∎

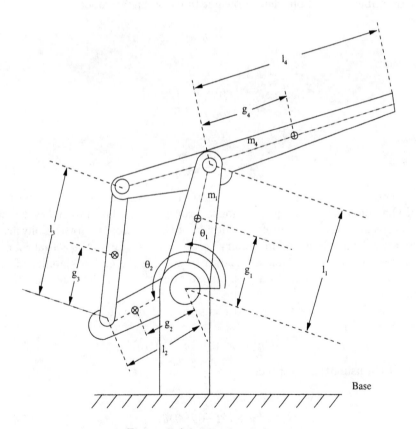

Figure 5.4.1: Five-bar linkage.

Some standard robotic structures may also be decoupled by design. Studies have been carried out to partially or totally decouple robots up to six links. The interested reader is referred to [Yang and Tzeng 1986], [Asada and Youcef-Toumi 1987], and [Kazerooni 1989] for good discussions of this topic.

Imaginary Robot Concept

The decoupled design alternative is very useful if the control engineer has access to, or can modify, the robot design at an early stage. It is more reasonable, however, to assume that the robot has already been constructed to satisfy may mechanical requirements before the control law is actually implemented. Thus a dynamics redesign is difficult if not impossible. The *imaginary robot* concept is presented as an alternative robust design methodology [Gu and Loh 1988]. The development of this approach is described next. Consider an output function of the robot given by

$$y = h(q), \qquad (5.4.1)$$

so that

$$\dot{y} = \left(\frac{\partial h}{\partial q}\right)^T \dot{q} = J\dot{q} \qquad (5.4.2)$$

and

$$\ddot{y} = J\ddot{q} + \dot{J}\dot{q}. \qquad (5.4.3)$$

The generalized output y may denote the coordinates of the end effector of the robot or the trajectory joint error $q_d - q$. The imaginary robot concept attempts to simplify the design of the control law for the physical robot, by controlling an "imaginary" robot that is close to the actual robot. This choice of the controller is shown to achieve the global stability of an imaginary robot whose joint positions are described by the components of the vector y.

The methodology starts by decomposing $M(q)$ as follows:

$$M(q) = J^T(q)J(q) + \tilde{M}(q) \qquad (5.4.4)$$

and then using the controller

$$\tau = J^T \ddot{y} + (\hat{M} - J^T J)\ddot{q} \qquad (5.4.5)$$
$$\ddot{y} = \ddot{y}_d + K_p(y_d - y) + K_v(\dot{y}_d - \dot{y}). \qquad (5.4.6)$$

5.4 Dynamics Redesign

Since $M(q)$ is unknown, however, the actual $\tilde{M}$ is not available. The resulting controller is then simpler and may be applied to the physical robot to lead acceptable, if not optimal behavior.

The following theorem illustrates a controller to guarantee the boundedness of the error.

THEOREM 5.4-1: Let

$$\tau = J^T \ddot{y} + (\hat{M} - J^T J)\ddot{q} \quad (1)$$
$$\ddot{y} = \ddot{y}_d + K_p(y_d - y) + K_v(\dot{y}_d - \dot{y}). \quad (2)$$

The $\mathcal{L}_\infty$ stability of the closed-loop system is guaranteed if

$$\|JM^{-1}(\tilde{M} - \hat{M} + J^T J)J^{-1}\| = \|J(I - M^{-1}\hat{M})J^{-1}\| < \frac{1}{2}. \quad (3)$$

Proof:
This is an immediate result of Theorem 5.2.2. See [Gu and Loh 1988] for more detail and for illustrative examples. ∎

The controller is shown in Table 5.4.1.

Table 5.4.1: Computed-Torque-Like Robot Controllers.

Assumptions:

$$y = h(q)$$

$$\dot{y} = \left(\frac{\partial h}{\partial q}\right)^T \dot{q} = J\dot{q}$$

$$M(q) = J^T(q)J(q) + \tilde{M}(q)$$

$$\|JM^{-1}(\tilde{M} - \hat{M} + J^T J)J^{-1}\| = \|J(I - M^{-1}\hat{M})J^{-1}\| < \frac{1}{2}$$

Controller:

$$\tau = J^T \ddot{y} + (\hat{M} - J^T J)\ddot{q}$$
$$\ddot{y} = \ddot{y}_d + K_p(y_d - y) + K_v(\dot{y}_d - \dot{y})$$

Performance:
The errors e and $\dot{e}$ are bounded in the $\mathcal{L}_\infty$ norm. Larger gains K_p and K_v will decrease the bounds on e and $\dot{e}$.

5.5 Summary

The design of robust motion controllers of rigid robots was reviewed. There main designs were identified and explained. All controllers were robust with respect to a range of uncertain parameters and will guarantee the boundedness of the position-tracking error. In the presence of disturbance torques, a bounded error is best achievable outcome. The question of which robust control method to choose is difficult to answer analytically, but the following guidelines are suggested. The linear-multivariable approach is useful when linear performance specifications (percent overshoot, damping ratio, etc.) are available. The one-DOF dynamic compensators performed rather well, with little or no knowledge of the robot dynamics. They may, however, result in high-gain control laws in the attempt to achieve robustness. The passive controllers are easy to implement but do not provide easily quantifiable performance measures. The modified variable-structure controllers seem to preform well when using the physics of the robot without excessive torque effort. The saturation controllers are most useful when a short transient error can be tolerated, but ultimately, the error will have to be bounded.

A common thread throughout this chapter has been the fact that a high-gain controller will guarantee the robustness of the closed-loop system. The challenge is, however, to guarantee the robust stability of the robot without requiring excessive torques. The robustness of the motion controllers when nonzero initial errors or disturbances are present was also verified through some of the examples and is discussed in the problems at the end of the chapter. It is useful to note that although the robot's dynamics are highly nonlinear, most successful controllers have exploited their physics and their very special structure. In the next chapter we describe the design of adaptive controllers in the case of uncertain dynamical description of the robots.

REFERENCES

[Abdallah et al. 1991] C.T. Abdallah and D. Dawson and P. Dorato and M. Jamshidi. "Survey of Robust Control for Rigid Robots". *IEEE Control Syst. Mag.*, vol. 11, number 2, pp. 24–30, 1991.

[Anderson 1989] R.J. Anderson. "A Network Approach to Force Control in Robotics and Teleoperation". Ph.D. Thesis, Departemnt of Electrical & Computer Engineering. *University of Illinois at Urbana-Champaign*, 1989.

[Arimoto and Miyazaki 1985] S. Arimoto and F. Miyazaki. "Stabiliy and Robustness of PID Feedback Control for Robot Manipulators of Sensory Capability". *Proc. Third Int. Symp. Robot. Res.*, Gouvieux, France. July, 1985.

[Asada and Youcef-Toumi 1987] H. Asada and K. Youcef-Toumi. "Direct-Drive Robots: Theory and Practice". *MIT Press*, Cambridge, MA, 1987.

[Becker and Grimm 1988] N. Becker and W.M. Grimm. "On L_2 and L_∞ Stability Approaches for the Robust Control of Robot Manipulators". *IEEE Trans. Autom. Control.* vol. 33, number 1, pp. 118–122, January, 1988.

[Canudas de Wit and Fixot 1991] C. Canudas de Wit and N. Fixot. "Robot Control via Robust Estimated State Feedback". *IEEE Trans. Autom. Control.* vol. 36, number 12, pp. 1497–1501, December, 1991.

[Chen 1989] Y. Chen. "Replacing a PID Controller by a Lag-Lead Compensator for a Robot: A Frequency Response Approach". *IEEE Trans. Robot. Autom.* vol. 5, number 2, pp. 174–182, April, 1989.

[Chen et al. 1990] Y-F Chen and T. Mita and S. Wahui. "A New and simple Algorithm for Sliding Mode Control of Robot Arms". *IEEE Trans. Autom. Control.* vol. 35, number 7, pp. 828–829, 1990.

[Corless 1989] M. Corless. "Tracking Controllers for Uncertain Systems: Application to a Manutec R3 Robot". *J. Dyn. Syst. Meas. Control.* vol. 111, pp. 609–618, December, 1989.

[Craig 1988] J.J. Craig. "Adaptive Control of Mechanical Manipulators".*Addison-Wesley*, Reading, MA, 1988.

[Dawson et al. 1990] D.M. Dawson et. al. "Robust Control for the Tracking of Robot Motion". *Int. J. Control.* vol. 52, number 3, pp. 581–595, 1990.

[DeCarlo et al. 1988] R.A. DeCarlo and S.H. Zak and G.P. Matthews. "Variable Structure Control of Nonlinear Multivariable Systems". *IEEE Proc.* vol. 76, number 3, pp. 212–232, March, 1988.

[Freund 1982] E. Freund. "Fast Nonlinear Control with Arbitrary Pole-Placement for Industrial Robots and Manipulators". *Int. J. Robot. Res.* vol. 1, number 1, pp. 65–78, 1982.

[Gu and Loh 1988] Y-L. Gu and N.K. Loh. "Dynamic Modeling and Control by Utilizing an Imaginary Robot Model". *IEEE Trans. Rob. Aut.* vol. 4, number 5, 1988.

[Kazerooni 1989] H. Kazerooni. "Design and Analysis of a Statically Balanced Direct-Drive Manipulator". *IEEE Control Syst. Mag.* vol. 9, number 2, pp. 30–34, February, 1989.

[Luh 1983] J.Y.S. Luh. "Conventional Controller Design for Industrial Robots: A Tutorial". *IEEE Trans. Syst. Man Cybern.* vol. 13, pp. 298–316, May-June, 1983.

[Ortega and Spong 1988] R. Ortega and M.W. Spong. "Adaptive Motion Control of Rigid Robots: A Tutorial". *Proc. IEEE Conf. Decision Control*, Austin, TX, December, 1988.

[Slotine 1985] J-J.E. Slotine. "The Robust Control of Robot Manipulators". *Int. J. Rob. Res.* vol. 4, number 4, pp. 49–64, 1985.

[Slotine 1988] J-J.E. Slotine. "Putting Physics Back in Control: The Example of Robotics". *IEEE Contr. Syst. Mag.* vol. 8, number 7, pp. 12–17, December, 1988.

[Spong and Vidyasagar 1987] M.W. Spong and M. Vidyasagar. "Robust Linear Compensator Design for Nonlinear Robotic Control". *IEEE Trans. Robot. Autom.* vol. 3, number 4, pp. 345–351, August, 1987.

REFERENCES

[Spong et al. 1987] M.W. Spong and J.S. Thorp and J.M. Kleinwaks. "Robust Microprocessor Control of Robot Manipulators". *Automatica.* vol. 23, number 3, pp. 373-379, 1987.

[Sugie et al. 1988] T. Sugie et. al. "Robust Controller Design for Robot Manipulators". *Trans. ASME Dyn. Syst. Meas. Control.* vol. 110, number 1, pp. 94–96, March, 1988.

[Tarn et al. 1984] T.J. Tarn and A.K. Bejczy and A. Isidori and Y. Chen. "Nonlinear Feedback in Robot Arm Control". *Proc. IEEE Conf. Decision Control*, Las Vegas, NV. December, 1984.

[Utkin 1977] V.I. Utkin. "Variable Structure Systems with Sliding Modes". *IEEE TRans. Autom. Control.* vol. 22, pp. 212–222, April, 1977.

[Vidyasagar 1985] M. Vidyasagar. "Control Systems Synthesis: A Factorization Approach". *MIT Press*, Cambridge, MA. 1985.

[Yang and Tzeng 1986] D.C-H. Yang and S.W. Tzeng. "Simplification and Linearization of Manipulator Dynamics by the Design of Inertia Distribution". *Int. J. Rob. Res.* vol. 5, number 3, pp. 120–128, 1986.

[Yeung and Chen 1988] K.S. Yeung and Y.P. Chen. "A New Controller Design for Manipulators Using the Theory of Variable Structure Systems". *IEEE Trans. Autom. Control.* vol. 33, number 2, pp. 200–206, February, 1988.

[Young 1978] K-K.D. Young. "Controller Design for a Manipulator Using theory of Variale Structure Systems". *IEEE Trans. Syst. Man Cybern.* vol. 8, number 2, pp. 210–218, February, 1978.

PROBLEMS

Section 5.2

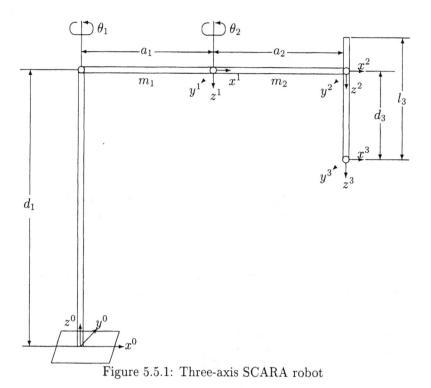

Figure 5.5.1: Three-axis SCARA robot

5.2-1 WE consider the three-axis SCARA robot shown in Figure 5.5.1, where all links are assumed to be thin, homogeneous rods of mass m_i and length a_i, $i = 1, 2, 3$. Then the dynamics are given by:

$$\begin{aligned} \tau_1 = & \left[\left(\frac{m_1}{3} + m_2 + m_3\right)a_1^2 + (m_2 + 2m_3)a_1a_2C_2 + \left(\frac{m_2}{3} + m_3\right)a_2^2\right]\ddot{q}_1 \\ & - \left[\left(\frac{m_2}{3} + m_3\right)a_1a_2C_2 + \left(\frac{m_2}{3} + m_3\right)a_2^2\ddot{q}_2\right] \\ & - a_1a_2S_2\left[(m_2 + 2m_3)\dot{q}_1\dot{q}_2 - \left(\frac{m_2}{3} + m_3\right)\dot{q}_2^2\right] \end{aligned}$$

$$\tau_2 = -\left[\left(\frac{m_2}{3} + m_3\right)a_1 a_2 C_2 + \left(\frac{m_2}{3} + m_3\right)a_2^2\right]\ddot{q}_1$$
$$+ \left(\frac{m_2}{3} + m_3\right)a_2^2 \ddot{q}_2$$
$$+ \left(\frac{m_2}{3} + m_3\right)a_1 a_2 S_2 \dot{q}_1^2$$

$$\tau_3 = m_3 \ddot{q}_3 - m_3 g$$

and let

$$m_1 = 2\text{kg}, \quad m_2 = 2\text{kg}, \quad m_3 = 1\text{kg},$$
$$a_1 = 2\text{m}, \quad a_2 = 1\text{m}, \quad l_3 = 1\text{m},$$
$$g = 9.8 m/s^2.$$

Note that the first two joints are decoupled from the last one. Find α, β_i, and μ_i defined in (5.2.8)-(5.2.11), assuming that $m_{12}(q) = m_{21}(q) = 0$ and that $n_1(q, \dot{q}) = n_2(q, \dot{q}) = 0$.

5.2-2 Choose a desired set point $q_{1d} = 45$ deg, $q_{2d} = 90$ deg, and $q_3 = \frac{1}{2}$ for the robot in Problem 5.2-1. Design an SPR controller as described in (5.2.14)-(5.2.19). Also, find a value of a to satisfy Theorem 5.2.1.

5.2-3 Let $q_{1d} = 10 \sin t$, $q_{2d} = 10 \cos t$, and $q_{3d} = \frac{1}{2}$ for the robot in Problem 5.2-1.

1. Design an SPR controller as described in (5.2.14)-(5.2.19).

2. Let the desired trajectory now be $q_{1d} = \sin t$, $q_{2d} = \cos t$, and $q_{3d} = \frac{1}{2}$. Study the effect that decreasing q_d has on a and on the trajectory error.

5.2-4 Choose a value of k_v that satisfies condition (5.2.28) for the robot in Problem 5.2-1 and the trajectory in Problem 5.2-3(b). What happens to k_v and to the trajectory errors if m_1, m_2, m_3 increase to 20?

5.2-5 Design a dynamic controller similar to that of Example 5.2.3 for the robot in Problem 5.2-1 and the trajectory in Problem 5.2-3(a). Compare the performance for $k = 10$ and $k = 50$.

5.2-6 Find the value of k that satisfies inequality (1) for you design in Problem 5.2-5. What happens to your conditions if both the velocity terms and gravity terms are available to feedback (i.e., $\beta_i = 0$)?

5.2-7 Consider the robot in Problem 5.2-1 and the set point of Problem 5.2-2. Also, assume that all $\beta_i = 0$ so that velocity and gravity terms are available to feedback. Find a gain k to satisfy (5.2.36) and implement the resulting controller.

5.2-8 Repeat Problem 5.2-7 with the trajectory of Problem 5.2-3(a).

5.2-9 Design a controller similar to the one in Example 5.2.5 for the robot of Problem 5.2-1 to follow the trajectory of Problem 5.2-3(a). Choose the same parameters used in that example and compare the resulting behavior to a set of parameters of your choosing.

Section 5.3

5.3-1 Design a controller similar to the one in Example 5.3.1 for the robot of Problem 5.2-1 to follow the trajectory of Problem 5.2-3(a). Choose the same parameters used in that example and compare the resulting behavior to a set of parameters of your choosing.

5.3-2 Consider the robot of Problem 5.2-1 with the desired set point of Problem 5.2-2. Design a variable-structure controller as described in Theorem 5.3.2. You may want to start your design with the parameter values in Example 5.3.2.

5.3-3 Repeat Problem 5.3-2 for the trajectory described in Problem 5.2-3(a).

5.3-4 Consider the robot of Problem 5.2-1 with the desired set point of Problem 5.2-2. Design a variable-structure controller as described in Theorem 5.3.3. You may want to start your design with the parameter values in Example 5.3.3.

5.3-5 Repeat Problem 5.3-4 for the trajectory described in Problem 5.2-3(a).

5.3-6 Consider the robot of Problem 5.2-1 with the desired set point of Problem 5.2-2. Design a saturation-type controller as described in Theorem 5.3.4. You may want to start your design with the parameter values in Example 5.3.5.

5.3-7 Repeat Problem 5.3-6 for the trajectory described in Problem 5.2-3(a).

5.3-8 Consider the robot of Problem 5.2-1 with the desired set point of Problem 5.2-2. Design a saturation-type controller as described in Theorem 5.3.5. You man want to start your design with the parameter values in Example 5.3.6.

5.3-9 Repeat Problem 5.3-8 for the trajectory described in Problem 5.2-3(a).

Chapter 6

Adaptive Control of Robotic Manipulators

In this chapter adaptive controllers are formulated by separating unknown constant parameters from known functions in the robot dynamic equation. The type of stability for each adaptive control strategy is discussed at length to motivate the formulation of the controllers. Some issues regarding parameter error convergence, persistency of excitation, and robustness are also discussed.

6.1 Introduction

The problem of designing adaptive control laws for rigid-robot manipulators that ensure asymptotic trajectory tracking has interested researchers for many years. The development of effective adaptive controllers represents an important step toward high-speed/precision robotic applications. Even in a well-structured industrial facility, robots may face uncertainty regarding the parameters describing the dynamic properties of the grasp load (e.g., unknown moments of inertia). Since these parameters are difficult to compute or measure, they limit the potential for robots to manipulate accurately objects of considerable size and weight. It has recently been recognized that the accuracy of conventional approaches in high-speed applications is greatly affected by parametric uncertainties.

To compensate for this parametric uncertainty, many researchers have proposed adaptive strategies for the control of robotic manipulators. An advantage of the adaptive approach over the robust control strategies discussed in Chapter 5 is that the accuracy of a manipulator carrying unknown loads improves with time because the adaptation mechanism continues extracting information from the tracking error. Therefore, adaptive controllers can give consistent performance in the face of load variations.

It is only recently that adaptive control results have included rigorous proofs for global convergence of the tracking error. Now that the existence of globally convergent adaptive control laws has been established, it is difficult to justify control schemes based on approximate models, local linearization techniques, or slowly time varying assumptions. In the control literature there also seems to be no general agreement as to what constitutes an adaptive control algorithm; therefore, in this chapter, the discussion will be limited to control schemes that explicitly incorporate parameter estimation in the control law.

6.2 Adaptive Control by a Computed-Torque Approach

In Chapter 3 we motivated the use of the computed-torque control law for controlling the robotic manipulator dynamics given by

$$\tau = M(q)\ddot{q} + V_m(q,\dot{q})\dot{q} + G(q) + F(\dot{q}). \qquad (6.2.1)$$

This motivation was actually quite simple. Specifically, if there are some nonlinear dynamics which one does not wish to deal with, one can change the nonlinear control problem to a linear control problem by directly canceling the nonlinearities. There is a wealth of available knowledge for controlling linear systems; therefore, if exact knowledge of the robot model is available, there is not much need for sophisticated nonlinear control techniques.

Approximate Computed-Torque Controller

Of course, in reality, we never have exact knowledge of the robot model due to many problems associated with model formulation. Two common uncertainties that do not allow exact model cancellation in robotic applications are unknown link masses due to payload disturbances and unknown friction coefficients. One way of dealing with these types of *parametric* uncertainties would be to use the computed-torque controller given in Chapter 3 with some fixed estimate of the unknown parameters in place of the actual parameters. This approximate computed-torque controller would have the form

$$\tau = \hat{M}(q)(\ddot{q}_d + K_v\dot{e} + K_p e) + \hat{V}_m(q,\dot{q})\dot{q} + \hat{G}(q) + \hat{F}(\dot{q}). \qquad (6.2.2)$$

6.2 Adaptive Control by a Computed-Torque Approach

where the superscript " $\hat{}$ " denotes the estimated dynamics with the unknown actual parameters replaced by the parameter estimates, K_v and K_p are control gain matrices, q_d is used to denote the desired trajectory, and the tracking error e is defined by

$$e = q_d - q.$$

We now illustrate the approximate computed-torque controller by examining an example.

EXAMPLE 6.2-1: Approximate Computed-Torque Controller

We wish to design and simulate an approximate computed-torque controller for the two-link arm given in Figure 6.2.1 (see Chapter 2 for the two-link revolute robot arm dynamics). Assuming that the friction is negligible, the link lengths are exactly known, and the masses m_1 and m_2 are known to be in the regions 0.8 ± 0.05 kg and 2.3 ± 0.1 kg, respectively, a possible approximated computed-torque controller can be written as

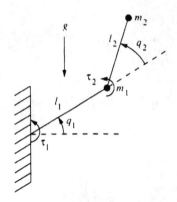

Figure 6.2.1: Two-link planar arm.

$$\tau_1 = \left(2\hat{m}_2 l_1 l_2 c_2 + \hat{m}_2 l_2^2 + (\hat{m}_1 + \hat{m}_2) l_1^2\right)(\ddot{q}_{d1} + k_{v1}\dot{e}_1 + k_{p1}e_1)$$
$$+ \left(\hat{m}_2 l_1 l_2 c_2 + \hat{m}_2 l_2^2\right)(\ddot{q}_{d2} + k_{v2}\dot{e}_2 + k_{p2}e_2) - \hat{m}_2 l_1 l_2 s_2 \dot{q}_2^2$$
$$- 2\hat{m}_2 l_1 l_2 s_2 \dot{q}_1 \dot{q}_2 + \hat{m}_2 l_2 g c_{12} + (\hat{m}_1 + \hat{m}_2) l_1 g c_1 \qquad (1)$$

$$\tau_2 = \left(\hat{m}_2 l_2^2 + \hat{m}_2 l_1 l_2 c_2\right)(\ddot{q}_{d1} + k_{v1}\dot{e}_1 + k_{p1}e_1) + \hat{m}_2 l_2 g c_{12}$$
$$+ \hat{m}_2 l_2^2 (\ddot{q}_{d2} + k_{v2}\dot{e}_2 + k_{p2}e_2) + \hat{m}_2 l_1 l_2 s_2 \dot{q}_1^2, \qquad (2)$$

where $l_1 = l_2 = 1$ m and g is the gravitational constant. We choose $\hat{m}_1 = 0.85$ kg and $\hat{m}_2 = 2.2$ kg since the actual values are assumed to be unknown. After substituting the control law above into the two-link robot dynamics, we can form the error system

$$\ddot{e} + K_v \dot{e} + K_p e = \hat{M}^{-1}(q) W(q, \dot{q}, \ddot{q}) \tilde{\varphi}, \qquad (3)$$

where $\hat{M}^{-1}(q)$ is the inverse of the inertia matrix $M(q)$ with m_1 and m_2 replaced by $\hat{m}_1$, and $\hat{m}_2$, respectively. The matrix $W(q, \dot{q}, \ddot{q})$, sometimes called the *regression* matrix [Craig 1985], is a 2×2 matrix given by

$$W(q, \dot{q}, \ddot{q}) = \begin{bmatrix} W_{11} & W_{12} \\ W_{21} & W_{22} \end{bmatrix}, \qquad (4)$$

where

$$W_{11} = l_1^2 \ddot{q}_1 + l_1 g c_1,$$
$$W_{12} = l_2^2 (\ddot{q}_1 + \ddot{q}_2) + l_1 l_2 c_2 (2\ddot{q}_1 + \ddot{q}_2) + l_1^2 \ddot{q}_1 - l_1 l_2 s_2 \dot{q}_2^2$$
$$\quad - 2l_1 l_2 s_2 \dot{q}_1 \dot{q}_2 + l_2 g c_{12} + l_1 g c_1,$$
$$W_{21} = 0,$$
$$W_{22} = l_1 l_2 c_2 \ddot{q}_1 + l_1 l_2 s_2 \dot{q}_1^2 + l_2 g c_{12} + l_2^2 (\ddot{q}_1 + \ddot{q}_2).$$

The vector $\tilde{\varphi}$, called the parameter error vector, is a 2×1 vector given by

$$\tilde{\varphi} = \begin{bmatrix} \tilde{\varphi}_1 \\ \tilde{\varphi}_2 \end{bmatrix}, \qquad (5)$$

where

$$\tilde{\varphi}_1 = m_1 - \hat{m}_1$$

and

$$\tilde{\varphi}_2 = m_2 - \hat{m}_2.$$

The associated tracking error 2×1 vector and 2×2 gain matrices in (3) are given by

6.2 Adaptive Control by a Computed-Torque Approach

$$e = \begin{bmatrix} e_1 \\ e_2 \end{bmatrix}, \quad K_v = \begin{bmatrix} k_{v1} & 0 \\ 0 & k_{v2} \end{bmatrix}, \quad \text{and} \quad K_p = \begin{bmatrix} k_{p1} & 0 \\ 0 & k_{p2} \end{bmatrix}.$$

We cannot use the error system (3) to select the values for K_v and K_p with standard linear control methods since the right-hand side of the error system is composed of nonlinear functions of q, $\dot{q}$, and $\ddot{q}$. For the approximate computed-torque controller, simulation can be used to select the appropriate values for K_v and K_p. A good starting point for selecting the values of K_p and K_v for the approximate computed-torque controller is to use the same values, as the computed-torque scheme would dictate, for a given desired damping ratio and natural frequency.

For $m_1 = 0.8$ kg and $m_2 = 2.3$ kg, the approximate computed torque controller (1)-(2) was simulated with $q(0) = \dot{q}(0) = 0$, with the controller gains set at

$$K_p = K_v = \begin{bmatrix} 4 & 0 \\ 0 & 4 \end{bmatrix} \qquad (6)$$

and with a desired trajectory of

$$q_{d1} = q_{d2} = \sin t. \qquad (7)$$

The tracking error is depicted in Figure 6.2.2. As illustrated by these figures, the tracking error remains bounded rather than going to zero. The reason for this type of bounded tracking error performance is that the error system given by (3) is constantly being excited by the dynamics on the right-hand side of (3).

■

Adaptive Computed-Torque Controller

For unknown parametric quantities such as link masses or friction coefficients, the approximate computed-torque controller simply substitutes a fixed estimate $\hat{\varphi}$ for the unknown parametric quantities. From Equation (3) in Example 6.2.1, one can clearly see that if the parameter error vector $\tilde{\varphi}$ is equal to zero, the tracking error can be shown to be asymptotically stable. For many applications, we cannot assume that $\tilde{\varphi}$ is equal to zero. For example, in the case of an unknown payload mass attached to the end effector of a robot, there will always be an unknown parametric quantity related to the payload mass that would appear in the parameter error vector.

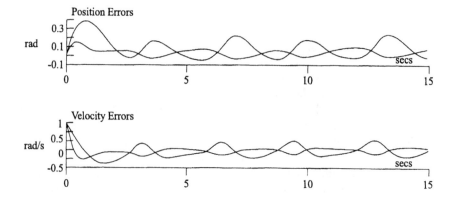

Figure 6.2.2: Simulation of approximate computed-torque controller.

The adaptive control strategy can, heuristically, be motivated by reasoning that one could expect better tracking performance if the parameter estimate was adjusted as the robot manipulator moves instead of always being a fixed quantity. That is, it seems reasonable to attempt to change our parameter estimates based on an adaptive update rule that would be a function of the robot configuration and the tracking error. The question then becomes: How do we formulate an adaptive control strategy, and how does this adaptive update rule affect the stability of the tracking error? The answer to both of these questions is that the adaptive update rule is formulated from the stability analysis of the tracking error system. That is, we ensure stability of the tracking error system by formulating the adaptive update rule and by analyzing the stability of the tracking error system at the same time.

The first adaptive control strategy that we will examine is the method outlined in [Craig 1985]. The adaptive computed-torque controller is the same as the approximate computed-torque controller (6.2.2) with the addition of an adaptive update rule for adjusting the parameter estimates. This adaptive controller is based on the fact that the parameters appear linearly in the robot model (see Chapter 2). That is, the robot dynamics (6.2.1) can be written in the form

$$W(q, \dot{q}, \ddot{q}) \varphi = M(q) \ddot{q} + V_m(q, \dot{q}) \dot{q} + G(q) + F(\dot{q}), \qquad (6.2.3)$$

where $W(q, \dot{q}, \ddot{q})$ is an $n \times r$ matrix of known time functions and φ is an $r \times 1$ vector of unknown constant parameters. This property is crucial for the type of adaptive control that Craig formulated in that it illustrates the separation of unknown parameters and the known time functions. The

6.2 Adaptive Control by a Computed-Torque Approach

reason that the robot dynamics can be separated in this form is that the robot dynamics are linear in the parameters expressed in the vector form φ. This separation of unknown parameters and known time functions will be used in the formulation of the adaptive update rule and also in the stability analysis of the tracking error system.

The first step in the study of the adaptive computed-torque controller is to form the tracking error system. Note that, by using (6.2.3), we may write the robot dynamic Equation given by (6.2.1) as

$$\tau = W(q, \dot{q}, \ddot{q})\varphi. \qquad (6.2.4)$$

From [Craig 1985], the adaptive computed-torque controller is given by

$$\tau = \hat{M}(q)(\ddot{q}_d + K_v\dot{e} + K_p e) + \hat{V}_m(q, \dot{q})\dot{q} + \hat{G}(q) + \hat{F}(\dot{q}). \qquad (6.2.5)$$

It is easy to see from our definition of the tracking error how (6.2.5) can be written as

$$\tau = \hat{M}(q)(\ddot{e} + K_v\dot{e} + K_p e) + \hat{M}(q)\ddot{q} + \hat{V}_m(q, \dot{q})\dot{q} + \hat{G}(q) + \hat{F}(\dot{q}). \qquad (6.2.6)$$

By utilizing (6.2.3), (6.2.6) can be written as

$$\tau = \hat{M}(q)(\ddot{e} + K_v\dot{e} + K_p e) + W(q, \dot{q}, \ddot{q})\hat{\varphi}, \qquad (6.2.7)$$

where $\hat{\varphi}$ is an $n \times 1$ vector used to represent a time-varying estimate of the unknown constant parameters. Substituting (6.2.7) into (6.2.4), we can form the *tracking error system*

$$\ddot{e} + K_v\dot{e} + K_p e = \hat{M}^{-1}(q) W(q, \dot{q}, \ddot{q}) \tilde{\varphi}, \qquad (6.2.8)$$

where the *parameter error* is

$$\tilde{\varphi} = \varphi - \hat{\varphi}. \qquad (6.2.9)$$

Now for convenience, rewrite (6.2.8) in the state-space form

$$\dot{\mathbf{e}} = A\mathbf{e} + BM^{-1}(q) W(q, \dot{q}, \ddot{q}) \tilde{\varphi}, \qquad (6.2.10)$$

where the *tracking error vector* is

$$\mathbf{e} = \begin{bmatrix} e \\ \dot{e} \end{bmatrix}$$

and

$$B = \begin{bmatrix} O_n \\ I_n \end{bmatrix}, \quad A = \begin{bmatrix} O_n & I_n \\ -K_p & -K_v \end{bmatrix},$$

with I_n being the $n \times n$ identity matrix, and O_n being the $n \times n$ zero matrix.

Now that the tracking error system has been formed, we use Lyapunov stability analysis (see Chapter 1) to show that the tracking error vector **e** is asymptotically stable with the right choice of adaptive update law. We first select the positive-definite Lyapunov-like function

$$V = \mathbf{e}^T P \mathbf{e} + \tilde{\varphi}^T \Gamma^{-1} \tilde{\varphi}, \tag{6.2.11}$$

where P is a $2n \times 2n$ positive-definite, constant, symmetric matrix, and Γ is a diagonal, positive-definite $r \times r$ matrix. That is, Γ can be written as

$$\Gamma = diag(\gamma_1, \gamma_2, \ldots, \gamma_r),$$

where the γ_i's are positive scalar constants.

Differentiating (6.2.11) with respect to time yields

$$\dot{V} = \mathbf{e}^T P \dot{\mathbf{e}} + \dot{\mathbf{e}}^T P \mathbf{e} + 2\tilde{\varphi}^T \Gamma^{-1} \dot{\tilde{\varphi}}. \tag{6.2.12}$$

It is important to note in (6.2.12) that we have used the fact that

$$\left[\tilde{\varphi}^T \Gamma^{-1} \dot{\tilde{\varphi}} \right]^T = \tilde{\varphi}^T \Gamma^{-1} \dot{\tilde{\varphi}} \tag{6.2.13}$$

since $\Gamma = \Gamma^T$. (Note that a scalar quantity can always be transposed.) Now, substituting for $\dot{\mathbf{e}}$ from (6.2.10) into (6.2.12) yields

$$\dot{V} = \mathbf{e}^T P \left(A\mathbf{e} + B\hat{M}^{-1}(q) W(\cdot) \tilde{\varphi} \right) \\ + \left(A\mathbf{e} + B\hat{M}^{-1}(q) W(\cdot) \tilde{\varphi} \right)^T P \mathbf{e} + 2\tilde{\varphi}^T \Gamma^{-1} \dot{\tilde{\varphi}}. \tag{6.2.14}$$

Combining terms in (6.2.14) and using the scalar transportation property gives

$$\dot{V} = -\mathbf{e}^T Q \mathbf{e} + 2\tilde{\varphi}^T \left(\Gamma^{-1} \dot{\tilde{\varphi}} + W^T(\cdot) \hat{M}^{-1}(q) B^T P \mathbf{e} \right), \tag{6.2.15}$$

where Q is a positive-definite, symmetric matrix that satisfies the Lyapunov equation

$$A^T P + PA = -Q. \tag{6.2.16}$$

6.2 Adaptive Control by a Computed-Torque Approach

From Chapter 1 we note that for stability, it is always desirable to have $\dot{V}$ at least negative semidefinite; therefore, the choice of adaptation update rule becomes obvious. Specifically, by substituting

$$\dot{\tilde{\varphi}} = -\Gamma W^T(\cdot)\hat{M}^{-1}(q) B^T P \mathbf{e}, \qquad (6.2.17)$$

(6.2.15) becomes

$$\dot{V} = -\mathbf{e}^T Q \mathbf{e}. \qquad (6.2.18)$$

To determine explicitly the type of stability, we must do further analysis; however, note first that (6.2.17) gives the adaptive update rule for the parameter estimate vector $\hat{\varphi}$ since $\dot{\varphi}$ is equal to zero. That is, by recalling that the actual unknown parameters are *constant*, we can substitute (6.2.9) into (6.2.17) to obtain the *adaptive update rule*:

$$\dot{\hat{\varphi}} = \Gamma W^T(\cdot)\hat{M}^{-1}(q) B^T P \mathbf{e} \qquad (6.2.19)$$

for the parameter estimate vector $\hat{\varphi}$.

Before examining the type of stability for the tracking error system, we note that Craig modified the adaptation update rule in (6.2.19) to prevent a circular argument in the stability analysis [Craig 1985]. Specifically, the parameter estimates are forced to remain within some known region. That is, if parameter estimates drift out of a known region, they are reset to within a known region. By resetting the parameter estimates in "software," we are guaranteed that the parameter estimates $\hat{\varphi}$ remain bounded.

We now detail the stability for the tracking error. Since $\dot{V}$ is negative semidefinite and V is lower bounded by zero, V remains upper bounded in the time interval $[O, \infty)$; furthermore,

$$\lim_{t \to \infty} V = V_\infty, \qquad (6.2.20)$$

where V_∞ is a positive scalar constant. Since V is upper bounded, it is obvious from the definition of V given in (6.2.11) that $\mathbf{e}$ and $\tilde{\varphi}$ are bounded, which also means that q, $\dot{q}$, and $\hat{\varphi}$ are bounded. Note that we have already assumed that $\hat{\varphi}$ is bounded, and we will always assume that the desired trajectory and its first two derivatives are bounded.

Now, from the robot Equation (6.2.1), it is clear that

$$\ddot{q} = M^{-1}(q)(\tau - V_m(q,\dot{q})\dot{q} - G(q) - F(\dot{q})); \qquad (6.2.21)$$

therefore, $\ddot{q}$ is bounded since $\ddot{q}$ and τ depend only on the bounded quantities q, $\dot{q}$, and $\hat{\varphi}$. If $\ddot{q}$ is bounded, (6.2.10) shows that $\dot{\mathbf{e}}$ is bounded. Since $\dot{\mathbf{e}}$ is bounded, we can state from (6.2.18) that $\ddot{V}$ is bounded. Therefore, since

V is lower bounded by zero, $\dot{V}$ is negative semidefinite, and $\ddot{V}$ is bounded, then by Barbalat's lemma (see Chapter 1),

$$\lim_{t \to \infty} \dot{V} = 0$$

which means that by the Rayleigh-Ritz Theorem (see Chapter 1)

$$\lim_{t \to \infty} \lambda_{\min}\{Q\} \|\mathbf{e}\|^2 = 0 \quad \text{which means that} \quad \lim_{t \to \infty} \mathbf{e} = 0. \quad (6.2.22)$$

Table 6.2.1: Adaptive Computed-Torque Controller

Torque Controller:
$$\tau = \hat{M}(q)(\ddot{q}_d + K_v \dot{e} + K_p e) + \hat{V}_m(q, \dot{q})\dot{q} + \hat{G}(q) + \hat{F}(\dot{q})$$

Update Rule:
$$\dot{\hat{\varphi}} = \Gamma W^T(q, \dot{q}, \ddot{q}) \hat{M}^{-1}(q) B^T P \mathbf{e}$$

where

$$\mathbf{e} = \begin{bmatrix} e \\ \dot{e} \end{bmatrix}, \quad B = \begin{bmatrix} O_n \\ I_n \end{bmatrix}, \quad A = \begin{bmatrix} O_n & I_n \\ -K_p & -K_v \end{bmatrix}$$

$$W(q, \dot{q}, \ddot{q})\hat{\varphi} = \hat{M}(q)\ddot{q} + \hat{V}_m(q, \dot{q})\dot{q} + \hat{G}(q) + \hat{F}(\dot{q})$$

$$A^T P + PA = -Q$$

for some positive-definite, symmetric matrices P and Q.

Stability:

Tracking error vector $\mathbf{e}$ is asymptotically stable.

Restrictions:

Parameter resetting method is required. Measurement of $\ddot{q}$ is required.

The information given in (6.2.22) informs us that the tracking error vector $\mathbf{e}$ is asymptotically stable. But what of the parameter error $\tilde{\varphi}$? Does it also converge to zero? From our analysis, all we can say about the parameter error is that it remains bounded if $\hat{M}^{-1}(q)$ exists. Indeed, this places a restriction on the parameter update law given in (6.2.19). That is, we must use the parameter resetting method discussed earlier to

6.2 Adaptive Control by a Computed-Torque Approach

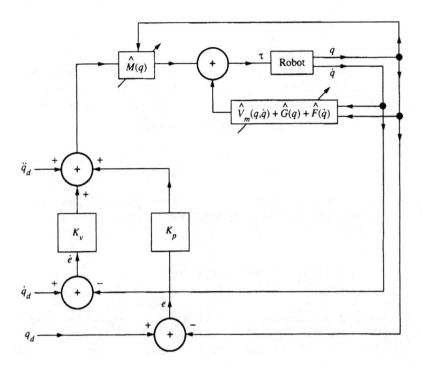

Figure 6.2.3: Block diagram of the adaptive computed-torque controller.

ensure that poor parameter estimates do not cause the inverse of $\hat{M}(q)$ to explode. In [Craig 1985] a possible method for ensuring that the parameter estimates and $\hat{M}^{-1}(q)$ remain bounded is outlined; furthermore, he shows how this method does not interfere with the stability result delineated by (6.2.22). We will not discuss this parameter estimate resetting method and the resulting stability proof since we will show later in this chapter how Slotine and Li used a more judicious choice of Lyapunov function to remove the need for resetting the parameter estimates, and at the same time remove the need for acceleration measurements in the regression matrix $W(q, \dot{q}, \ddot{q})$! The adaptive computed-torque controller is summarized in Table 6.2.1 and depicted in Figure 6.2.3.

We now present an example to illustrate how Table 6.2.1 can be used to generate adaptive controllers for robotic manipulators.

EXAMPLE 6.2-2: Adaptive Computed-Torque Controller

It is desired to design and simulate the adaptive computed-torque controller given in Table 6.2.1 for the two-link arm given in Figure 6.2.1.

Assuming that the friction is negligible and that the link lengths are exactly known, the adaptive computed-torque controller can be written in the same form as that given in Example 6.2.1, with the exception that we must find the update rules for $\hat{m}_1$ and $\hat{m}_2$. That is, we use Equations (1) and (2) in Example 6.2.1 for the joint torque control and then formulate the update rule for m_1 and m_2 according to Table 6.2.1.

For simplicity, in this example we select the servo gains as

$$K_v = k_v I_n \quad \text{and} \quad K_p = k_p I_n, \tag{1}$$

where k_v and k_p are positive, scalar constants and for this case I_n is the 2×2 identity matrix. We propose that the matrix P in Table 6.2.1 be selected as

$$P = \begin{bmatrix} P_1 I_n & P_2 I_n \\ P_2 I_n & P_3 I_n \end{bmatrix} = \frac{1}{2} \begin{bmatrix} (K_p + \frac{1}{2}k_v) I_n & \frac{1}{2}I_n \\ \frac{1}{2}I_n & I_n \end{bmatrix}. \tag{2}$$

Note that P is symmetric, and that it is positive definite if k_v is selected to be greater than 1 (see the Gerschgorin Theorem in Chapter 1). To see if our selection of P gives a positive-definite Q, perform the matrix operation

$$A^T P + PA = -Q \tag{3}$$

$$Q = \begin{bmatrix} \frac{1}{2}k_p I_n & O_n \\ O_n & (K_v + \frac{1}{2}) I_n \end{bmatrix}. \tag{4}$$

Since we have already restricted $k_v > 1$, it can be verified that Q is a positive definite, symmetric matrix. We note here that the process of finding a positive definite, symmetric P and Q for the general Lyapunov approach is not always an easy task.

Now that we have found an appropriate P, we can formulate the adaptive update rule given in Table 6.2.1. The associated parameter estimate vector is

$$\hat{\varphi} = \begin{bmatrix} \hat{m}_1 \\ \hat{m}_2 \end{bmatrix}$$

with update rules

6.3 Adaptive Control by an Inertia-Related Approach

$$\dot{\hat{m}}_1 = \gamma_1[(W_{11}MI_{11} + W_{21}MI_{21})(P_2 e_1 + P_3 \dot{e}_1) \\ + (W_{11}MI_{21} + W_{21}MI_{22})(P_2 e_2 + P_3 \dot{e}_2)] \quad (5)$$

and

$$\dot{\hat{m}}_2 = \gamma_2[(W_{12}MI_{11} + W_{22}MI_{21})(P_2 e_1 + P_3 \dot{e}_1) \\ + (W_{12}MI_{21} + W_{22}MI_{22})(P_2 e_2 + P_3 \dot{e}_2)], \quad (6)$$

where

$$MI_{11} = \frac{1}{\Delta}\left(\hat{m}_2 l_2^2\right),$$

$$MI_{21} = -\frac{1}{\Delta}\left(\hat{m}_2 l_1 l_2 c_2 + \hat{m}_2 l_2^2\right),$$

$$MI_{22} = \frac{1}{\Delta}\left(2\hat{m}_2 l_1 l_2 c_2 + \hat{m}_2 l_2^2 + (\hat{m}_1 + \hat{m}_2) l_1^2\right),$$

$$\Delta = \left(2\hat{m}_2 l_1 l_2 c_2 + \hat{m}_2 l_2^2 + (\hat{m}_1 + \hat{m}_2) l_1^2\right)\left(\hat{m}_2 l_2^2\right) - \left(\hat{m}_2 l_2^2 + \hat{m}_2 l_1 l_2 c_2\right)^2,$$

the P_i's are defined in (2), and the quantities forming the regression matrix W_{ii}'s are found in Example 6.2.1.

For $m_1 = 0.8$ kg and $m_2 = 2.3$ kg, this adaptive computed-torque controller was simulated with $k_v = 50$, $k_p = 125$, $\gamma_1 = 500$, $\gamma_2 = 500$, $\hat{m}_1(0) = 0.85$, $\hat{m}_2(0) = 2.2$, and with the same desired trajectory and initial joint conditions used in Example 6.2.1. The tracking error and mass estimates are depicted in Figure 6.2.4. As illustrated by this figure, the tracking error goes to zero, and the parameter estimates remain bounded as predicted by the theory. Note that we did not take any special precautions in this simulation to ensure the existence of $\hat{M}^{-1}(q)$. We could have developed some sort of procedure to guarantee that $\hat{M}^{-1}(q)$ existed; however, this is not really needed since we show in the next section how to eliminate this restriction. ∎

6.3 Adaptive Control by an Inertia-Related Approach

In Section 6.2 we showed how adaptive control can be used to compensate for parametric uncertainties. This led Craig to develop the adaptive

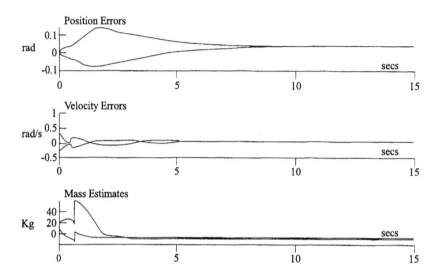

Figure 6.2.4: Simulation of the adaptive computed-torque controller.

computed-torque controller. We also gave the two restrictions required for the implementation of the adaptive computed-torque controller [i.e., the need for measuring $\ddot{q}$ and ensuring that $\hat{M}^{-1}(q)$ exists]. Both of these restrictions can be quite cumbersome. For example, most industrial robots have only position and velocity sensors, and since differentiation of velocity is not desirable in general, we must add additional costly sensors for measuring $\ddot{q}$. Furthermore, if large, unknown payloads relative to the manipulator's weight are being lifted by the manipulator, it may be extremely difficult to ensure that $\hat{M}^{-1}(q)$ exists.

After researchers reexamined the structure of the adaptive computed-torque controller, they began to wonder if all the available information about the robot manipulator was being used in designing adaptive control schemes. That is, are all the properties inherent to a mechanical manipulator being exploited? In [Arimoto and Miyazaki 1986] a proportional-derivative (PD) feedback controller with gravity compensation is proposed. It should be noted that this controller is not a product of the feedback linearization approach. Rather, this controller employs an inertia-related Lyapunov function in the stability analysis which utilizes physical properties inherent to a mechanical manipulator. After reexamining the adaptive computed-torque controller, one can see that the Lyapunov function used in the stability analysis is not inertia-related but is somewhat arbitrary.

6.3 Adaptive Control by an Inertia-Related Approach

Examination of a PD Plus Gravity Controller

One method [Slotine 1988] of motivating a PD plus gravity controller [Arimoto and Miyazaki 1986] is to write the manipulator dynamics in the conservation-of-energy form

$$\frac{1}{2}\frac{d}{dt}\left[\dot{q}^T M(q) \dot{q}\right] = \dot{q}^T \left(\tau - G(q) - F(\dot{q})\right), \qquad (6.3.1)$$

where the left-hand side of (6.3.1) is the derivative of the manipulator kinetic energy, and the right-hand side of (6.3.1) represents the power supplied from the actuators minus the power dissipated due to gravity and friction. Note that the Coriolis and centripetal terms are accounted for in (6.3.1) since these terms are related to the time derivative of the inertia matrix.

Suppose that we now want to design a constant set-point controller (i.e., $\dot{q}_d = 0$) for the system given in the conservation of energy form (6.3.1). To begin, we select the inertia-related Lyapunov function

$$V = 1/2 \dot{q}^T M(q) \dot{q} + 1/2 e^T K_p e. \qquad (6.3.2)$$

Since a Lyapunov function can be thought of heuristically as an energy function, this Lyapunov function seems quite reasonable. That is, the Lyapunov function is composed of the kinetic energy of the robotic manipulator system ($\frac{1}{2}\dot{q}^T M(q)\dot{q}$) and an additional energy damping term ($\frac{1}{2}e^T K_p e$). This energy damping term can be thought of as using physical springs so that the manipulator will be better behaved.

Differentiating (6.3.2) with respect to time yields

$$\dot{V} = \frac{1}{2}\frac{d}{dt}\left[\dot{q}^T M(q) \dot{q}\right] + e^T K_p \dot{e}. \qquad (6.3.3)$$

Substituting (6.3.1) into (6.3.3), we have

$$\dot{V} = \dot{q}^T \left(\tau - G(q) - F(\dot{q}) - K_p e\right), \qquad (6.3.4)$$

since we are solving the set-point control problem. Since it is desirable for $\dot{V}$ to be at least negative semidefinite, the control

$$\tau = G(q) + F(\dot{q}) + K_p e - K_v \dot{q} \qquad (6.3.5)$$

is motivated by the form of (6.3.4). That is, substituting (6.3.5) into (6.3.4) yields

$$\dot{V} = -\dot{q}^T K_v \dot{q}. \qquad (6.3.6)$$

The analysis above illustrates that V is decreasing for all time except for $\dot{q}$ equal to zero. We now use this information to illustrate that the desired set point q_d is achieved. That is, if $\dot{V} = 0$, then $\dot{q} = 0$, and hence $\ddot{q} = 0$; therefore, for $\dot{q} = \ddot{q} = 0$, we can utilize (6.2.1) and (6.3.5) to write the closed-loop system in the form $K_p e = 0$, which implies that $e = 0$. From Chapter 1, LaSalle's Theorem can now be used to show that the tracking error e is asymptotically stable.

Adaptive Inertia-Related Controller

Although the controller given in (6.3.5) utilizes the conservation of energy property, it has two disadvantages. First, the controller merely ensures that the manipulator reaches a desired set point. In general, a robot control designer must ensure that the manipulator tracks a desired time-varying trajectory. Second, the controller requires exact knowledge of any parameters associated with the robot manipulator model since the gravity and friction terms are included in the control law (6.3.5).

In [Slotine and Li 1985(a)] the conservation of energy formulation given in (6.3.1) is exploited to design an adaptive controller for the trajectory-following problem. This controller can be motivated in much the same way as our treatment of the PD plus gravity controller. In other words, we use the stability analysis to guide us in finding an adaptive controller. Since we are designing an adaptive trajectory-following controller, we should select a Lyapunov function that is a function of the tracking error and the parameter error. Slotine selected the inertia-related Lyapunov-like function

$$V = 1/2 r^T M(q)^r + 1/2 \tilde{\varphi}^T \Gamma^{-1} \tilde{\varphi}, \qquad (6.3.7)$$

where

$$r = \Lambda e + \dot{e}, \qquad (6.3.8)$$

with Γ defined as in (6.2.11), Λ defined as positive-definite, diagonal matrix such that

$$\Lambda = \operatorname{diag}(\lambda_1, \lambda_2, \ldots, \lambda_n),$$

and $\tilde{\varphi}$ defined as in (6.2.9). The auxiliary signal $r(t)$ given in (6.3.8) may be considered as a filtered tracking error.

After differentiating (6.3.7) with respect to time, we have

$$\dot{V} = r^T M(q) \dot{r} + 1/2 r^T \dot{M}(q) r + \tilde{\varphi}^T \Gamma^{-1} \dot{\tilde{\varphi}}. \qquad (6.3.9)$$

From (6.3.9) it is clear that we must substitute for the variable $\dot{r}$; therefore, we must write the robot equation in terms of the variable r. Using (6.2.1), the robot dynamics can be rewritten as

6.3 Adaptive Control by an Inertia-Related Approach

$$M(q)\dot{r} = Y(\cdot)\varphi - \tau - V_m(q,\dot{q})r, \qquad (6.3.10)$$

where

$$Y(\cdot)\varphi = M(q)(\ddot{q}_d + \Lambda\dot{e}) + V_m(q,\dot{q})(\dot{q}_d + \Lambda e) + G(q) + F(\dot{q}), \qquad (6.3.11)$$

and $Y(\cdot)$ is an $n \times r$ matrix of known time functions. This is the same type of parametric separation that was used in the formulation of the adaptive computed-torque controller; however, note that $Y(\cdot)$ is not a function of joint acceleration $\ddot{q}$!

Substituting (6.3.10) into (6.3.9) gives

$$\dot{V} = r^T \left(Y(\cdot)\varphi - \tau \right) + r^T \left(\tfrac{1}{2}\dot{M}(q) - V_m(q,\dot{q}) \right) r + \tilde{\varphi}^T \Gamma^{-1} \dot{\tilde{\varphi}}. \qquad (6.3.12)$$

Applying the skew-symmetric property (see Chapter 2), we can write (6.3.12) as

$$\dot{V} = r^T (Y(\cdot)\varphi - \tau) + \tilde{\varphi}^T \Gamma^{-1} \dot{\tilde{\varphi}}. \qquad (6.3.13)$$

Again, the stability analysis has guided us to a choice of torque controller and adaptive update rule. That is, if we select the torque control to be

$$\tau = Y(\cdot)\hat{\varphi} + K_v r, \qquad (6.3.14)$$

(6.3.13) becomes

$$\dot{V} = -r^T K_v r + \tilde{\varphi}^T \left(\Gamma^{-1}\dot{\tilde{\varphi}} + Y^T(\cdot) r \right). \qquad (6.3.15)$$

By selecting the adaptive update rule as

$$\dot{\hat{\varphi}} = -\dot{\tilde{\varphi}} = \Gamma Y^T(\cdot) r, \qquad (6.3.16)$$

(6.3.15) becomes

$$\dot{V} = -r^T K_v r. \qquad (6.3.17)$$

We now detail the type of stability for the tracking error. First, since $\dot{V}$ in (6.3.17) is negative semidefinite, we can state that V in (6.3.7) is upper bounded. Using the facts that V is upper bounded and that $M(q)$ is a positive-definite matrix (see Chapter 2), we can state that r and $\tilde{\varphi}$ are bounded. From the definition of r given in (6.3.8), we can use standard linear control arguments to state that e and $\dot{e}$ (and hence q and $\dot{q}$) are bounded. Since e, $\dot{e}$, r, and $\tilde{\varphi}$ are bounded, we can use (6.3.10) and

(6.3.14) to show that $\dot{r}$ (and hence $\ddot{V}$ obtained by differentiating (6.3.17)) is bounded. Second, note that since $M(q)$ is lower bounded, we can state that V given in (6.3.7) is lower bounded. Since V is lower bounded, $\dot{V}$ is negative semidefinite, and $\ddot{V}$ is bounded, we can use Barbalat's lemma (see Chapter 1) to state that

$$\lim_{t\to\infty} \dot{V} = 0, \qquad (6.3.18)$$

which means that by the Rayleigh-Ritz Theorem (see Chapter 1)

$$\lim_{t\to\infty} \lambda_{\min}\{K_v\}\|r\|^2 = 0 \quad \text{or} \quad \lim_{t\to\infty} r = 0. \qquad (6.3.19)$$

Note that (6.3.8) is a stable first-order differential equation driven by the "input" r; therefore, by standard linear control arguments and (6.3.19), we can write

$$\lim_{t\to\infty} e = 0 \quad \text{and} \quad \lim_{t\to\infty} \dot{e} = 0. \qquad (6.3.20)$$

This result informs us that the tracking errors e and $\dot{e}$ are asymptotically stable. Again, from the analysis above, all we can say about the parameter error is that it remains bounded. Later, we will show that the parameter error also converges to zero under certain conditions on the desired trajectory. The adaptive controller derived above is summarized in Table 6.3.1 and depicted in Figure 6.3.1.

After glancing through Table 6.3.1, we can see that the restrictions with regard to the adaptive computed-torque controller have been removed. Specifically, for the adaptive inertia-related controller, we do not require measurements of acceleration or any ad hoc adjustment of the parameter estimates to ensure that $\hat{M}^{-1}(q)$ exists.

We now present an example to illustrate how Table 6.3.1 can be used to design adaptive controllers for robotic manipulators.

EXAMPLE 6.3-1: Adaptive Inertia-Related Controller

We wish to design and simulate the adaptive inertia-related controller given in Table 6.3.1 for the two-link arm given in Figure 6.2.1. Assuming that the friction is negligible and the link lengths are exactly known, the adaptive inertia-related torque controller can be written as

$$\tau_1 = Y_{11}\hat{m}_1 + Y_{12}\hat{m}_2 + k_{v1}\dot{e}_1 + k_{v1}\lambda_1 e_1 \qquad (1)$$

and

6.3 Adaptive Control by an Inertia-Related Approach

Table 6.3.1: Adaptive Inertia-Related Controller

Torque Controller:
$$\tau = Y(\cdot)\hat{\varphi} + K_v \dot{e} + K_v \Lambda e$$

Update Rule:
$$\dot{\hat{\varphi}} = \Gamma Y^T(\cdot)(\Lambda e + \dot{e})$$

where
$$Y(\cdot)\hat{\varphi} = \hat{M}(q)(\ddot{q}_d + \Lambda \dot{e}) + \hat{V}_m(q,\dot{q})(\dot{q}_d + \Lambda e) + \hat{G}(q) + \hat{F}(\dot{q})$$

Stability:

Tracking error e and $\dot{e}$ are asymptotically stable. Parameter estimate $\hat{\varphi}$ is bounded.

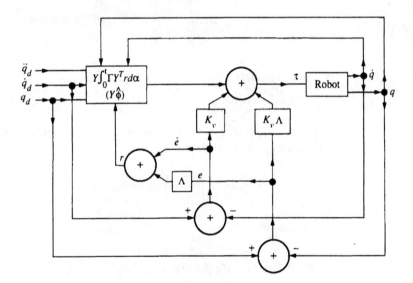

Figure 6.3.1: Block diagram of the adaptive inertia-related controller.

$$\tau_2 = Y_{21}\hat{m}_1 + Y_{22}\hat{m}_2 + k_{v2}\dot{e}_2 + k_{v2}\lambda_2 e_2. \qquad (2)$$

In the expression for the control torques, the regression matrix $Y(\cdot)$ is given by

$$Y(\ddot{q}_d, \dot{q}_d, q_d, q, \dot{q}) = \begin{bmatrix} Y_{11} & Y_{12} \\ Y_{21} & Y_{22} \end{bmatrix}, \quad (3)$$

where

$$Y_{11} = l_1^2 (\ddot{q}_{d1} + \lambda_1 \dot{e}_1) + l_1 g c_1, \quad (4)$$

$$\begin{aligned} Y_{12} &= (l_2^2 + 2l_1 l_2 c_2 + l_1^2)(\ddot{q}_{d1} + \lambda_1 \dot{e}_1) \\ &+ (l_2^2 + l_1 l_2 c_2)(\ddot{q}_{d2} + \lambda_2 \dot{e}_2) - l_1 l_2 s_2 \dot{q}_2 (\dot{q}_{d1} + \lambda_1 e_1) \\ &- l_1 l_2 s_2 (\dot{q}_1 + \dot{q}_2)(\dot{q}_{d2} + \lambda_2 e_2) + l_2 g c_{12} + l_1 g c_1, \end{aligned} \quad (5)$$

$$Y_{21} = 0, \quad (6)$$

and

$$\begin{aligned} Y_{22} &= (l_1 l_2 c_2 + l_2^2)(\ddot{q}_{d1} + \lambda_1 \dot{e}_1) + l_2^2 (\ddot{q}_{d2} + \lambda_2 \dot{e}_2) \\ &- l_1 l_2 s_2 \dot{q}_1 (\dot{q}_{d1} + \lambda_1 e_1) + l_2 g c_{12}. \end{aligned} \quad (7)$$

Formulating the adaptive update rule as given in Table 6.3.1, the associated parameter estimate vector is

$$\hat{\varphi} = \begin{bmatrix} \hat{m}_1 \\ \hat{m}_2 \end{bmatrix}$$

with the adaptive update rules

$$\dot{\hat{m}}_1 = \gamma_1 [Y_{11}(\lambda_1 e_1 + \dot{e}_1) + Y_{21}(\lambda_2 e_2 + \dot{e}_2)] \quad (8)$$

and

$$\dot{\hat{m}}_2 = \gamma_2 [Y_{12}(\lambda_1 e_1 + \dot{e}_1) + Y_{22}(\lambda_2 e_2 + \dot{e}_2)]. \quad (9)$$

For $m_1 = 0.8$ kg and $m_2 = 2.3$ kg, the adaptive inertia-related controller was simulated with $k_{v1} = k_{v2} = 10$, $\lambda_1 = \lambda_2 = 2.5$, $\gamma_1 = \gamma_2 =$

6.4 Adaptive Controllers Based on Passivity

20, $\hat{m}_1(0) = 0$, $\hat{m}_2(0) = 0$, and with the same desired trajectory and initial joint conditions as given in Example 6.2.1 . The tracking error and mass estimates are depicted in Figure 6.3.2. As illustrated by the figure, the tracking error is asymptotically stable, and the parameter estimates remain bounded. ∎

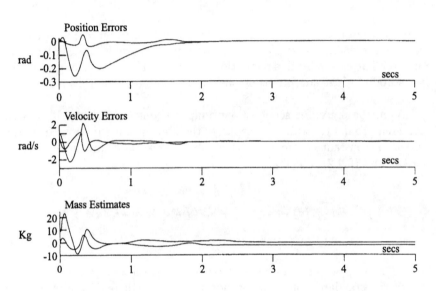

Figure 6.3.2: Simulation of the adaptive inertia-related controller.

6.4 Adaptive Controllers Based on Passivity

In recent years, many authors have developed adaptive control schemes that are different with regard to the torque control law or the adaptive update rule. To unify some of the approaches, general adaptive control strategies have been developed based on the passivity approach (see [Ortega and Spong 1988] and [Brogliato et al. 1990]). In this section we illustrate how the passivity approach can be used to develop a class of torque control laws and adaptive update rules for the control of robot manipulators.

Passive Adaptive Controller

First, we define an auxiliary filtered tracking error variable that is similar to that defined for the adaptive inertia-related controller. That is, we define our tracking variable to be

$$r(s) = H^{-1}(s) e(s), \qquad (6.4.1)$$

where

$$H^{-1}(s) = \left[sI_n + \frac{1}{s}K(s)\right] \qquad (6.4.2)$$

and s is the Laplace transform variable. In (6.4.2), the $n \times n$ gain matrix $K(s)$ is chosen such that $H(s)$ is a strictly proper, stable transfer function matrix. The reason for this restriction on $H(s)$ will be clear after we analyze the stability of the adaptive controller that is presented later in this section.

As in the preceding sections, our adaptive control strategies have been centered around the ability to separate the known time functions from the unknown constant parameters. Therefore, we use the expressions given in (6.4.1) and (6.4.2) to define

$$Z(\cdot)\varphi = M(q)(\ddot{q}_d + K(s)e) + V_m(q,\dot{q})\left(\dot{q}_d + \frac{1}{s}K(s)e\right)$$
$$+ G(q) + F(\dot{q}), \qquad (6.4.3)$$

where in this control formulation, $Z(\cdot)$ is a known $n \times r$ regression matrix. [Note the standard abuse of notation in (6.4.3), where $K(s)e$ is used to represent the inverse Laplace transform of $K(s)$ convolved with $e(t)$.] It is important to note that $K(s)$ can be selected so that $Z(\cdot)$ and r do not depend on measurement of $\ddot{q}$. Indeed, if $K(s)$ is selected such that $H(s)$ has a relative degree of 1 [Kailath 1980], $Z(\cdot)$ and r will not depend on $\ddot{q}$.

The adaptive control formulation given in this section is called the *passivity approach* because the mapping of $-r \to Z(\cdot)\tilde{\varphi}$ is constructed to be a passive mapping. That is, we construct an adaptive update rule such that

$$\int_0^t -r^T(\sigma) Z(\sigma) \tilde{\varphi}(\sigma) \, d\sigma \geq -\beta \qquad (6.4.4)$$

is satisfied for all time and for some positive scalar constant β. This passivity concept is used in analyzing the stability of the error system, as we shall show. However, for now let us illustrate the use of (6.4.4) in generating an adaptive update rule.

6.4 Adaptive Controllers Based on Passivity

EXAMPLE 6.4-1: Adaptive Update Rule by Passivity

Let us show that the adaptive update rule

$$\dot{\tilde{\varphi}} = -\dot{\hat{\varphi}} = -\Gamma Z^T(\cdot) r \tag{1}$$

satisfies the inequality given by (6.4.4). Note that Γ is defined as in (6.2.11).

First rewrite (1) in the form

$$\dot{\tilde{\varphi}}^T \Gamma^{-1} = -r^T Z(\cdot), \tag{2}$$

where we have used the fact that Γ is a diagonal matrix. Substituting (2) into (6.4.4) gives

$$\int_0^t \dot{\tilde{\varphi}}^T(\sigma) \Gamma^{-1} \tilde{\varphi}(\sigma) \, d\sigma \geq -\beta. \tag{3}$$

Since Γ is a constant matrix, we can use the product rule to rewrite (3) as

$$\tfrac{1}{2} \int_0^t \frac{d}{d\sigma} \left(\tilde{\varphi}^T(\sigma) \Gamma^{-1} \tilde{\varphi}(\sigma) \right) d\sigma \geq -\beta \tag{4}$$

or

$$\tfrac{1}{2} \tilde{\varphi}^T(t) \Gamma^{-1} \tilde{\varphi}(t) - \tfrac{1}{2} \tilde{\varphi}^T(0) \Gamma^{-1} \tilde{\varphi}(0) \geq -\beta. \tag{5}$$

From (5) it is now obvious that if β is selected as

$$\beta = \tfrac{1}{2} \tilde{\varphi}^T(0) \Gamma^{-1} \tilde{\varphi}(0), \tag{6}$$

then the passivity integral given in (6.4.4) is satisfied for the adaptive update rule given in (1).
■

Now that we have a feeling for how the passivity integral (6.4.4) can be used to generate adaptive update rules, we use the concept of passivity to analyze the stability of a class of adaptive controllers. For this class of adaptive controllers, the torque control is given by

$$\tau = Z(\cdot) \hat{\varphi} + K_v r. \tag{6.4.5}$$

Note that many types of torque controllers can be generated from (6.4.5) by selecting different transfer function matrices for $K(s)$ in the definition of r. That is, for different $K(s)$, we have different types of feedback because the feedback term $K_v r$ will change accordingly.

To form the error system, rewrite the robot dynamics (6.2.1) in terms of the tracking error variable r and the regression matrix $Z(\cdot)$ as

$$M(q)\dot{r} = -\tau - V_m(q,\dot{q})r + Z(\cdot)\varphi. \tag{6.4.6}$$

Substituting (6.4.5) into (6.4.6) yields the tracking error system

$$M(q)\dot{r} = -K_v r - V_m(q,\dot{q})r + Z(\cdot)\hat{\varphi}. \tag{6.4.7}$$

For analyzing the stability of this system, we use the Lyapunov-like function

$$V = {}^1\!/_2 r^T M(q) r + \beta - \int_0^t r^T(\sigma) Z(\sigma) \tilde{\varphi}(\sigma) \, d\sigma \tag{6.4.8}$$

[Ortega and Spong 1988]. Note that V is positive since the parameter estimate update rule is constructed to guarantee (6.4.4). That is, if (6.4.4) is satisfied, then

$$\beta - \int_0^t r^T(\sigma) Z(\sigma) \tilde{\varphi}(\sigma) \, d\sigma \geq 0; \tag{6.4.9}$$

therefore, V is a positive scalar function. Differentiating (6.4.8) with respect to time gives

$$\dot{V} = r^T M(q) \dot{r} + {}^1\!/_2 r^T \dot{M}(q) r - r^T Z(\cdot)\tilde{\varphi}. \tag{6.4.10}$$

Substituting (6.4.7) into (6.4.10) yields

$$\dot{V} = -r^T K_v r + r^T \left({}^1\!/_2 \dot{M}(q) - V_m(q,\dot{q}) \right) r. \tag{6.4.11}$$

Utilizing the skew-symmetric property (see Chapter 2) allows one to write

$$\dot{V} = -r^T K_v r. \tag{6.4.12}$$

We now detail the type of stability for the tracking error. First note from (6.4.12) that we can place the new upper bound on $\dot{V}$:

$$\dot{V} \leq -\lambda_{\min}\{K_v\} \|r\|^2, \tag{6.4.13}$$

which can also be written as

6.4 Adaptive Controllers Based on Passivity

$$\int_0^\infty \dot{V}(\sigma)\, d\sigma \leq -\lambda_{\min}\{K_v\} \int_0^\infty \|r(\sigma)\|^2\, d\sigma. \tag{6.4.14}$$

Multiplying (6.4.14) by -1 and integrating the left-hand side of (6.4.14) yields

$$V(0) - V(\infty) \geq \lambda_{\min}\{K_v\} \int_0^\infty \|r(\sigma)\|^2\, d\sigma. \tag{6.4.15}$$

Since $\dot{V}$ is negative semidefinite as delineated by (6.4.12), we can state that V is a nonincreasing function that is upper bounded by $V(0)$. By recalling that $M(q)$ is lower bounded, as delineated by the positive-definite property of the inertia matrix (see Chapter 2), we can state that V given in (6.4.8) is lower bounded by zero. Since V is nonincreasing, upper bounded by $V(0)$, and lower bounded by zero, we can write (6.4.15) as

$$\lambda_{\min}\{K_v\} \int_0^\infty \|r(\sigma)\|^2\, d\sigma < \infty \tag{6.4.16}$$

or

$$\sqrt{\int_0^\infty \|r(\sigma)\|^2\, d\sigma} < \infty. \tag{6.4.17}$$

The bound delineated by (6.4.17) informs us that $r \in L_2^n$ (see Chapter 1), which means that the filtered tracking r is bounded in the "special" way given by (6.4.17).

To establish a stability result for the position tracking error e, we establish the transfer function relationship between the position tracking error and the filtered tracking error r. From (6.4.1) we can state that

$$e(s) = H(s)\, r(s), \tag{6.4.18}$$

where $H(s)$ is as defined in (6.4.2). Since $H(s)$ is a strictly proper, asymptotically stable transfer function matrix and $r \in L_2^n$, we can use Theorem 1.4.7 in Chapter 1 to state that

$$\lim_{t \to \infty} e = 0. \tag{6.4.19}$$

The result above informs us that the position tracking error is asymptotically stable. In accordance with the theoretical development above, all we can say about the velocity tracking error is that it is bounded.

The passivity-based controller is summarized in Table 6.4.1. From this table we can see that the passivity approach gives a general class of torque control laws. We illustrate this concept with some examples that unify some of the research in adaptive control.

EXAMPLE 6.4-2: Passivity of the Adaptive Inertia-Related Controller

In this example we show how the adaptive inertia-related controller can be derived using passivity concepts. First, note that by defining

$$K(s) = \Lambda s \tag{1}$$

and

$$H(s) = (sI_n + \Lambda)^{-1} \tag{2}$$

in Table 6.4.1, we obtain the torque control law

$$\tau = Z(\cdot)\hat{\varphi} + K_v \dot{e} + K_v \Lambda e, \tag{3}$$

where

$$Z(\cdot)\hat{\varphi} = \hat{M}(q)(\ddot{q}_d + \Lambda \dot{e}) + \hat{V}_m(q,\dot{q})(\dot{q}_d + \Lambda e) + \hat{G}(q) + \hat{F}(\dot{q}). \tag{4}$$

This corresponds to the definition given in (6.3.11); therefore, using (2), we have obtained the adaptive inertia-related torque controller as given in Table 6.3.1.

The last item to check is whether the adaptive inertia-related update rule satisfies the passivity integral given in Table 6.4.1. From Table 6.3.1 the adaptive inertia-related update rule can be written as

$$\dot{\hat{\varphi}} = -\Gamma Z^T(\cdot)(\dot{e} + \Lambda e) = -\dot{\tilde{\varphi}} \tag{5}$$

for the choice of $K(s)$ given in (1). After reexamining Example 6.4.1 it is now obvious that we have derived the adaptive inertia-related controller with the passivity approach. ∎

6.4 Adaptive Controllers Based on Passivity

Table 6.4.1: Passive Class of Adaptive Controllers

Torque Controller:

$$\tau = Z(\cdot)\hat{\varphi} + K_v r$$

where

$$Z(\cdot)\hat{\varphi} = \hat{M}(q)(\ddot{q}_d + K(s)e) + \hat{V}_m(q,\dot{q})\left(\dot{q}_d + \tfrac{1}{s}K(s)e\right) \\ + \hat{G}(q) + \hat{F}(\dot{q}),$$

$$r(s) = H^{-1}(s)e(s), \quad H^{-1}(s) = \left[sI_n + \tfrac{1}{s}K(s)\right],$$

and the gain matrix $K(s)$ is chosen such that $H(s)$ is a strictly proper, stable transfer function matrix with relative degree 1.

Update Rule Must Satisfy:

$$\int_0^t -r^T(\sigma)Z(\sigma)\tilde{\varphi}(\sigma)\,d\sigma \geq -\beta$$

Stability:

Position tracking error e is asymptotically stable.

EXAMPLE 6.4-3: PID Torque Control Law

In Example 6.4.2 the torque control law was shown to be

$$\tau = Z(\cdot)\hat{\varphi} + K_v \dot{e} + K_v \Lambda e. \tag{1}$$

This torque controller is a proportional-derivative (PD) feedback controller since we are using e and $\dot{e}$ in the feedback portion of the control. It is now desired to find the choice of $K(s)$ in Table 6.4.1 to give a proportional-integral-derivative (PID) type of torque controller of the form

$$\tau = Z(\cdot)\hat{\varphi} + K_v \dot{e} + K_p e + K_I \int e\,dt. \tag{2}$$

Using Table 6.4.1, we can select

$$K(s) = \Lambda s + \Psi, \tag{3}$$

where

$$\Lambda = \text{diag}\ \{\lambda_1, \lambda_2, \ldots, \lambda_r\},$$
$$\Psi = \text{diag}\ \{\Psi_1, \Psi_2, \ldots, \Psi_r\},$$

λ_i's are positive scalar constants, ψ_i's are positive scalar constants, $K_p = K_v\Lambda$, and $K_I = K_v\Psi$. Now we must check to verify that $H(s)$ is indeed a strictly proper, stable transfer function matrix. For this choice of $K(s)$ in Table 6.4.1, we can write

$$H(s) = \left(sI_n + \Lambda + \frac{1}{s}\Psi\right)^{-1}. \tag{4}$$

Note that since Λ and Ψ have been selected to be diagonal positive-definite matrices, $H(s)$ is a decoupled transfer function matrix. That is, the transfer function for the ith system is

$$h_i(s) = \frac{s}{s^2 + \lambda_i s + \Psi_i}. \tag{5}$$

Since the λ_i's and ψ's are positive, $H(s)$ is a strictly proper, stable transfer function matrix.

∎

General Adaptive Update Rule

As mentioned earlier, the adaptive control scheme outlined in Table 6.4.1 allows one to formulate different adaptive update laws by ensuring that the proposed update satisfies the passivity integral given in (6.4.4). Landau proposed the general update rule (which satisfies the passivity integral)

$$\hat{\varphi} = \int_0^t F_I(t - \sigma) Z^T(\sigma) r(\sigma)\ d\sigma + F_p Z^T(\cdot) r, \tag{6.4.20}$$

where F_p is an $r \times r$ positive definite, constant matrix, and $F_I(t)$ is an $r \times r$ positive definite matrix kernel whose Laplace transform is a positive real transfer function matrix with a pole at $s = 0$ [Landau 1979].

By utilizing this general update law, many types of adaptation may be designed. All we need to keep in mind is that the conditions on $F_I(t)$ and F_p must be met. One possible adaptive scheme that comes directly from (6.4.20) is the proportional + integral (PI) adaptation scheme. The PI update law is the same as that given by (6.4.20), with

$$F_I(t) = K_1, \qquad (6.4.21)$$

where K_1 is a diagonal, constant positive-definite matrix. It has been pointed out in [Landau 1979] that with regard to adaptive model following, PI adaptation has shown a significant improvement over integral adaptation. Therefore, this type of adaptation might be beneficial for the tracking control of robot manipulators.

6.5 Persistency of Excitation

For the adaptive controllers presented in the previous sections the tracking error has been shown to be asymptotically stable; however, all that could be said about the parameter error was that it was bounded. In general, parameter identification will occur in adaptive control systems only if certain conditions on the regression matrix can be established. Specifically, several researchers [Morgan and Narendra 1977], [Anderson 1977] have studied the asymptotic stability of adaptive control systems similar to the ones we have presented in this chapter. For example, parameter error convergence can be established for the adaptive inertia-related controller if the regression matrix $Y(\cdot)$ satisfies

$$\alpha I_r \leq \int_{t_0}^{t_0+\rho} Y^T(q, \dot{q}, q_d, \dot{q}_d, \ddot{q}_d)\, Y(q, \dot{q}, q_d, \dot{q}_d, \ddot{q}_d)\, dt \leq \beta I_r \qquad (6.5.1)$$

for all t_0, where α, β, and ρ are all positive scalars. Furthermore, since the tracking error is asymptotically stable, we can rewrite (6.5.1) as

$$\alpha I_r \leq \int_{t_0}^{t_0+\rho} Y^T(q_d, \dot{q}_d, \ddot{q}_d)\, Y(q_d, \dot{q}_d, \ddot{q}_d)\, dt \leq \beta I_r, \qquad (6.5.2)$$

where the arguments q and $\dot{q}$ have been replaced by q_d and $\dot{q}_d$, respectively.

The condition given in (6.5.2) informs us that if $Y(\cdot)$ varies sufficiently over the interval given by ρ so that the entire r-dimensional parameter space is spanned, we know the parameter error converges to zero. This amounts to a condition on the desired trajectory such that all parameters will be identified after a sufficient learning interval. This condition can be helpful in formulating desired trajectories to ensure that parameters such as friction coefficients or payload masses are identified. We now illustrate the meaning of a persistently exciting trajectory with some examples.

EXAMPLE 6.5-1: Lack of Persistency of Excitation for a One-Link Robot Arm

We wish to investigate the persistency of excitation conditions for the one-link robot arm given in Figure 6.5.1. The dynamics of this robot arm will be taken to be

$$\tau = m\ddot{q} + b\dot{q}, \tag{1}$$

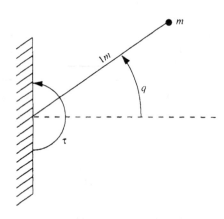

Figure 6.5.1: One-link revolute arm.

where the term b is used to denote the positive scalar representing the dynamic coefficient of friction. We assume that this robot arm is in the plane not affected by the gravitational force and that m and b are unknown positive constants.

a. Adaptive Controller

By using Table 6.3.1, the adaptive inertia-related controller for the dynamics (1) can be shown to be given by

$$\tau = Y_{11}\hat{m} + Y_{12}\hat{b} + k_v\dot{e} + k_v\lambda e. \tag{2}$$

In the expression above for the control torque, the regression matrix $Y(\cdot)$ is given by

6.5 Persistency of Excitation

$$Y(q_d, \dot{q}_d, \ddot{q}_d, q, \dot{q}) = \begin{bmatrix} Y_{11} & Y_{12} \end{bmatrix} \tag{3}$$

where

$$Y_{11} = (\ddot{q}_d + \lambda \dot{e}) \quad \text{and} \quad Y_{12} = \dot{q}.$$

The corresponding parameter estimate vector is given by

$$\hat{\varphi} = \begin{bmatrix} \hat{m} \\ \hat{b} \end{bmatrix}.$$

We can formulate the adaptive update rule given in Table 6.3.1 as

$$\dot{\hat{m}} = \gamma_1 Y_{11} (\lambda e + \dot{e}) \tag{4}$$

and

$$\dot{\hat{b}} = \gamma_2 Y_{12} (\lambda e + \dot{e}). \tag{5}$$

b. Persistency of Excitation

With this adaptive controller it is desired to show analytically that $q_d = 1 - e^{-2t}$ is not persistently exciting. From (6.5.2), the integrand of the persistently exciting condition for this example is given by

$$Y^T(q_d, \dot{q}_d, \ddot{q}_d) Y(q_d, \dot{q}_d, \ddot{q}_d) = \begin{bmatrix} \ddot{q}_d^2 & \dot{q}_d \ddot{q}_d \\ \dot{q}_d \ddot{q}_d & \dot{q}_d^2 \end{bmatrix}$$

or

$$Y^T(\cdot) Y(\cdot) = \begin{bmatrix} 16e^{-4t} & -8e^{-4t} \\ -8e^{-4t} & 4e^{-4t} \end{bmatrix}. \tag{6}$$

Multiplying the first column of $Y^T(\cdot)Y(\cdot)$ by $\frac{1}{2}$ and adding it to the second column of $Y^T(\cdot)Y(\cdot)$ gives

$$R_{YY} = \begin{bmatrix} 16e^{-4t} & 0 \\ -8e^{-4t} & 0 \end{bmatrix}. \tag{7}$$

Since the matrix R_{YY} has the same range space as the matrix $Y^T(\cdot)Y(\cdot)$, we can see from (7) that the range space of the matrix $Y^T(\cdot)Y(\cdot)$ will always be one-dimensional; therefore, the persistent excitation condition does not hold for

$$q_d = 1 - e^{-2t}.$$

∎

EXAMPLE 6.5-2: Persistency of Excitation for a One-Link Arm

a. Desired Trajectory with Single Frequency

With the adaptive controller in Example 6.5.1, we would like to show by simulation that $q_d = \sin t$ is not persistently exciting. For $m = 1$ kg and $b = 1$ N-m-s, the adaptive inertia-related controller was simulated with

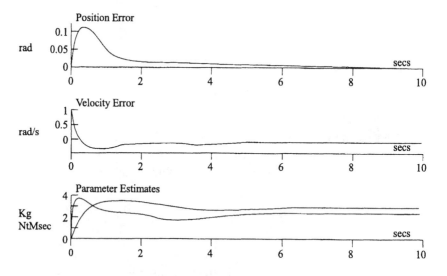

Figure 6.5.2: Lack of persistency of excitation.

$$k_v = 10, \quad \lambda = 2.5, \quad \gamma_1 = \gamma_2 = 20, \tag{1}$$

with initial conditions of

$$q(0) = \dot{q}(0) = \hat{m}_1(0) = \hat{m}_2(0) = 0. \tag{2}$$

The tracking error and the parameter estimates are depicted in Figure 6.5.2. As anticipated, the tracking error is asymptotically stable, and the parameter estimates remain bounded. Note that $\tilde{b}$ and $\tilde{m}$ do

6.6 Composite Adaptive Controller

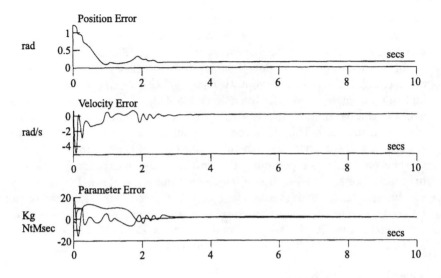

Figure 6.5.3: Persistency of excitation.

not go to zero since the desired trajectory is not persistently exciting.

b. Desired Trajectory with Multiple Frequencies

With the adaptive controller in Example 6.5.1, we desire to show by simulation that $q_d = \sin t + \cos 3t$ is persistently exciting. The adaptive controller given in Example 6.5.1 should be simulated under the same conditions as those given in part (a) of this example except for the change in the desired trajectory. The tracking error and the parameter error are depicted in Figure 6.5.3. As illustrated by the figure, the tracking error is asymptotically stable, and the parameter estimates remain bounded. Note that $\tilde{b}$ and $\tilde{m}$ converge to the exact values of b and m, respectively. This is because the desired trajectory is persistently exciting.

■

6.6 Composite Adaptive Controller

In both the adaptive computed-torque and the adaptive inertia-related control strategies, we have shown that the tracking error is asymptotically stable and the parameter error is bounded. It was then illustrated that if a persistency of excitation condition holds, the parameter error $\tilde{\varphi}$ converges to zero. In some robotic applications it may not be practical to utilize a persistently exciting trajectory; therefore, we are motivated to redesign the adaptive control strategy to achieve parameter identification.

In this section we show how the adaptive controller given in Section 6.3 can be modified to ensure asymptotic convergence of both the tracking error and the parameter error [Slotine and Li 1985(b)]. The asymptotic convergence of the parameter error is shown to hold if a condition on the filtered regression matrix holds. This condition, often called the *infinite integral condition*, is less restrictive than the persistency of excitation condition.

The procedure for designing the new adaptive controller can be outlined as follows. First, a filtered regression matrix is formed from torque measurements. Second, it is shown how this filtered regression matrix can be used to formulate a least-squares estimator for estimating the unknown parameters. Finally, the adaptive update rule in Table 6.3.1 is modified to include an additional least-squares estimator term.

Torque Filtering

We now show how the regression matrix that is formed from a torque measurement can be filtered to eliminate the need for acceleration measurements. From Section 6.2 we can write the robot Equation (6.2.1) in the following form:

$$\tau = M(q)\ddot{q} + V_m(q, \dot{q})\dot{q} + G(q) + F(\dot{q}) = W(q, \dot{q}, \ddot{q})\varphi. \tag{6.6.1}$$

The middle expression in (6.6.1) is written in the form

$$\tau = \dot{h} + g, \tag{6.6.2}$$

where

$$\dot{h} = \frac{d}{dt}(M(q)\dot{q}) \tag{6.6.3}$$

and

$$g = -\dot{M}(q)\dot{q} + V_m(q, \dot{q})\dot{q} + G(q) + F(\dot{q}). \tag{6.6.4}$$

The reason for writing the robot equation in the form given by (6.6.2) is that this equation has now been separated in a way that allows $\ddot{q}$ to be filtered out or removed. That is, by filtering (6.6.2), we have

$$\tau_f = f * \tau = f * \dot{h} + f * g, \tag{6.6.5}$$

6.6 Composite Adaptive Controller

where f is the impulse response of a linear stable, strictly proper filter, and the $*$ is used to denote the convolution operation. For example, we could use the first-order filter given by

$$f(s) = \frac{a}{s+a}, \qquad (6.6.6)$$

where a is a positive scalar constant. By using the property of convolution

$$f * \dot{h} = \dot{f} * h + f(0)h - fh(0), \qquad (6.6.7)$$

we can rewrite (6.6.5) as

$$\tau_f = f * \tau = \dot{f} * h + f(0)h - fh(0) + f * g. \qquad (6.6.8)$$

After substituting the expressions for h and g, note that $\ddot{q}$ has been filtered out. That is, the explicit expression for if is given by

$$\tau_f = \dot{f} * (M(q)\dot{q}) + f(0)M(q)\dot{q} - fM(q(0))\dot{q}(0)$$
$$+ f * \left(-\dot{M}(q)\dot{q} + V_m(q,\dot{q})\dot{q} + G(q) + F(\dot{q})\right), \qquad (6.6.9)$$

where $\dot{f}$ is the impulse response of a proper, stable filter; for example,

$$sf(s) = \frac{as}{s+a}. \qquad (6.6.10)$$

By linearity, the unknown parameters can still be separated out with regard to (6.6.9). That is, (6.6.9) can be rewritten as

$$\tau_f = W_f(q,\dot{q})\varphi, \qquad (6.6.11)$$

where $W_f(\cdot)$ is an $n \times r$ filtered regression matrix, and φ is an $r \times 1$ vector of unknown parameters. We now use an example to show how torque filtering can be used to eliminate the need for acceleration measurements.

EXAMPLE 6.6-1: Torque Filtering of a One-Link Robot Arm

Using the dynamics of the one-link robot arm given in Example 6.5.1, it is desired to find the filter regression matrix $W_f(q,\dot{q})$ given in (6.6.11), where the linear filter is given by

$$f(s) = \frac{1}{s+1}, \qquad (1)$$

or, in the time domain,

$$f(t) = e^{-t}. \tag{2}$$

Using (6.6.9), the filtered torque expression for the one-link arm is given by

$$\tau_f = \dot{f} * m\dot{q} + m\dot{q} - fm\dot{q}(0) + f * b\dot{q}, \tag{3}$$

where

$$\dot{f}(s) = \frac{s}{s+1}. \tag{4}$$

The expression in (3) is used to separate the known functions from the unknown constants into the form

$$\tau_f = W_f(q, \dot{q}) \varphi, \tag{5}$$

where the filtered regression matrix and parameter vector are given by

$$W_f(q, \dot{q}) = \left[\left(\dot{f} * \dot{q} \right) + \dot{q} - f\dot{q}(0) \quad \dot{f} * \dot{q} \right] \tag{6}$$

and

$$\varphi = \begin{bmatrix} m & b \end{bmatrix}^T. \tag{7}$$

∎

The important concept to realize in the regression matrix formulation given by (6.6.11) is that the quantities τ_f and $W_f(q, \dot{q})$ are known or assumed to be measurable; however, φ is unknown. To estimate φ, we define the estimate of the filtered torque based on the estimate of the unknown parameters, that is,

$$\hat{\tau}_f = W_f(\cdot) \hat{\varphi}. \tag{6.6.12}$$

We can now define the measurable quantity

$$\tilde{\tau}_f = W_f(\cdot) \tilde{\varphi}, \tag{6.6.13}$$

where

$$\tilde{\tau}_f = \tau_f - \hat{\tau}_f. \tag{6.6.14}$$

6.6 Composite Adaptive Controller

The use of $\tilde{\tau}_f$ is crucial in the development of the least-squares estimator that is developed in the next section. We can easily see how $\tilde{\tau}_f$ is measurable by writing (6.6.14) in the form

$$\tilde{\tau}_f = \tau_f - W_f(\cdot)\hat{\varphi}. \qquad (6.6.15)$$

As explained earlier, $\tilde{\tau}_f$ and $W_f(\cdot)$ are assumed to be known or measurable; therefore, all we need is the parameter estimate term [i.e., $\hat{\varphi}$ in (6.6.15)] for if to be known. Later, we generate an adaptive update rule that will give us the parameter estimate term $\hat{\varphi}$ in (6.6.15). So for now we will assume that $\tilde{\tau}_f$ is known.

Least-Squares Estimation

Least-squares estimation methods have been used in many types of parameter identification schemes [Astrom and Wittenmark 1989]. It turns out that this type of estimation method extracts the maximum amount of parametric information even when the desired trajectory is not persistently exciting. This is an important fact to realize when designating adaptive control systems for robot manipulators because in many robot applications, the persistency of excitation condition will not be valid. Therefore, least-squares estimation offers an attractive solution to the design of adaptive controllers for robot manipulators.

We now show how the least-squares estimation method can be used to generate an adaptive update rule. First, define the least-squares update rule

$$\dot{\hat{\varphi}} = P W_f^T(\cdot)\tilde{\tau}_f, \qquad (6.6.16)$$

where

$$\dot{P} = -P W_f^T(\cdot) W_f(\cdot) P \qquad (6.6.17)$$

and P is an $r \times r$ time-varying symmetric matrix.

With this least-squares estimation method, if an "infinite integral" condition holds, the parameter error converges to zero. Specifically, if

$$\lim_{t\to\infty} \lambda_{\min}\left\{\int_0^t W_f^T(\sigma) W_f(\sigma)\, d\sigma\right\} = \infty \qquad (6.6.18)$$

holds, then

$$\lim_{t\to\infty} \tilde{\varphi} = 0. \qquad (6.6.19)$$

As pointed out in [Slotine and Li 1985(b)], this infinite integral condition is a weaker condition than the persistency of excitation condition. That is, in practical robot applications, (6.6.18) can often be validated.

EXAMPLE 6.6-2: Least-Squares Estimator for a One-Link Robot Arm

Using the dynamics of the one-link robot arm given in Example 6.5.1, it is desired to find the least-squares estimator given by (6.6.16) and (6.6.17). Since the number of unknown parameters is two, define the matrix P to be

$$P = \begin{bmatrix} P_1 & P_2 \\ P_2 & P_3 \end{bmatrix}. \tag{1}$$

Utilizing the filtered regression matrix from Example 6.6.1, we have

$$W_f(q, \dot{q}) = \begin{bmatrix} W_{f11} & W_{f12} \end{bmatrix}, \tag{2}$$

where

$$W_{f11} = \left(\dot{f} * \dot{q}\right) + \dot{q} - f\dot{q}(0) \quad \text{and} \quad W_{f12} = \dot{f} * \dot{q}.$$

Using (6.6.17), it is easy to see that the matrix P should be updated in the following manner:

$$\dot{P}_1 = -\left(W_{f11}P_1 + W_{f12}P_2\right)^2 \tag{3}$$

$$\dot{P}_2 = -W_{f11}^2 P_1 P_2 - W_{f12}^2 P_2 P_3 - W_{f11}W_{f12}\left(P_1 P_3 + P_2^2\right), \tag{4}$$

and

$$\dot{P}_3 = -\left(W_{f11}P_2 + W_{f12}P_3\right)^2. \tag{5}$$

Now using (6.6.16), the parameter update rules are

$$\dot{\hat{m}} = (P_1 W_{f11} + P_2 W_{f12})\tilde{\tau}_f \tag{6}$$

and

6.6 Composite Adaptive Controller

$$\dot{\hat{b}} = (P_2 W_{f11} + P_3 W_{f12}) \tilde{\tau}_f, \qquad (7)$$

where, from (6.6.16), $\tilde{\tau}_f$ is given by

$$\tilde{\tau}_f = \tau_f - W_{f11}\hat{m} - W_{f12}\hat{b}. \qquad (8)$$

■

For insight into how the least-squares estimation method extracts parameter information, we now show how (6.6.18) is obtained. Utilizing (6.6.13) and the fact that the parameters are constant, we write (6.6.16) as

$$\dot{\tilde{\varphi}} = -P W_f^T(\cdot) W_f(\cdot) \tilde{\varphi}. \qquad (6.6.20)$$

Using the matrix identity $\dot{P} = -P\dot{P}^{-1}P$ we can write (6.6.17) as

$$\dot{P}^{-1} = W_f^T(\cdot) W_f(\cdot). \qquad (6.6.21)$$

Substituting (6.6.21) into (6.6.20) yields the differential equation

$$\dot{\tilde{\varphi}} = -P\dot{P}^{-1}\tilde{\varphi}. \qquad (6.6.22)$$

We claim that

$$\tilde{\varphi} = -PP^{-1}(0)\tilde{\varphi}(0) \qquad (6.6.23)$$

is the solution to (6.6.22). This fact can be verified by substituting (6.6.23) into the right-hand and left-hand sides of (6.6.22). That is, we obtain

$$-\dot{P}P^{-1}(0)\tilde{\varphi}(0) = P\dot{P}^{-1}PP^{-1}(0)\tilde{\varphi}(0); \qquad (6.6.24)$$

therefore, (6.6.23) is the solution. Now from (6.6.21) it is easy to see that the solution for P is given by

$$P = \left\{ P^{-1}(0) + \int_0^t W_f^T(\sigma) W_f(\sigma) d\sigma \right\}^{-1}. \qquad (6.6.25)$$

After examining (6.6.25), we can intuitively see that if the infinite integral condition is satisfied, then

$$\lim_{t \to \infty} \lambda_{\max}\{P\} = 0 \qquad (6.6.26)$$

and

$$\lim_{t \to \infty} \lambda_{\min}\{P^{-1}\} = \infty. \qquad (6.6.27)$$

Now if (6.6.26) holds, we can see from (6.6.23) that the parameter error converges to zero. This proof is detailed in [Li and Slotine 1987].

Composite Adaptive Controller

The composite adaptive controller is the same as the controller given in Table 6.3.1, with the exception of a modification to the adaptive update rule. This modification is given by

$$\dot{\hat{\varphi}} = -\dot{\tilde{\varphi}} = PY^T(\cdot)r + PW_f^T(\cdot)\tilde{\tau}_f. \tag{6.6.28}$$

To prove that the tracking error and the parameter error both converge to zero, start with the Lyapunov-like function,

$$V = 1/2 r^T M(q) r + 1/2 \tilde{\varphi}^T P^{-1} \tilde{\varphi}. \tag{6.6.29}$$

Differentiating (6.6.29) with respect to time yields

$$\dot{V} = r^T M(q) \dot{r} + 1/2 r^T \dot{M}(q) r + \tilde{\varphi}^T P^{-1} \dot{\tilde{\varphi}} + 1/2 \tilde{\varphi}^T \dot{P}^{-1} \tilde{\varphi}. \tag{6.6.30}$$

From the control law given in Table 6.3.1 and the development in Section 6.3, we can form the tracking error system

$$M(q)\dot{r} = Y(\cdot)\tilde{\varphi} - K_v r - V_m(q,\dot{q})r. \tag{6.6.31}$$

Substituting (6.6.31) into (6.6.30) yields

$$\dot{V} = -r^T K_v r + \tilde{\varphi}^T \left(P^{-1}\dot{\tilde{\varphi}} + Y^T(\cdot)r \right) + 1/2 \tilde{\varphi}^T \dot{P}^{-1} \tilde{\varphi}. \tag{6.6.32}$$

After substituting $\dot{\tilde{\varphi}}$ in (6.6.28), $\dot{P}^{-1}$ in (6.6.21), and $\tilde{\tau}_f$ in (6.6.13) into (6.6.32), we have

$$\dot{V} = -r^T K_v r - 1/2 \tilde{\varphi}^T W_f^T(\cdot) W_f(\cdot) \tilde{\varphi}. \tag{6.6.33}$$

We now detail the type of stability for the tracking error and the parameter error. First, since $\dot{V}$ in (6.6.33) is at least negative semidefinite in the form

$$\dot{V} \leq -\lambda_{\min}\{K_v\} \|r\|^2, \tag{6.6.34}$$

we can state that V in (6.6.29) is bounded. Since V is bounded, $M(q)$ is a positive-definite matrix, and P^{-1} satisfies the condition given by (6.6.27), we can state that r and $\tilde{\varphi}$ are bounded. Furthermore, from the definition of r given in (6.3.8), we can use standard linear control arguments to state

6.6 Composite Adaptive Controller

that e and $\dot{e}$ (and hence q and $\dot{q}$) are bounded. We can now use the same arguments presented in Section 6.4 to show that $r \in L_2^n$. Given that $r \in L_2^n$, we can determine a stability result for the position tracking error by establishing the transfer function relationship between the position tracking error and the filtered tracking error r. From (6.3.8), we can state that

$$e(s) = G(s) r(s), \qquad (6.6.35)$$

where s is the Laplace transform variable,

$$G(s) = (sI + \Lambda)^{-1}, \qquad (6.6.36)$$

I is the $n \times n$ identity matrix, and Λ is an $n \times n$ positive-definite matrix. Since $G(s)$ is a strictly proper, asymptotically stable transfer function and $r \in L_2^n$, we can use Theorem 1.4.7 in Chapter 1 to state that

$$\lim_{t \to \infty} e = 0. \qquad (6.6.37)$$

Second, since $\dot{V}$ is at least negative semidefinite, we know that V must be nonincreasing, and hence V is upper bounded by $V(0)$. Furthermore, by the infinite integral assumption, we have concluded in (6.6.27) that

$$\lim_{t \to \infty} \lambda_{\min}\{P^{-1}\} = \infty. \qquad (6.6.38)$$

Since the term

$$\tfrac{1}{2}\tilde{\varphi}^T P^{-1} \tilde{\varphi} \qquad (6.6.39)$$

in V given in (6.6.29) is upper bounded by $V(0)$, we can see that for (6.6.38) to hold, we must have

$$\lim_{t \to \infty} \tilde{\varphi} = 0.$$

Therefore, from the argument above, the position tracking error and the parameter error are asymptotically stable for the composite adaptive controller outlined in Table 6.6.1. In accordance with the theoretical development above, all we can say about the velocity tracking error is that it is bounded; however, Barbalat's lemma can be invoked to illustrate that the velocity tracking error is also asymptotically stable (see Problem 6.6-3). We now use an example to illustrate how the composite adaptive controller is formulated.

Table 6.6.1: Composite Adaptive Controller

Controller:

$$\tau = Y(\cdot)\hat{\varphi} + K_v\dot{e} + K_v\Lambda e$$

Update Rule:

$$\dot{\hat{\varphi}} = PY^T(\cdot)(\dot{e} + \Lambda e) + PW_f^T(\cdot)(f*\tau - W_f(\cdot)\hat{\varphi})$$

where

$$Y(\cdot)\hat{\varphi} = \hat{M}(q)(\ddot{q}_d + \Lambda\dot{e}) + \hat{V}_m(q,\dot{q})(\dot{q}_d + \Lambda e) + \hat{G}(q) + \hat{F}(\dot{q})$$

$$\dot{P} = -PW_f^T(\cdot)W_f(\cdot)P$$

$$W_f(\cdot)\hat{\varphi} = f * \left[\hat{M}(q)\ddot{q} + \hat{V}_m(q,\dot{q})\dot{q} + \hat{G}(q) + \hat{F}(\dot{q})\right]$$

f is the impulse response of a strictly proper stable filter.

Stability:

Tracking error e is asymptotically stable. Parameter error $\tilde{\varphi}$ is asymptotically stable if the infinite integral condition is satisfied.

EXAMPLE 6.6-3: Composite Adaptive Controller for a One-Link Robot Arm

It is desired to formulate the composite adaptive controller for the one-link robot arm given in Figure 6.5.1. The torque control law is the same as that given by Equation (2) in Example 6.5.1; therefore, all that need be done is the formulation of the composite adaptive update rule. From Table 6.6.1, the composite parameter update rules are

$$\dot{\hat{m}} = (P_1Y_{11} + P_2Y_{12})r + (P_1W_{f11} + P_2W_{f12})\tilde{\tau}_f \qquad (1)$$

and

$$\dot{\hat{b}} = (P_2Y_{11} + P_3Y_{12})r + (P_2W_{f11} + P_3W_{f12})\tilde{\tau}_f, \qquad (2)$$

where r is defined in (6.3.8), Y_{11}, Y_{12} are defined in Example 6.5.1, and $P_1, P_2, P_3, W_{f11}, W_{f12}, \tilde{\tau}_f$ are defined in Example 6.6.2. ∎

6.7 Robustness of Adaptive Controllers

All of the adaptive control schemes discussed ensure asymptotic tracking of a desired reference trajectory for the robot manipulator dynamics; however, in reality we know that there will always be disturbances in any electromechanical system. A simplistic way to take into account some sort of disturbance effect is to add a bounded disturbance term to the manipulator dynamic equation. With this additive disturbance term the robot equation becomes

$$\tau = M(q)\ddot{q} + V_m(q,\dot{q})\dot{q} + G(q) + F(\dot{q}) + T_d, \qquad (6.7.1)$$

where T_d is an $n \times 1$ vector that represents an additive disturbance.

Applying the adaptive inertia-related control strategy and ignoring the term T_d in (6.7.1) gives the adaptive control scheme of Table 6.3.1. However, if we reexamine the stability analysis given in Section 6.3 for the adaptive inertia-related controller, we can see that a bounded disturbance term gives us a different type of stability result for the tracking error. Specifically, with the addition of the bounded disturbance term in (6.7.1), the derivative of the Lyapunov function in (6.3.7) becomes

$$\dot{V} = -r^T K_v r + r^T T_d. \qquad (6.7.2)$$

From (6.7.2) it is obvious that $\dot{V}$ can no longer be taken to be negative semidefinite. From our previous experience with Lyapunov stability theory, it was desired to have $\dot{V}$ be "negative"; therefore, it stands to reason that it would be advantageous to find the region where $\dot{V}$ is negative in (6.7.2). By the use of the Rayleigh-Ritz Theorem (see Chapter 1), we can write (6.7.2) as

$$\dot{V} \leq -\lambda_{\min}\{K_v\} \|r\|^2 + \|r\| \|T_d\|. \qquad (6.7.3)$$

From (6.7.3), a sufficient condition on the negativity of $\dot{V}$ can be obtained. That is, $\dot{V}$ will be negative if

$$\|r\| > \frac{\|T_d\|}{\lambda_{\min}\{K_v\}}. \qquad (6.7.4)$$

If (6.7.4) is satisfied, $\dot{V}$ is negative and V will decrease. If V decreases, then by our definition of the Lyapunov function given in (6.3.7), r must eventually decrease. However, if r decreases such that

$$\|r\| \leq \frac{\|T_d\|}{\lambda_{\min}\{K_v\}}. \tag{6.7.5}$$

then $\dot{V}$ may become positive, which means that V will start to increase. If V starts to increase, we gain insight into the problem by examining two possibilities. One, the increase in V causes r to increase such that (6.7.4) is satisfied. This means that V will start to decrease and hence r will eventually decrease. If r increased and decreased in this fashion continually, then r and $\tilde{\varphi}$ both remain bounded. The other possibility is that the increase in V causes $\tilde{\varphi}$ to increase while r stays small enough such that (6.7.5) is satisfied. For this case, $\dot{V}$ remains positive; therefore, $\tilde{\varphi}$ could continue to increase. If V continues to increase in this fashion, r is bounded; however, $\tilde{\varphi}$ and hence $\hat{\varphi}$ both become unbounded.

The argument above reveals that the parameter estimate in the adaptive inertia-related control law *may* go unstable in the presence of a bounded disturbance. That is, the parameter estimate may diverge under the assumption that the robot model is given by (6.7.1). If the parameter estimate becomes too large, we can see from Table 6.3.1 that the input torque will start to grow and possibly saturate the joint motors; therefore, it would be desirable to modify the adaptive controller to eliminate the possibility of torque saturation.

EXAMPLE 6.7-1: Effects of Disturbance on Adaptive Control

In this example we simulate the same adaptive controller given in Example 6.3.1 with the same control parameters, initial conditions, and desired trajectory; however, we have added the disturbance term

$$T_d = \begin{bmatrix} 2\sin(10t) \\ 2\sin(10t) \end{bmatrix} \tag{1}$$

to the two-link manipulator dynamics. The tracking error and parameter estimates are illustrated in Figure 6.7.1. From the figure, note that for the disturbance given by (1), the parameter estimates do not become unbounded; however, the tracking error is no longer asymptotically stable.

∎

Torque-Based Disturbance Rejection Method

To reject an additive disturbance term in the robot model, we illustrate how the parameter estimates remain bounded if the torque control is modified to be

6.7 Robustness of Adaptive Controllers

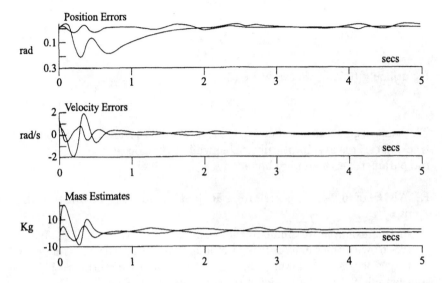

Figure 6.7.1: Simulation of adaptive controller in the pressence of disturbance.

$$\tau = \tau_a + k_d \operatorname{sgn}(r) \tag{6.7.6}$$

where τ_d is the torque control given in Table 6.3.1,

$$\operatorname{sgn}(r) = [\operatorname{sgn}(r_1), \operatorname{sgn}(r_2), \ldots, \operatorname{sgn}(r_n)]^T, \tag{6.7.7}$$

sgn($\cdot$) is used to denote the signum function, and k_d is a scalar constant that satisfies

$$k_d > \max\{|T_{di}|\} \tag{6.7.8}$$

with T_{di}; representing the ith component of the $n \times 1$ vector T_d [Slotine and Li 1985(c)].

Applying the disturbance rejection controller given in (6.7.6), the derivative of the Lyapunov function in (6.3.7) becomes

$$\dot{V} = -r^T K_v r + r^T T_d - k_d r^T \operatorname{sgn}(r). \tag{6.7.9}$$

By noting that

$$|r_i| = r_i \operatorname{sgn}(r_i), \tag{6.7.10}$$

(6.7.9) can be written as

$$\dot{V} \leq -r^T K_v r + \sum_{i=1}^{n} |r_i| \left(|T_{di}| - k_d \right). \qquad (6.7.11)$$

By utilizing (6.7.8), we can write (6.7.11) as

$$\dot{V} \leq -r^T K_v r. \qquad (6.7.12)$$

The same arguments as in Section 6.6 can be used to show that the position tracking error is asymptotically stable while the velocity tracking error and the parameter estimate are bounded.

EXAMPLE 6.7-2: Disturbance Rejection for a Two-Link Robot Arm

In this example we simulate the modified adaptive controller given in (6.7.6) with the same control parameters, initial conditions, and desired trajectory as in Example 6.3.1 and also with the disturbance given in Example 6.7.1. The modified torque controller is given by

$$\tau_1 = \tau_{a1} + k_d \text{sgn}\left(\lambda_1 e_1 + \dot{e}_1\right) \qquad (1)$$

and

$$\tau_2 = \tau_{a2} + k_d \text{sgn}\left(\lambda_2 e_2 + \dot{e}_2\right), \qquad (2)$$

where τ_{a1}, and τ_{a2} are the same adaptive torque controllers given in Example 6.3.1, λ_1, and λ_2 are the same scalar constants defined in Example 6.3.1, and

$$k_d = 2.2. \qquad (3)$$

Note that k_d has been chosen to satisfy the condition given in (6.7.8) and that the update laws are the same as those given in Example 6.3.1. The tracking error and parameter estimates are illustrated in Figure 6.7.2. From the figure we note that the parameter estimates remain bounded; furthermore, the tracking error is now asymptotically stable even in the presence of a disturbance. It is important to note that for the theoretical development given for the torque-based disturbance rejection method above, we only guaranteed the velocity tracking error to be bounded.

∎

Estimator-Based Disturbance Rejection Method

In [Reed and Ioannou 1988] a modified version of the σ-modification [Ioannou and Tsakalis 1985] to the adaptive inertia-related control algorithm [Slotine and Li 1985(a)] was introduced to compensate for unmodeled dynamics and bounded disturbances in the robot model. In this method the torque control is the same as that given by Table 6.3.1; however, the adaptive update rule is modified to be

$$\dot{\hat{\varphi}} = \Gamma Y^T(\cdot) r - \sigma_s \hat{\varphi}, \tag{6.7.13}$$

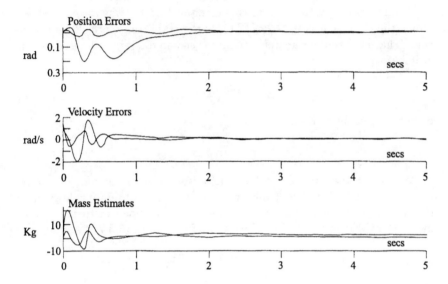

Figure 6.7.2: Simulation of adaptive controller with disturbance rejection.

where

$$\sigma_s = \begin{cases} 0 & \text{if } \|\hat{\varphi}\| < \varphi_0 \\ \frac{\|\hat{\varphi}\|}{\varphi_0} - 1 & \text{if } \varphi_0 \leq \|\hat{\varphi}\| \leq 2\varphi_0 \\ 1 & \text{if } \|\hat{\varphi}\| > 2\varphi_0 \end{cases} \tag{6.7.14}$$

and

$$\varphi_0 > \|\varphi\|. \tag{6.7.15}$$

With the update rule (6.7.13), Reed showed that the tracking error could be confined to a residual set and that all closed-loop signals are bounded.

As we have shown, a bounded disturbance can cause the parameter estimates to go unstable. The update rule given by (6.7.13) is intended to remedy this problem by regulating on-line the size of the parameter estimates. This is done by the scalar design constant φ_0. That is, by checking the size of the parameter estimates against φ_0, the parameter estimates are forced to remain bounded by using this new update rule. One can see clearly that if $\|\hat{\varphi}\| < \varphi_0$, the update rule is the same as that given by Table 6.3.1. In other words, if the parameter estimates do not get too large, the controller is the same as the adaptive inertia-related controller. On the other hand, if parameter estimates get too large, the adaptive update rule is modified to ensure that the parameter estimates remain bounded. How this σ-*modification* accomplishes this task is now discussed.

To motivate how the update rule given in (6.7.13) was formulated, examine the case when $\|\hat{\varphi}\| > 2\varphi_0$. This is the stabilizing part of the update rule. That is, if the parameter estimates become too large, the update rule switches to

$$\dot{\hat{\varphi}} = \Gamma Y^T(\cdot) r - \hat{\varphi}, \qquad (6.7.16)$$

or in terms of the parameter error,

$$\dot{\tilde{\varphi}} = -\Gamma Y^T(\cdot) r - \tilde{\varphi} + \varphi. \qquad (6.7.17)$$

We now reexamine the stability analysis given in Section 6.3 for the adaptive inertia-related controller with the parameter update rule given by (6.7.16). Specifically, with the addition of the bounded disturbance term in (6.7.1), the derivative of the Lyapunov function in (6.3.7) becomes

$$\dot{V} = -r^T K_v r - \tilde{\varphi}^T \Gamma^{-1} \tilde{\varphi} + r^T T_d + \tilde{\varphi}^T \Gamma^{-1} \varphi \qquad (6.7.18)$$

or

$$\dot{V} = -x^T K^* x + x^T T_d^*, \qquad (6.7.19)$$

where

$$x = \begin{bmatrix} r \\ \tilde{\varphi} \end{bmatrix}, \quad K^* = \begin{bmatrix} K_v & 0 \\ 0 & \Gamma^{-1} \end{bmatrix}, \quad \text{and} \quad T_d^* = \begin{bmatrix} T_d \\ \Gamma^{-1}\varphi \end{bmatrix}. \qquad (6.7.20)$$

By the use of the Rayleigh-Ritz Theorem (see Chapter 1), we can write (6.7.19) as

$$\dot{V} \leq -\lambda_{\min}\{K^*\} \|x\|^2 + \|x\| \|T_d^*\|. \qquad (6.7.21)$$

From (6.7.21), $\dot{V}$ will be negative if

6.8 Summary

$$\|x\| > \frac{\|T_d^*\|}{\lambda_{\min}\{K^*\}}. \tag{6.7.22}$$

It is important to note that the right-hand side of (6.7.22) is a constant; therefore, if (6.7.22) is satisfied, $\dot{V}$ is negative, which causes V to decrease. If V decreases, then by our definition of the Lyapunov function given in (6.3.7), x must eventually decrease. However, if x decreases such that

$$\|x\| \leq \frac{\|T_d^*\|}{\lambda_{\min}\{K^*\}}, \tag{6.7.23}$$

then $\dot{V}$ may be positive, which means that V will start to increase. The increase in V causes x to increase such that (6.7.22) is satisfied. This means that V now starts to decrease again and hence x eventually decreases. This argument illustrates how x is bounded. If x is bounded, then from (6.7.20), r and $\tilde{\varphi}$ are bounded. Since r is bounded, standard linear control arguments can be used to show that e and $\dot{e}$ are bounded.

One last point is now discussed regarding the region $\varphi_0 \leq \|\hat{\varphi}\| \leq 2\varphi_0$. for the adaptive update rule given in (6.7.13). This part of the adaptive update rule is used to ensure that there is a smooth transition between the adaptive inertia-related update rule and the stabilizing portion of the update rule given by (6.7.16). That is, this ensures that we do not obtain any discontinuities in the parameter estimates, which could cause a large discontinuity in the input torque. A large discontinuity is undesirable in the input torque signal since this type of signal could cause the robot manipulator to jerk violently.

6.8 Summary

In this chapter an account of several of the most recent adaptive control results for rigid robots has been given. The intent has been to lend some perspective to the growing list of adaptive control results for robot manipulators. Some research areas, such as transient behavior, digital implementation, and robustness to unmodeled dynamics, will no doubt be addressed in the future. An issue that remains to be investigated is the comparison of the advantages and disadvantages of the different servo and adaptive laws.

Some excellent adaptive control work with regard to robot manipulators by other researchers is outlined in [Ortega and Spong 1988]. Since this is such a well-studied field and there is limited space available in this chapter, we apologize to anyone who has been left out.

REFERENCES

[Anderson 1977] Anderson, B., "Exponential stability of linear equations arising in adaptive identification," *IEEE Trans. Autom. Control*, Feb. 1977, pp. 83-88.g

[Arimoto and Miyazaki 1986] Arimoto, S., and F. Miyazaki, "Stability and robustness of PD feedback control with gravity compensation for robot manipulators," *Robot. Theory Pract.*, DSC Vol. 3, pp. 67-72, *ASME Winter Annual Meeting*, Dec. 1986.

[Åström and Wittenmark 1989] K.Åström and B. Wittenmark, *Adaptive Control.* Reading, MA: Addison-Wesley, 1989.

[Brogliato et al. 1990] Brogliato, B., I. Landau, and R. Lozano-Leal, "Adaptive motion control of robot manipulators: a unified approach based on passivity," *Proc. IEEE Am. Controls Conf*, pp. 2259-2264, San Diego, CA, May 1990.

[Craig 1985] Craig, J., *Adaptive Control of Mechanical Manipulators.* Reading, MA.: AddisonWesley, 1985.

[Ioannou and Tsakalis 1985] Ioannou, P., and K. Tsakalis, "A robust direct adaptive controller," *IEEE Trans. Autom. Control*, vol. AC-31, no. 11, pp. 1033-1043, 1985.

[Kailath 1980] Kailath, T, *Linear Systems.* Englewood Cliffs, NJ: Prentice Hall, 1980.

[Landau 1979] Landau, Y., *Adaptive Control: The Model Reference Approach.* New York: Marcel Dekker, 1979.

[Li and Slotine 1987] Li, W., and J. Slotine, "Parameter estimation strategies for robotic applications," *ASME Winter Annual Meeting*, Boston, 1987.

[Morgan and Narendra 1977] Morgan, A., and K. Narendra, "On the uniform asymptotic stability of certain linear nonautonomous differential equations," *SIAM J. Control Optim.*, 1977, p. 15.

[Ortega and Spong 1988] Ortega, R., and M. Spong, "Adaptive motion control of rigid robots: a tutorial," *Proc. IEEE Conf Decision Control*, Austin, TX, 1988.

[Reed and Ioannou 1988] Reed, J., and P. Ioannou, "Instability analysis and robust adaptive control of robotic manipulators," *Proc. IEEE Conf Decision Control*, Austin, TX, 1988.

[Slotine 1988] Slotine, J., (1988), "Putting physics in control: the example of robotics," *Control Syst. Mag.*, Dec. 1988, Vol. 8, pp 12-15.

[Slotine and Li 1985(a)] Slotine, J., and W. Li, "Theoretical issues in adaptive control," *5th Yale Workshop on Applications of Adaptive Systems Theory*, Yale University, New Haven, CT, 1985(a).

[Slotine and Li 1985(b)] Slotine, J., and W. Li, "Adaptive robot control: a new perspective," *Proc. IEEE Conf Decision Control*, Los Angeles, 1985(b).

[Slotine and Li 1985(c)] Slotine, J., and W. Li, "Adaptive strategies in constrained manipulation," *Proc. IEEE Int. Conf Robot. Autom.*, Raleigh, NC, Mar. 1985(c), pp. 595-601.

[Vidyasagar 1978] Vidyasagar, M., *Nonlinear Systems Analysis*. Englewood Cliffs, NJ: Prentice Hall, 1978.

PROBLEMS

Section 6.2
6.2-1 Design and simulate the adaptive computed-torque controller given in Table 6.2.1 for the two-link polar robot arm given in Chapter 2.

6.2-2 Find different positive-definite, symmetric matrices, P and Q from that given in Example 6.2.2 that satisfy

$$A^T P + PA = -Q,$$

where

$$A = \begin{bmatrix} O_n & I_n \\ -K_p & -K_v \end{bmatrix},$$

K_p, K_v are diagonal positive-definite matrices, and O_n, I_n represent the $n \times n$ zero matrix and $n \times n$ identity matrix, respectively.

6.2-3: With the P and Q found in Problem 6.2-3, redo Problem 6.2-1 and report the differences in the tracking error performance.

Section 6.3
6.3-1 Design and simulate the adaptive inertia-related controller given in Table 6.3.1 for the two-link polar robot arm given in Chapter 2.

6.3-2 For the simulation given in Problem 6.3-1, run several simulations with different values of the control parameters (ie., Λ, K_v, Γ), and report the effects on tracking error performance.

6.3-3 Enumerate the the advantages of the adaptive controller given in Table 6.3.1 over the adaptive controller given in Table 6.2.1.

6.3-4 As given in (6.3.8), the filtered tracking error is defined by

$$r = \Lambda e + \dot{e},$$

where Λ is a positive-definite diagonal matrix. Show that if

$$\lim_{t \to \infty} r(t) = 0, \text{ then } \lim_{t \to \infty} e(t) = 0 \text{ and } \lim_{t \to \infty} \dot{e}(t) = 0.$$

Section 6.4
6.4-1 Design and simulate an adaptive controller for the two-link revolute arm given in Example 6.3.1 with the PID servo law given in Example 6.4.3 and the adaptation law given in Example 6.4.1. Report any

differences from that given in Example 6.3.1.

6.4-2 Redo Problem 6.4-1 with the proportional + integral adaptation law given by Equations (6.4.20) and (6.4.21).

Section 6.5

6.5-1 Show analytically that $q_d = \sin t$ in Example 6.5.2 is not persistently exciting.

Section 6.6

6.6-1 Show that

$$f(t) * \dot{h}(t) = \dot{f}(t) * h(t) + f(0)h(t) - f(t)h(0).$$

6.6-2 Simulate the composite adaptive controller given in Example 6.6.3 and report the effects on the tracking error of using different values for $P(0)$ and a (i.e., the pole of the filter used for the filtered regression matrix).

6.6-3 Show how Barbalat's lemma given in Chapter 1 can be used in the proof of the composite adaptive controller to yield

$$\lim_{t \to \infty} \dot{e}(t) = 0.$$

Section 6.7

6.7-1 Redo Problem 6.3-1 with the additive bounded disturbance

$$T_d = \begin{bmatrix} 0.75 \sin(3t) \\ 0.25 \cos(2t) \end{bmatrix},$$

added to two-link polar robot arm dynamics given in Chapter 2.

6.7-2 Redo Problem 6.3-1 with the additive bounded disturbance given in Problem 6.7-1 and with the term

$$k_d \text{sgn}(r),$$

added to the adaptive controller. Run several simulations with different values of k_d, and report the effects on tracking error performance.

Chapter 7

Advanced Control Techniques

In this chapter some advanced control techniques for the tracking control of robot manipulators are discussed. The controllers that are developed in this chapter address computational issues and the effects of actuator dynamics. The analytical concepts and the control developments presented in this chapter are in general more complex than those presented in the previous chapters; therefore, it is highly recommended that the previous chapters be studied before examining this new material.

7.1 Introduction

As research in robot control has progressed over the last couple of years, many robot control researchers have begun to focus on implementational issues. That is, implementational concerns, such as the reduction of on-line computation and the effects of actuator dynamics, are causing researchers to rethink the previous theoretical development of robot controllers so that these concerns are addressed. This constant retooling of the previous control development to coincide with the implementational restrictions is how previous progress in robot control research has proceeded. Utilizing this concept of forcing the theoretical development to satisfy implementational restrictions, we illustrate how some researchers have begun to address problems such as reducing on-line computation and compensating for the effects of actuator dynamics.

7.2 Robot Controllers with Reduced On-Line Computation

In this section we examine the robot controllers designed by Sadegh and coworkers [Sadegh and Horowitz 1990], [Sadegh et al. 1990]. We separate these controllers from related work since this work addresses the extremely relevant implementation issue of on-line controller computation. Specifically, this adaptive controller reduces on-line computation as opposed to other control techniques, such as the adaptive controllers presented in Chapter 6. Following the development of the adaptive controller research, a "repetitive" controller is also presented. This repetitive controller also reduces online computation.

Desired Compensation Adaptation Law

One of the disadvantages of the adaptive controllers in Chapter 6 is that the regression matrix (e.g., the matrix $Y(\cdot)$ in the adaptive inertia-related controller) used as feedforward compensation must be calculated on-line. The regression matrix must be calculated on-line since it depends on the measurements of the joint position and velocity (i.e., q and $\dot{q}$). For the simple two-link robot controller given in Example 6.3.1, it is evident that online calculation of $Y(\cdot)$ is computationally intensive. As one can imagine, on-line computation of the regression matrix can be very computationally intensive if one desires to control a robot manipulator with many degrees of freedom.

To eliminate the need for on-line computation of the regression matrix, we will now examine the desired compensation adaptation law (DCAL) [Sadegh and Horowitz 1990]. The DCAL eliminates the need for on-line computation of the regression matrix by replacing q and $\dot{q}$ with the desired joint position and velocity (i.e., q_d and $\dot{q}_d$). That is, the DCAL regression matrix only depends on desired trajectory information; therefore, the DCAL regression matrix can be calculated *a priori* off-line. Of course, this modification of the regression matrix forces us to reexamine the adaptive control design and the corresponding stability analysis.

For purposes of control design in this section, we assume that the robotic manipulator is a revolute manipulator with dynamics given by

$$\tau = M(q)\ddot{q} + V_m(q, \dot{q})\dot{q} + G(q) + F_d\dot{q}, \qquad (7.2.1)$$

where F_d is a $n \times n$ positive-definite, diagonal matrix that is used to represent the dynamic coefficients of friction, and all other quantities are as defined in Chapter 3. As in other chapters, we define the joint tracking error to be

7.2 Robot Controllers with Reduced On-Line Computation

$$e = q_d - q. \tag{7.2.2}$$

As explained in Chapter 6, adaptive control of robot manipulators involves separating the known time functions from the unknown constant parameters. For example, recall that this separation of parameters from time functions for the adaptive inertia-related controller is given by

$$Y(\cdot)\varphi = M(q)(\ddot{q}_d + \dot{e}) + V_m(q,\dot{q})(\dot{q}_d + e) + G(q) + F_d\dot{q}, \tag{7.2.3}$$

where $Y(\cdot)$ is an $n \times r$ regression matrix that depends only on known time functions of the actual and desired trajectory, and φ is an $r \times 1$ vector of unknown constant parameters. (Note that Λ defined Table 6.3.1 is taken to be the identity matrix.)

In the DCAL, this separation of parameters from time functions is given by

$$Y(\cdot)\varphi = M(q_d)\ddot{q}_d + V_m(q_d,\dot{q}_d)\dot{q}_d + G(q_d) + F_d\dot{q}_d, \tag{7.2.4}$$

where $Y_d(\cdot)$ is an $n \times r$ regression matrix that depends only on known functions of the *desired trajectory*. Note that if we substitute q_d and $\dot{q}_d$ for q and $\dot{q}$, respectively, into (7.2.3), the regression matrix formulation given by (7.2.3) is equivalent to that given by (7.2.4).

Utilizing the regression matrix formulation given in (7.2.4), the DCAL is formulated as

$$\tau = Y_d(\cdot)\hat{\varphi} + k_v r + k_p e + k_a \|e\|^2 r, \tag{7.2.5}$$

where k_v, k_p, k_a are scalar, constant, control gains, $\hat{\varphi}$ is the $r \times 1$ vector of parameter estimates, and the filtered tracking error is defined as

$$r = e + \dot{e}. \tag{7.2.6}$$

The corresponding DCAL parameter adaptive update law is

$$\dot{\hat{\varphi}} = -\dot{\tilde{\varphi}} = \Gamma Y_d^T(\cdot) r, \tag{7.2.7}$$

where Γ is an $r \times r$ positive definite, diagonal, constant, adaptive gain matrix, and the parameter error is defined by

$$\tilde{\varphi} = \varphi - \hat{\varphi}. \tag{7.2.8}$$

Note that the DCAL given by (7.2.5) is quite similar to adaptive controllers discussed in Chapter 6 with the exception of the term $k_a\|e\|^2 r$ in (7.2.5). It turns out that this additional term is used to compensate for the difference between $Y(\cdot)\varphi$ and $Y_d(\cdot)\varphi$ given in (7.2.3) and (7.2.4), respectively. As shown in [Sadegh and Horowitz 1990], this difference between the actual regression matrix and the desired regression matrix formulations can be quantified as

$$\left\|\tilde{Y}\right\| \leq \zeta_1 \left\|e\right\| + \zeta_2 \left\|e\right\|^2 + \zeta_3 \left\|r\right\| + \zeta_4 \left\|r\right\| \left\|e\right\|, \qquad (7.2.9)$$

where

$$\tilde{Y} = Y(\cdot)\varphi - Y_d(\cdot)\varphi \qquad (7.2.10)$$

and ζ_1, ζ_2, ζ_3, and ζ_4 are positive bounding constants that depend on the desired trajectory and the physical properties of the specific robot configuration (i.e., link mass, link length, friction coefficients, etc.).

To analyze the stability of the controller given by (7.2.5), we must form the corresponding error system. First, we rewrite (7.2.1) in terms of $Y(\cdot)\varphi$ and r defined in (7.2.3) and (7.2.6), respectively. That is, we have

$$M(q)\dot{r} = -V_m(q,\dot{q})r + Y(\cdot)\varphi - \tau. \qquad (7.2.11)$$

Adding and subtracting the term $Y_d(\cdot)\varphi$ on the right-hand side of (7.2.11) yields

$$M(q)\dot{r} = -V_m(q,\dot{q})r + Y_d(\cdot)\varphi + \tilde{Y} - \tau, \qquad (7.2.12)$$

where $\tilde{Y}$ is defined in (7.2.10). Substituting the control given by (7.2.5) into (7.2.12) yields the error system

$$M(q)\dot{r} = -V_m(q,\dot{q})r - k_v r - k_p e - k_a \left\|e\right\|^2 r + Y_d(\cdot)\tilde{\varphi} + \tilde{Y}, \qquad (7.2.13)$$

where $\tilde{\varphi}$ is defined in (7.2.8).

We now analyze the stability of the error system given by (7.2.13) with the Lyapunov-like function

$$V = 1/2 r^T M(q) r + 1/2 k_p e^T e + 1/2 \tilde{\varphi}^T \Gamma^{-1} \tilde{\varphi}. \qquad (7.2.14)$$

Differentiating (7.2.14) with respect to time yields

$$\dot{V} = 1/2 r^T \dot{M}(q) r + r^T M(q) \dot{r} + k_p e^T \dot{e} + \tilde{\varphi}^T \Gamma^{-1} \dot{\tilde{\varphi}} \qquad (7.2.15)$$

since scalar quantities can be transposed. Substituting (7.2.13) into (7.2.15) fields

7.2 Robot Controllers with Reduced On-Line Computation

$$\dot{V} = k_p e^T \dot{e} - k_v r^T r - k_p r^T e - k_a \|e\|^2 r^T r + r^T \tilde{Y}$$
$$+ \tfrac{1}{2} r^T \left(\dot{M}(q) - 2V_m(q,\dot{q}) \right) r + \tilde{\varphi}^T \Gamma^{-1} \dot{\tilde{\varphi}} + r^T Y_d(\cdot) \tilde{\varphi}. \quad (7.2.16)$$

By utilizing the skew-symmetric property (see Chapter 3) and the update law in (7.2.7), it is easy to see that the second line in (7.2.16) is equal to zero. therefore, by invoking the definition of r given in (7.2.6), (7.2.16) simplifies to

$$\dot{V} = -k_p e^T e - k_v r^T r - k_a \|e\|^2 r^T r + r^T \tilde{Y}. \quad (7.2.17)$$

From (7.2.17), we can place an upper bound on $\dot{V}$ in the following manner:

$$\dot{V} \leq -k_p \|e\|^2 - k_v \|r\|^2 - k_a \|e\|^2 \|r\|^2 + \|r\| \left\| \tilde{Y} \right\|. \quad (7.2.18)$$

A new upper bound on $\dot{V}$ can be obtained by substituting (7.2.9) into (7.2.18) to yield

$$\dot{V} \leq - k_p \|e\|^2 - k_v \|r\|^2 - k_a \|e\|^2 \|r\|^2 + \zeta_1 \|e\| \|r\|$$
$$+ \zeta_2 \|e\|^2 \|r\| + \zeta_3 \|r\|^2 + \zeta_4 \|r\|^2 \|e\|. \quad (7.2.19)$$

By rearranging the second line of (7.2.19), it can be written as

$$\dot{V} \leq - k_p \|e\|^2 - k_v \|r\|^2 - k_a \|e\|^2 \|r\|^2 + \zeta_1 \|e\| \|r\|$$
$$- \zeta_2 \|e\|^2 [1/2 - \|r\|]^2 - \zeta_4 \|r\|^2 [1/2 - \|e\|]^2$$
$$+ (\zeta_2 + \zeta_4) \|e\|^2 \|r\|^2 + (\zeta_2/4) \|e\|^2 + (\zeta_3 + \zeta_4/4) \|r\|^2. \quad (7.2.20)$$

After collecting common terms in (7.2.20), it can be rewritten as

$$\dot{V} \leq - (k_p - \zeta_2/4) \|e\|^2 - (k_v - \zeta_3 - \zeta_4/4) \|r\|^2 + \zeta_1 \|e\| \|r\|$$
$$- \zeta_2 \|e\|^2 [1/2 - \|r\|]^2 - \zeta_4 \|r\|^2 [1/2 - \|e\|]^2$$
$$- (k_a - \zeta_2 - \zeta_4) \|e\|^2 \|r\|^2. \quad (7.2.21)$$

By noting that if the control gain k_a is adjusted in accordance with

$$k_a > \zeta_2 + \zeta_4, \quad (7.2.22)$$

we can see that the terms on the second line of (7.2.21) will all be negative; therefore, we can obtain the new upper bound on $\dot{V}$:

$$\dot{V} \leq -(k_p - \zeta_2/4)\|e\|^2 - (k_v - \zeta_3 - \zeta_4/4)\|r\|^2 + \zeta_1 \|e\| \|r\|. \qquad (7.2.23)$$

By rewriting (7.2.23) in the matrix form

$$\dot{V} \leq -x^T \bar{Q} x, \qquad (7.2.24)$$

where

$$\bar{Q} = \begin{bmatrix} k_p - \zeta_2/4 & -\zeta_1/2 \\ -\zeta_1/2 & k_v - \zeta_3 - \zeta_4/4 \end{bmatrix} \quad \text{and} \quad x = \begin{bmatrix} \|e\| \\ \|r\| \end{bmatrix},$$

we can establish sufficient conditions on k_p and k_v such that the matrix $\overline{Q}$ in (7.2.24) is positive definite. Specifically, by using the Gerschgorin theorem (see Chapter 2), we can see that if

$$k_p > \zeta_1/2 + \zeta_2/4 \qquad (7.2.25)$$

and

$$k_v > \zeta_1/2 + \zeta_3 + \zeta_4/4, \qquad (7.2.26)$$

the matrix $\overline{Q}$ defined in (7.2.24) will be positive definite; therefore, $\dot{V}$ will be negative semidefinite.

We now detail the type of stability for the tracking error. First, since $\dot{V}$ is negative semidefinite, we can state that V is upper bounded. Using the fact that V is upper bounded, we can state that e, $\dot{e}$, r, and $\tilde{\varphi}$ are bounded. Since e, $\dot{e}$, r, and $\tilde{\varphi}$ are bounded, we can use (7.2.13) to show that $\dot{r}$, $\ddot{q}$, and hence $\ddot{V}$ in (7.2.17) are bounded. Second, note that since $M(q)$ is lower bounded as delineated by the positive-definite property of the inertia matrix (see Chapter 3), we can state that V given in (7.2.14) is lower bounded. Since V is lower bounded, $\dot{V}$ is negative semidefinite, and $\ddot{V}$ is bounded, we can use Barbalat's lemma (see Chapter 2) to state that

$$\lim_{t \to \infty} \dot{V} = 0.$$

Therefore, from the argument above and (7.2.24), we know that

$$\lim_{t \to \infty} \begin{bmatrix} e \\ r \end{bmatrix} = 0. \qquad (7.2.27)$$

From (7.2.27), we can also determine the stability result for the velocity tracking error. Specifically, from (7.2.6), note that r is defined to be a

7.2 Robot Controllers with Reduced On-Line Computation

stable first-order differential equation in terms of the variable e; therefore, by standard linear control arguments, we can write

$$\lim_{t \to \infty} \dot{e} = 0. \tag{7.2.28}$$

This result informs us that if the controller gains are selected according to (7.2.22), (7.2.25), and (7.2.26), the tracking errors e and $\dot{e}$ are asymptotically stable. From the analysis above, all we can say about the parameter error is that it remains bounded. The adaptive controller just derived is summarized in Table 7.2.1 and depicted in Figure 7.2.1.

Table 7.2.1: DCAL Controller

Torque Controller:

$$\tau = Y_d(\cdot)\hat{\varphi} + k_v r + k_p e + K_a \|e\|^2 r$$

where

$$Y_d(\cdot)\varphi = M(q_d)\ddot{q}_d + V_m(q_d, \dot{q}_d)\dot{q}_d + G(q_d) + F_d\dot{q}_d$$

$$r = e + \dot{e}.$$

Update Rule:

$$\dot{\hat{\varphi}} = \Gamma Y_d^T(\cdot) r$$

Stability:

Tracking error e and $\dot{e}$ are asymptotically stable. Parameter estimate $\hat{\varphi}$ is bounded.

Comments: Controller gains k_a, k_p, and k_v must be sufficiently large.

After glancing through Table 7.2.1, we can see that as opposed to the adaptive inertia-related controller, the DCAL has the obvious advantage of reduced on-line calculations. Specifically, the regression matrix $Y_d(\cdot)$ depends only on the desired trajectory; therefore, the regression matrix can be calculated off-line. We now present an example to illustrate how Table 7.2.1 can be used to design adaptive controllers for robotic manipulators.

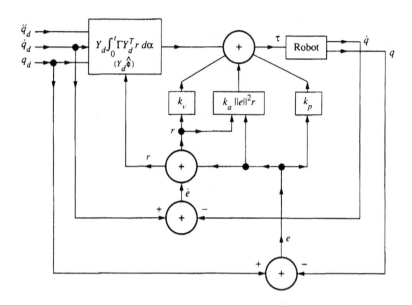

Figure 7.2.1: Block diagram of DCAL controller.

EXAMPLE 7.2-1: DCAL for the Two-Link Arm

We wish to design and simulate the DCAL given in Table 7.2.1 for the two-link arm given in Figure 6.2.1. (The dynamics for this robot arm are given in Chapter 3.) Assuming that the friction is negligible and the link lengths are exactly known to be of length 1 m each, the DCAL can be written as

$$\tau_1 = Y_{11}\hat{m}_1 + Y_{12}\hat{m}_2 + k_v r_1 + k_p e_1 + k_a \|e\|^2 r_1 \tag{1}$$

and

$$\tau_2 = Y_{21}\hat{m}_1 + Y_{22}\hat{m}_2 + k_v r_2 + k_p e_2 + k_a \|e\|^2 r_2, \tag{2}$$

where $r_1 = e_1 + \dot{e}_1$, $r_2 = e_2 + \dot{e}_2$, and $\|e\|^2 = e_1^2 + e_2^2$.

In the expression for the control torques, the regression matrix $Y_d(\cdot)$ is given by

$$Y_d(\ddot{q}_d, \dot{q}_d, q_d) = \begin{bmatrix} Y_{11} & Y_{12} \\ Y_{21} & Y_{22} \end{bmatrix}, \tag{3}$$

7.2 Robot Controllers with Reduced On-Line Computation

where

$$Y_{11} = l_1^2 \ddot{q}_{d1} + l_1 g \cos(q_{d1}), \tag{4}$$

$$Y_{12} = \left(l_2^2 + 2l_1 l_2 \cos(q_{d2}) + l_1^2\right) \ddot{q}_{d1} + \left(l_2^2 + l_1 l_2 \cos(q_{d2})\right) \ddot{q}_{d2}$$
$$- l_1 l_2 \sin(q_{d2}) \dot{q}_{d2} \dot{q}_{d1} - l_1 l_2 \sin(q_{d2}) (\dot{q}_{d1} + \dot{q}_{d2}) \dot{q}_{d2}$$
$$+ l_2 g \cos(q_{d1} + q_{d2}) + l_1 g \cos(q_{d1}), \tag{5}$$

$$Y_{21} = 0, \tag{6}$$

$$Y_{22} = \left(l_1 l_2 \cos(q_{d2}) + l_2^2\right) \ddot{q}_{d1} + l_2^2 \ddot{q}_{d2} + l_1 l_2 \sin(q_{d2}) \dot{q}_{d1}^2$$
$$+ l_2 g \cos(q_{d1} + q_{d2}). \tag{7}$$

Formulating the adaptive update rule as given in Table 7.2.1, the associated parameter estimate vector is

$$\hat{\varphi} = \begin{bmatrix} \hat{m}_1 \\ \hat{m}_2 \end{bmatrix}$$

with the adaptive update rules

$$\dot{\hat{m}}_1 = \gamma_1 \left[Y_{11} r_1 + Y_{21} r_2\right] \tag{8}$$

and

$$\dot{\hat{m}}_2 = \gamma_2 \left[Y_{12} r_1 + Y_{22} r_2\right]. \tag{9}$$

For $m_1 = 0.8$ kg and $m_2 = 2.3$ kg, the DCAL was simulated with

$$k_a = k_v = k_p = 50, \quad \gamma_1 = \gamma_2 = 20,$$

$$\hat{m}_1(0) = \hat{m}_2(0) = q_1(0) = q_2(0) = \dot{q}_1(0) = \dot{q}_2(0) = 0,$$

and

$$q_{d1} = q_{d2} = \sin t.$$

The tracking error and mass estimates are depicted in Figure 7.2.2. As illustrated by the figure, the tracking error is asymptotically stable, and the parameter estimates remain bounded.

Repetitive Control Law

In many industrial applications, robot manipulators are used to perform the same task repeatedly. For example, a robot may be required to paint the same assembly-line part over and over again. As one can imagine, the desired trajectory for this painting operation would be a periodic function. That is, after the desired trajectory has been generated for painting the first part (i.e., the first "trial"), the same desired trajectory should be followed in a repetitive fashion for painting the next part (i.e., the next "trial").

If a control strategy does not take into account the nature of a repetitive operation, mistakes made along the first trajectory will be repeated from trial to trial. Therefore, one is motivated to design a controller that utilizes the tracking error measurements in the present trial to improve the tracking performance in the next trial. These types of controllers are often referred to as "learning" or repetitive controllers. The term "learning controller" is used to emphasize the fact that the controller attempts to learn the repeatable part of the manipulator dynamics.

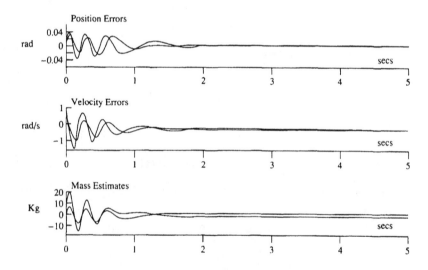

Figure 7.2.2: Simulation of DCAL controller.

To motivate the design of the repetitive control law (RCL) [Sadegh et al. 1990], we note that the dynamics given by

$$u_d(t) = M(q_d)\ddot{q}_d + V_m(q_d, \dot{q}_d)\dot{q}_d + G(q_d) + F_d\dot{q}_d \tag{7.2.29}$$

7.2 Robot Controllers with Reduced On-Line Computation

are repeatable if the desired trajectory is periodic. That is, even though there may be unknown constant parametric quantities in (7.2.29), the signal represented by the $n \times 1$ vector $u_d(t)$ will be periodic or repeatable. Therefore, in the subsequent discussion, we assume that the desired trajectory is periodic with period T. This periodic assumption on the desired trajectory allows us to write

$$u_d(t) = u_d(t - T) \tag{7.2.30}$$

since the dynamics represented by $u_d(t)$ depend only on periodic quantities.

Utilizing the repeatability of the dynamics given by (7.2.29), the RCL is formulated as

$$\tau = \hat{u}_d(t) + k_v r + k_p e + k_a \|e\|^2 r, \tag{7.2.31}$$

where the $n \times 1$ vector $\hat{u}_d(t)$ is a learning term that is used to compensate for the repeatable dynamics $u_d(t)$, and all other quantities are the same as those defined for the DCAL. The learning term $\hat{u}_d(t)$ is updated from trial to trial by the *learning update rule*

$$\hat{u}_d(t) = \hat{u}_d(t - T) + k_L r, \tag{7.2.32}$$

where k_L is a positive scalar control gain.

As done similarly in the adaptive control development, we will write the learning update rule given in (7.2.32) in terms of the *learning error*, which is defined as

$$\tilde{u}_d(t) = u_d(t) - \hat{u}_d(t). \tag{7.2.33}$$

Specifically, multiplying (7.2.32) by -1 and then adding $u_d(t)$ to both sides of (7.2.32) yields

$$u_d(t) - \hat{u}_d(t) = u_d(t) - \hat{u}_d(t - T) - k_L r. \tag{7.2.34}$$

By utilizing the periodic assumption given by (7.2.30), we can write (7.2.34) as

$$u_d(t) - \hat{u}_d(t) = u_d(t - T) - \hat{u}_d(t - T) - k_L r, \tag{7.2.35}$$

which gives the learning error update rule

$$\tilde{u}_d(t) = \tilde{u}_d(t - T) - k_L r, \tag{7.2.36}$$

where $\tilde{u}_d(t)$ is defined in terms of (7.2.33).

Before we analyze the stability of the controller given in (7.2.31), we will form the corresponding error system. First, we rewrite (7.2.1) in terms of r defined in (7.2.6). That is, we have

$$M(q)\dot{r} = -V_m(q,\dot{q})r + u_a(t) - \tau, \tag{7.2.37}$$

where the $n \times 1$ vector $u_a(t)$ is used to represent the "actual manipulator dynamics" given by

$$u_a(t) = M(q)(\ddot{q}_d + \dot{e}) + V_m(q,\dot{q})(\dot{q}_d + e) + G(q) + F_d\dot{q}. \tag{7.2.38}$$

Adding and subtracting the term $u_d(t)$ on the right-hand side of (7.2.37) yields

$$M(q)\dot{r} = -V_m(q,\dot{q})r + u_d(t) + \tilde{U} - \tau, \tag{7.2.39}$$

where $\tilde{U}$ is defined as

$$\tilde{U} = u_a(t) - u_d(t). \tag{7.2.40}$$

As shown similarly in [Sadegh and Horowitz 1990], this difference between the actual manipulator dynamics (i.e., $u_a(t)$) and the repeatable manipulator dynamics (i.e., $u_d(t)$) can be quantified as

$$\left\|\tilde{U}\right\| \leq \zeta_1 \|e\| + \zeta_2 \|e\|^2 + \zeta_3 \|r\| + \zeta_4 \|r\| \|e\|, \tag{7.2.41}$$

where ζ_1, ζ_2, ζ_3, and ζ_4 are positive bounding constants that depend on the desired trajectory and the physical properties of the specific robot configuration (i.e., link mass, link length, friction coefficients, etc.).

The last step in forming the error system is to substitute the control given by (7.2.31) into (7.2.39) to yield

$$M(q)\dot{r} = -V_m(q,\dot{q})r - k_v r - k_p e - k_a \|e\|^2 r + \tilde{u}_d(t) + \tilde{U}. \tag{7.2.42}$$

We now analyze the stability of the error system given by (7.2.42) with the Lyapunov-like function

$$V = \frac{1}{2}r^T M(q) r + \frac{1}{2}k_p e^T e + \frac{1}{2k_L}\int_{t-T}^{t} \tilde{u}_d^T(\sigma)\tilde{u}_d(\sigma)\,d\sigma. \tag{7.2.43}$$

Differentiating (7.2.43) with respect to time yields

7.2 Robot Controllers with Reduced On-Line Computation

$$\dot{V} = \frac{1}{2} r^T \dot{M}(q) r + r^T M(q) \dot{r} + k_p e^T \dot{e}$$
$$+ \frac{1}{2k_L} \left(\tilde{u}_d^T(t) \tilde{u}_d(t) - \tilde{u}_d^T(t-T) \tilde{u}_d(t-T) \right). \quad (7.2.44)$$

Substituting the error system given by (7.2.42) into (7.2.44) yields

$$\dot{V} = k_p e^T \dot{e} - k_v r^T r - k_p r^T e - k_a \|e\|^2 r^T r + r^T \tilde{U}$$
$$+ \frac{1}{2} r^T \left(\dot{M}(q) - 2V_m(q,\dot{q}) \right) r + r^T \tilde{u}_d(t)$$
$$+ \frac{1}{2k_L} \left(\tilde{u}_d^T(t) \tilde{u}_d(t) - \tilde{u}_d^T(t-T) \tilde{u}_d(t-T) \right). \quad (7.2.45)$$

By utilizing the skew-symmetric property and the learning error update law in (7.2.36), it is easy to show that the second line in (7.2.45) is equal to

$$-\tfrac{1}{2} k_L r^T r.$$

Therefore, by invoking the definition of r given in (7.2.6), (7.2.45) simplifies to

$$\dot{V} = -k_p e^T e - (k_v + \tfrac{1}{2} k_L) r^T r - k_a \|e\|^2 r^T r + r^T \tilde{U}. \quad (7.2.46)$$

From (7.2.46) we can place an upper bound on $\dot{V}$ in the following manner:

$$\dot{V} \leq -k_p \|e\|^2 - (k_v + \tfrac{1}{2} k_L) \|r\|^2 - k_a \|e\|^2 \|r\|^2 + \|r\| \left\| \tilde{U} \right\|. \quad (7.2.47)$$

The rest of the stability argument is a modification of the stability argument presented in the preceding section for the DCAL. Specifically, we first note that (7.2.18) and (7.2.47) are almost identical since $\tilde{Y}$ in (7.2.18) and $\tilde{U}$ in (7.2.47) are bounded by the same scalar function. After one retraces the steps of the DCAL stability argument, we can see that the controller gains k_a and k_p should still be adjusted according to (7.2.22) and (7.2.25), respectively. However, the controller gain k_v, is adjusted in conjunction with the controller gain k_L to satisfy

$$k_v + \tfrac{1}{2} k_L > \zeta_1/2 + \zeta_3 + \zeta_4/4 \quad (7.2.48)$$

where $\zeta_1, \zeta_2, \zeta_3$, and ζ_4 are defined (7.2.41). If the controller gains are adjusted according to (7.2.22), (7.2.25), and (7.2.48), then from the analytical development given for the DCAL (i.e., (7.2.18) to (7.2.24)) and (7.2.47), we can place the new upper bound on $\dot{V}$:

$$\dot{V} \leq -\lambda_3 \|x\|^2, \qquad (7.2.49)$$

where λ_3 is a positive scalar constant given by $\lambda_{min}\{Q_0\}$,

$$Q_o = \begin{bmatrix} k_p - \zeta_2/4 & -\zeta_1/2 \\ -\zeta_1/2 & k_v + {}^1\!/\!_2 k_L - \zeta_3 - \zeta_4/4 \end{bmatrix}, \text{ and } x = \begin{bmatrix} \|e\| \\ \|r\| \end{bmatrix}.$$

We now detail the type of stability for the tracking error. First note that from (7.2.49), we can place the new upper bound on $\dot{V}$:

$$\dot{V} \leq -\lambda_3 \|r\|^2, \qquad (7.2.50)$$

which implies that

$$\int_0^\infty \dot{V}(\sigma)\,d\sigma \leq -\lambda_3 \int_0^\infty \|r(\sigma)\|^2 d\sigma. \qquad (7.2.51)$$

Multiplying (7.2.51) by -1 and integrating the left-hand side of (7.2.51) yields

$$V(0) - V(\infty) \geq \lambda_3 \int_0^\infty \|r(\sigma)\|^2 d\sigma. \qquad (7.2.52)$$

Since $\dot{V}$ is negative semidefinite as delineated by (7.2.49), we can state that V is a nonincreasing function and therefore is upper bounded by $V(0)$. By recalling that $M(q)$ is lower bounded as delineated by the positive-definite property of the inertia matrix, we can state that V given in (7.2.43) is lower bounded by zero. Since V is nonincreasing, upper bounded by $V(0)$, and lower bounded by zero, we can write (7.2.52) as

$$\lambda_3 \int_0^\infty \|r(\sigma)\|^2 d\sigma < \infty \qquad (7.2.53)$$

or

$$\sqrt{\int_0^\infty \|r(\sigma)\|^2 d\sigma} < \infty. \qquad (7.2.54)$$

The bound delineated by (7.2.54) informs us that $r \in L_2^n$ (see Chapter 2), which means that the filtered tracking error r is bounded in the "special" way given by (7.2.54).

7.2 Robot Controllers with Reduced On-Line Computation

To establish a stability result for the position tracking error e, we establish the transfer function relationship between the position tracking error and the filtered tracking error r. From (7.2.6), we can state that

$$e(s) = G(s) r(s), \qquad (7.2.55)$$

where s is the Laplace transform variable,

$$G(s) = (sI + I)^{-1}, \qquad (7.2.56)$$

and I is the $n \times n$ identity matrix. Since $G(s)$ is a strictly proper, asymptotically stable transfer function and $r \in L_2^n$, we can use Theorem 2.4.7 in Chapter 2 to state that

$$\lim_{t \to \infty} e = 0. \qquad (7.2.57)$$

Therefore, if the controller gains are selected according to (7.2.22), (7.2.25), and (7.2.48), the position tracking error e is asymptotically stable. In accordance with the theoretical development presented in this section, all we can say about the velocity tracking error $\dot{e}$ is that it is bounded. It should be noted that if the learning estimate $\hat{u}_d(t)$ in (7.2.32) is "artificially" kept from growing, we can conclude that the velocity tracking error is asymptotically stable [Sadegh et al. 1990]. The stability proof for this modification is a straightforward application of the adaptive control proofs presented in Chapter 6.

The repetitive controller examined in this section is summarized in Table 7.2.2 and depicted in Figure 7.2.3. After glancing through Table 7.2.2, we can see that the RCL requires very little information about the robot being controlled as opposed to adaptive controllers that required the formulation of regression-type matrices. Another obvious advantage of the RCL is that it requires very little on-line computation. We now present an example to illustrate how Table 7.2.2 can be used to design repetitive controllers for robot manipulators.

EXAMPLE 7.2-2: RCL for the Two-Link Arm

We wish to design and simulate the RCL given in Table 7.2.2 for the two-link arm given in Figure 6.2.1. (The dynamics for this robot arm are given in Chapter 3.) From Table 7.2.2, the RCL can be written as

$$\tau_1 = \hat{u}_{d1} + k_v r_1 + k_p e_1 + k_a \|e\|^2 r_1 \qquad (1)$$

and

Table 7.2.2: RCL Controller

Torque Controller:

$$\tau = \hat{u}_d + k_v r + k_p e + k_a \|e\|^2 r$$

where

$$r = e + \dot{e}.$$

Learning Update Rule:

$$\hat{u}_d(t) = \hat{u}_d(t - T) + k_L r$$

Stability:

Tracking error e is asymptotically stable. Tracking error $\dot{e}$ is bounded.

Comments: Desired trajectory must be periodic with period T, and the controller gains k_a, k_p, k_L, and k_v must be sufficiently large.

$$\tau_2 = \hat{u}_{d2} + k_v r_2 + k_p e_2 + k_a \|e\|^2 r_2, \tag{2}$$

where $r_1 = e_1 + \dot{e}_1$, $r_2 = e_2 + \dot{e}_2$, and $\|e\|^2 = e_1^2 + e_2^2$.

Formulating the learning update rule as given in Table 7.2.2 yields

$$\hat{u}_{d1}(t) = \hat{u}_{d1}(t - T) + k_L r_1 \tag{3}$$

and

$$\hat{u}_{d2}(t) = \hat{u}_{d2}(t - T) + k_L r_2. \tag{4}$$

For $m_1 = 0.8$ kg, $m_2 = 2.3$ kg, and link lengths of 1 m, the RCL was simulated with

$$k_a = k_v = k_p = k_L = 50, \quad T = 2\pi,$$

$$q_1(0) = q_2(0) = \dot{q}_1(0) = \dot{q}_2(0) = 0,$$

and

7.3 Adaptive Robust Control

$$q_{d1} = q_{d2} = \sin t.$$

The position and velocity tracking error is depicted in Figure 7.2.4. As illustrated by the figure, the position and velocity tracking error are both asymptotically stable; however, in accordance with the theoretical development in this subsection, we are only guaranteed that the position tracking error will be asymptotically stable. ∎

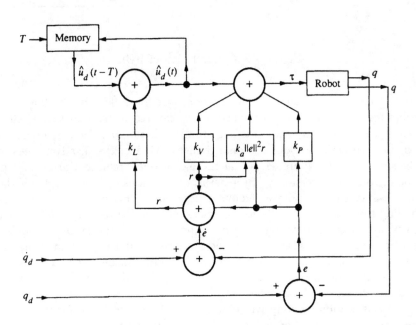

Figure 7.2.3: Block diagram of RCL.

7.3 Adaptive Robust Control

In Chapter 6 we discussed the use of adaptive controllers for the tracking control of robot manipulators. One of the attractive features of the adaptive controllers is that the control implementation does not require *a priori* knowledge of unknown constant parameters such as payload masses or friction coefficients. Two disadvantages of the adaptive controllers are that large amounts of on-line calculation are required, and the lack of robustness to additive bounded disturbances.

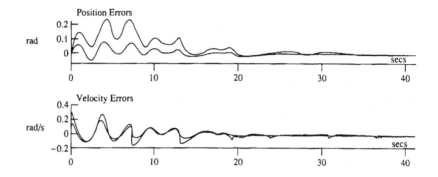

Figure 7.2.4: Simulation of RCL.

In Chapter 5 we discussed the use of robust controllers for the control of robot manipulators. Two of the attractive features of the robust controllers are that on-line computation is kept to a minimum and their inherent robustness to additive bounded disturbances. One of the disadvantages of the robust control approach is that these controllers require *a priori* known bounds on the uncertainty. In general, calculations of the bounds on the uncertainty can be quite a tedious process since this calculation involves finding the maximum values for the mass and friction related constants for each link of the robot manipulator. Another disadvantage of the robust control approach is that even in the absence of additive bounded disturbances, we cannot guarantee asymptotic stability of the tracking error. In general, it would be desirable to obtain at least a "theoretical" asymptotic stability result for the tracking error.

In this section an adaptive robust controller is developed for the tracking control of robot manipulators. The adaptive robust controller can be thought of as combining the best qualities of the adaptive controller and the robust controller. This control approach has the advantages of reduced online calculations (compared to the adaptive control method), robustness to additive bounded disturbances, no *a priori* knowledge of system uncertainty, and asymptotic tracking error performance.

For purposes of control design in this section, we assume that the robotic manipulator is a revolute manipulator with dynamics given by

$$\tau = M(q)\ddot{q} + V_m(q,\dot{q})\dot{q} + G(q) + F_d\dot{q} + F_s(\dot{q}) + T_d, \tag{7.3.1}$$

where F_d is a $n \times n$ positive definite, diagonal matrix that is used to represent the dynamic coefficients of friction, $F_s(\dot{q})$ is a $n \times 1$ vector containing the

7.3 Adaptive Robust Control

static friction terms, T_d is a $n \times 1$ vector representing an unknown bounded disturbance, and all other quantities are as defined in Chapter 3.

The adaptive robust controller is very similar to the robust control strategies discussed in Chapter 5 in that an auxiliary controller is used to "bound" the uncertainty. Recall from Chapter 5 that the robust controllers bounded the uncertainty by using a scalar function that was composed of tracking error norms and positive bounding constants. For example, suppose that the dynamics given by

$$w = M(q)(\ddot{q}_d + \dot{e}) + V_m(q, \dot{q})(\dot{q}_d + e) + G(q) + F_d \dot{q} + F_s(\dot{q}) + T_d, \quad (7.3.2)$$

represent the uncertainty for a given robot controller. That is, the dynamics given by (7.3.2) are uncertain in that payload masses, coefficients of friction, and disturbances are not known exactly. It is assumed; however, that a positive scalar function ρ can be used to bound the uncertainty as follows:

$$\rho \geq \|w\|. \quad (7.3.3)$$

As delineated in [Dawson et al. 1990], the physical properties of the robot manipulator can be used to show that the dynamics given by (7.3.2) can be bounded as

$$\rho = \delta_0 + \delta_1 \|\mathbf{e}\| + \delta_2 \|\mathbf{e}\|^2 \geq \|w\|, \quad (7.3.4)$$

where

$$\mathbf{e} = \begin{bmatrix} e \\ \dot{e} \end{bmatrix} \quad (7.3.5)$$

and δ_0, δ_1, and δ_2 are positive bounding constants that are based on the largest possible payload mass, link mass, friction coefficients, disturbances, and so on.

In general, the robust controllers presented in Chapter 5 required that the bounding constants defined in (7.3.4) be formulated *a priori*. The adaptive robust controller that will be developed in this section "learns" these bounding constants on-line as the manipulator moves. That is, in the control implementation, we do not require knowledge of the bounding constants; rather, we only require the existence of the bounding constants defined in (7.3.4).

Similar to the general development presented in [Corless and Leitmann 1983], the adaptive robust controller has the form

$$\tau = K_v r + v_R, \quad (7.3.6)$$

where K_v is a $n \times n$ diagonal, positive-definite matrix, r (the filtered tracking error) is defined as in (7.2.6), and v_R is a $n \times 1$ vector representing an auxiliary controller. The auxiliary controller v_R in (7.3.6) is defined by

$$v_R = \frac{r\hat{\rho}^2}{\hat{\rho}\|r\| + \varepsilon}, \qquad (7.3.7)$$

where

$$\dot{\varepsilon} = -k_\varepsilon \varepsilon, \quad \varepsilon(0) > 0, \qquad (7.3.8)$$

k_ε is a positive scalar control constant, $\hat{\rho}$ is a scalar function defined as

$$\hat{\rho} = \hat{\delta}_0 + \hat{\delta}_1 \|\mathbf{e}\| + \hat{\delta}_2 \|\mathbf{e}\|^2, \qquad (7.3.9)$$

and $\hat{\delta}_0$, $\hat{\delta}_1$, and $\hat{\delta}_2$ are the dynamic estimates of the corresponding bounding constants δ_0, δ_1, and δ_2 defined in (7.3.4). The bounding estimates denoted by " $\hat{\ }$ " are changed on-line based on an adaptive update rule. Before giving the update rule, we write (7.3.9) in the more convenient form

$$\hat{\rho} = S\hat{\theta}, \qquad (7.3.10)$$

where

$$S = \begin{bmatrix} 1 & \|\mathbf{e}\| & \|\mathbf{e}\|^2 \end{bmatrix} \quad \text{and} \quad \hat{\theta} = \begin{bmatrix} \hat{\delta}_0 & \hat{\delta}_1 & \hat{\delta}_2 \end{bmatrix}^T.$$

The actual bounding function ρ given in (7.3.3) can also be written in the matrix form

$$\rho = S\theta, \qquad (7.3.11)$$

where

$$\theta = \begin{bmatrix} \delta_0 & \delta_1 & \delta_2 \end{bmatrix}^T.$$

Note the similarity between the regression matrix formulation in the adaptive approach (see Chapter 6) and the formulation given by (7.3.10). Specifically, the 1×3 matrix S resembles a "regression matrix," and the 3×1 vector $\hat{\theta}$ resembles a "parameter estimate vector."

The bounding estimates defined in (7.3.10) are updated on-line by the relation

$$\dot{\hat{\theta}} = \gamma S^T \|r\|, \qquad (7.3.12)$$

where r is defined in (7.2.6), S is defined in (7.3.10), and γ is a positive scalar control constant. For convenience, we also note that since δ_0, δ_1, and δ_2 defined in (7.3.4) are constants, (7.3.12) can be written as

7.3 Adaptive Robust Control

$$\dot{\hat{\theta}} = -\gamma S^T \|r\| \tag{7.3.13}$$

since we will define the difference between θ and $\hat{\theta}$ as

$$\tilde{\theta} = \theta - \hat{\theta}. \tag{7.3.14}$$

We now turn our attention to analyzing the stability of the corresponding error system for the controller given in (7.3.6). Substituting the controller (7.3.6) into the robot Equation (7.3.1) gives the error system

$$M(q)\dot{r} = -V_m(q,\dot{q})r - K_v r + w - v_R, \tag{7.3.15}$$

where w is defined in (7.3.2).

We now analyze the stability of the error system given by (7.3.15) with the Lyapunov-like function

$$V = 1/2 r^T M(q) r + 1/2 \tilde{\theta}^T \gamma^{-1} \tilde{\theta} + k_\varepsilon^{-1} \varepsilon. \tag{7.3.16}$$

Differentiating (7.3.16) with respect to time yields

$$\dot{V} = 1/2 r^T \dot{M}(q) r + r^T M(q) \dot{r} + \tilde{\theta}^T \gamma^{-1} \dot{\tilde{\theta}} + k_\varepsilon^{-1} \dot{\varepsilon} \tag{7.3.17}$$

since scalar quantities can be transposed. Substituting (7.3.13) and (7.3.15) into (7.3.17) yields

$$\dot{V} = -r^T K_v r - S\tilde{\theta} \|r\| + r^T (w - v_R) + k_\varepsilon^{-1} \dot{\varepsilon} \\ + 1/2 r^T \left(\dot{M}(q) - 2V_m(q,\dot{q}) \right) r. \tag{7.3.18}$$

By utilizing the skew-symmetric property, it is easy to see that the second line in (7.3.18) is equal to zero. From (7.3.18), we can use (7.3.3) and (7.3.11) to place an upper bound on $\dot{V}$ in the following manner:

$$\dot{V} \leq -r^T K_v r - S\tilde{\theta} \|r\| + S\theta \|r\| - r^T v_R + k_\varepsilon^{-1} \dot{\varepsilon}. \tag{7.3.19}$$

Substituting (7.3.7), (7.3.8), (7.3.10) and (7.3.14) into (7.3.19), we obtain

$$\dot{V} \leq -r^T K_v r - \varepsilon + S\hat{\theta} \|r\| - \frac{r^T r \left(S\hat{\theta}\right)^2}{S\hat{\theta} \|r\| + \varepsilon}, \tag{7.3.20}$$

which can be written as

$$\dot{V} \leq -r^T K_v r - \varepsilon + S\hat{\theta}\,\|r\| - \frac{\|r\|^2 \left(S\hat{\theta}\right)^2}{S\hat{\theta}\,\|r\| + \varepsilon}, \qquad (7.3.21)$$

Obtaining a common denominator for the last two terms in (7.3.21) enables us to write (7.3.21) as

$$\dot{V} \leq -r^T K_v r - \varepsilon + \frac{\varepsilon S\hat{\theta}\,\|r\|}{S\hat{\theta}\,\|r\| + \varepsilon}, \qquad (7.3.22)$$

Since the sum of the last two terms in (7.3.22) is always less than zero, we can place the new upper bound on $\dot{V}$:

$$\dot{V} \leq -r^T K_v r. \qquad (7.3.23)$$

We now detail the type of stability for the tracking error. First, note from (7.3.23) that we can place the new upper bound on $\dot{V}$:

$$\dot{V} \leq -\lambda_{\min}\{K_v\}\,\|r\|^2. \qquad (7.3.24)$$

As illustrated for the RCL in the preceding section, we can use (7.3.24) to show that all signals are bounded and that $r \in L_2^n$ (see Chapter 2). Following the RCL stability analysis, we can use (7.2.6) to show that the position tracking error (e) is related to the filtered tracking error r by the transfer function relationship

$$e(s) = G(s)\,r(s), \qquad (7.3.25)$$

where s is the Laplace transform variable and $G(s)$ is a strictly proper, asymptotically stable transfer function. Therefore, we can use Theorem 2.4.7 in Chapter 2 to state that

$$\lim_{t \to \infty} e = 0. \qquad (7.3.26)$$

The result above informs us that the position tracking error e is asymptotically stable. In accordance with the theoretical development presented in this section, we can only state that the velocity tracking error $\dot{e}$ and the bounding estimates $\hat{\theta}$ are bounded. It should be noted that in [Corless and Leitmann 1983], a more complex theoretical development is presented that proves the velocity tracking error is asymptotically stable. However, in the interest of brevity this additional information is left for the reader to pursue.

7.3 Adaptive Robust Control

Table 7.3.1: Adaptive Robust Controller

Torque Controller:

$$\tau = K_v r + \frac{r\hat{\rho}^2}{\hat{\rho}\|r\|+\varepsilon}$$

where

$$\hat{\rho} = S\hat{\theta} = \left[1 \, \left\|\begin{bmatrix} e \\ \dot{e} \end{bmatrix}\right\| \, \left\|\begin{bmatrix} e \\ \dot{e} \end{bmatrix}\right\|^2 \right] \begin{bmatrix} \hat{\delta}_0 & \hat{\delta}_1 & \hat{\delta}_2 \end{bmatrix}^T$$

$$r = e + \dot{e}, \quad \text{and} \quad \dot{\varepsilon} = -k_\varepsilon \varepsilon$$

Bounding Estimate Update Rule:

$$\dot{\hat{\theta}} = \gamma S^T \|r\|$$

Stability:

Position tracking error e is asymptotically stable. Bounding estimate $\hat{\theta}$ and velocity tracking error $\dot{e}$ are bounded.

Comments: The adaptive robust controller can compensate for bounded disturbances with no modification.

The adaptive robust controller derived in this section is summarized in Table 7.3.1 and depicted in Figure 7.3.1. We now present an example to illustrate how Table 7.3.1 can be used to design adaptive robust controllers for robot manipulators.

EXAMPLE 7.3-1: Adaptive Robust Controller for the Two-Link Arm

We wish to design and simulate the adaptive robust controller given in Table 7.3.1 for the two-link arm given in Figure 6.2.1. (The dynamics for this robot arm are given in Chapter 3.) To model friction and disturbances, the dynamics

$$2\dot{q}_1 + 0.5\,\text{sgn}\,(\dot{q}_1) + 0.2\sin(3t) \tag{1}$$

and

$$2\dot{q}_2 + 0.5\text{sgn}\,(\dot{q}_2) + 0.2\sin(3t) \qquad (2)$$

were added to τ_1 and τ_2, respectively, in the two-link robot model.

We can now use Table 7.3.1 to formulate the adaptive robust controller as

$$\tau_1 = k_v r_1 + r_1 \hat{\rho}^2 \frac{1}{\hat{\rho}\,\|r\| + \varepsilon} \qquad (3)$$

and

$$\tau_2 = k_v r_2 + r_2 \hat{\rho}^2 \frac{1}{\hat{\rho}\,\|r\| + \varepsilon}, \qquad (4)$$

where $K_v = k_v I$, $r_1 = e_1 + \dot{e}_1$, $r_2 = e_2 + \dot{e}_2$, $\dot{\varepsilon} = -k_\varepsilon \varepsilon$, and $\|r\| = \sqrt{r_1^2 + r_2^2}$.

In the expression above for the control torques, the bounding function $\hat{\rho}$ is given by

$$\hat{\rho} = S\hat{\theta} = \begin{bmatrix} 1 & \|e\| & \|e\|^2 \end{bmatrix} \begin{bmatrix} \hat{\delta}_0 & \hat{\delta}_1 & \hat{\delta}_2 \end{bmatrix}^T, \qquad (5)$$

where $\|e\| = \sqrt{e_1^2 + e_2^2 + \dot{e}_1^2 + \dot{e}_2^2}$. From Table 7.3.1, the associated bounding estimates are updated in the fashion

$$\dot{\hat{\delta}}_0 = \gamma\,(r), \quad \dot{\hat{\delta}}_1 = \gamma\,\|e\|\,\|r\|, \quad \text{and} \quad \dot{\hat{\delta}}_2 = \gamma\,\|e\|^2\,\|r\|. \qquad (6)$$

For $m_1 = 0.8$ kg, $m_2 = 2.3$ kg, and link lengths of 1 m each, the adaptive robust controller was simulated with the control parameters, initial conditions, and desired trajectory given by

$$k_v = 50, \quad \gamma = 5, \quad \varepsilon\,(0) = 1, \quad k_\varepsilon = 1, \quad \hat{\delta}_0\,(0) = 20,$$

$$q_1\,(0) = q_2\,(0) = \dot{q}_1\,(0) = \dot{q}_2\,(0) = \hat{\delta}_1\,(0) = \hat{\delta}_2\,(0) = 0,$$

and

$$q_{d1} = q_{d2} = \sin t.$$

7.4 Compensation for Actuator Dynamics

The tracking error and mass estimates are depicted in Figure 7.3.2. As illustrated by the figure, the position and the velocity tracking error are both asymptotically stable, and the bounding estimates remain bounded. It should be noted that from the theoretical development given in this section, we are only guaranteed that the position tracking error is asymptotically stable while all other signals remain bounded. ∎

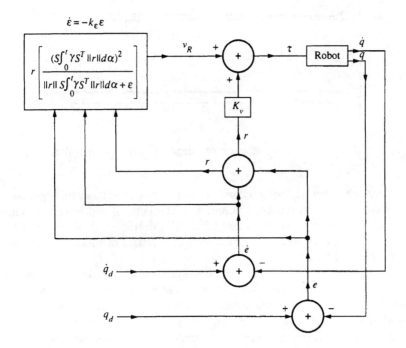

Figure 7.3.1: Block diagram of adaptive robust controller.

7.4 Compensation for Actuator Dynamics

Throughout this book we have discussed controllers that are designed at the "torque input level." That is, any dynamics associated with the actuators have been neglected. The reason for bringing up this point is not to denigrate the control development discussed previously, since this research has been involved with solving a very difficult problem, namely, the global tracking control of a highly nonlinear system in the presence of uncertainty. We mention this deficiency in previous approaches to highlight the fact that

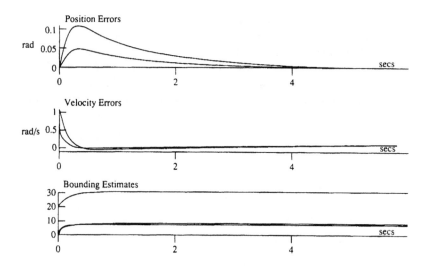

Figure 7.3.2: Simulation of adaptive robust controller.

in many robot control researchers' opinions, it is now time to begin to include the effects of actuator dynamics in the control synthesis. Recently, several researchers have postulated that the detrimental effects of actuators are preventing high-speed motion/force control of robot manipulators [Eppinger and Seering 1987].

In this section we illustrate how a systematic approach can be used to compensate for actuator dynamics in the form of electrical effects and joint flexibilities. Using this approach and the assumption of exact model knowledge, controllers are developed that yield a global asymptotic stability result for the link tracking error. Although we assume exact knowledge of the model, it is important to realize that in some cases, it may be possible to formulate adaptive and robust nonlinear tracking controllers to compensate for "uncertainty." The compensation of uncertain systems in the presence of actuator dynamics is currently being researched [Ghorbel and Spong 1990].

Electrical Dynamics

In this subsection, we illustrate how a "corrective" controller [Kokotovic et al. 1986] can be synthesized that ensures asymptotic link tracking despite the electrical dynamics that a motor will add to the overall system dynamics. The terminology corrective controller is used to emphasize the fact that the controller corrects for the electrical dynamics. The class of robots studied in this subsection will be referred to as rigid-link electrically driven (RLED) robots. For simplicity, we assume that the ac-

7.4 Compensation for Actuator Dynamics

tuator is a direct-current (dc) motor; however, the following analysis, with some modifications, can be used for more complicated motors such as the switched-reluctance motor [Taylor 1989].

The model [Tarn et al. 1991] for the RLED robot is taken to be

$$\mathbf{M}(q)\ddot{q} + N(q,\dot{q}) = K_T I \qquad (7.4.1)$$

$$L_a \dot{I} + R(I,\dot{q}) = u_E, \qquad (7.4.2)$$

where

$$\mathbf{M}(q) = M(q) + J \qquad (7.4.3)$$

$M(q)$ is a $n \times n$ link inertia matrix, $N(q,\dot{q})$ is a $n \times I$ vector containing the centripetal, Coriolis, gravity, damping, and friction terms, J is a $n \times n$ constant, diagonal, positive-definite matrix used to represent the actuator inertia, $I(t)$ is an $n \times 1$ vector used to denote the current in each actuator, K_T is a constant diagonal $n \times n$ matrix used to represent the conversion between torque and current, L_a is a $n \times n$ constant positive-definite diagonal matrix used to represent the electrical inductance, $R(I,\dot{q})$ is a $n \times 1$ vector used to represent the electrical resistance and the motor back-electromotive force, and $u_E(t)$ is an $n \times 1$ control vector used to represent the input motor voltage.

Throughout the book, a good deal of emphasis has been placed on the utilization of physical properties of robot manipulators to aid us in the stability analysis. In this tradition we note that the composite inertia matrix $\mathbf{M}(q)$ defined in (7.4.3) is symmetric, positive definite, and is uniformly bounded as a function of q; therefore, we can state for any $n \times 1$ vector x that

$$\mathbf{m}_1 \|x\|^2 = \lambda_{\min}\{\mathbf{M}(q)\}\|x\|^2 \le x^T \mathbf{M}(q) x, \qquad (7.4.4)$$

where $\mathbf{m}_1$ is a positive scalar constant that depends on the mass properties of the specific robot (see Chapter 3). From (7.4.4), it can also be established that

$$x^T \mathbf{M}^{-1}(q) x \le \lambda_{\max}\{\mathbf{M}^{-1}(q)\}\|x\|^2 = \frac{1}{\mathbf{m}_1}\|x\|^2 = \left\|\mathbf{M}^{-1}(q)\right\|_{i2}\|x\|^2, \qquad (7.4.5)$$

where $\|\cdot\|_{i2}$ is used to denote the induced 2-norm (see Chapter 2).

As discussed many times before, we are interested in the performance of the link tracking error. To avoid confusion, we restate that the *tracking error* is defined to be

$$e = q_d - q, \tag{7.4.6}$$

where q_d represents the desired link trajectory. We will assume that q_d and its first, second, and third derivatives are all bounded as functions of time. We also assume that the first derivative of the link dynamics on the left-hand side of (7.4.1) exists. These assumptions on the "smoothness" of the desired trajectory and the link dynamics ensure that the controller, which will be developed later, remains bounded.

The control objective will be to obtain asymptotic link tracking despite the electrical dynamics. To accomplish this objective, we first rewrite (7.4.1) in terms of the tracking error given by (7.4.6) to yield

$$\mathbf{M}(q)\ddot{q}_d - \mathbf{M}(q)\ddot{e} + N(q, \dot{q}) = K_T I. \tag{7.4.7}$$

The error system given by (7.4.7) can also be written in the state-space form

$$\dot{\mathbf{e}} = A_o \mathbf{e} + B\left[\ddot{q}_d + \mathbf{M}^{-1}(q) N(q, \dot{q}) - \mathbf{M}^{-1}(q) K_T I\right], \tag{7.4.8}$$

where

$$A_o = \begin{bmatrix} 0_{n \times n} & I_{n \times n} \\ 0_{n \times n} & 0_{n \times n} \end{bmatrix}, \quad B = \begin{bmatrix} 0_{n \times n} \\ I_{n \times n} \end{bmatrix}, \quad \mathbf{e} = \begin{bmatrix} e \\ \dot{e} \end{bmatrix},$$

$0_{n \times n}$ is the $n \times n$ zero matrix, and $I_{n \times n}$ is the $n \times n$ identity matrix.

As one can plainly see, there is no control input in (7.4.8); therefore, we will add and subtract the term $B\mathbf{M}^{-1}(q)u_L$ on the right-hand side of (7.4.8) to yield

$$\begin{aligned}\dot{\mathbf{e}} =& A_o \mathbf{e} + B\left[\ddot{q}_d + \mathbf{M}^{-1}(q) N(q, \dot{q}) - \mathbf{M}^{-1}(q) u_L\right] \\ &+ B\left[\mathbf{M}^{-1}(q)(u_L - K_T I)\right], \end{aligned} \tag{7.4.9}$$

where u_L is an $n \times 1$ vector representing a "fictitious" $n \times 1$ control input. As it turns out, the controller u_L is the computed-torque controller that ensures asymptotic link tracking error if the electrical dynamics were not present. As we will see later, the fictitious controller u_L is actually embedded inside the overall control strategy, which is designed at the voltage control input u_E.

Continuing with the error system development, we define u_L for RLED robots to be the computed-torque controller

$$u_L = \mathbf{M}(q)\left(\ddot{q}_d + K_{Lv}\dot{e} + K_{Lp}e\right) + N(q, \dot{q}), \tag{7.4.10}$$

7.4 Compensation for Actuator Dynamics

where K_{Lv} and K_{Lp} are defined to be $n \times n$ positive-definite diagonal matrices. Substituting (7.4.10) for only the first u_L term in (7.4.9) yields the link tracking error system

$$\dot{\mathbf{e}} = A_L \mathbf{e} + B_L \mathbf{M}^{-1}(q)\eta_E, \qquad (7.4.11)$$

where

$$A_L = \begin{bmatrix} O_{n \times n} & I_{n \times n} \\ -K_{Lp} & -K_{Lv} \end{bmatrix} \quad \text{and} \quad \eta_E = u_L - K_T I. \qquad (7.4.12)$$

With regard to the link tracking error system given by (7.4.11), if η_E could be guaranteed to be zero for all time, we could easily show that the tracking error would be asymptotically stable since A_L defined in (7.4.12) has stable eigenvalues. Therefore, one can view the control objective as forcing the "perturbation" η_E to go to zero.

To design a control law for η_E, we must first establish its dynamic characteristics. From (7.4.12), the derivative of η_E with respect to time is given by

$$\dot{\eta}_E = \dot{u}_L - K_T \dot{I}. \qquad (7.4.13)$$

To obtain the the dynamic characteristics of η_E, we substitute for $\dot{I}$ in (7.4.13) from (7.4.2) to yield

$$\dot{\eta}_E = \dot{u}_L - K_T L_a^{-1}(u_E - R(I, \dot{q})). \qquad (7.4.14)$$

We can now use (7.4.14) to design a control law at the input u_E to force η_E to go to zero. The fact that η_E should go to zero motivates the corrective control law

$$u_E = L_a K_T^{-1}(\dot{u}_L + K_{Ep}\eta_E) + R(I, \dot{q}), \qquad (7.4.15)$$

where K_{Ep} is defined to be a $n \times n$ positive-definite diagonal matrix. Substituting (7.4.15) into (7.4.14) yields

$$\dot{\eta}_E = -K_{Ep}\eta_E. \qquad (7.4.16)$$

The dynamic equations given by (7.4.11) and (7.4.16) can be thought of as two interconnected systems representing the overall closed-loop dynamics. As one would expect, it would be desirable to determine the type of stability of the overall closed-loop system. To determine the type of stability of (7.4.11) and (7.4.16), we will utilize the Lyapunov function

$$V = \mathbf{e}^T P_L \mathbf{e} + 1/2 \eta_E^T \eta_E, \qquad (7.4.17)$$

where

$$P_L = 1/2 \begin{bmatrix} K_{Lp} + 1/2 K_{Lv} & 1/2 I_{n \times n} \\ 1/2 I_{n \times n} & I_{n \times n} \end{bmatrix}.$$

If the sufficient condition given by

$$\lambda_{\min}\{K_{Lv}\} > 1 \tag{7.4.18}$$

is satisfied, then by the Gerschgorin theorem (see Chapter 2), it is obvious that P_L is a positive-definite matrix, and hence V is a Lyapunov function. The condition given by (7.4.18) simply means that the smallest velocity controller gain should be larger than 1.

Differentiating (7.4.17) with respect to time yields

$$\dot{V} = \mathbf{e}^T P_L \dot{\mathbf{e}} + \dot{\mathbf{e}}^T P_L \mathbf{e} + 1/2 \eta_E^T \dot{\eta}_E + 1/2 \dot{\eta}_E^T \eta_E. \tag{7.4.19}$$

Substituting (7.4.11) and (7.4.16) into (7.4.19) yields

$$\dot{V} = -\mathbf{e}^T Q_L \mathbf{e} - \eta_E^T K_{Ep} \eta_E + 2\mathbf{e}^T P_L B \mathbf{M}^{-1}(q) \eta_E, \tag{7.4.20}$$

where

$$Q_L = -\left(A_L^T P_L + P_L A_L\right) = \begin{bmatrix} 1/2 K_{Lp} & O_{n \times n} \\ O_{n \times n} & K_{Lv} - 1/2 I_{n \times n} \end{bmatrix}. \tag{7.4.21}$$

Note that if the sufficient condition given by (7.4.18) holds, it is obvious that the matrix Q_L is positive definite. Using the fact that Q_L is positive definite allows us to place an upper bound on $\dot{V}$ given in (7.4.20). This upper bound is given by

$$\dot{V} \leq -\lambda_{\min}\{Q_L\} \|\mathbf{e}\|^2 - \lambda_{\min}\{K_{Ep}\} \|\eta_E\|^2 + 2\beta_o \|\mathbf{e}\| \|\eta_E\|, \tag{7.4.22}$$

where

$$\beta_o = 1/\mathbf{m}_1 > \|P_L B \mathbf{M}^{-1}(q)\|_{i2} = 1/2 \left\| \begin{bmatrix} 1/2 \mathbf{M}^{-1}(q) \\ \mathbf{M}^{-1}(q) \end{bmatrix} \right\|_{i2},$$

where $\mathbf{m}_l$ is defined in (7.4.5).

To determine the sufficient conditions on the controller gains for asymptotic stability, we rewrite (7.4.22) in the matrix form

$$\dot{V} \leq -x_o^T Q_o x_o, \tag{7.4.23}$$

where

7.4 Compensation for Actuator Dynamics

$$Q_o = \begin{bmatrix} \lambda_{\min}\{Q_L\} & -\beta_o \\ -\beta_o & \lambda_{\min}\{K_{Ep}\} \end{bmatrix} \quad \text{and} \quad x_o = \begin{bmatrix} \|e\| \\ \|\eta_E\| \end{bmatrix}.$$

By the Gerschgorin Theorem, the matrix Q_0 defined in (7.4.23) will be positive definite if the sufficient condition

$$\min\{\lambda_{\min}\{Q_L\}, \lambda_{\min}\{K_{Ep}\}\} > \beta_o \qquad (7.4.24)$$

holds. Therefore, if the controller gains satisfy the conditions given by (7.4.18) and (7.4.24), we can use standard Lyapunov stability arguments (see Chapter 2) to state that the vector x_0 defined in (7.4.23) and hence $\|e\|$, e, and $\dot{e}$ are all asymptotically stable. It is easy to show that if the sufficient condition

$$\min\{\lambda_{\min}\{K_{Lp}\}, \lambda_{\min}\{K_{Lv}\}, \lambda_{\min}\{K_{Ep}\}\} > 2/\mathbf{m}_1 + 1 \qquad (7.4.25)$$

holds, the conditions given by (7.4.18) and (7.4.24) are always satisfied.

It should be noted that the control given by (7.4.15) depends on the measurement of u_L, $\dot{u}_L$, and I. At first one might be tempted to state that this controller requires measurements of q, $\dot{q}$, $\ddot{q}$, and I; however, since we have assumed exact knowledge of the dynamic model given by (7.4.1) and (7.4.2), we can use this information to eliminate the need for measuring $\ddot{q}$. That is, by differentiating (7.4.10) with respect to time, $\dot{u}_L$ can be written as

$$\dot{u}_L = \dot{\mathbf{M}}(q)(\ddot{q}_d + K_{Lv}\dot{e} + K_{Lp}e) + \dot{N}(q, \dot{q})$$
$$+ \mathbf{M}(q)\left[\frac{d}{dt}\ddot{q}_d + K_{Lv}(\ddot{q}_d - \ddot{q}) + K_{Lp}\dot{e}\right], \qquad (7.4.26)$$

where $\ddot{q}$ is found from (7.4.1) to be

$$\ddot{q} = \mathbf{M}^{-1}(q)[K_T I - N(q, \dot{q})].$$

After substituting for $\ddot{q}$ in (7.4.26), u_L will depend only on the measurement of q, $\dot{q}$, and I. The actual control that would be implemented at the control input u_E can be found by making the appropriate substitution into (7.4.15). That is, the corrective control given by (7.4.15) can be written as

$$u_E = L_a K_T^{-1} \dot{u}_L + R(I, \dot{q}) + L_a K_T^{-1} K_{Ep}$$
$$[\mathbf{M}(q)(\ddot{q}_d + K_{Lv}\dot{e} + K_{Lp}e) + N(q, \dot{q}) - K_T I], \qquad (7.4.27)$$

where $\dot{u}_L$ would be given by (7.4.26).

After examining the functional dependence of $\dot{u}_L$ given in (7.4.26), it is now obvious why we have assumed that the desired trajectory and the link dynamics be sufficiently smooth. Specifically, we can see from (7.4.26) that the corrective controller requires that the first, second, and third time derivatives of the desired trajectory to be bounded while requiring the existence of the first derivative of the link dynamics. These assumptions on the desired trajectory and the link dynamics ensure that the control input will remain bounded.

The corrective controller derived above is summarized in Table 7.4.1 and depicted in Figure 7.4.1. We now present an example to illustrate how Table 7.4.1 can be used to design corrective controllers for RLED robots.

EXAMPLE 7.4-1: Corrective Controller for the One-Link RLED Arm

We wish to design and simulate a corrective controller using Table 7.4.1 for the one-link motor-driven robot arm given in Figure 7.4.2. The dynamics for the system are taken to be

$$\left(mL^2 + J\right)\ddot{q} + mLg\sin q + f_d\dot{q} = K_T I \tag{1}$$

and

$$L_a \dot{I} + RI + k_b \dot{q} = u_E, \tag{2}$$

where $m = 1$ kg, $K_T = 2$ N/A, $k_b = 0.3$ V-S, $f_d = 3$ kg-m/s, $L = 1$ m, $L_a = 0.1$ H, $R = 5$ Ω, g is the gravitational coefficient, $J = 0.2$ kg-m^2, I is the motor current, and u_E is the motor input voltage.

Assuming that the model given by (1) and (2) is known exactly, we can use Table 7.4.1 to formulate the corrective controller

$$\begin{aligned}u_E =& L_a K_T^{-1} \dot{u}_L + RI + k_b \dot{q} + L_a K_T^{-1} K_{Ep} \\ & \left[\left(mL^2 + J\right)\left(\ddot{q}_d + K_{Lv}\dot{e} + K_{Lp}e\right) + mLg\sin q + f_d\dot{q} - K_T I\right],\end{aligned} \tag{3}$$

where

$$\begin{aligned}\dot{u}_L =& mLg\dot{q}\cos q + f_d\ddot{q} + \left(mL^2 + J\right) \\ & \left[\frac{d}{dt}\ddot{q}_d + K_{Lv}\left(\ddot{q}_d - \ddot{q}\right) + K_{Lp}\dot{e}\right],\end{aligned} \tag{4}$$

7.4 Compensation for Actuator Dynamics

Table 7.4.1: RLED Corrective Controller

Voltage Controller:

$$u_E = L_a K_T^{-1} \dot{u}_L + R(I, \dot{q})$$
$$+ L_a K_T^{-1} K_{Ep} \left[\mathbf{M}(q)(\ddot{q}_d + K_{Lv}\dot{e} + K_{Lp}e) + N(q, \dot{q}) - K_T I \right]$$

where

$$\dot{u}_L = \dot{\mathbf{M}}(q)(\ddot{q}_d + K_{Lv}\dot{e} + K_{Lp}e) + \dot{N}(q, \dot{q})$$
$$+ \mathbf{M}(q) \left[\tfrac{d}{dt}\ddot{q}_d + K_{Lv}(\ddot{q}_d - \ddot{q}) + K_{Lp}\dot{e} \right]$$
$$\ddot{q} = \mathbf{M}^{-1}(q)[K_T I - N(q, \dot{q})]$$

Stability:

Tracking error e and $\dot{e}$ are asymptotically stable.

Comments: Controller requires exact knowledge of system dynamics and the controller gain matrices K_{Lv}, K_{Lp}, and K_{Ep} mustbe sufficiently large. Desired trajectory must be sufficiently smooth.

---

where $\ddot{q}$ is found from (1) to be

$$\ddot{q} = (mL^2 + J)^{-1} [K_T I - mLg \sin q - f_d \dot{q}]. \tag{5}$$

The corrective controller was simulated with the control parameters, initial conditions, and desired trajectory given by

$$K_{Lv} = K_{Lp} = K_{Ep} = 5,$$

$$q(0) = \dot{q}(0) = I(0) = 0,$$

and

$$q_d = \sin t.$$

The tracking error and the control voltage are depicted in Figure 7.4.3. As illustrated by the figure, the tracking error is asymptotically stable.

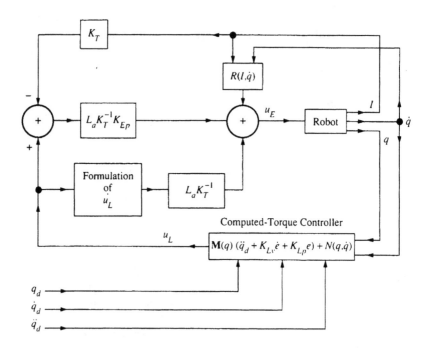

Figure 7.4.1: Block diagram of RLED corrective controller.

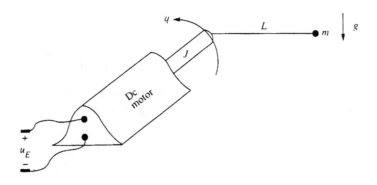

Figure 7.4.2: One-link RLED robot.

■

Joint Flexibilities

In this subsection we illustrate how a "corrective" controller can be synthesized that ensures asymptotic link tracking despite the joint flexibilities

7.4 Compensation for Actuator Dynamics

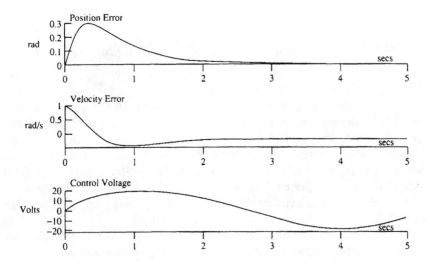

Figure 7.4.3: Simulation of RLED corrective controller.

that a drive or gearing will add to the overall system dynamics. The terminology corrective controller is used to emphasize the fact that the controller corrects for the dynamics that are used to represent the effects of joint flexibilities. The class of robots studied in this subsection will be referred to as rigid-link flexible-joint (RLFJ) robots.

The model [Spong 1987] for the RLFJ robot is taken to be

$$M(q)\ddot{q} + \mathbf{N}(q, \dot{q}) = K q_m \qquad (7.4.28)$$

$$J\ddot{q}_m + B_m(q_m, \dot{q}_m, q) = u_F, \qquad (7.4.29)$$

where

$$\mathbf{N}(q, \dot{q}) = N(q, \dot{q}) + Kq, \qquad (7.4.30)$$

$q_m(t)$ is a $n \times 1$ vector representing the motor displacement, K is a constant, diagonal, positive-definite $n \times n$ joint flexibility matrix, $B_m(q_m, \dot{q}_m, q)$ is an $n \times 1$ vector that represents the motor damping and flexibility effects, $u_F(t)$ is an $n \times 1$ control vector used to represent the input torque, and all other quantities are defined as in the preceding subsection. With regard to the rigid link model given in (7.4.28), we note from Chapter 3 that for any $n \times 1$ vector x

$$x^T M^{-1}(q) x \leq \lambda_{\max}\{M^{-1}(q)\} \|x\|^2 = \frac{1}{m_1} \|x\|^2 = \|M^{-1}(q)\|_{i2} \|x\|^2, \tag{7.4.31}$$

where m_1 is a positive scalar constant.

As in the preceding subsection, we are interested in the performance of the link tracking error defined in (7.4.6). For the control of RLFJ robots, we will assume that q_d and its first, second, third, and fourth derivatives are all bounded as functions of time. We also assume that the first and second derivatives of the link dynamics on the left-hand side of (7.4.28) exist. These assumptions on the "smoothness" of the desired trajectory and the link dynamics ensure that the controller, developed later, remains bounded.

Following the same analytical development given in the previous sections, we write (7.4.28) in terms of the tracking error given by (7.4.6) to yield the state-space form

$$\dot{\mathbf{e}} = A_o \mathbf{e} + B \left[\ddot{q}_d + M^{-1}(q) \mathbf{N}(q,\dot{q}) - M^{-1}(q) K q_m \right], \tag{7.4.32}$$

where A_0, $\mathbf{e}$, and B are defined as in (7.4.8). Again, since there is no control input in (7.4.32); we add and subtract the term $BM^{-1}(q)u_L$ on the right-hand side of (7.4.32) to yield

$$\begin{aligned}\dot{\mathbf{e}} = &A_o \mathbf{e} + B\left[\ddot{q}_d + M^{-1}(q)\mathbf{N}(q,\dot{q}) - M^{-1}(q)u_L\right] \\ &+ B\left[M^{-1}(q)(u_L - Kq_m)\right],\end{aligned} \tag{7.4.33}$$

where u_L is again used to represent a fictitious $n \times 1$ control input. As before, the fictitious controller u_L will be embedded inside the overall control strategy, which is designed at the control input u_F.

Continuing with the error system development, we define u_L for RLFJ robots to be the computed-torque controller

$$u_L = M(q)\left(\ddot{q}_d + K_{Lv}\dot{\mathbf{e}} + K_{Lp}\mathbf{e}\right) + \mathbf{N}(q,\dot{q}), \tag{7.4.34}$$

where K_{Lv} and K_{Lp} are defined as in (7.4.10). Substituting (7.4.34) into (7.4.33) yields the link tracking error system

$$\dot{\mathbf{e}} = A_L \mathbf{e} + BM^{-1}(q) C\eta_F, \tag{7.4.35}$$

where A_L is defined as in (7.4.12),

7.4 Compensation for Actuator Dynamics

$$C = \begin{bmatrix} I_{n\times n} & O_{n\times n} \end{bmatrix}, \text{ and } \eta_F = \begin{bmatrix} \eta_F \\ \dot{\eta}_F \end{bmatrix} = \begin{bmatrix} u_L - Kq_m \\ \dot{u}_L - K\dot{q}_m \end{bmatrix}. \quad (7.4.36)$$

The reason for defining η_F in terms of $(u_L - Kq_m)$ and its derivative is that the dynamics given by (7.4.29) are second-order dynamics. That is, since the actuator dynamics are second order, we force η_F and its derivative (i.e., $\dot{\eta}_F$) to zero to ensure that the link tracking error (**e**) goes to zero.

To design a control law for η_F, we must first establish its dynamic characteristics. From (7.4.36), the derivative of η_F is given by

$$\dot{\eta}_F = \begin{bmatrix} \dot{u}_L - K\dot{q}_m \\ \ddot{u}_L - K\ddot{q}_m \end{bmatrix}. \quad (7.4.37)$$

To obtain the the dynamic characteristics of η_F, we substitute for $\ddot{q}_m$ in (7.4.37) from (7.4.29) to yield

$$\dot{\eta}_F = \begin{bmatrix} \dot{u}_L - K\dot{q}_m \\ \ddot{u}_L - KJ^{-1}(u_F - B_m(q_m, \dot{q}_m, q)) \end{bmatrix}. \quad (7.4.38)$$

We can now use (7.4.38) to design a control law at the input u_F to force η_F to go to zero. The fact that η_F should go to zero motivates the control law

$$u_F = JK^{-1}(\ddot{u}_L + K_{Fv}\dot{\eta}_F + K_{Fp}\eta_F) + B_m(q_m, \dot{q}_m, q), \quad (7.4.39)$$

where K_{Fv} and K_{Fp} are defined to be $n \times n$ positive-definite diagonal matrices. Substituting (7.4.39) into (7.4.38) yields

$$\dot{\eta}_F = A_F \eta_F, \quad (7.4.40)$$

where

$$A_F = \begin{bmatrix} O_{n\times n} & I_{n\times n} \\ -K_{Fp} & -K_{Fv} \end{bmatrix}.$$

The dynamic equations given by (7.4.35) and (7.4.40) can be thought of as two interconnected systems representing the overall closed-loop dynamics. To determine the type of stability for the closed-loop dynamics, we will utilize the Lyapunov function

$$V = \mathbf{e}^T P_L \mathbf{e} + \eta_F^T P_F \eta_F, \quad (7.4.41)$$

where P_L is defined in (7.4.17) and

$$P_F = \tfrac{1}{2} \begin{bmatrix} K_{Fp} + \tfrac{1}{2}K_{Fv} & \tfrac{1}{2}I_{n\times n} \\ \tfrac{1}{2}I_{n\times n} & I_{n\times n} \end{bmatrix}.$$

If the sufficient condition given by

$$\min\{\lambda_{\min}\{K_{Lv}\}, \lambda_{\min}\{K_{Fv}\}\} > 1 \qquad (7.4.42)$$

is satisfied, then by the Gerschgorin theorem it is obvious that the matrices P_L and P_F given in (7.4.41) are positive-definite matrices, and hence V is a Lyapunov function.

Differentiating (7.4.41) with respect to time yields

$$\dot{V} = \mathbf{e}^T P_L \dot{\mathbf{e}} + \dot{\mathbf{e}}^T P_L \mathbf{e} + \eta_F^T P_F \dot{\eta}_F + \dot{\eta}_F^T P_F \eta_F. \qquad (7.4.43)$$

Substituting (7.4.35) and (7.4.40) into (7.4.43) yields

$$\dot{V} = -\mathbf{e}^T Q_L \mathbf{e} - \eta_F^T Q_F \eta_F + 2\mathbf{e}^T P_L B M^{-1}(q) C \eta_F, \qquad (7.4.44)$$

where Q_L is defined in (7.4.21) and

$$Q_F = -\left(A_F^T P_F + P_F A_F\right) = \begin{bmatrix} \tfrac{1}{2}K_{Fp} & O_{n\times n} \\ O_{n\times n} & K_{Fv} - \tfrac{1}{2}I_{n\times n} \end{bmatrix}. \qquad (7.4.45)$$

Note that if the sufficient condition given by (7.4.42) holds, it is obvious that the matrices Q_L and Q_F defined in (7.4.44) are positive-definite matrices. Using the fact. that Q_L and Q_F are positive definite allows us to place an upper bound on $\dot{V}$ given in (7.4.44). This upper bound is given by

$$\dot{V} \leq -\lambda_{\min}\{Q_L\}\|\mathbf{e}\|^2 - \lambda_{\min}\{Q_F\}\|\eta_F\|^2 + 2\beta_1\|\mathbf{e}\|\|\eta_F\|, \qquad (7.4.46)$$

where

$$\beta_1 = 1/m_1 > \left\|P_L B M^{-1}(q) C\right\|_{i2} = \tfrac{1}{2}\left\|\begin{bmatrix} \tfrac{1}{2}M^{-1}(q) & O_{n\times n} \\ M^{-1}(q) & O_{n\times n} \end{bmatrix}\right\|_{i2},$$

where m_1 is defined in (7.4.31).

To determine the sufficient conditions on the controller gains for asymptotic stability, we rewrite (7.4.46) in the matrix form

$$\dot{V} \leq -x_1^T Q_1 x_1, \qquad (7.4.47)$$

where

7.4 Compensation for Actuator Dynamics

$$Q_1 = \begin{bmatrix} \lambda_{\min}\{Q_L\} & -\beta_1 \\ -\beta_1 & \lambda_{\min}\{Q_F\} \end{bmatrix} \quad \text{and} \quad x_1 = \begin{bmatrix} \|e\| \\ \|\eta_F\| \end{bmatrix}.$$

By the Gerschgorin theorem, the matrix Q_1 defined in (7.4.47) will be positive definite if the sufficient condition

$$\min\{\lambda_{\min}\{Q_L\}, \lambda_{\min}\{Q_F\}\} > \beta_1 \tag{7.4.48}$$

holds. Therefore, if the controller gains satisfy the conditions given by (7.4.42) and (7.4.48), we can use standard Lyapunov stability arguments to state that the vector x_1 defined in (7.4.47), and hence $\|e\|$, e, and $\dot{e}$, are all asymptotically stable. It is easy to show that if the sufficient condition

$$\min\{\lambda_{\min}\{K_{Lp}\}, \lambda_{\min}\{K_{Lv}\}, \lambda_{\min}\{K_{Fv}\}, \lambda_{\min}\{K_{Fp}\}\} > 2/m_1 + 1 \tag{7.4.49}$$

holds, the conditions given by (7.4.42) and (7.4.48) are always satisfied.

It should be noted that the corrective controller given by (7.4.39) depends on the measurement of u_L, $\dot{u}_L$, $\ddot{u}_L$, q_m, and $\dot{q}_m$. Again, it seems that this control would require measurements of $\ddot{q}$ and its derivative; however, since we have assumed exact knowledge of the dynamic model, we can use this information to eliminate the need for measuring $\ddot{q}$ and its derivative. That is, $\dot{u}_L$ can be written as

$$\dot{u}_L = \dot{M}(q)(\ddot{q}_d + K_{Lv}\dot{e} + K_{Lp}e) + \dot{N}(q, \dot{q})$$
$$+ M(q)\left[\frac{d}{dt}\ddot{q}_d + K_{Lv}(\ddot{q}_d - \ddot{q}) + K_{Lp}\dot{e}\right], \tag{7.4.50}$$

where $\ddot{q}$ is found from (7.4.28) to be

$$\ddot{q} = M^{-1}(q)[Kq_m - N(q, \dot{q})]. \tag{7.4.51}$$

Substituting (7.4.51) into (7.4.50), we see that $\dot{u}_L$ can be written as a function of q, $\dot{q}$, q_m, and time (i.e., t) since the desired trajectory can be explicitly written as a function of time. We can delineate this functional dependence by use of the equation

$$\dot{u}_L = f(t, q, \dot{q}, q_m), \tag{7.4.52}$$

where $f(t, q, \dot{q}, q_m)$ is an $n \times 1$ vector given by the right-hand side of (7.4.50). We can obtain the functional dependence for $\ddot{u}_L$ by differentiating (7.4.52) with respect to time to yield

$$\ddot{u}_L = \dot{f}(t, q, \dot{q}, q_m) = g(t, q, \dot{q}, \ddot{q}, q_m, \dot{q}_m), \qquad (7.4.53)$$

where $g(t, q, \dot{q}, \ddot{q}, q_m, \dot{q}_m)$ is an $n \times 1$ vector. Note that (7.4.51) can be used again to eliminate the need for measurement of $\ddot{q}$; therefore, the expressions given for $\dot{u}_L$ and $\ddot{u}_L$ in (7.4.50) and (7.4.53), respectively, depend only on the measurements of q, $\dot{q}$, q_m, and $\dot{q}_m$. The actual control that would be implemented at the control input u_F can be found by making the appropriate substitution into (7.4.39). That is, the corrective control given by (7.4.39) can be written as

$$\begin{aligned} u_F =& JK^{-1}\ddot{u}_L + B_m(q_m, \dot{q}_m, q) + JK^{-1}K_{Fv}(\dot{u}_L - K\dot{q}_m) \\ &+ JK^{-1}K_{Fp} \\ &[M(q)(\ddot{q}_d + K_{Lv}\dot{e} + K_{Lp}e) + \mathbf{N}(q, \dot{q}) - Kq_m], \qquad (7.4.54) \end{aligned}$$

where $\dot{u}_L$ and $\ddot{u}_L$ would be given by (7.4.50) and (7.4.53), respectively.

After closely examining the functional dependence of $\dot{u}_L$ and $\ddot{u}_L$ given in (7.4.50) and (7.4.53), respectively, it is obvious why we have assumed that the desired trajectory and link dynamics be sufficiently smooth. Specifically, we can see from (7.4.50) and (7.4.53) that the corrective controller given in (7.4.39) will require that the first, second, third, and fourth time derivative of the desired trajectory be bounded while also requiring the existence of the first and second derivatives of the link dynamics. These assumptions on the desired trajectory and the link dynamics ensure that the control input will remain bounded.

The corrective controller derived above is summarized in Table 7.4.2 and depicted in Figure 7.4.4. We now present an example to illustrate how Table 7.4.2 can be used to design corrective controllers for RLFJ robots.

EXAMPLE 7.4-2: Corrective Controller for the One-Link RLFJ Arm

We wish to design and simulate a corrective controller using Table 7.4.2 for the one-link flexible joint robot arm given in Figure 7.4.5. The dynamics for the system are taken to be

$$mL^2\ddot{q} + mLg\sin q + f_d\dot{q} + Kq = Kq_m \qquad (1)$$

and

$$J\ddot{q}_m + B\dot{q}_m + K(q_m - q) = u_F, \qquad (2)$$

7.4 Compensation for Actuator Dynamics

Table 7.4.2: RLED Corrective Controller

Torque Controller:

$$u_F = JK^{-1}\ddot{u}_L + B_m(q_m, \dot{q}_m, q) + JK^{-1}K_{Fv}(\dot{u}_L - K\dot{q}_m)$$
$$+ JK^{-1}K_{Fp}\left[M(q)(\ddot{q}_d + K_{Lv}\dot{e} + K_{Lp}e) + \mathbf{N}(q, \dot{q}) - Kq_m\right]$$

where

$$\dot{u}_L = \dot{M}(q)(\ddot{q}_d + K_{Lv}\dot{e} + K_{Lp}e) + \dot{\mathbf{N}}(q, \dot{q})$$
$$+ M(q)\left[\tfrac{d}{dt}\ddot{q}_d + K_{Lv}(\ddot{q}_d - \ddot{q}) + K_{Lp}\dot{e}\right]$$
$$\ddot{q} = M^{-1}(q)[Kq_m - \mathbf{N}(q, \dot{q})]$$
$$\ddot{u}_L = \tfrac{d}{dt}\dot{u}_L \quad \text{utilizing } \ddot{q}$$

Stability:

Tracking error e and $\dot{e}$ are asymptotically stable.

*Comments:*Controller requires exact knowledge of system dynamics and the controller gain matrices K_{Lv}, K_{Lp}, K_{Fv}, and K_{Fp} must be sufficiently large. Desired Trajectory must be sufficiently smooth.

---

where $m = 1$ kg, $K = 10$ N, $B = 5$ kg-m/s, $f_d = 3$ kg-m/s, $L = 1$ m, g is the gravitational coefficient, $J = 0.2$ kg-m^2, q_m is the motor displacement measured in radians, and u_F is the input torque.

Assuming that the model given by (1) and (1) is exactly known, we can use Table 7.4.2 to formulate the corrective controller

$$u_F = JK^{-1}\ddot{u}_L + B\dot{q}_m + K(q_m - q) + JK^{-1}K_{Fv}(\dot{u}_L - K\dot{q}_m)$$
$$+ JK^{-1}K_{Fp}$$
$$\left[mL^2(\ddot{q}_d + K_{Lv}\dot{e} + K_{Lp}e) + mLg\sin q + f_d\dot{q} + Kq - Kq_m\right], (3)$$

where

$$\dot{u}_L = mLg\dot{q}\cos q + f_d\ddot{q} + K\dot{q} + mL^2\left[\frac{d}{dt}\ddot{q}_d + K_{Lv}(\ddot{q}_d - \ddot{q}) + K_{Lp}\dot{e}\right], (4)$$

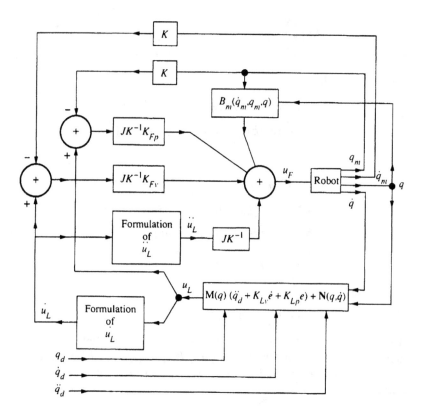

Figure 7.4.4: Block diagram of RLFJ corrective controller.

and $\ddot{q}$ is found from (1) to be

$$\ddot{q} = \left(mL^2\right)^{-1} \left[Kq_m - mLg\sin q - f_d\dot{q} - Kq\right], \qquad (5)$$

To obtain an expression for $\ddot{u}_L$ given in (3), we substitute (5) into (4) to yield

$$\begin{aligned}\dot{u}_L =& mLg\dot{q}\cos q + f_d\left(mL^2\right)^{-1}\left[Kq_m - mLg\sin q - f_d\dot{q} - Kq\right] \\ &+ K\dot{q} + mL^2 \\ &\left[\frac{d}{dt}\ddot{q}_d + K_{Lv}\left[\ddot{q}_d - \left(mL^2\right)^{-1}\left[Kq_m - mLg\sin q - f_d\dot{q} - Kq\right]\right] + K_{Lp}\dot{e}\right]\end{aligned}$$
$$(6)$$

7.4 Compensation for Actuator Dynamics

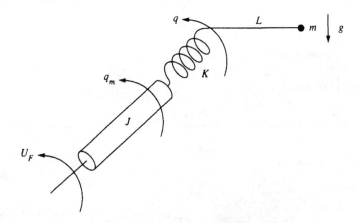

Figure 7.4.5: One-link RLFJ robot.

Differentiating (6) with respect to time yields

$$\ddot{u}_L = mLg\ddot{q}\cos q - mLg\dot{q}^2 \sin q + f_d \left(mL^2\right)^{-1}$$
$$[K\dot{q}_m - mLg\dot{q}\cos q - f_d\ddot{q} - K\dot{q}] + K\ddot{q} + mL^2$$
$$\left[\frac{d^2}{dt^2}\ddot{q}_d + K_{Lv}\left[\frac{d}{dt}\ddot{q}_d - \left(mL^2\right)^{-1}[K\dot{q}_m - mLg\dot{q}\cos q - f_d\ddot{q} - K\dot{q}]\right] + K_{Lp}\left(\ddot{q}_d - \ddot{q}\right)\right] \tag{7}$$

where $\ddot{q}$ is found from (5).

The corrective controller was simulated with the control parameters, initial conditions, and desired trajectory given by

$$K_{Lv} = K_{Lp} = K_{Fv} = K_{Fp} = 5,$$

$$q(0) = \dot{q}(0) = q_m(0) = \dot{q}_m(0) = 0,$$

and

$$q_d = \sin t.$$

The tracking error and the control torque are depicted in Figure 7.4.6. As illustrated by the figure, the tracking error is asymptotically stable.

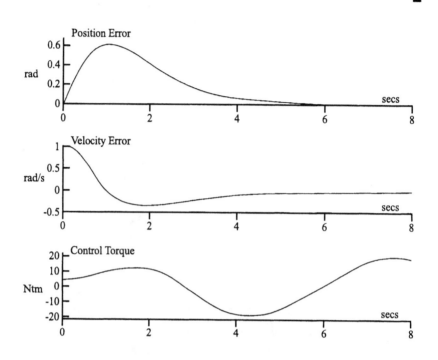

Figure 7.4.6: Simulation of RLFJ corrective controller.

7.5 Summary

In this chapter an account of several of the more advanced control techniques for the control of robot manipulators has been given. The intent of this chapter has been to study controllers that reduce online computation and controllers that compensate for actuator dynamics. Some current research issues involve the integration of force controllers with advanced motion controllers and the corresponding digital implementation.

REFERENCES

[Corless and Leitmann 1983] Corless, M., and G. Leitmann, "Adaptive control of systems containing uncertain functions and unknown functions with uncertain bounds," *J. Optim. Theory Appl.*, Jan. 1983.

[Dawson et al. 1990] Dawson, D. M., Z. Qu, F. L. Lewis, and J. F. Dorsey, "Robust control for the tracking of robot motion," *Int. J. Control*, vol. 52, pp. 581-595, 1990.

[Eppinger and Seering 1987] Eppinger, S., and W. Seering, "Introduction to Dynamic models for robot force control," *IEEE Control Syst. Mag.*, vol. 7, no. 2, pp. 48-52, Apr. 1987.

[Ghorbel and Spong 1990] Ghorbel, F:, and M. Spong, "Stability analysis of adaptively controlled flexible joint robots," *Proc. IEEE Conf. Decision Control*, Honolulu, pp 2538-2544, 1990.

[Kokotovic et al. 1986] Kokotovic, P. V., H. Khalil, and J. O'Reilly, *Singular Perturbation Methods in Control Analysis and Design*. New York: Academic Press, 1986.

[Sadegh and Horowitz 1990] Sadegh, N., and R. Horowitz, "Stability and robustness analysis of a class of adaptive controllers for robotic manipulators," *Int. J. Robot. Res.*, vol. 9, no. 3, pp. 74-92, June 1990.

[Sadegh et al. 1990] Sadegh, N., R. Horowitz, W. Kao, and M. Tomizuka, "A unified approach to the design of adaptive and repetitive controllers for robotic manipulators," *Trans. ASME*, vol. 112, pp. 618-629, Dec. 1990.

[Spong 1987] Spong, M., "Modeling and control of elastic joint robots," *J. Dyn. Syst., Meas. Control*, vol. 109, pp. 310-319, Dec. 1987.

[Tarn et al. 1991] Tarn, T., A. Bejczy, X. Yun, and Z. Li, "Effect of motor dynamics on nonlinear feedback robot arm control," *IEEE Trans. Robot. Autom.*, vol. 7, pp. 114-122, Feb. 1991.

[Taylor 1989] Taylor, D., "Composite control of direct-drive robots," *Proc. IEEE Conf. Decision Control*, pp. 1670-1675, Dec. 1989.

PROBLEMS

Section 7.2

7.2-1 Illustrate how the DCAL stability analysis can be modified if the filter tracking error is defined as

$$r = \Lambda e + \dot{e},$$

where Λ is a positive-definite diagonal matrix.

7.2-2 Design and simulate the DCAL controller given in Table 7.2.1 for the two-link polar robot arm given in Chapter 3. (Ignore the fact that this robot has a prismatic link.)

7.2-3 Can the DCAL stability analysis (and consequently, the controller itself) be modified to account for the prismatic link robot given in Problem 7.2-2? If so, explain how.

7.2-4 Illustrate how the RCL stability analysis can be modified to show that the velocity tracking error is asymptotically stable if the learning term $\hat{u}_d(t)$ is forced to remain within the a priori bounds

$$\hat{u}_{d\min} \leq \hat{u}_{di}(t) \leq \hat{u}_{d\max},$$

where the subscript i is used to denote the ith component of the $n \times 1$ vector $\hat{u}_d(t)$, and $\hat{u}_{d\min}$, $\hat{u}_{d\max}$ are scalar constants.

7.2-5 Design and simulate the RCL controller given in Table 7.2.2 for the two-link polar robot arm given in Chapter 3. (Ignore the fact that this robot has a prismatic link.)

7.2-6 Can the RCL stability analysis (and consequently, the controller itself) be modified to account for the prismatic link robot given in Problem 7.2-5? If so, explain how.

Section 7.3

7.3-1 Design and simulate the adaptive robust controller given in Table 7.3.1 for the two-link polar robot arm given in Chapter 3. (Ignore the fact that this robot has a prismatic link.)

7.3-2 Can the adaptive robust controller stability analysis (and consequently, the controller itself) be modified to account for the prismatic link given in Problem 7.3-1? If so, explain how.

7.3-3 Show how Barbalat's lemma (see Chapter 2) can be used to modify the stability analysis for the adaptive robust controller to guarantee that the velocity tracking error is also asymptotically stable.

Section 7.4

7.4-1 For a constant, symmetric, positive-definite, $n \times n$ matrix A, show that

(a)
$$\|A\|_{i2} = \lambda_{\max}\{A\}$$

(b)
$$\lambda_{\max}\{A\} > \frac{1}{2} \left\| \begin{bmatrix} \frac{1}{2}A \\ A \end{bmatrix} \right\|_{i2}$$

Section 7.4

7.4-2 Design and simulate the RLED corrective controller given in Table 7.4.1 for the two-link revolute robot arm given in Chapter 3. Assume that both motors can be modeled as the motor given in Example 7.4-1.

7.4-3 For a constant, symmetric, positive-definite, $n \times n$ matrix A show that

$$\lambda_{\max}\{A\} > \frac{1}{2} \left\| \begin{bmatrix} \frac{1}{2}A & 0_{n \times n} \\ A & 0_{n \times n} \end{bmatrix} \right\|_{i2},$$

where $O_{n \times n}$ is the $n \times n$ zero matrix.

7.4-4 Design and simulate the RLFJ corrective controller given in Table 7.4.2 for the two-link revolute robot arm given in Chapter 3. Assume that both joints can be modeled similar to the joint given in Example 7.4-2.

Chapter 8

Neural Network Control of Robots

A framework is given for controller design using neural networks. These structures possess a universal approximation property that allows them to be used in feedback control of unknown systems without requirements for linearity in the system parameters or finding a regression matrix. Feedback control topologies and weight tuning algorithms are given here that guarantee closed-loop stability and bounded weights.

8.1 Introduction

In recent years, there has been a great deal of effort to design feedback control systems that mimic the functions of living biological systems [White and Sofge 1992], [Miller et al. 1991]. There has been great interest recently in 'universal model-free controllers' that do not need a mathematical model of the controlled plant, but mimic the functions of biological processes to learn about the systems they are controlling on-line, so that performance improves automatically. Techniques include fuzzy logic control, which mimics linguistic and reasoning functions, and artificial neural networks, which are based on biological neuronal structures of interconnected nodes. Neural networks (NNs) have achieved great success in classification, pattern recognition, and other open-loop applications in digital signal processing and elsewhere. Rigorous analyses have shown how to select NN topologies and weights, for instance, to discriminate between specified exemplar patterns. By now, the theory and applications of NN in open-loop applications are well understood, so that NN have become an important tool in the repertoire of the signal processor and computer scientist.

There is a large literature on NN for feedback control of unknown plants. Until the 1990's, design and analysis techniques were ad hoc, with no repeatable design algorithms or proofs of stability and guaranteed performance. In spite of this, simulation results appearing in the literature showed good performance. Most of the early approaches used standard backpropagation weight tuning [Werbos 1992] since rigorous derivations of tuning algorithms suitable for closed-loop control purposes were not available. Many NN design techniques mimicked adaptive control approaches, where rigorous analysis results were available [Åström and Wittenmark 1989], [Landau 1979], [Goodwin et al. 1984], proposing NN feedback control topologies such as indirect identification-based control, inverse dynamics control, series-parallel techniques, etc.

In keeping with the philosophy of those working in control system theory since Maxwell, Lyapunov, A.N. Whitehead, von Bertalanffy, and others, to address such issues it is necessary to begin with the knowledge available about the system being controlled. Narendra [Narendra 1991], [Narendra and Parthasarathy 1991] and others [Miller et al. 1991], [White and Sofge 1992], [Werbos 1992] pioneered rigorous NN controls applications by studying the dynamical behavior of NN in closed-loop systems, including computation of the gradients needed for backprop tuning. Other groups providing rigorous analysis and design techniques for closed-loop NN controllers included [Sanner and Slotine 1991] who showed how to use radial basis function NN in feedback control, [Chen and Khalil 1992], [Chen and Liu 1994] who provided NN tuning algorithms based on deadzone methods, [Polycarpou and Ioannou 1991], [Polycarpou 1996] who used projection methods, [Lewis et al. 1993], [Lewis et al. 1995] who used an e-mod approach, [Sadegh 1993] who provided NN controllers for discrete-time systems, and [Rovithakis and Christodoulou 1994] who used dynamic NN for feedback control.

In this chapter is given an approach to the design and analysis of *neural network* controllers based on several PhD dissertations and a body of published work, including [Lewis et al. 1999], [Kim and Lewis 1998]. The control structures discussed here are *multiloop controllers* with NN in some of the loops and an outer tracking unity-gain PD loop. There are repeatable design algorithms and guarantees of system performance including both small tracking errors and bounded NN weights. It is shown that as uncertainty about the controlled system increases or performance requirements increase, the NN controllers require more and more structure. An exposition of this material in greater depth appears in [Lewis 1999], which also shows relationships with Fuzzy Logic Systems.

8.2 Background in Neural Networks

In this section is provided the background in neural network structures required for feedback control. For more details see [Haykin 1994], [Lewis et al. 1999], [Lewis and Kim 1998]. Two key features that make NN useful for feedback control are their *universal approximation property* and their *learning capability*, which arises due to the fact that their weights are tunable parameters that can be updated to improve controller performance. The universal approximation property is the main feature that makes nonlinear network structures more suitable for robot control than adaptive robot controllers, which have generally relied upon the determination of a regression matrix, which in turn requires linearity in the tunable parameters (LIP).

Multilayer Neural Networks

A multilayer neural network is shown in Figure 8.2.1 . This NN has two layers of adjustable weights, and is called here a 'two-layer' net. This NN has no internal feedback connections and so is termed *feedforward*, and no internal dynamics and so is said to be *static*. The NN output y is a vector with m components that are determined in terms of the n components of the input vector x by the recall equation

$$y_i = \sum_{j=1}^{L} \left[w_{ij}\sigma \left(\sum_{k=1}^{n} v_{jk}x_k + \theta_{v_j} \right) + \theta_{w_i} \right]; \quad i = 1, \ldots, m \qquad (8.2.1)$$

where $\sigma(\cdot)$ are the *activation functions* and L is the number of *hidden-layer neurons*. The first-layer interconnections weights are denoted v_{jk} and the second-layer interconnection weights by w_{ij}. The threshold offsets are denoted by $\theta_{v_j}, \theta_{w_i}$.

Many different activation functions $\sigma(\cdot)$ are in common use. For feedback control using multilayer NN it is required that $\sigma(\cdot)$ be smooth enough so that at least its first derivative exists. Suitable choices include the sigmoid

$$\sigma(x) = \frac{1}{1 + e^{-x}}, \qquad (8.2.2)$$

the hyperbolic tangent

$$\sigma(x) = \frac{e^x - e^{-x}}{e^x + e^{-x}}, \qquad (8.2.3)$$

and other logistic-curve-type functions, as well as the gaussian and an assortment of other functions.

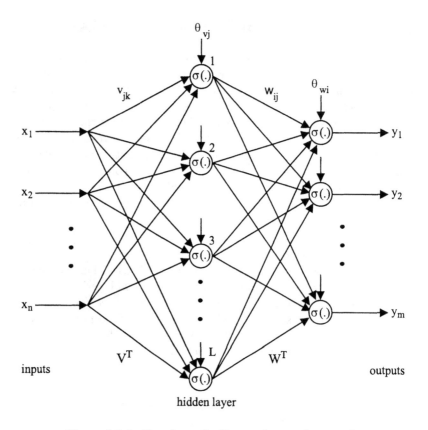

Figure 8.2.1: Two-layer feedforward neural network.

By collecting all the NN weights v_{jk}, w_{ij} into matrices of weights V^T, W^T, the NN recall equation may be written in terms of vectors as

$$y = W^T \sigma(V^T x). \qquad (8.2.4)$$

The thresholds are included as the first columns of the weight matrices W^T, V^T; to accommodate this, the vectors x and $\sigma(\cdot)$ need to be augmented by placing a '1' as their first element (e.g. $x \equiv [1 \; x_1 \; x_2 \; \ldots \; x_n]^T$). In this equation, to represent (8.2.1) one has sufficient generality if $\sigma(\cdot)$ is taken as a diagonal function from $\Re^L$ to $\Re^L$, that is $\sigma(z) \equiv diag\{\sigma(z_j)\}$ for a vector $z = [z_1 \; z_2 \ldots z_L]^T \in \Re^L$.

Note that the recall equation is *nonlinear* in the first-layer weights and thresholds V.

8.2 Background in Neural Networks

Universal Function Approximation Property. NN satisfy many important properties. A main one of concern for feedback control purposes is the *universal function approximation property* [Hornik and Stinchombe 1989]. Let $f(x)$ be a general smooth function from $\Re^n$ to $\Re^m$. Then, it can be shown that, as long as x is restricted to a compact set S of $\Re^n$, there exist weights and thresholds such that one has

$$f(x) = W^T \sigma(V^T x) + \epsilon \qquad (8.2.5)$$

for some number of hidden-layer neurons L. This holds for a large class of activation functions, including those just mentioned. The value ϵ is called the *NN functional approximation error*, and it generally decreases as the net size L increases. In fact, for any choice of a positive number ϵ_N, one can find a feedforward NN such that

$$\|\epsilon\| < \epsilon_N \qquad (8.2.6)$$

for all x in S. The selection of the net size L for a prescribed accuracy ϵ_N is an open question for general unstructured fully-connected NN. This problem can be solved for CMAC, FL nets, and other structured nonlinear nets as subsequently described.

The ideal NN weights in matrices W, V that are needed to best approximate a given nonlinear function $f(x)$ are difficult to determine. In fact, they may not even be unique. However, all one needs to know for controls purposes is that, for a specified value of ϵ_N some ideal approximating NN *weights exist*. Then, an *estimate* of $f(x)$ can be given by

$$\hat{f}(x) = \hat{W}^T \sigma(\hat{V}^T x) \qquad (8.2.7)$$

where $\hat{W}, \hat{V}$ are *estimates* of the ideal NN weights that are provided by some on-line weight tuning algorithms subsequently to be detailed. Note that all 'hat' quantities are known, since they may be computed using measurable signals and current NN weight estimates.

The assumption that there exist ideal weights such that the approximation property holds is very much like various similar assumptions in adaptive control [Åström and Wittenmark 1989], [Landau 1979], including *Erzberger's assumptions* and *linearity in the parameters*. The very important difference is that in the NN case, *the universal approximation property always holds*, while in adaptive control such assumptions often do not hold in practice, and so they imply restrictions on the form of the systems that can be controlled.

Overcoming the NLIP Problem. Multilayer NN are nonlinear in the weights V and so weight tuning algorithms that yield guaranteed stability and bounded weights in closed-loop feedback systems have been difficult to discover until a few years ago. The NLIP problem is easily overcome if a correct and rigorous approach is taken. One approach is the following appropriate use of the Taylor series [Lewis et al. 1996].

Define the functional estimation error

$$\tilde{f} = f - \hat{f}, \tag{8.2.8}$$

the weight estimation errors

$$\tilde{W} = W - \hat{W}, \quad \tilde{V} = V - \hat{V}, \tag{8.2.9}$$

and the hidden-layer output error

$$\tilde{\sigma} = \sigma - \hat{\sigma} \equiv \sigma(V^T x) - \sigma(\hat{V}^T x). \tag{8.2.10}$$

For any z one may write the Taylor series

$$\sigma(z) = \sigma(\hat{z}) + \sigma'(\hat{z})\tilde{z} + O(\tilde{z})^2, \tag{8.2.11}$$

where σ' is the jacobian and the last term means terms of order $\tilde{z}^2$. Therefore,

$$\tilde{\sigma} = \sigma'(\hat{V}^T x)\tilde{V}^T x + O(\tilde{V}^T x)^2 \equiv \hat{\sigma}' \tilde{V}^T x + O(\tilde{V}^T x)^2. \tag{8.2.12}$$

This key equation allows one to write the functional estimation error as

$$\begin{aligned}
\tilde{f} =& f - \hat{f} = W^T \sigma\left(V^T x\right) - \hat{W}^T \sigma\left(\hat{V}^T x\right) + \varepsilon \\
=& \tilde{W}^T \sigma(\hat{V}^T x) + W^T[\sigma(V^T x) - \sigma(\hat{V}^T x)] + \epsilon \\
=& \tilde{W}^T \hat{\sigma} + W^T[\hat{\sigma}' \tilde{V}^T x + O(\tilde{V}^T x)^2] + \epsilon \\
=& \tilde{W}^T [\hat{\sigma} - \hat{\sigma}' \hat{V}^T x] + \hat{W}^T \hat{\sigma}' \tilde{V}^T x + \tilde{W}^T \hat{\sigma}' V^T x \\
& + W^T O(\tilde{V}^T x)^2 + \epsilon.
\end{aligned} \tag{8.2.13}$$

The first term has $\tilde{W}$ multiplied by a known quantity (in square braces), and the second term has $\tilde{V}$ multipled by a known quantity. When used subsequently in the closed-loop error dynamics, this form allows one to derive tuning laws for $\hat{V}$ and $\hat{W}$ that guarantee closed-loop stability.

8.2 Background in Neural Networks

Linear-in-the-parameter neural nets

If the first-layer weights and thresholds V in (8.2.4) are fixed and only the second-layer weights and thresholds W are tuned, then the NN has only one layer of tunable weights. One may then define the fixed function $\phi(x) = \sigma(V^T x)$ so that such a 1-layer NN has the recall equation

$$y = W^T \phi(x), \qquad (8.2.14)$$

where $x \in \Re^n, y \in \Re^m, \phi(\cdot) : \Re^n \to \Re^L$, and L is the number of hidden-layer neurons. This NN is linear in the tunable parameters W, so that it is far easier to tune. In fact, standard adaptive control proofs can be used to derive suitable tuning algorithms.

Though LIP, these NN still offer an enormous advantage over standard adaptive control approaches that require the determination of a regression matrix since they satisfy a universal approximation property if the functions $\phi(\cdot)$ are correctly chosen. Note that adaptive controllers are linear in the *system* parameters, while 1-layer NN are linear in the *NN weights*. An advantage of 1-layer NN over 2-layer NN is that firm results exist for the former on the selection of the number of hidden layer neurons L for a specified approximation accuracy. A disadvantage of LIP networks is that Barron [Barron 1993] has shown a lower bound on the approximation accuracy.

Functional-Link Basis Neural Networks. More generality is gained if $\sigma(\cdot)$ is not diagonal, but $\phi(\cdot)$ is allowed to be a general function from $\Re^n$ to $\Re^L$. This is called a *functional-link neural net (FLNN)* [Sadegh 1993]. For LIP NN, the functional approximation property does not generally hold. However, a 1-layer NN can still approximate functions as long as the activation functions $\phi(\cdot)$ are selected as a *basis*, which must satisfy the following two requirements on a compact simply-connected set S of $\Re^n$:

1. A constant function on S can be expressed as (8.2.14) for a finite number L of hidden-layer neurons.

2. The functional range of (8.2.14) is dense in the space of continuous functions from S to $\Re^m$ for countable L.

If $\phi(\cdot)$ provides a basis, then a smooth function $f(x)$ from $\Re^n$ to $\Re^m$ can be approximated on a compact set S of $\Re^n$, by

$$f(x) = W^T \phi(x) + \epsilon. \qquad (8.2.15)$$

for some ideal weights and thresholds W and some number of hidden layer neurons L. In fact, for any choice of a positive number ϵ_N, one can find a feedforward NN such that $\|\epsilon\| < \epsilon_N$ for all x in S.

The approximation for $f(x)$ is given by

$$\hat{f}(x) = \hat{W}^T \phi(x), \tag{8.2.16}$$

where $\hat{W}$ are the NN weight and threshold estimates given by the tuning algorithm. For LIP NN, the functional estimation error is given by

$$\begin{aligned}\tilde{f} &= f - \hat{f} = W^T \phi(x) - \hat{W}^T \phi(x) + \epsilon \\ &= \tilde{W}^T \phi(x) + \epsilon,\end{aligned} \tag{8.2.17}$$

which, when used subsequently in the error dynamics, directly yields tuning algorithms for $\hat{W}$ that guarantee closed-loop stability.

Barron [Barron 1993] has shown that for all LIP approximators there is a fundamental lower bound, so that ϵ is bounded below by terms on the order of $1/L^{2/n}$. Thus, as the number of NN inputs n increases, increasing L to improve the approximation accuracy becomes less effective. As shown subsequently, this is not a major limitation on adaptive controllers designed using 1-layer NN. This lower bound problem does not occur in the NLIP multilayer nonlinear nets.

A special FLNN is now discussed. We often use $\sigma(\cdot)$ in place of $\phi(\cdot)$, with the understanding that, for LIP nets, this activation function vector is not diagonal, but is a general function from $\Re^n$ to $\Re^L$.

Gaussian or Radial Basis Function (RBF) Networks. The selection of a basis set of activation functions is considerably simplified in various sorts of *structured nonlinear networks*, including radial basis function, CMAC, and fuzzy logic nets. It will be shown here that the key to the design of such structured nonlinear nets lies in a *more general set of NN thresholds* than allowed in the standard equation (8.2.1), and in their appropriate selection.

A NN activation function often used is the gaussian or radial basis function (RBF) [Sanner and Slotine 1991] given when x is a scalar as

$$\sigma(x) = e^{-(x-\bar{x})^2/2p}, \tag{8.2.18}$$

where $\bar{x}$ is the mean and p the variance. RBF NN can be written as (8.2.4), but have an advantage over the usual sigmoid NN in that the n-dimensional gaussian function is well understood from probability theory, Kalman filtering, and elsewhere, so that n-dimensional RBF are easy to conceptualize.

The j-th activation function can be written as

8.2 Background in Neural Networks

$$\sigma_j(x) = e^{-\frac{1}{2}(x-\bar{x}_j)^T P_j^{-1}(x-\bar{x}_j)} \quad (8.2.19)$$

with $x, \bar{x}_j \in \Re^n$. Define the vector of activation functions as $\sigma(x) \equiv [\sigma_1(x) \; \sigma_2(x) \; ... \; \sigma_L(x)]^T$. If the covariance matrix is diagonal so that $P_j = diag\{p_{jk}\}$, then (8.2.19) becomes *separable* and may be decomposed into components as

$$\sigma_j(x) = e^{-\frac{1}{2}\sum_{k=1}^{n} -(x_k-\bar{x}_{jk})^2/p_{jk}} = \prod_{k=1}^{n} e^{-\frac{1}{2}(x_k-\bar{x}_{jk})^2/p_{jk}}, \quad (8.2.20)$$

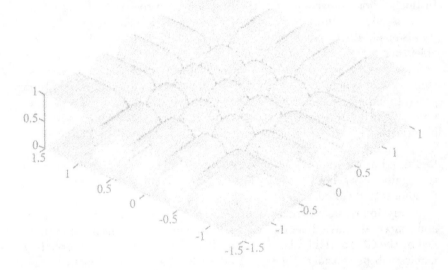

Figure 8.2.2: 2-dimensional separable gaussian functions for an RBF NN.

where $x_k, \bar{x}_{jk}$ are the k-th components of $x, \bar{x}_j$. Thus, the n-dimensional activation functions are the product of n scalar functions. Note that this equation is of the form of the activation functions in (8.2.1), but with *more general thresholds*, as a threshold is required for each different component of x at each hidden layer neuron j; that is, the threshold at each hidden-layer neuron in Figure 8.2.1 is a *vector*. The RBF variances p_{jk} and offsets $\bar{x}_{jk}$ are usually selected in designing the RBF NN and left fixed; only the output-layer weights W^T are generally tuned. Therefore, the RBF NN is a special sort of FLNN (8.2.14) (where $\phi(x) = \sigma(x)$).

Figure 8.2.2 shows separable gaussians for the case $x \in \Re^2$. In this figure, all the variances p_{jk} are identical, and the mean values $\bar{x}_{jk}$ are chosen in a special way that spaces the activation functions at the node points of a 2-D grid. To form an RBF NN that approximates functions over the region $\{-1 < x_1 \leq 1, -1 < x_2 \leq 1\}$ one has here selected $L = 5 \times 5 = 25$ hidden-layer neurons, corresponding to 5 cells along x_1 and 5 along x_2. Nine of these neurons have 2-D gaussian activation functions, while those along the boundary require the illustrated 'one-sided' activation functions.

8.3 Tracking Control Using Static Neural Networks

In this section is discussed feedback tracking control design using nonlinear nets and assuming full state-variable feedback. For more details see [Lewis et al. 1999] and other listed references. If full state feedback is available, then a static feedforward NN suffices for control. A control topology and net tuning algorithms are provided here that guarantee closed-loop stability and bounded weights. The techniques to be discussed apply for general nonlinear nets including both neural networks and fuzzy logic systems [Wang 1997], [Lewis et al. 1997], so that the abbreviation NN might henceforth be construed as meaning 'nonlinear network'. The resulting multiloop control topology has an *outer PD tracking loop* and an inner NN feedback linearization or *action generating loop*. It is found that backprop tuning does not generally suffice, but modified tuning algorithms are needed for guaranteed closed-loop performance.

Many industrial mechanical systems, as well as automobiles, aircraft, and spacecraft, have dynamics in the *Lagrangian form*, which are exemplified by the class of rigid robot systems. Therefore, the Lagrangian robot dynamics will be considered [Lewis et al. 1993], [Lewis et al. 1996]. The NN control techniques presented may also be applied to other unknown systems including certain important classes of nonlinear systems [Lewis et al. 1999].

Robot Arm Dynamics and Error System

The dynamics of rigid Lagrangian systems, including robot arms, have some important physical, structural, and passivity properties [Craig 1988], [Lewis et al. 1993], [Slotine 1988], [Spong and Vidyasagar 1989] that make it very natural to use NN in their control. These properties should be taken into account in the design of any controller— in fact, they provide the foundation for rigorous design algorithms for NN controllers.

The dynamics of an n-link rigid (i.e. no flexible links or high-frequency

8.3 Tracking Control Using Static Neural Networks

joint/motor dynamics) robot manipulator may be expressed in the Lagrange form

$$M(q)\ddot{q} + V_m(q,\dot{q})\dot{q} + G(q) + F(\dot{q}) + \tau_d = \tau \qquad (8.3.1)$$

with $q(t) \in \Re^n$ the joint variable vector, whose entries are the robot arm joint angles or link extensions. $M(q)$ is the inertia matrix, $V_m(q,\dot{q})$ the coriolis/centripetal matrix, $G(q)$ the gravity vector, and $F(\dot{q})$ the friction. Bounded unknown disturbances (including e.g. unstructured unmodelled dynamics) are denoted by τ_d, and the control input torque is $\tau(t)$. The robot dynamics have the following standard properties:

Property 1: $M(q)$ is a positive definite symmetric matrix bounded by $m_1 I < M(q) < m_2 I$, with m_1, m_2 positive constants.

Property 2: The norm of the matrix $V_m(q,\dot{q})$ is bounded by $v_b(q)\|\dot{q}\|$, for some function $v_b(q)$.

Property 3: The matrix $\dot{M} - 2V_m$ is skew-symmetric. This is equivalent to the fact that the internal forces do no work.

Property 4: The unknown disturbance satisfies $\|\tau_d\| < d_B$, with b_D a positive constant.

Given a desired arm trajectory $q_d(t) \in \Re^n$, the tracking error is

$$e(t) = q_d(t) - q(t) \qquad (8.3.2)$$

and the filtered tracking error is

$$r = \dot{e} + \Lambda e \qquad (8.3.3)$$

where Λ is a symmetric positive definite design parameter matrix, usually selected diagonal. If a controller is found such that $r(t)$ is bounded, then $e(t)$ is also bounded; in fact $\|e\| \leq \|r\|/\lambda_{min}(\Lambda)$ and $\|\dot{e}\| \leq \|r\|$, with $\lambda_{min}(\Lambda)$ the minimum singular value of Λ.

Differentiating $r(t)$ and using (8.3.1), the arm dynamics may be written in terms of the filtered tracking error as

$$M\dot{r} = -V_m r - \tau + f + \tau_d \qquad (8.3.4)$$

where the nonlinear robot function is

$$f(x) = M(q)(\ddot{q}_d + \Lambda \dot{e}) + V_m(q,\dot{q})(\dot{q}_d + \Lambda e) + G(q) + F(\dot{q}). \qquad (8.3.5)$$

The vector x required to compute $f(x)$ can be defined, for instance, as

$$x \equiv [e^T \; \dot{e}^T \; q_d^T \; \dot{q}_d^T \; \ddot{q}_d^T]^T, \quad (8.3.6)$$

which can be measured. Function $f(x)$ contains potentially unknown robot parameters including payload mass and complex forms of friction.

A suitable control input for trajectory following is given by the computed-torque-like control

$$\tau = \hat{f} + K_v r - v \quad (8.3.7)$$

with $K_v = K_v^T > 0$ a gain matrix, generally chosen diagonal, and $\hat{f}(x)$ an estimate of the robot function $f(x)$ that is provided by some means. The robustifying signal $v(t)$ is needed to compensate for unmodelled unstructured disturbances. Using this control, the closed-loop error dynamics is

$$M\dot{r} = -(K_v + V_m)r + \tilde{f} + \tau_d + v. \quad (8.3.8)$$

In computing the control signal, the estimate $\hat{f}$ can be provided by several techniques, including adaptive control or neural or fuzzy networks. The auxiliary control signal $v(t)$ can be selected by several techniques, including sliding-mode methods and others under the general aegis of robust control methods.

The desired trajectory is assumed bounded so that

$$\left\| \begin{array}{c} q_d(t) \\ \dot{q}_d(t) \\ \ddot{q}_d(t) \end{array} \right\| \leq q_B, \quad (8.3.9)$$

with q_B a known scalar bound. It is easy to show that for each time t, $x(t)$ is bounded by

$$\|x\| \leq c_1 + c_2 \|r\| \leq q_B + c_0 \|r(0)\| + c_2 \|r\| \quad (8.3.10)$$

for computable positive constants c_0, c_1, c_2.

Adaptive Control

Standard adaptive control techniques in robotics [Craig 1988], [Slotine 1988], [Spong and Vidyasagar 1989], [Lewis et al. 1993] use the assumption that the nonlinear robot function is linear in the tunable parameters so that

$$f(x) = \Theta(x)p + \epsilon \quad (8.3.11)$$

8.3 Tracking Control Using Static Neural Networks

where p is a vector of unknown parameters and $\Theta(x)$ is a known *regression matrix*. The control is selected as

$$\tau = K_v r + \Theta(x)\hat{p} \qquad (8.3.12)$$

and tuning algorithms are determined for the parameter estimates $\hat{p}$. This is facilitated by the LIP assumption. If the approximation error ϵ is nonzero, then the tuning algorithm must be modified in order to guarantee bounded parameter estimates. Various robustifying techniques may be used for this including the σ-mod [Ioannou and Kokotovic 1984], the e-modification [Narendra and Annaswamy 1987], or the dead-zone techniques [Kreisselmeier and Anderson 1986]. There are adaptive control techniques by now that do not require LIP or the determination of a regression matrix [Colbaugh and Seraji 1994]. It is interesting to compare the complexity of these to the NN controllers to be discussed herein.

Neural Net Feedback Tracking Controller

A NN will be used in the control $\tau(t)$ in (8.3.7) to provide the estimate $\hat{f}$ for the unknown robot function $f(x)$. The NN approximation property assures us that there always exists a NN that can accomplish this within a given accuracy ϵ_N. For 2-layer NLIP NN the approximation is

$$\hat{f}(x) = \hat{W}^T \sigma(\hat{V}^T x), \qquad (8.3.13)$$

while for 1-layer LIP NN it is

$$\hat{f}(x) = \hat{W}^T \phi(x), \qquad (8.3.14)$$

with $\phi(\cdot)$ selected as a basis.

The structure of the NN controller appears in Figure 8.3.1, where $\underline{e} \equiv [e^T \; \dot{e}^T]^T, \underline{q} \equiv [q^T \; \dot{q}^T]^T$. The neural network that provides the estimate for $f(x)$ appears in an *inner control loop*, and there is an *outer tracking loop* provided by the PD term $K_v r$. In control theory terminology, the inner loop is a *feedback linearization* controller [Slotine and Li 1991], while in computer science terms it is the *action generating loop* [Werbos 1992], [Miller et al. 1991], [White and Sofge 1992]. This *multiloop intelligent control structure* is derived naturally from robot control notions, and is not ad hoc. As such, it is immune to philosophical deliberations concerning suitable NN control topologies including the common discussions on feedforward vs. feedback, direct vs. indirect, and so on. It is to be noted that the static feedforward NN in this diagram is turned into a *dynamic NN* by closing a feedback loop around it (c.f. [Narendra 1991]).

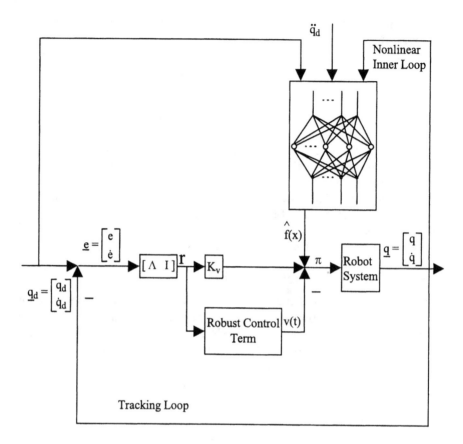

Figure 8.3.1: Neural net robot controller, showing nonlinear neural network loop and outer tracking loop.

Advantage of NN over LIP Adaptive Control. Each robotic system has its own regression matrix, so that a different $\Theta(x)$ must be determined for each system. This regression matrix is often complicated and determining it can be time consuming. One notes that the regression matrix effectively provides a basis for function approximation for a *specific* given system. On the other hand, the universal approximation property of nonlinear networks shows that NN provide a basis for *all* sufficiently smooth systems. Therefore, NN can be used to approximate smooth $f(x)$ for *all rigid robotic systems*, effectively allowing the design of a single tunable controller for a class of systems. No regression matrix need be determined, for the same NN activation functions suffice for the whole class of plants.

Even the 1-layer NN, which is LIP, has this advantage. Note that the 1-layer NN is linear not in the *system parameters*, but in the NN weights. Even the LIP NN has a universal approximation property so that no regression matrix is needed.

Initial Tracking Errors and Initial NN Weights. It is now required to determine how to tune the NN weights to yield guaranteed closed-loop stability. Several cases are considered for NN controller design in this section. All of them need the following construction.

Since the NN approximation property holds on a compact set, one must define an *allowable initial condition set* as follows. Let the NN approximation property hold for the function $f(x)$ given in (8.3.5) with a given accuracy ϵ_N in (8.2.6) for all x inside the ball of radius $b_x > q_B$. Define the set of allowable initial tracking errors as

$$S_r = \{r \mid \|r\| < (b_x - q_B)/(c_0 + c_2)\}. \tag{8.3.15}$$

Note that the approximation accuracy of the NN determines size of S_r. For a larger NN (i.e. more hidden-layer units), ϵ_N is small for a larger radius b_x. Thus, the allowed initial condition set S_r is larger. On the other hand, a more active desired trajectory (e.g. containing higher frequency components) results in a larger acceleration $\ddot{q}_d(t)$, which yields a larger bound q_B, thereby decreasing S_r. It is important to note the dependence of S_r on the PD design ratio Λ— both c_0 and c_2 depend on Λ.

A key feature of our the Initial Condition Requirement is its *independence of the NN initial weights*. This is in stark contrast to other techniques in the literature where the proofs of stability depend on selecting some initial stabilizing NN weights, which is very difficult to do.

8.4 Tuning Algorithms for Linear-in-the-Parameters NN

Suppose now that a LIP FLNN is used to approximate the nonlinear robot function (8.3.5) according to (8.2.15) with $\|\epsilon\| \leq \epsilon_N$ on a compact set, the ideal approximating weights W constant, and $\phi(x)$ selected as a basis. An estimate of $f(x)$ is given by (8.3.14). Then the control law (8.3.7) becomes

$$\tau = \hat{W}^T \phi(x) + K_v r - v \tag{8.4.1}$$

and the closed-loop filtered error dynamics (8.3.8) are

$$M\dot{r} = -(K_v + V_m)r + \tilde{W}^T \phi(x) + (\epsilon + \tau_d) + v. \tag{8.4.2}$$

The next theorem derives the tuning law for a NN controller that is augmented by an e-mod term as in [Narendra and Annaswamy 1987].

It must now be assumed that the ideal weights W are constant and bounded so that

$$\|W\|_F \leq W_B, \tag{8.4.3}$$

with the bound W_B known. In [Polycarpou 1996] it is shown how to use standard adaptive robust techniques to avoid the assumption that the bound W_B is known.

THEOREM 8.4-1: *(NN Weight Tuning Algorithm).*

Let the desired trajectory $q_d(t)$ be bounded by q_B and the initial tracking error $r(0)$ be in S_r. Let the NN reconstruction error bound ϵ_N and the disturbance bound d_B be constants. Assume the ideal NN target weights are bounded by W_B as in (8.4.3). Let the control input for the robot arm be given by (8.4.1) with $v(t) = 0$ and gain

$$K_{v_{min}} > \frac{(\kappa W_B^2/4 + \varepsilon_N + d_B)(c_0 + c_2)}{b_x - q_B}. \tag{1}$$

Let the weight tuning be

$$\dot{\hat{W}} = F\phi r^T - \kappa F\|r\|\hat{W}, \tag{2}$$

with $F = F^T > 0$ and $\kappa > 0$ a small design parameter. Then the filtered tracking error $r(t)$ and the NN weight estimates $\hat{W}(t)$ are UUB with practical bounds given respectively by the right-hand sides of (8), (9). Moreover, the tracking error may be made as small as desired by increasing the tracking gain K_v.

Proof:

Let the NN approximation property (8.2.15) hold for the function $f(x)$ given in (8.3.5) with a given accuracy ε_N for all x in the compact set $S_x \equiv \{x \mid \|x\| < b_x\}$ with $b_x > q_B$. Let $r(0) \in S_r$. Then the approximation property holds at time 0.

Define the Lyapunov function candidate

$$L = \frac{1}{2}r^T M r^T + \frac{1}{2}tr\{\tilde{W}^T F^{-1}\tilde{W}\} \tag{3}$$

Differentiating yields

$$\dot{L} = r^T M\dot{r} + \frac{1}{2}r^T \dot{M} r + tr\{\tilde{W}^T F^{-1}\dot{\tilde{W}}\} \tag{4}$$

8.4 Tuning Algorithms for Linear-In-The-Parameters NN

whence substitution from (8.4.2) yields

$$\dot{L} = -r^T K_v r + \frac{1}{2} r^T (\dot{M} - 2V_m) r + tr\{\tilde{W}^T (F^{-1} \dot{\tilde{W}} + \phi r^T)\} + r^T (\epsilon + \tau_d). \quad (5)$$

Then, using tuning rule (2) yields

$$\dot{L} = -r^T K_v r + \kappa \|r\| tr\{\tilde{W}^T (W - \tilde{W})\} + r^T (\varepsilon + \tau_d). \quad (6)$$

Since

$$tr\{\tilde{W}^T (W - \tilde{W})\} = <\tilde{W}, W>_F - \|\tilde{W}\|_F^2 \le \|\tilde{W}\|_F \|W\|_F - \|\tilde{W}\|_F^2,$$

there results

$$\dot{L} \le -K_{v_{min}} \|r\|^2 + \kappa \|r\| \cdot \|\tilde{W}\|_F (W_B - \|\tilde{W}\|_F) + (\varepsilon_N + d_B) \|r\|$$
$$= -\|r\|[K_{v_{min}} \|r\| + \kappa \|\tilde{W}\|_F (\|\tilde{W}\|_F - W_B) - (\varepsilon_N + d_B)], \quad (7)$$

which is negative as long as the term in braces is positive. Completing the square yields

$$K_{v_{min}} \|r\| + \kappa \|\tilde{W}\|_F (\|\tilde{W}\|_F - W_B) - (\varepsilon_N + d_B)$$
$$= \kappa (\|\tilde{W}\|_F - W_B/2)^2 - \kappa W_B^2/4 + K_{v_{min}} \|r\| - (\varepsilon_N + d_B)$$

which is guaranteed positive as long as

$$\|r\| > \frac{\kappa W_B^2/4 + (\varepsilon_N + d_B)}{K_{v_{min}}} \equiv b_r \quad (8)$$

or

$$\|\tilde{W}\|_F > W_B/2 + \sqrt{\kappa W_B^2/4 + (\varepsilon_N + d_B)/\kappa} \equiv b_W. \quad (9)$$

Thus, $\dot{L}$ is negative outside a compact set. Selecting the gain according to (1) ensures that the compact set defined by $\|r\| \le b_r$ is contained in S_r, so that the approximation property holds throughout. This demonstrates the UUB of both $\|r\|$ and $\|\tilde{W}\|_F$. ∎

This proof is similar to [Narendra and Annaswamy 1987] but is modified to reflect the fact that the NN approximates only on a compact set. In this result one now requires the bound W_B on the ideal weights W, which appears in the PD gain condition (1). Now, bounds are discovered both for the tracking error $r(t)$ and the weights $\hat{W}(t)$. Note that no persistence of excitation condition is needed to establish the bounds on $\tilde{W}$ with the modified weight tuning algorithm.

The NN weight tuning algorithm (2) consists of a backpropagation through time term plus a new second term. The importance of the κ term, called the e-modification, is that it adds a quadratric term in $\|\tilde{W}\|_F$ in (7), so that it is possible to establish that $\dot{L}$ is negative outside a compact set in the $(\|r\|, \|\tilde{W}\|_F)$ plane [Narendra and Annaswamy 1987]. The e-mod term makes the tuning law *robust* to unmodelled dynamics so that the PE condition is not needed. In [Polycarpou and Ioannou 1991] a projection algorithm is used to keep the NN weights bounded. In [Chen and Khalil 1992], [Chen and Liu 1994] a deadzone technique is employed.

Weight Initialization and On-Line Tuning. In the NN control schemes derived in this paper there is *no preliminary off-line learning phase*. The weights are simply initialized at zero, for then Figure 8.3.1 shows that the controller is just a PD controller. Standard results in the robotics literature [Dawson et al. 1990] show that a PD controller gives bounded errors if K_v is large enough. Therefore, the closed-loop system remains stable until the NN begins to learn. The weights are tuned *on-line in real-time* as the system tracks the desired trajectory. As the NN learns $f(x)$, the tracking performance improves. This is a significant improvement over other NN control techniques where one must find some *initial stabilizing weights*, generally an impossible feat for complex nonlinear systems.

Bounds on the Tracking Error and NN Weight Estimation Errors. The right-hand side of (8) can be taken as a *practical bound on the tracking error* in the sense that $r(t)$ will never stray far above it. It is important to note from this equation that the tracking error increases with the NN reconstruction error ϵ_N and robot disturbances d_B, yet *arbitrarily small tracking errors may be achieved* by selecting large gains K_v. This is in spite of Barron's lower bound on ϵ.

Note that the NN weights $\hat{W}$ are not guaranteed to approach the ideal unknown weights W that give good approximation of $f(x)$. However, this is of no concern as long as $W - \hat{W}$ is bounded, as the proof guarantees. This guarantees bounded control inputs $\tau(t)$ so that the tracking objective can be achieved.

8.5 Tuning Algorithms for Nonlinear-in-the-Parameters NN

Now suppose that a 2-layer NN is used to approximate the robot function according to (8.2.5) with $\|\epsilon\| \leq \epsilon_N$ on a compact set, and the ideal approximating weights W, V constant. Then the control law (8.3.7) becomes

$$\tau = \hat{W}^T \sigma(\hat{V}^T x) + K_v r - v. \tag{8.5.1}$$

The proposed NN control structure is shown in Figure 8.3.1.

The NLIP NN controller is far more powerful than the 1-layer FLNN as it is not necessary to select a basis of activation functions. In effect, the weights $\hat{V}$ are adjusted on-line to automatically provide a suitable basis. The control is nonlinear in the first-layer weights $\hat{V}$, which presents complications in deriving suitable weight tuning laws. Auxiliary signal $v(t)$ is a function to be detailed subsequently that provides robustness in the face of higher-order terms in the Taylor series arising from the NLIP problem.

Using the result (8.2.13) to properly address the NLIP problem, the closed-loop error dynamics (8.3.8) can be written as

$$M\dot{r} = -(K_v + V_m)r + \tilde{W}^T(\hat{\sigma} - \hat{\sigma}'\hat{V}^T x) + \hat{W}^T\hat{\sigma}'\tilde{V}^T x + w + v \tag{8.5.2}$$

where the disturbance terms are

$$w(t) = \tilde{W}^T \hat{\sigma}' V^T x + W^T O(\tilde{V}^T x)^2 + \varepsilon + \tau_d \tag{8.5.3}$$

Define

$$Z \equiv \begin{bmatrix} W & 0 \\ 0 & V \end{bmatrix} \tag{8.5.4}$$

and $\hat{Z}, \tilde{Z}$ equivalently. It is not difficult to show that

$$\|w(t)\| \leq C_0 + C_1\|\tilde{Z}\|_F + C_2\|\tilde{Z}\|_F\|r\| \tag{8.5.5}$$

with C_i known positive constants.

It is assumed that the ideal weights are bounded so that

$$\|Z\|_F \leq Z_B \tag{8.5.6}$$

with Z_B known. (Standard adaptive robust control techniques can be used to lift the assumption that Z_B is known [Polycarpou 1996].)

The next theorem presents the most powerful NN controller in this chapter.

THEOREM 8.5-1: *(Augmented Backprop Weight Tuning for Multilayer NN Controller).*

Let the desired trajectory $q_d(t)$ be bounded by q_B and the initial tracking error $r(0)$ be in S_r. Let the ideal target NN weights be bounded by known Z_B. Take the control input for the robot dynamics (8.3.1) as (8.5.1) with PD gain satisfying

$$K_{v_{min}} > \frac{(C_0 + \kappa C_3^2/4)(c_0 + c_2)}{b_x - q_B}, \tag{1}$$

where C_3 is defined in the proof. Let the robustifying term be

$$v(t) = -K_z(\|\hat{Z}\|_F + Z_B)r \tag{2}$$

with gain

$$K_z > C_2. \tag{3}$$

Let NN weight tuning be provided by

$$\dot{\hat{W}} = F\hat{\sigma}r^T - F\hat{\sigma}'\hat{V}^T x r^T - \kappa F\|r\|\hat{W} \tag{4}$$

$$\dot{\hat{V}} = Gx(\hat{\sigma}'^T \hat{W} r)^T - \kappa G\|r\|\hat{V} \tag{5}$$

with any constant matrices $F = F^T > 0$, $G = G^T > 0$, and $\kappa > 0$ a small scalar design parameter. Then the filtered tracking error $r(t)$ and NN weight estimates $\hat{V}, \hat{W}$ are UUB, with the bounds given specifically by (10) and (11). Moreover, the tracking error may be kept as small as desired by increasing the gains K_v in (8.5.1).

Proof:

Let the NN approximation property (8.2.5) hold for the function $f(x)$ given in (8.3.5) with a given accuracy ε_N for all x in the compact set $S_x \equiv \{x \mid \|x\| < b_x\}$ with $b_x > q_B$. Let $r(0) \in S_r$. Then the approximation property holds at time 0.

Define the Lyapunov function candidate

$$L = \frac{1}{2}r^T M(q)r + \frac{1}{2}tr\{\tilde{W}^T F^{-1} \tilde{W}\} + \frac{1}{2}tr\{\tilde{V}^T G^{-1} \tilde{V}\} \tag{6}$$

Differentiating yields

$$\dot{L} = r^T M\dot{r} + \frac{1}{2}r^T \dot{M}r + tr\{\tilde{W}^T F^{-1}\dot{\tilde{W}}\} + tr\{\tilde{V}^T G^{-1}\dot{\tilde{V}}\}. \tag{7}$$

8.5 Tuning Algorithms for Nonlinear-In-The-Parameters NN

Substituting now from the error system (8.5.2) yields

$$\dot{L} = -r^T K_v r + \frac{1}{2} r^T (\dot{M} - 2V_m) r$$
$$+ tr\{\tilde{W}^T(F^{-1}\dot{\tilde{W}} + \hat{\sigma} r^T - \sigma' \hat{V}^T x r^T)\}$$
$$+ tr\{\tilde{V}^T(G^{-1}\dot{\tilde{V}} + x r^T \hat{W}^T \hat{\sigma}')\} \qquad (8)$$

The tuning rules give

$$\dot{L} = -r^T K_v r + \kappa \|r\| tr\{\tilde{W}^T(W - \tilde{W})\}$$
$$+ \kappa \|r\| tr\{\tilde{V}^T(V - \tilde{V})\} + r^T(w + v)$$
$$= -r^T K_v r + \kappa \|r\| tr\{\tilde{Z}^T(Z - \tilde{Z})\} + r^T(w + v).$$

Since

$$tr\{\tilde{Z}^T(Z - \tilde{Z})\} = <\tilde{Z}, Z> - \|\tilde{Z}\|_F^2 \le \|\tilde{Z}\|_F \|Z\|_F - \|\tilde{Z}\|_F^2,$$

there results

$$\dot{L} \le -r^T K_v r + \kappa \|r\| \cdot \|\tilde{Z}\|_F (Z_B - \|\tilde{Z}\|_F)$$
$$- K_Z (\|\hat{Z}\|_F + Z_B) \|r\|^2 + \|r\| \cdot \|w\|$$
$$\le - K_{v_{min}} \|r\|^2 + \kappa \|r\| \cdot \|\tilde{Z}\|_F (Z_B - \|\tilde{Z}\|_F)$$
$$- K_Z (\|\hat{Z}\|_F + Z_B) \|r\|^2$$
$$+ \|r\| \left[C_0 + C_1 \|\tilde{Z}\|_F + C_2 \|r\| \cdot \|\tilde{Z}\|_F \right]$$
$$\le - \|r\| \left\{ K_{v_{min}} \|r\| - \kappa \cdot \|\tilde{Z}\|_F (Z_B - \|\tilde{Z}\|_F) - C_0 - C_1 \|\tilde{Z}\|_F \right\}$$

where $K_{v_{min}}$ is the minimum singular value of K_v, and the last inequality holds due to (3).

$\dot{L}$ is negative as long as the term in braces is positive. Defining $C_3 = Z_B + C_1/\kappa$ and completing the square yields

$$K_{v_{min}} \|r\| - \kappa \cdot \|\tilde{Z}\|_F (Z_B - \|\tilde{Z}\|_F) - C_0 - C_1 \|\tilde{Z}\|_F$$
$$= \kappa(\|\tilde{Z}\|_F - C_3/2)^2 + K_{v_{min}} \|r\| - C_0 - \kappa C_3^2/4 \qquad (9)$$

which is guaranteed positive as long as either

$$\|r\| > \frac{C_0 + \kappa C_3^2/4}{K_{v_{min}}} \equiv b_r \tag{10}$$

or

$$\|\tilde{Z}\|_F > C_3/2 + \sqrt{C_0/\kappa + C_3^2/4} \equiv b_Z. \tag{11}$$

Thus, $\dot{L}$ is negative outside a compact set. According to the LaSalle extension [Lewis et al. 1993] this demonstrates the UUB of both $\|r\|$ and $\|\tilde{Z}\|_F$ as long as the control remains valid within this set. However, the PD gain condition (1) shows that the compact set defined by $\|r\| \leq b_r$ is contained in S_r, so that the approximation property holds throughout. ∎

The first terms in the tuning algorithms (4), (5) are nothing but backpropagation of the tracking error through time. The last terms are the Narendra e-mod, and the second term in (4) is a novel second-order term needed due to the NLIP of the NN. The robustifying term $v(t)$ is required also due to the NLIP problem.

The comments appearing after Theorem 8.4.1 are germane here. Specifically, there is no preliminary off-line tuning phase and NN weight tuning occurs on-line in real-time simultaneously with control action. Weight initialization is not an issue in the proof and it is not necessary to provide initial stabilizing weights; the weights are simply initialized so that the NN output is equal to zero, for then the PD loop keeps the system stable until the NN begins to learn. However, practical experiments [Gutierrez and Lewis 1998] show that it is important to initialize the weights suitably. A good choice is to select $\hat{V}(0)$ randomly and $\hat{W}(0)$ equal to zero.

In practice it is not necessary to compute the constants c_0, c_1, c_2, C_0, C_1, C_2, Z_B nor determine S_r. The size L of the NN and the PD gains K_v are simply selected large and a simulation is performed. The simulation is then repeated with a different L and K_v. Based on the difference in behavior between the two simulations, L and K_v can be modified.

Straight Backpropagation Weight Tuning. The backprop algorithm (usually in discrete-time form) has been proposed innumerable times in the literature for closed-loop control. It can be shown that if the NN approximation error ϵ is zero, the disturbances $\tau_d(t)$ are zero, and there are no higher-order terms in the Taylor series (8.2.12), then the straight backprop algorithm

$$\dot{\hat{W}} = F\hat{\sigma} r^T$$
$$\dot{\hat{V}} = Gx(\hat{\sigma}'^T \hat{W} r)^T \tag{8.5.7}$$

8.5 Tuning Algorithms for Nonlinear-In-The-Parameters NN

may be used for closed-loop control instead of (4), (5). PE is required to show that the NN weights are bounded using straight backprop. The conditions mentioned are very restrictive and do not occur in practice; they essentially require the robot arm to be linear (e.g. 1=link). Moreover, PE conditions are not easy to guarantee in NN.

Passivity Properties of NN Controllers

The NN used in this paper are static feedforward nets, but since they appear in a closed feedback loop and are tuned using differential equations, they are dynamical systems. Therefore, one can discuss the *passivity* of these NN. In general a NN cannot be guaranteed to be passive. However, the NN controllers in this paper have some important passivity properties that result in robust closed-loop performance.

The system $\dot{x} = f(x,u), y = h(x,u)$ is said to be *passive* if it verifies an equality of the so-called *power form* [Slotine and Li 1991]

$$\dot{L}(t) = y^T u - g(t) \tag{8.5.8}$$

for some lower-bounded $L(t)$ and some $g(t) \geq 0$. That is,

$$\int_0^T y^T(\tau)u(\tau)d\tau \geq \int_0^T g(\tau)d\tau - \gamma^2 \tag{8.5.9}$$

for all $T \geq 0$ and some $\gamma \geq 0$. The system is *dissipative* if it is passive and in addition

$$\int_0^\infty y^T(\tau)u(\tau)d\tau \neq 0 \text{ implies } \int_0^\infty g(\tau)d\tau > 0. \tag{8.5.10}$$

A special sort of dissipativity occurs if $g(t)$ is a quadratic function of $\|x\|$ with bounded coefficients. We call this *state strict passivity (SSP)*; then, the norm of the internal states is bounded in terms of the power delivered to the system. Somewhat surprisingly, the concept of SSP has not been extensively used in the literature [Lewis et al. 1993], [Seron et al. 1994], though see [Goodwin et al. 1984] where input and output strict pasivity are defined. SSP turns out to be pivotal in studying the passivity properties of NN controllers, and allows one to conclude some internal boundedness properties without any assumptions of persistence of excitation.

Passivity of the Robot Tracking Error Dynamics

The error dynamics in this paper (e.g. (8.5.2)) appear in Figure 8.5.1 and have the form

$$M\dot{r} = -(K_v + V_m)r + \xi_0 \tag{8.5.11}$$

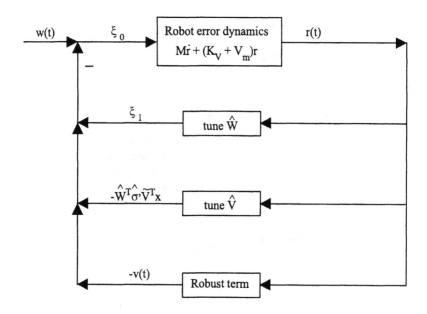

Figure 8.5.1: Two-layer neural net closed-loop error system.

where $r(t)$ is the filtered tracking error and $\xi_0(t)$ is appropriately defined. This system satisfies the following strong passivity property.

THEOREM 8.5-2: (SSP of Robot Error Dynamics).

The dynamics (8.5.11) from $\xi_0(t)$ to $r(t)$ are a state strict passive system.

Proof:

Take the nonnegative function

$$L = \frac{1}{2} r^T M(q) r$$

so that, using (8.5.11), one obtains

$$\dot{L} = r^T M \dot{r} + \frac{1}{2} r^T \dot{M} r = -r^T K_v r + \frac{1}{2} r^T (\dot{M} - 2V_m) r + r^T \xi_0$$

whence the skew-symmetry property yields the power form

$$\dot{L} = r^T \xi_0 - r^T K_v r.$$

8.5 Tuning Algorithms for Nonlinear-In-The-Parameters NN

This is the power delivered to the system minus a quadratic term in $\|r\|$, verifying state strict passivity. ∎

Passivity Properties of 2-layer NN Controllers

The closed-loop system is in Figure 8.3.1, where the NN appears in an inner feedback linearization loop. The error dynamics for the 2-layer NN controller are given by (8.5.2). The closed-loop error system appears in Figure 8.5.1, with the signal ξ_1 defined as

$$\xi_1(t) = -\tilde{W}^T(\hat{\sigma} - \hat{\sigma}'\hat{V}^T x) \qquad (8.5.12)$$

Note the role of the NN in the error dynamics, where it is decomposed into two effective blocks appearing in a typical feedback configuration.

We now reveal the passivity properties engendered by straight back-propagation tuning (8.5.7). To prove this algorithm, one uses the error dynaqmics in a different form than (8.5.2), so that in Figure 8.5.1 one has $\xi_1(t) = -\tilde{W}^T \hat{\sigma}$.

THEOREM 8.5-3: *(Passivity of Backprop NN Tuning Algorithm).*

The simple backprop weight tuning algorithm (8.5.7) makes the map from $r(t)$ to $-\tilde{W}^T \hat{\sigma}$, and the map from $r(t)$ to $-\tilde{W}^T \hat{\sigma}' \hat{V}^T x$, both passive maps.

Proof:
The dynamics with respect to $\tilde{W}$, $\tilde{V}$ are

$$\dot{\tilde{W}} = -F\hat{\sigma}r^T \qquad (1)$$

$$\dot{\tilde{V}} = -Gx(\hat{\sigma}'^T \hat{W} r)^T \qquad (2)$$

1. Selecting the nonnegative function

$$L = \frac{1}{2}tr\{\tilde{W}^T F^{-1} \tilde{W}\}$$

and evaluating $\dot{L}$ along the trajectories of (1) yields

$$\dot{L} = tr\{\tilde{W}^T F^{-1} \dot{\tilde{W}}\} = -tr\{\tilde{W}^T \hat{\sigma} r^T\} = r^T(-\tilde{W}^T \hat{\sigma}),$$

which is in power form (8.5.8).

2. Selecting the nonnegative function

$$L = \frac{1}{2}tr\{\tilde{V}^T G^{-1} \tilde{V}\}$$

and evaluating $\dot{L}$ along the trajectories of (2) yields

$$\dot{L} = tr\{\tilde{V}^T G^{-1}\dot{\tilde{V}}\} = -tr\{\tilde{V}^T x(\hat{\sigma}'^T \hat{W} r)^T\} = r^T(-\hat{W}^T \hat{\sigma}' \tilde{V}^T x).$$

which is in power form. ∎

Thus, the robot error system in Figure 8.5.1 is state strict passive (SSP) and the weight error blocks are passive; this guarantees the passivity of the closed-loop system. Using the passivity theorem [Slotine and Li 1991] one may now conclude that the input/output signals of each block are bounded as long as the external inputs are bounded.

Unfortunately, though passive, the closed-loop system cannot be guaranteed to be SSP, so that when disturbances are nonzero, this does not yield boundedness of the internal states of the weight blocks (i.e. $\tilde{W}$, $\tilde{V}$) unless those blocks are *observable*, that is persistently exciting (PE).

The next result shows why a PE condition is not needed with the modified weight update algorithm given in Theorem 8.5.1.

THEOREM 8.5-4: *(SSP of Augmented Backprop NN Tuning Algorithm).*

The modified weight tuning algorithms in Theorem 8.5.1 make the map from $r(t)$ to $-\tilde{W}^T(\hat{\sigma} - \hat{\sigma}'\hat{V}^T x)$, and the map from $r(t)$ to $-\hat{W}^T \hat{\sigma}' \tilde{V}^T x$, both state strict passive (SSP) maps.

Proof:

The dynamics relative to $\tilde{W}$, $\tilde{V}$ using the tuning algorithms in Theorem 8.5.1 are given by

$$\dot{\tilde{W}} = -F\hat{\sigma} r^T + F\hat{\sigma}'\hat{V}^T x r^T + \kappa F\|r\|\hat{W}$$

$$\dot{\tilde{V}} = -Gx(\hat{\sigma}'^T \hat{W} r)^T + \kappa G\|r\|\hat{V}.$$

1. Selecting the nonnegative function

$$L = \frac{1}{2} tr\{\tilde{W}^T F^{-1} \tilde{W}\}$$

and evaluating $\dot{L}$ yields

$$\begin{aligned}\dot{L} &= tr\{\tilde{W}^T F^{-1}\dot{\tilde{W}}\}\\ &= tr\left\{\left[-\tilde{W}^T(\hat{\sigma} - \hat{\sigma}'^T \hat{V}^T x)\right] r^T + \kappa\|r\|\tilde{W}^T \hat{W}\right\}\end{aligned}$$

Since

8.5 Tuning Algorithms for Nonlinear-In-The-Parameters NN

$$tr\{\tilde{W}^T(W - \tilde{W})\} = <\tilde{W}, W>_F - \|\tilde{W}\|_F^2$$
$$\leq \|\tilde{W}\|_F \cdot \|W\|_F - \|\tilde{W}\|_F^2,$$

there results

$$\dot{L} \leq r^T\left[-\tilde{W}^T(\hat{\sigma} - \hat{\sigma}'\hat{V}^T x)\right] - \kappa\|r\| \cdot (\|\tilde{W}\|_F^2 - \|\tilde{W}\|_F\|W\|_F)$$
$$\leq r^T\left[-\tilde{W}^T(\hat{\sigma} - \hat{\sigma}'\hat{V}^T x)\right] - \kappa\|r\| \cdot (\|\tilde{W}\|_F^2 - W_B\|\tilde{W}\|_F)$$

which is in power form with the last function quadratic in $\|\tilde{W}\|_F$.

2. Selecting the nonnegative function

$$L = \frac{1}{2}tr\{\tilde{V}^T G^{-1}\tilde{V}\}$$

and evaluating $\dot{L}$ yields

$$\dot{L} = tr\{\tilde{V}^T G^{-1}\dot{\tilde{V}}\}$$
$$= r^T(-\hat{W}^T\hat{\sigma}'\tilde{V}^T x) - \kappa\|r\|(\|\tilde{V}\|_F^2 - <\tilde{V}, V>_F)$$
$$\leq r^T(-\hat{W}^T\hat{\sigma}'\tilde{V}^T x) - \kappa\|r\|(\|\tilde{V}\|_F^2 - V_B\|\tilde{V}\|)$$

which is in power form with the last function quadratic in $\|\tilde{V}\|_F$. ∎

The SSP of both the robot dynamics and the weight tuning blocks does guarantee SSP of the closed-loop system, so that *the norms of the internal states are bounded in terms of the power delivered to each block.* Then, boundedness of input/output signals assures state boundedness without any sort of observability or PE requirement.

Definition of Passive NN and Robust NN. We define a dynamically tuned NN as *passive* if, in the error formulation, the tuning guarantees the passivity of the weight tuning subsystems. Then, an extra PE condition is needed to guarantee boundedness of the weights. This PE condition is generally in terms of the outputs of the hidden layers of the NN. We define a dynamically tuned NN as *robust* if, in the error formulation, the tuning guarantees the SSP of the weight tuning subsystem. Then, no extra PE condition is needed for boundedness of the weights. Note that (1) SSP of the open-loop plant error system is needed in addition for tracking stability, and (2) the NN passivity properties are dependent on the weight tuning algorithm used.

Passivity Properties of 1-Layer NN Controllers

In a similar fashion, it is shown that the FLNN controller tuning algorithm makes the system passive, so that an additional PE condition is needed to verify internal stability of the NN weights. On the other hand, the augmented tuning algorithm in Theorem 8.4.1 yields SSP, so that no PE is needed.

8.6 Summary

Neural network controller design algorithms were given for a general class of industrial Lagrangian motion systems characterized by the rigid robot arms. The design procedures are based on rigorous nonlinear derivations and stability proofs, and yield a *multiloop intelligent control structure* with NN in some of the loops. NN weight tuning algorithms were given that do not require complicated initialization procedures or any off-line learning phase, work on-line in real-time, and offer guaranteed tracking and bounded NN weights and control signals. The NN controllers given here are *model-free controllers* in that they work for any system in a prescribed class without the need for extensive modeling and preliminary analysis to find a 'regression matrix'. Unlike standard robot adaptive controllers, they do not require linearity in the parameters (see also [Colbaugh and Seraji 1994] which does not need LIP). Proper design allows the NN controllers to avoid requirements such as persistence of excitation and certainty equivalence. NN passivity and robustness properties are defined and studied here.

REFERENCES

[Åström and Wittenmark 1989] K.J. Åström and B. Wittenmark, *Adaptive Control*, Addision Wesley, Reading, MA, 1989.

[Barron 1993] Barron, A.R., "Universal approximation bounds for superpositions of a sigmoidal function," *IEEE Trans. Info. Theory*, vol. 39, no. 3, pp. 930-945, May 1993.

[Chen and Khalil 1992] F.-C. Chen and H.K. Khalil, "Adaptive control of nonlinear systems using neural networks," *Int. J. Control*, vol. 55, no. 6, pp. 1299-1317, 1992.

[Chen and Liu 1994] F.-C. Chen and C.-C. Liu, "Adaptively controlling nonlinear continuous-time systems using multilayer neural networks," *IEEE Trans. Automat. Control*, vol. 39, no. 6, pp. 1306-1310, June 1994.

[Colbaugh and Seraji 1994] R. Colbaugh, H. Seraji, and K. Glass, "A new class of adaptive controllers for robot trajectory tracking," *J. Robotic Systems*, vol. 11, no. 8, pp. 761-772, 1994.

[Craig 1988] Craig, J.J., *Adaptive Control of Mechanical Manipulators*, Reading, MA: Addison-Wesley, 1988.

[Dawson et al. 1990] Dawson, D.M., Z. Qu, F.L. Lewis, and J.F. Dorsey, "Robust control for the tracking of robot motion," *Int. J. Control*, vol. 52, no. 3, pp. 581-595, 1990.

[Goodwin et al. 1984] Goodwin, C.G., and K.S. Sin, *Adaptive Filtering, Prediction, and Control*, Prentice-Hall, New Jersey, 1984.

[Gorinevsky 1995] D. Gorinevsky, "On the persistency of excitation in radial basis function network identificaiton of nonlinear systems," *IEEE Trans. Neural Networks*, vol. 6, no. 5, pp. 1237-1244, Sept. 1995.

[Gutierrez and Lewis 1998] L.B. Gutierrez and F.L. Lewis, "Implementation of a neural net tracking controller for a single flexible link: comparison with PD and PID controllers," *IEEE Trans. Industrial Electronics*, to appear, April 1998.

[Haykin 1994] S. Haykin, *Neural Networks*, IEEE Press, New York, 1994.

[Hornik and Stinchombe 1989] K. Hornik, M. Stinchombe, and H. White, "Multilayer feedforward networks are universal approximators," *Neural Networks*, vol. 2, pp. 359-366, 1989.

[Ioannou and Kokotovic 1984] P.A. Ioannou and P.V. Kokotovic, "Instability analysis and improvement of robustness of adaptive control," *Automatica*, vol. 20, no. 5, pp. 583-594, 1984.

[Kim and Lewis 1998] Y.H. Kim and F.L. Lewis, *High-Level Feedback Control with Neural Networks*, World Scientific, Singapore, 1998.

[Kreisselmeier and Anderson 1986] Kreisselmeier, G., and B. Anderson, "Robust model reference adaptive control," *IEEE Trans. Automat. Control*, vol. AC-31, no. 2, pp. 127-133, Feb. 1986.

[Landau 1979] Y.D. Landau, *Adaptive Control*, Marcel Dekker, Basel, 1979.

[Lewis 1999] F.L. Lewis, "Nonlinear Network Structures for Feedback Control", *Asian Journal of Control*, vol. 1, no. 4, pp.205-228, December 1999.

[Lewis et al. 1993] F.L. Lewis, C.T. Abdallah, and D.M. Dawson, *Control of Robot Manipulators*, New York: Macmillam, 1993.

[Lewis et al. 1999] F.L. Lewis, S. Jagannathan, and A. Yeşildirek, *Neural Network Control of Robot Manipulators and Nonlinear Systems*, Taylor and Francis, London, 1999.

[Lewis and Kim 1998] F. L. Lewis and Y.H. Kim, "Neural Networks for Feedback Control," Encyclopedia of Electrical and Electronics Engineering, ed. J.G. Webster, New York: Wiley, 1998. To appear.

[Lewis et al. 1993] Lewis, F.L., K. Liu, and A. Yeşildirek, "Neural net robot controller with guaranteed tracking performance," *Proc. Int. Symp. Intelligent Control*, pp. 225-231, Chicago., Aug. 1993.

[Lewis et al. 1995] Lewis, F.L., K. Liu, and A. Yeşildirek, "Neural net robot controller with guaranteed tracking performance," *IEEE Trans. Neural Networks*, vol. 6, no. 3, pp. 703-715, 1995.

REFERENCES

[Lewis et al. 1997] F.L. Lewis, K. Liu, and S. Commuri, "Neural networks and fuzzy logic systems for robot control," in *Fuzzy Logic and Neural Network Applications*, ed. F. Wang, World Scientific Pub., to appear, 1997.

[Lewis et al. 1996] F.L. Lewis, A. Yeşildirek, and K. Liu, "Multilayer neural net robot controller: structure and stability proofs," *IEEE Trans. Neural Networks*, vol. 7, no. 2, pp. 1-12, Mar. 1996.

[Miller et al. 1991] W.T. Miller, R.S. Sutton, P.J. Werbos, ed., *Neural Networks for Control*, Cambridge: MIT Press, 1991.

[Narendra 1991] K.S. Narendra, "Adaptive Control Using Neural Networks," *Neural Networks for Control*, pp 115-142. ed. W.T. Miller, R.S. Sutton, P.J. Werbos, Cambridge: MIT Press, 1991.

[Narendra and Annaswamy 1987] K.S. Narendra and A.M. Annaswamy, "A new adaptive law for robust adaptive control without persistent excitation," *IEEE Trans. Automat. Control*, vol. 32, pp. 134-145, Feb., 1987.

[Narendra and Parthasarathy 1991] Narendra, K.S., and K. Parthasarathy, "Gradient methods for the optimization of dynamical systems containing neural networks," *IEEE Trans. Neural Networks*, vol. 2, no. 2, pp. 252-262, Mar. 1991.

[Polycarpou 1996] M.M. Polycarpou, "Stable adaptive neural control scheme for nonlinear systems," *IEEE Trans. Automat. Control*, vol. 41, no. 3, pp. 447-451, Mar. 1996.

[Polycarpou and Ioannou 1991] M.M. Polycarpou and P.A. Ioannou, "Identification and control using neural network models: design and stability analysis," *Tech. Report 91-09-01*, Dept. Elect. Eng. Sys., Univ. S. Cal., Sept. 1991.

[Rovithakis and Christodoulou 1994] G.A. Rovithakis and M.A. Christodoulou, "Adaptive control of unknown plants using dynamical neural networks," *IEEE Trans. Systems, Man, and Cybernetics*, vol. 24, no. 3, pp. 400-412, Mar. 1994.

[Sadegh 1993] N. Sadegh, "A perceptron network for functional identification and control of nonlinear systems," *IEEE Trans. Neural Networks*, vol. 4, no. 6, pp. 982-988, Nov. 1993.

[Sanner and Slotine 1991] R.M. Sanner and J.-J.E. Slotine, "Stable adaptive control and recursive identification using radial gaussian networks," *Proc. IEEE Conf. Decision and Control*, Brighton, 1991.

[Seron et al. 1994] Seron, M.M., D.J. Hill, and A.L. Fradkov, "Adaptive passification of nonlinear systems," *Proc. IEEE Conf. Decision and Control*, pp. 190-195, Dec. 1994.

[Slotine 1988] Slotine, J.-J.E., "Putting physics in control the example of robotics," *IEEE Control Systems Magazine*, pp. 12-17, Dec. 1988.

[Slotine and Li 1991] J.-J. Slotine and W. Li, *Applied Nonlinear Control*, Prentice Hall, New Jersey, 1991.

[Spong and Vidyasagar 1989] Spong, M.W., and M. Vidyasagar, *Robot Dynamics and Control*, New York: Wiley, 1989.

[Wang 1997] L.-X. Wang, *A course in fuzzy systems and control*, Prentice-Hall, New Jersey, 1997.

[Werbos 1989] P.J. Werbos, "Back propagation: past and future," *Proc. 1988 Int. Conf. Neural Nets*, vol. 1, pp. I343-I353, 1989.

[Werbos 1992] P.J. Werbos, "Neurocontrol and supervised learning: an overview and evaluation," in *Handbook of Intelligent Control*, ed. D.A. White and D.A. Sofge, New York: Van Nostrand Reinhold, 1992.

[White and Sofge 1992] D.A. White and D.A. Sofge, ed. *Handbook of Intelligent Control*, New York: Van Nostrand Reinhold, 1992.

Chapter 9

Force Control

In this chapter the fundamentals of position/force controllers are studied. The topics that are covered include stiffness control, hybrid position/force control, hybrid impedance control, and reduced-state position/force control. Emphasis is placed on controller development, stability, and implementation issues.

9.1 Introduction

For tasks performed by robot manipulators, such as moving payloads or painting objects, position controllers give adequate performance because these types of tasks only require the robot to follow a desired trajectory. However, during grinding or an assembly task, the robot manipulator comes in contact with the environment; therefore, interaction forces develop between the robot manipulator and the environment. Consequently, these interaction forces, as well as the position of the end effector, must be controlled.

To motivate the need for using a combination of force and position control, consider the problem of controlling a manipulator to write a sentence on a blackboard. To form the letters in the sentence, we must certainly control the end-effector position or, equivalently, the position of the chalk. As anyone who has written on a blackboard knows, the force with which one presses on the blackboard must also be controlled. That is, pressing too lightly can result in letters that are not easily readable, while pressing too hard can result in broken chalk. This example clearly illustrates that many robotic applications will require that a desired positional trajectory and a desired force trajectory must be specified. In this chapter we present some general control strategies that control not only the robot end-effector position but also the force that the end effector exerts on the environment. It should be noted that throughout this chapter, we assume that the desired velocity and force trajectories, which are commanded by the controllers, are consistent with the model of the environment [Lipkin and Duffy 1988]. If this is not the case, it may be possible to modify the desired velocity and force trajectories to be consistent with the model of the environment. The interested reader is referred to [Lipkin and Duffy 1988] for information on this modifying or "kinestatic filtering" of the desired trajectories.

9.2 Stiffness Control

Since the first robot manipulators involved in industrial processes were required to perform positional tasks (e.g., spray painting), robot manipulators were manufactured to be very rigid. This rigid design allowed the robot control designer to obtain reasonable positional accuracy by utilizing simple control laws. As one might expect, force control applications (e.g., grinding or sanding) are extremely difficult to accomplish with such a "stiff" robot. Therefore, if the robot manipulator "stiffness" could be controlled, force control applications could be accomplished more easily. In this section the concept of stiffness control is formulated for a simple single-degree-of-freedom example. The robot manipulator equation is then modified to account for the forces exerted on the environment. Using this new model, the stiffness control concept [Salisbury and Craig 1980] is then generalized to an n-link robot manipulator.

Stiffness Control of a Single-Degree-of-Freedom Manipulator

To motivate the concept of stiffness control, consider the problem of force control for the system depicted by Figure 9.2.1. Here the manipulator with mass m is assumed to be in contact with the environment, which is located at the static position x_e. The control problem is to specify an input force (i.e., τ) so that the manipulator moves to a desired constant position (i.e., x_d). In this system, we also assume that if the position of the manipulator (i.e., x) is greater than x_e, the force (i.e., f) exerted on the environment is given by

$$f = k_e (x - x_e), \qquad (9.2.1)$$

where k_e is a positive constant used to denote the environmental stiffness. That is, we are assuming that the environmental stiffness can be modeled as a linear spring with a spring constant denoted by k_e. From this assumption

9.2 Stiffness Control

we can visualize the single-degree-of-freedom system as the mass-spring diagram given by Figure 9.2.2.

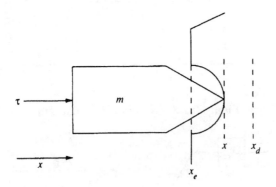

Figure 9.2.1: Single-degree-of-freedom system.

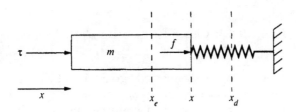

Figure 9.2.2: Mass-spring diagram.

Assuming that gravity and friction are negligible, the equation of motion for the system given in Figure 9.2.2 is given by

$$\tau = m\ddot{x} + k_e(x - x_e). \qquad (9.2.2)$$

The system block diagram for (9.2.2) is given by Figure 9.2.3. Note that in Figure 9.2.3, we have used the variable s to denote the Laplace transform variable.

The form of the dynamics given by (9.2.2) motivates the simple PD control law

$$\tau = -k_v \dot{x} + k_p(x_d - x), \qquad (9.2.3)$$

where k_v and k_p are positive scalar control gains. After substituting (9.2.3) into (9.2.2), we obtain the closed-loop system

$$m\ddot{x} + k_v \dot{x} + (k_p + k_e)x = k_p x_d + k_e x_e, \qquad (9.2.4)$$

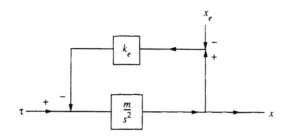

Figure 9.2.3: System block diagram.

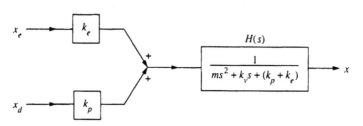

Figure 9.2.4: Closed-loop system block diagram.

which can be represented by the block diagram given by Figure 9.2.4. From Figure 9.2.4 we know that the closed-loop system is stable since we have defined the constants m, k_v, k_p, and k_e to be positive. That is, we can show that the poles of the transfer function

$$H(s) = \frac{1}{ms^2 + k_v s + (k_p + k_e)}$$

are in the open left-half s-plane.

To investigate how the PD control given in (9.2.3) controls the force exerted on the environment, we examine the system in steady-state conditions. Because x_d and x_e are constant, the Laplace transform of x can be found from (9.2.4) to be

$$x(s) = \frac{k_p x_d + k_e x_e}{s \left(ms^2 + k_v s + (k_p + k_e) \right)}. \tag{9.2.5}$$

Therefore, the steady-state manipulator position (i.e., $\bar{x}$) can easily be shown to be

$$\bar{x} = \lim_{s \to 0} s\, x(s) = \frac{k_p x_d + k_e x_e}{k_p + k_e}. \tag{9.2.6}$$

9.2 Stiffness Control

The steady-state manipulator position can now be used to calculate the steady-state force (i.e., $\bar{f}$) exerted on the environment. Specifically, upon substituting (9.2.6) into (9.2.1), the steady-state force is given by

$$\bar{f} = \frac{k_p k_e (x_d - x_e)}{k_p + k_e}. \tag{9.2.7}$$

As one would expect, the spring constant of the environment is often considered to be large because the robot is pushing on a nearly rigid surface in most robot force control applications. Thus, if we assume that $k_e \gg k_p$, we can approximate the steady-state force in (9.2.7) as

$$\bar{f} \cong k_p (x_d - x_e). \tag{9.2.8}$$

From the discussion above, we can see that the position control strategy given by (9.2.3) does indeed exert a force on the environment. Specifically, this force is created by commanding a desired trajectory that is slightly inside the contact surface. In attempting to eliminate the position error, the position controller exerts a steady-state force on the surface. From the approximate steady-state force given by (9.2.8), the position gain (i.e., k_p) can be thought of as representing the desired "stiffness" of the manipulator. That is, the manipulator can be visualized as a spring, with spring constant k_p, exerting a force on the environment. Hence the term "stiffness control" has often been associated with the PD control given by (9.2.3) since the stiffness of the manipulator can be set by the adjustment of k_p.

The Jacobian Matrix and Environmental Forces

Before the stiffness controller can be generalized to an n-link robot manipulator, we must define some notation with regard to the forces that the robot exerts on the environment. As explained in [Spong and Vidyasagar 1989], the forces are commonly transformed into the joint-space via a Jacobian matrix. In this chapter we define the Jacobian matrix in terms of a *task space* coordinate system which is defined for the specific robot application in question. That is, for a certain application, we may wish to have the end effector apply forces along a particular set of directions while moving along other directions. This concept is illustrated in Figure 9.2.5, which depicts a manipulator moving along a slanted surface. The task space coordinate system is given by the directions u and v since we wish to move the end effector along the surface in the direction v while applying a force normal to the surface along the direction u. An appropriate task space coordinate system is usually defined in most robotic applications from this line of reasoning.

Following this logic, let x be the $n \times 1$ task space vector defined by

$$x = h(q), \qquad (9.2.9)$$

where $h(q)$ is found from the manipulator kinematics and the appropriate relationships between the joint and task spaces. The derivative of x is defined as

$$\dot{x} = J(q)\dot{q}, \qquad (9.2.10)$$

where the $n \times n$ *task space Jacobian* matrix $J(q)$ [Spong and Vidyasagar 1989] is defined as

$$J(q) = \begin{bmatrix} I & 0 \\ 0 & T \end{bmatrix} \frac{\partial h(q)}{\partial q} \qquad (9.2.11)$$

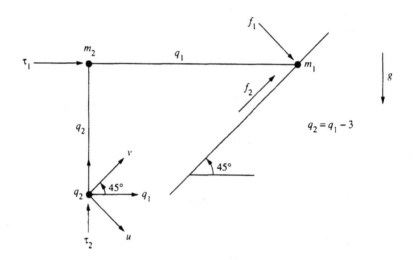

Figure 9.2.5: Manipulator moving along slanted surface.

with the identity matrix I, the zero matrix 0, and the transformation matrix T having dimensions dependent on the task space coordinate system selected. The transformation matrix T is typically used when converting joint velocities to the derivatives of the roll, pitch, and yaw angles associated with end-effector orientation. For brevity we assume that the robot manipulators discussed in this chapter are nonredundant and are always in a nonsingular configuration; therefore, the Jacobian matrix is a nonsingular square matrix.

Using the task space coordinate concept, we now examine how the robot equation must be modified for the purposes of force control. If the end effector of the manipulator is in contact with the environment, force interactions

9.2 Stiffness Control

will exist between the end effector and the environment. If the interaction forces are measured in the joint space, the manipulator dynamic equation can be written as

$$\tau = M(q)\ddot{q} + V_m(q,\dot{q})\dot{q} + G(q) + F(\dot{q}) + \tau_e, \qquad (9.2.12)$$

where τ_e is an $n \times 1$ vector in joint space coordinates, which denotes the force exerted on the environment. The dynamic equation given by (9.2.12) makes sense because if the manipulator is not moving (i.e., $\ddot{q} = \dot{q} = 0$), then (9.2.12) reduces to

$$\tau = G(q) + \tau_e. \qquad (9.2.13)$$

That is, if the robot manipulator is in static operation, the actuator force is equal to the force exerted on the environment plus the force needed to withstand the gravitational forces. Note that in (9.2.13), we have assumed that static friction can be neglected.

A joint space representation for the force exerted on the environment is not the standard notation in the robotics literature; rather, the robot manipulator equation is usually given by

$$\tau = M(q)\ddot{q} + V_m(q,\dot{q})\dot{q} + G(q) + F(\dot{q}) + J^T(q)f, \qquad (9.2.14)$$

where f is the $n \times 1$ vector of contact forces and torques in task space.

To understand the origin of (9.2.14), equate the right-hand sides of (9.2.12) and (9.2.14) to yield

$$\tau_e = J^T(q)f. \qquad (9.2.15)$$

This equation can be shown to be true quite easily by using a conservation of energy argument. Specifically, by conservation of energy, we know that

$$\dot{q}^T \tau_e = \dot{x}^T f. \qquad (9.2.16)$$

Substituting (9.2.10) into (9.2.16) yields

$$\dot{q}^T \tau_e = \dot{q}^T J^T(q) f.$$

Since the relationship above must hold for all $\dot{q}$, we can see that (9.2.15) obviously represents a true statement. To shed some light on the process of developing the Jacobian and the task space formulation, two examples are now discussed.

EXAMPLE 9.2-1: Task Space Formulation for a Slanted Surface

We want to find the manipulator dynamics for the Cartesian manipulator system (i.e., both joints are prismatic) given in Figure 9.2.5 and to decompose the forces exerted on the surface into a normal force and a tangent force. First, the motion portion of the dynamics can easily be determined when the robot is not constrained by the surface. After removing the surface and the interaction forces f_1, and f_2, the manipulator dynamics can be shown to be

$$\tau = M\ddot{q} + G + F(\dot{q}), \qquad (1)$$

where

$$\tau = \begin{bmatrix} \tau_1 \\ \tau_2 \end{bmatrix}, \, q = \begin{bmatrix} q_1 \\ q_2 \end{bmatrix}, \, G = \begin{bmatrix} 0 \\ (m_1 + m_2)g \end{bmatrix},$$

$$M = \begin{bmatrix} m_1 & 0 \\ 0 & m_1 + m_2 \end{bmatrix},$$

and $F(\dot{q})$ is the 2×1 vector $[F_1(\dot{q}_1) F_2(\dot{q}_2)]^T$ that models the friction as discussed in Chapter 2.

To account for the interaction forces, let x be the 2×1 task space vector defined by

$$x = \begin{bmatrix} u \\ v \end{bmatrix}, \qquad (2)$$

where u and v define a fixed coordinate system such that u represents the normal distance to the surface, and v represents the tangent distance along the surface. As in (9.2.9), the task space coordinates can be expressed in terms of the joint space coordinates by

$$x = h(q), \qquad (3)$$

where $h(q)$ is found from the geometry of the problem to be

$$h(q) = \frac{1}{\sqrt{2}} \begin{bmatrix} q_1 - q_2 \\ q_1 + q_2 \end{bmatrix}. \qquad (4)$$

9.2 Stiffness Control

The task space Jacobian matrix is found from (9.2.11) by utilizing the fact that T is the identity matrix for this problem because we do not have to concern ourselves with any end-effector angles of orientation. That is, $J(q)$ is given as

$$J = \frac{\partial h(q)}{\partial q} = \frac{1}{\sqrt{2}} \begin{bmatrix} 1 & -1 \\ 1 & 1 \end{bmatrix}. \tag{5}$$

Following (9.2.14), the robot manipulator equation is given by

$$\tau = M\ddot{q} + G + F(\dot{q}) + J^T f, \tag{6}$$

where

$$f = \begin{bmatrix} f_1 \\ f_2 \end{bmatrix}.$$

It is important to realize that the normal force (i.e., f_1) and the tangent force (i.e., f_2) are drawn in the direction of the task space coordinate system given by (2) (see Fig. 9.2.5).

■

EXAMPLE 9.2-2: Task Space Formulation for an Elliptical Surface

We wish to find the manipulator dynamics for the Cartesian manipulator system given in Figure 9.2.6 and to decompose the forces exerted on the surface into a normal force and a tangent force. The motion portion of the dynamics is the same as in Example 9.2.1; however, due to the change in the environmental surface, a new task space coordinate system must be defined. Specifically, let x be the 2×1 task space vector defined by

$$x = \begin{bmatrix} u \\ v \end{bmatrix}, \tag{1}$$

where u and v define a rotating coordinate system such that u represents the normal distance to the surface and v represents the tangent distance along the surface. As in (9.2.9), the task space coordinates can be expressed in terms of the joint space coordinates by

$$x = h(q), \tag{2}$$

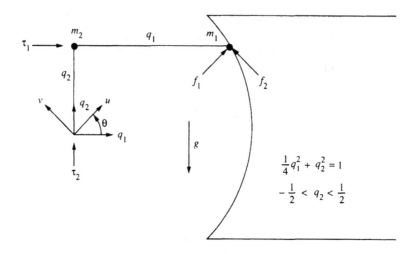

Figure 9.2.6: Manipulator moving along elliptical surface.

where $h(q)$ is found to be

$$h(q) = \begin{bmatrix} \bar{u} \cdot \bar{q}_1 & \bar{u} \cdot \bar{q}_2 \\ \bar{v} \cdot \bar{q}_1 & \bar{v} \cdot \bar{q}_2 \end{bmatrix} \begin{bmatrix} q_1 \\ q_2 \end{bmatrix}, \quad (3)$$

with $\bar{u}$, $\bar{v}$, $\bar{q}_1$ and $\bar{q}_2$ being appropriately defined unit vectors used in the dot product notation given in (3). The unit vectors $\bar{q}_1$ and $\bar{q}_2$ are defined in terms of the fixed coordinate set given by q_1 and q_2. These unit vectors are defined as

$$\bar{q}_1 = \begin{bmatrix} q_1 \\ q_2 \end{bmatrix} = \begin{bmatrix} 1 \\ 0 \end{bmatrix} \quad \text{and} \quad \bar{q}_2 = \begin{bmatrix} q_1 \\ q_2 \end{bmatrix} = \begin{bmatrix} 0 \\ 1 \end{bmatrix}. \quad (4)$$

To find the unit vectors $\bar{u}$ and $\bar{v}$, we first use the function of the surface

$$1/4 q_1^2 + q_2^2 = 1 \quad (5)$$

to parameterize the surface in terms of one variable (i.e., q_2) as follows:

$$\begin{bmatrix} q_1 \\ q_2 \end{bmatrix} = \begin{bmatrix} 2\sqrt{1 - q_2^2} \\ q_2 \end{bmatrix}. \quad (6)$$

9.2 Stiffness Control

The partial derivative of (6) with respect to q_2 divided by the length of the vector yields a unit vector (i.e., $\bar{v}$) that is always tangent to the surface. That is, $\bar{v}$ is given by

$$\bar{v} = \begin{bmatrix} \frac{\partial q_1}{\partial q_2} \\ \frac{\partial q_2}{\partial q_2} \end{bmatrix} = \frac{1}{\Delta} \begin{bmatrix} -2q_2 \left[1 - q_2^2\right]^{-1/2} \\ 1 \end{bmatrix}, \qquad (7)$$

where

$$\Delta = \sqrt{1 + 4q_2^2 \left(1 - q_2^2\right)^{-1}}.$$

By using (5) again, the expression for $\bar{v}$ can be simplified to yield

$$\bar{v} = \frac{1}{\Delta} \begin{bmatrix} -4q_2/q_1 \\ 1 \end{bmatrix}, \qquad (8)$$

where

$$\Delta = \sqrt{1 + 16q_2^2/q_1^2}.$$

Because the vectors $\bar{u}$ and $\bar{v}$ must be orthogonal (i.e., $\bar{u} \cdot \bar{v} = 0$), via (8) and the geometry of the problem, we know that

$$\bar{u} = \frac{1}{\Delta} \begin{bmatrix} 1 \\ 4q_2/q_1 \end{bmatrix}. \qquad (9)$$

Substituting (4), (8), and (9) into (3) yields

$$h(q) = \frac{1}{\Delta} \begin{bmatrix} q_1 + 4q_2^2/q_1 \\ -3q_2 \end{bmatrix}. \qquad (10)$$

The task space Jacobian matrix is found from (9.2.11) by utilizing the fact that T is the identity matrix. That is, $J(q)$ is given as

$$J(q) = \begin{bmatrix} J_{11} & J_{12} \\ J_{21} & J_{22} \end{bmatrix}, \qquad (11)$$

where

$$J_{11} = {}^1\!/\!_2 q_1 \left[{}^1\!/\!_4 q_1^2 + 7 q_2^2\right] \left[{}^1\!/\!_4 q_1^2 + 4 q_2^2\right]^{-3/2},$$
$$J_{12} = q_2 \left[8 q_2^2 - q_1^2\right] \left[{}^1\!/\!_4 q_1^2 + 4 q_2^2\right]^{-3/2},$$
$$J_{21} = -6 q_2^3 \left[{}^1\!/\!_4 q_1^2 + 4 q_2^2\right]^{-3/2},$$
$$J_{22} = -{}^3\!/\!_8 q_1^3 \left[{}^1\!/\!_4 q_1^2 + 4 q_2^2\right]^{-3/2},$$

Following (9.2.14), the robot manipulator equation is given by

$$\tau = M\ddot{q} + G + F(\dot{q}) + J^T(q) f, \qquad (12)$$

where τ, M, q, G, f, and $F(\dot{q})$ are as defined in Example 9.2.1. It is important to note that the normal force (i.e., f_1) and the tangent force (i.e., f_2) are drawn in the direction of the task space coordinate system given by (1) (see Fig. 9.2.6).

■

Stiffness Control of an N-Link Manipulator

Now that we have the robot manipulator dynamics in a form which includes the environmental interaction forces, the stiffness controller for the n-link robot manipulator can be formulated. As before, the force exerted on the environment is defined as

$$f = K_e (x - x_e), \qquad (9.2.17)$$

where K_e is an $n \times n$ diagonal, positive semi-definite, constant matrix used to denote the environmental stiffness, and x_e is an $n \times 1$ vector measured in task space that is used to denote the static location of the environment. Note that if the manipulator is not constrained in a particular task space direction, the corresponding diagonal element of the matrix K_e is assumed to be zero. Also, the environmental surface friction is typically neglected in the stiffness control formulation.

The multidimensional stiffness controller is the PD-type controller

$$\tau = J^T(q) \left(-K_v \dot{x} + K_p \tilde{x}\right) + G(q) + F(\dot{q}), \qquad (9.2.18)$$

where K_v and K_p are $n \times n$ diagonal, constant, positive-definite matrices and the task space tracking error is defined as

$$\tilde{x} = x_d - x.$$

9.2 Stiffness Control

As before, x_d is used to denote the desired constant end-effector position that we wish to move the robot manipulator to; however, x_d is now an $n \times 1$ vector. Substituting (9.2.17) and (9.2.18) into (9.2.14) yields the closed-loop dynamics

$$M(q)\ddot{q} + V_m(q,\dot{q})\dot{q} = J^T(q)\left(-K_v \dot{x} + K_p \tilde{x} - K_e(x - x_e)\right). \quad (9.2.19)$$

To analyze the stability of the system given by (9.2.19), we utilize the Lyapunov-like function

$$V = \tfrac{1}{2}\dot{q}^T M(q)\dot{q} + \tfrac{1}{2}\tilde{x}^T K_p \tilde{x} + \tfrac{1}{2}(x - x_e)^T K_e(x - x_e). \quad (9.2.20)$$

Differentiating (9.2.20) with respect to time and utilizing (9.2.10) yields

$$\begin{aligned}\dot{V} =\ & \tfrac{1}{2}\dot{q}^T \dot{M}(q)\dot{q} - \dot{q}^T M(q)\ddot{q} - \dot{q}^T J^T(q) K_p \tilde{x} \\ & + \dot{q}^T J^T(q) K_e(x - x_e).\end{aligned} \quad (9.2.21)$$

Note that in (9.2.21) we have used the fact that x_e and x_d are constant and that the transpose of either a scalar function or a diagonal matrix is equal to that function or matrix, respectively. Substituting (9.2.19) into (9.2.21) and utilizing (9.2.10) yields

$$\dot{V} = \dot{q}^T \left(\tfrac{1}{2}\dot{M}(q) - V_m(q,\dot{q})\right)\dot{q} - \dot{q}^T J^T(q) K_v J(q)\dot{q}. \quad (9.2.22)$$

Applying the skew-symmetric property (see Chapter 2) to (9.2.22) yields

$$\dot{V} = -\dot{q}^T J^T(q) K_v J(q)\dot{q}, \quad (9.2.23)$$

which is nonpositive. Since the matrices $J(q)$ and hence $J^T(q)K_v J(q)$ are nonsingular, $\dot{V}$ can only remain zero along trajectories where $\dot{q} = 0$ and hence $\ddot{q} = 0$ (see LaSalle's theorem in Chapter 1). Substituting $\dot{q} = 0$ and $\ddot{q} = 0$ into (9.2.19) and utilizing (9.2.10) yields

$$\lim_{t \to \infty} [K_p \tilde{x} - K_e(x - x_e)] = 0 \quad (9.2.24)$$

or, equivalently,

$$\lim_{t \to \infty} x_i = (K_{pi} + K_{ei})^{-1}(K_{pi} x_{di} + K_{ei} x_{ei}), \quad (9.2.25)$$

where the subscript i is used to denote the ith component of the vectors x, x_d, x_e, and the ith diagonal element of the matrices K_p and K_e.

The stability analysis above can be interpreted to mean that the robot manipulator will stop moving when the task space coordinates are given by (9.2.25). That is, the final position or steady-state position of the end effector is given by (9.2.25), which in the single-degree-of-freedom case is equivalently given by (9.2.6). To obtain the ith component of the steady-state force exerted on the environment, we substitute (9.2.25) into the ith component of (9.2.17) to yield

$$\lim_{t\to\infty} f_i = K_{ei}\left(K_{pi} + K_{ei}\right)^{-1} K_{pi}\left(x_{di} - x_{ei}\right). \quad (9.2.26)$$

Thus the steady-state force exerted by the end effector on the environment is given by (9.2.26), which in the single-degree-of-freedom case is equivalently given by (9.2.7). As in the single-degree-of-freedom case, we assume that K_{ei} is much larger than K_{pi} for the task space directions that are to be force controlled. That is, the steady-state force in (9.2.26) can be approximated by

Table 9.2.1: Stiffness Controller

Torque Controller:

$$\tau = J^T\left(q\right)\left(-K_v \dot{x} + K_p \tilde{x}\right) + G\left(q\right) + F\left(\dot{q}\right)$$

where $J\left(q\right)$ is the task space Jacobian.

Stability:

Nonconstrained directions: set-point positional control

$$\lim_{t\to\infty} x_i\left(t\right) = x_{di}$$

Constrained directions: steady-state force control approximated by

$$\lim_{t\to\infty} f_i\left(t\right) \cong K_{pi}\left(x_{di} - x_{ei}\right)$$

Comments: Control gains K_{pi} are used to adjust stiffness of the manipulator.

$$\lim_{t\to\infty} f_i \cong K_{pi}\left(x_{di} - x_{ei}\right); \quad (9.2.27)$$

9.2 Stiffness Control

therefore, K_{pi} can interpreted as specifying the stiffness of the manipulator in these task space directions.

If the manipulator is not constrained in a task space direction, the corresponding stiffness constant K_{ei} is equal to zero. Substituting $K_{ei} = 0$ into (9.2.25) yields

$$\lim_{t \to \infty} x_i = x_{di}. \tag{9.2.28}$$

This means that for the nonconstrained task space directions, we obtain set-point control; therefore, in steady state the desired position set point is reached. The stiffness controller along with the corresponding stability result are both summarized in Table 9.2.1. We now illustrate the concept of stiffness control with an example.

EXAMPLE 9.2-3: Stiffness Controller for a Cartesian Manipulator

We want to design and simulate a stiffness controller for the robot manipulator system given in Figure 9.2.5. The control objective is to move the end effector to a desired final position of $v_d = 3$ m while exerting a final desired normal force of $f_{d1} = 2$ N. We neglect the surface friction (i.e., f_2) and joint friction, and assume that the normal force (i.e., f_1) satisfies the relationship

$$f_1 = k_e (u - u_e), \tag{1}$$

where $u_e = 3/\sqrt{2}$ m and $k_e = 1000$ N/m. The robot link masses are assumed to be unity, and the initial end-effector position is given by

$$\nu(0) = 5\,\text{m} \quad \text{and} \quad u(0) = 3/\sqrt{2}\,\text{m}. \tag{2}$$

To accomplish the control objective, the stiffness controller from Table 9.2.1 is given by

$$\tau = J^T(q)\left(-K_v \dot{x} + K_p \tilde{x}\right) + G(q), \tag{3}$$

where

$$\tilde{x} = \begin{bmatrix} u_d - u \\ \nu_d - \nu \end{bmatrix},$$

τ, J, G, and x are as defined in Example 9.2.1, u_d is defined as the desired normal position, and the gain matrices K_v and K_p have been taken to be $K_v = k_v I$ and $K_p = k_p I$. For this example we select $k_v = k_p = 10$, which will guarantee that $k_p \ll k_e$ as required in the stiffness control formulation. To satisfy the control objective that $f_{d1} = 2$ N, we utilize (9.2.27) to determine the desired normal position. Specifically, substituting the values of f_{d1}, k_e, and u_e into

$$f_{d1} = k_p (u_d - u_e) \tag{4}$$

yields $u_d = (0.2 + 3/\sqrt{2})$ m.

The simulation of the stiffness controller given by (3) for the robot manipulator system (Figure 9.2.5) is given in Figure 9.2.7. As indicated by the simulation, the desired tangential position and normal force are reached in about 4 s.

■

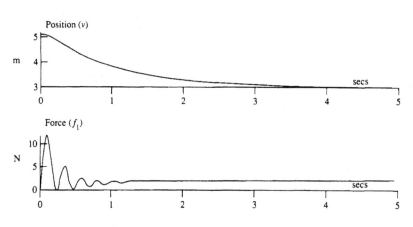

Figure 9.2.7: Simulation of stiffness controller.

9.3 Hybrid Position/Force Control

A major disadvantage of the stiffness controller given in Section 9.2 is that it can only be used for set-point control; in other words, the desired end effector manipulator position and the desired force exerted on the environment must be constant. In many robotic applications, such as grinding, the end effector must track a desired positional trajectory along the object

9.3 Hybrid Position/Force Control

surface while tracking a desired force trajectory exerted onto the object surface. In this type of application, a stiffness controller will not perform adequately; therefore, another control approach must be utilized.

The so-called hybrid position/force controller [Chae et al. 1988] and [Raibert and Craig 1981] can be used for tracking position and force trajectories simultaneously. The basic concept of the hybrid position/force controller is to decouple the position and force control problems into subtasks via a task space formulation. As we have seen, the task space formulation is valuable in determining which directions should be force or position controlled. That is, the position and force control subtasks are easily determined from the task space formulation. After the control subtasks have been identified, separate position and force controllers can then be developed.

Hybrid Position/Force Control of a Cartesian Two-Link Arm

To illustrate this concept of hybrid position/force control, consider the robot manipulator system given by Figure 9.3.1. For this application, the position along the surface and the normal force exerted on the surface should both be controlled; therefore, one must determine which variables should be force controlled and which should be position controlled.

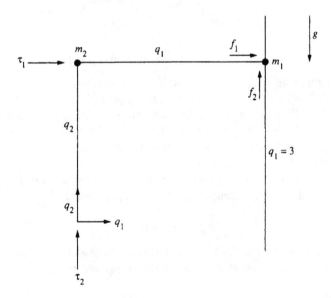

Figure 9.3.1: Manipulator moving along perpendicular surface.

Following the task space concept given in Section 9.2, the task space formulation for the manipulator system given in Figure 9.3.1 is

$$x = \begin{bmatrix} u \\ v \end{bmatrix} = h(q) = \begin{bmatrix} q_1 \\ q_2 \end{bmatrix}, \qquad (9.3.1)$$

with the task-space Jacobian matrix given by

$$J = \frac{\partial h(q)}{\partial q} = \begin{bmatrix} 1 & 0 \\ 0 & 1 \end{bmatrix}.$$

As illustrated by (9.3.1), the task space and the joint space are equivalent for this problem; therefore, we will refer to joint variables as task-space variables throughout this problem.

To design the position/force controller for the manipulator system, we must first determine the dynamic equations for the task space formulation given by (9.3.1). Using this task space formulation and neglecting joint friction, the manipulator dynamics can be shown to be

$$\tau = M\ddot{q} + G + f, \qquad (9.3.2)$$

where τ, M, G, and f are as defined in Example 9.2.1. The two dynamic equations given in the matrix form represented by (9.3.2) are

$$\tau_1 = m_1 \ddot{q}_1 + f_1 \qquad (9.3.3)$$

and

$$\tau_2 = (m_1 + m_2)\ddot{q}_2 + (m_1 + m_2)g + f_2. \qquad (9.3.4)$$

In formulating a hybrid position/force controller, we design separate controllers for the dynamics given by (9.3.3) and (9.3.4). As illustrated by Figure 9.3.1, the position along the task space direction q_2 should be position controlled; therefore, we should use (9.3.4) for designing the position controller. [This is obvious because the dynamics given by (9.3.3) do not contain the task space variable q_2.]

Because we are designing a position controller to track a desired trajectory, we will define the "tangent space" tracking error to be

$$\tilde{x} = q_{d2} - q_2, \qquad (9.3.5)$$

where q_{d2} represents the desired position trajectory along or tangent to the surface. The position controller will be the computed-torque controller (see Chapter 3)

$$\tau_2 = (m_1 + m_2)a_T + (m_1 + m_2)g + f_2, \qquad (9.3.6)$$

9.3 Hybrid Position/Force Control

where

$$a_T = \ddot{q}_{d2} + k_{Tv}\dot{\tilde{x}} + k_{Tp}\tilde{x}, \tag{9.3.7}$$

with k_{Tv} and k_{Tp} being positive control gains. Substituting (9.3.6) into (9.3.4) gives the position tracking error system

$$\ddot{\tilde{x}} + k_{Tv}\dot{\tilde{x}} + k_{Tp}\tilde{x} = 0. \tag{9.3.8}$$

By using the fact that k_{Tv} and k_{Tp} are positive, we can apply standard linear control results to (9.3.8) to yield

$$\lim_{t \to \infty} \tilde{x} = 0;$$

therefore, asymptotic positional tracking is guaranteed with the controller given by (9.3.6). Note that the position controller requires measurement of the joint position, joint velocity, and surface friction force; therefore, from an implementation point of view, force measurements are required in the position controller.

The position controller given in (9.3.6) will ensure good position tracking along the surface of the environment; however, we also want to control the force exerted on the environment. For the manipulator system given in Figure 9.3.1, the task space direction normal to the surface is q_1; therefore, we will assume that in this direction the environment can be modeled as a spring. Specifically, the normal force f_1 exerted on the environment is given by

$$f_1 = k_e (q_1 - q_e), \tag{9.3.9}$$

where k_e represents the environment stiffness and $q_e = 3$. Taking the second derivative of (9.3.9) with respect to time gives the expression

$$\ddot{q}_1 = \frac{1}{k_e}\ddot{f}_1, \tag{9.3.10}$$

where the normal task space acceleration is written in terms of the second derivative of the normal force. Substituting (9.3.10) into (9.3.3) yields the force dynamic equation

$$\tau_1 = \frac{m_1}{k_e}\ddot{f}_1 + f_1. \tag{9.3.11}$$

We can now use the force dynamic equation given in (9.3.11) to design a force controller to track a desired force trajectory. First, define the force tracking error to be

$$\tilde{f} = f_{d1} - f_1, \qquad (9.3.12)$$

where f_{d1} represents the desired normal force that is to be exerted on the environment. Similar to the position controller, the force controller will be the computed-torque controller

$$\tau_1 = \frac{m_1}{k_e} a_N + f_1. \qquad (9.3.13)$$

where

$$a_N = \ddot{f}_{d1} + k_{Nv}\dot{\tilde{f}} + k_{Np}\tilde{f}, \qquad (9.3.14)$$

with k_{Nv} and k_{Np} being positive control gains. Substituting (9.3.13) into (9.3.11) gives the force tracking error system,

$$\ddot{\tilde{f}} + k_{Nv}\dot{\tilde{f}} + k_{Np}\tilde{f} = 0. \qquad (9.3.15)$$

Using the fact k_{Nv} and k_{Np} are positive in (9.3.15) yields

$$\lim_{t \to \infty} \tilde{f} = 0;$$

therefore, asymptotic force tracking is guaranteed with the controller given by (9.3.13). It is important to realize that the force controller requires measurement of the normal force and the derivative of the normal force. Because the force derivative is often not available for measurement, it is manufactured from (9.3.9), that is,

$$\dot{f}_1 = k_e \dot{q}_1; \qquad (9.3.16)$$

therefore, the stiffness of the environment and the normal task space velocity are used to simulate the derivative of the force.

Hybrid Position/Force Control of an N-Link Manipulator

The hybrid position/force controller given in the preceding section can easily be extended to the multidegree case by using the task space formulation concept. Specifically, one can develop a feedback-linearizing control that will globally linearize the robot manipulator equation and then develop linear controllers to track the desired force and position trajectories.

First, the control designer selects a task space formulation

$$x = h(q) \qquad (9.3.17)$$

9.3 Hybrid Position/Force Control

such that the normal and tangent surface motions are decomposed as discussed in Section 9.2. The robot dynamics given in (9.3.14) are then written in terms of the task space acceleration by differentiating (9.3.17) twice with respect to time to obtain

$$\ddot{x} = J(q)\ddot{q} + \dot{J}(q)\dot{q}, \qquad (9.3.18)$$

where $J(q)$ is the task space Jacobian defined in (9.2.11). Solving (9.3.18) for $\ddot{q}$ yields

$$\ddot{q} = J^{-1}(q)\left(\ddot{x} - \dot{J}(q)\dot{q}\right) \qquad (9.3.19)$$

Substituting (9.3.19) into (9.2.14) yields

$$\tau = M(q)J^{-1}(q)\left(\ddot{x} - \dot{J}(q)\dot{q}\right) + V_m(q,\dot{q})\dot{q} + G(q) + F(\dot{q}) + J^T(q)f. \qquad (9.3.20)$$

The corresponding feedback linearizing control for the dynamics given by (9.3.20) is given by

$$\tau = M(q)J^{-1}(q)\left(\mathrm{a} - \dot{J}(q)\dot{q}\right) + V_m(q,\dot{q})\dot{q} + G(q) + F(\dot{q}) + J^T(q)f, \qquad (9.3.21)$$

where a is an $n \times 1$ vector used to represent the linear position and force control strategies, which will be discussed later. After substituting (9.3.21) into (9.3.20), we have

$$\ddot{x} = \mathrm{a}. \qquad (9.3.22)$$

From (9.3.22), we can see that the task space motion has been globally linearized and decoupled; therefore, we can design the position and force controllers independently in a method similar to that of the preceding section. Specifically, linear position controllers can be designed for the task space variables that represent tangent motion; moreover, linear force controllers can be designed for the task space variables that represent normal force.

Because the dynamics given in (9.3.20) have been decoupled in the task space, we will define the tangent space components of x as x_{Ti}, where the subscript T is used to denote the tangent space, and the subscript i is used to denote the ith component of x_T. From this notation the tangent space components of (9.3.22) are given as

$$\ddot{x}_{Ti} = \mathrm{a}_{Ti}, \qquad (9.3.23)$$

where $\mathbf{a}_{Ti}$ is the ith linear tangent space position controller. For the purpose of feedback control, we define the tangent space tracking error to be

$$\tilde{x}_{Ti} = x_{Tdi} - x_{Ti}, \qquad (9.3.24)$$

where x_{Tdi} represents the ith desired position trajectory tangent to the environment surface. As in the preceding section, the corresponding linear controller is then given as

$$\mathbf{a}_{Ti} = \ddot{x}_{Tdi} + k_{Tvi}\dot{\tilde{x}}_{Ti} + k_{Tpi}\tilde{x}_{Ti}, \qquad (9.3.25)$$

with k_{Tvi} and k_{Tpi} being the ith positive control gains. Substituting (9.3.25) into (9.3.23) gives the position tracking error system

$$\ddot{\tilde{x}}_{Ti} + k_{Tvi}\dot{\tilde{x}}_{Ti} + k_{Tpi}\tilde{x}_{Ti} = 0. \qquad (9.3.26)$$

Using the fact that k_{Tvi} and k_{Tpi} are positive in (9.3.26) yields

$$\lim_{t \to \infty} \tilde{x}_{Ti} = 0;$$

therefore, asymptotic positional tracking is guaranteed.

For purposes of force control, we define the normal space components of x as x_{Nj} where the subscript N is used to denote normal space, and the subscript j is used to denote the jth component of x_N. From this notation, the normal space components of (9.3.22) are given as

$$\ddot{x}_{Nj} = \mathbf{a}_{Nj}, \qquad (9.3.27)$$

where $\mathbf{a}_{Nj}$ is the jth linear normal space force controller. As in the preceding section, we assume that the environment can be modeled as a spring. Specifically, the normal force f_{Nj} exerted on the environment is given by

$$f_{Nj} = k_{ej}(x_{Nj} - x_{ej}), \qquad (9.3.28)$$

where k_{ej} is the jth component of environmental stiffness, and x_{ej} is used to represent the static location of the environment in the direction of the normal space x_{Nj}.

As done for the single-degree-of-freedom robot in the preceding subsection, we must formulate the force dynamics before we can develop the force controller. Taking the second derivative of (9.3.28) with respect to time gives the expression

$$\ddot{x}_{Nj} = \frac{1}{k_{ej}} \ddot{f}_{Nj}, \qquad (9.3.29)$$

9.3 Hybrid Position/Force Control

where the normal task space acceleration is written in terms of the second derivative of the normal force. Substituting (9.3.29) into (9.3.27) yields the force dynamics

$$\frac{1}{k_{ej}} \ddot{f}_{Nj} = a_{Nj}. \tag{9.3.30}$$

For the purpose of feedback control, we define the force tracking error to be

$$\tilde{f}_{Nj} = f_{Ndj} - f_{Nj}, \tag{9.3.31}$$

where f_{Ndj} represents the jth component of the desired force exerted normal to the environment. As in the preceding section, the corresponding linear controller is then given by

$$a_{Nj} = \frac{1}{k_{ej}} \left(\ddot{f}_{Ndj} + k_{Nvj} \dot{\tilde{f}}_{Nj} + k_{Npj} \tilde{f}_{Nj} \right), \tag{9.3.32}$$

with k_{Nvj} and k_{Npj} being the jth positive control gains. Substituting (9.3.32) into (9.3.30) gives the force tracking error system

$$\ddot{\tilde{f}}_{Nj} + k_{Nvj} \dot{\tilde{f}}_{Nj} + k_{Npj} \tilde{f}_{Nj} = 0 \tag{9.3.33}$$

Using the fact that k_{Nvj} and k_{Npj} are positive in (9.3.33) yields

$$\lim_{t \to \infty} \tilde{f}_{Nj} = 0;$$

therefore, asymptotic force tracking is guaranteed.

The hybrid position/force controller and the corresponding stability result are both summarized in Table 9.3.1. We now illustrate the concept of hybrid position/force control with an example.

EXAMPLE 9.3-1: Hybrid Position/Force Control Along a Slanted Surface

We want to design and simulate a hybrid position/force controller for the robot manipulator system given in Figure 9.2.5. The control objective is to move the end effector with a desired surface trajectory of $v_d = \sin(t)$ m while exerting a normal force trajectory of $f_{d1} = 1 - e^{-t} N$. We neglect joint friction and assume that the normal force (i.e., f_1) satisfies the relationship

$$f_1 = k_e (u - u_e), \tag{1}$$

Table 9.3.1: Hybrid Position/Force Controller

Torque Controller:

$$\tau = M(q)J^{-1}(q)\left(\mathbf{a} - \dot{J}(q)\dot{q}\right) + V_m(q,\dot{q})\dot{q} + G(q)$$
$$+ F(\dot{q}) + J^T(q)f$$

where $J(q)$ is the task space Jacobian.

Position control:

$$\mathrm{a}_{Ti} = \ddot{x}_{Tdi} + k_{Tvi}\dot{\tilde{x}}_{Ti} + k_{Tpi}\tilde{x}_{Ti}$$

Force control:

$$\mathrm{a}_{Nj} = \frac{1}{k_{ej}}\left(\ddot{f}_{Ndj} + k_{Nvj}\dot{\tilde{f}}_{Nj} + k_{Npj}\tilde{f}_{Nj}\right)$$

Stability:

Nonconstrained directions: position tracking control

$$\lim_{t\to\infty} x_{Ti}(t) = x_{Tdi}(t)$$

Constrained directions: force tracking control

$$\lim_{t\to\infty} f_{Nj}(t) = f_{Ndj}(t)$$

Comments: Environment is modeled as a spring.

---

where $u_e = 3/\sqrt{2}$ m and $k_e = 1000$ N/m. The robot link masses are assumed to be unity, and the initial end-effector position is given by

$$v(0) = 0 \text{ m} \quad \text{and} \quad u(0) = 3/\sqrt{2} \text{ m}. \tag{2}$$

To accomplish the control objective, the hybrid position/force controller from Table 9.3.1 is given by

$$\tau = MJ^{-1}\mathbf{a} + G + J^T f, \tag{3}$$

where $\mathbf{a}$ is a 2×1 vector representing the linear position and force controllers with τ, J, G, and f as defined in Example 9.2.1. The controller

9.3 Hybrid Position/Force Control

given by (3) decouples the robot dynamics in the task space as follows:

$$\ddot{x} = \begin{bmatrix} \ddot{u} \\ \ddot{v} \end{bmatrix} = a. \qquad (4)$$

From Figure 9.2.5, we can see that the task space variable u represents the normal space, and the task space variable v represents the tangent space; therefore, (4) may rewritten in the notation given in Table 9.3.1 as

$$\ddot{x} = \begin{bmatrix} \ddot{u} \\ \ddot{v} \end{bmatrix} = \begin{bmatrix} \ddot{x}_{N1} \\ \ddot{x}_{T1} \end{bmatrix} = \begin{bmatrix} a_{N1} \\ a_{T1} \end{bmatrix} = a. \qquad (5)$$

From Table 9.3.1, the corresponding linear position and force controllers are then given by

$$a_{T1} = \ddot{x}_{Tdl} + k_{Tv1}\dot{\tilde{x}}_{T1} + k_{Tp1}\tilde{x}_{T1} \qquad (6)$$

and

$$a_{N1} = \frac{1}{k_{e1}} \left(\ddot{f}_{Nd1} + k_{Nv1}\dot{\tilde{f}}_{N1} + k_{Np1}\tilde{f}_{N1} \right) \qquad (7)$$

where $x_{Tdl} = \sin t$, $f_{Nd1} = 1 - e^{-t}$, and $k_{e1} = 1000$.

The simulation of the hybrid position/force controller given by (3), (6), and (7) for the robotic manipulator system (Figure 9.2.5) is given in Figure 9.3.2. The controller gains were selected as

$$k_{Nv1} = k_{Np1} = k_{Tv1} = k_{Tp1} = 10.$$

As indicated by the simulation, the position and force tracking error go to zero in about 4 s.

∎

Implementation Issues

After reexamining the hybrid position/force controller, one can see that the task space forces are needed for implementation of the control law. Often a wrist-mounted force sensor is used for measuring the end-effector forces; however, in general, these forces will not be the task space forces that are needed for control implementation. Fortunately, a transformation can be

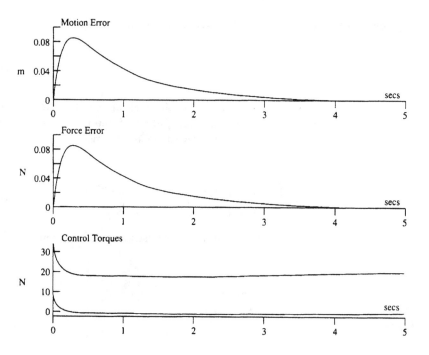

Figure 9.3.2: Simulation of hybrid position/force controller.

used to obtain the task space forces for any particular force application. That is, the task space forces are related to the sensor forces by

$$J^T(q) f = J_s^T(q) f_s, \qquad (9.3.34)$$

where $J_s(q)$ is an $n \times n$ Jacobian sensor matrix, and f_s is an $n \times 1$ vector of sensor forces. Using (9.3.34), the task space forces are given by

$$f = J^{-T}(q) J_s^T(q) f_s. \qquad (9.3.35)$$

As we mentioned earlier, the force control law requires measurement of the task space force derivative; however, this signal is not usually available. Often, the stiffness equation (9.3.28) is used to obtain the jth task space force derivative as

$$\dot{f}_{Nj} = k_{ej} \dot{x}_{Nj}, \qquad (9.3.36)$$

which is in terms of the measurable jth task space normal velocity.

9.4 Hybrid Impedance Control

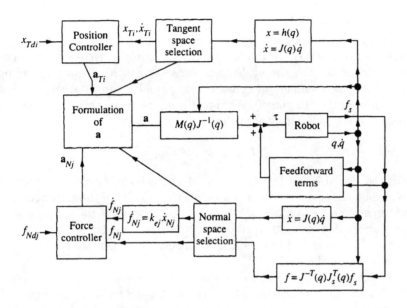

Figure 9.3.3: Hybrid position/force controller.

With these implementational concerns in mind, the overall hybrid position/force control strategy is depicted in Figure 9.3.3. The feedforward terms in the block diagram are used to represent the terms

$$-M(q) J^{-1}(q) \dot{J}(q) \dot{q} + V_m(q, \dot{q}) \dot{q} + G(q) + F(\dot{q}) + J^T(q) f \quad (9.3.37)$$

in the control law given in Table 9.3.1.

9.4 Hybrid Impedance Control

Impedance control is based on the concept that the controller should be used to regulate the dynamic behavior between the robot manipulator motion and the force exerted on the environment [Hogan 1987] rather than considering the motion and force control problems separately. Using this concept, the control designer specifies the desired dynamic behavior between the motion of the manipulator and the force exerted on the environment. This desired behavior is sometimes referred to as the target impedance because it is used to represent an Ohm's law type of relationship between motion and force.

Modeling the Environment

As pointed out in [Hogan 1987], the environmental model is central to any force control strategy. In the force control strategies discussed previously, the environment has simply been modeled as a spring; however, as one might imagine, a simple spring model may not adequately describe all types of environments. To classify the many types of environments, we use the linear transfer function relationship

$$f(s) = Z_e(s)\dot{x}(s), \tag{9.4.1}$$

where the variable s is the Laplace transform variable, f represents the force exerted on the environment, $\dot{x}$ represents the velocity of the manipulator at the environmental contact point, and $Z_e(s)$ represents the environmental impedance. For now, all quantities are assumed to be scalar functions; however, at the end of this section we generalize impedance control to the multidimensional case.

The quantity $Z_e(s)$ is called an impedance because (9.4.1) represents an Ohm's law type of relationship between motion and force. As in circuit theory, environmental impedances can be separated into different categories. To further our impedance control discussion, we now give three commonly used categories which are used to classify environmental impedances.

DEFINITION 9.4-1 *An impedance is inertial if and only if* $|Z(0)| = 0$. ∎

An illustration of an inertial environment is given in Figure 9.4.1a. This figure depicts a robot manipulator moving a payload of mass h with velocity $\dot{x}$. The corresponding interaction force is given by

$$f = h\ddot{x};$$

therefore, utilizing (9.4.1) yields an inertial environmental impedance of

$$Z_e(s) = hs. \tag{9.4.2}$$

We can easily verify that this impedance is indeed inertial by applying Definition 9.4.1.

DEFINITION 9.4-2 *An impedance is resistive if and only if* $|Z(0)| = c$ *where* $0 < c < \infty$. ∎

An illustration of a resistive environment is given in Figure 9.4.1b. This figure depicts a robot manipulator moving through a liquid medium with

9.4 Hybrid Impedance Control

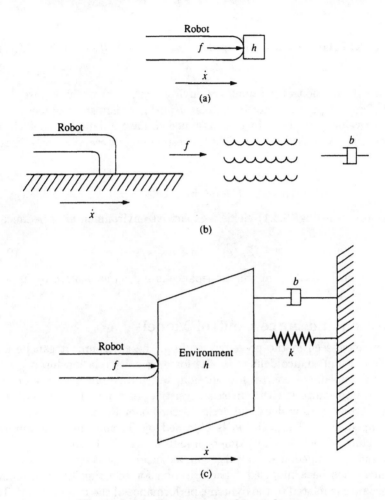

Figure 9.4.1: Environmental impedances.

velocity $\dot{x}$. The liquid medium is assumed to have a damping coefficient of b. The corresponding interaction force is given by

$$f = b\dot{x};$$

therefore, utilizing (9.4.1) yields a resistive environmental impedance of

$$Z_e(s) = b. \qquad (9.4.3)$$

We can easily verify that this impedance is indeed resistive by applying Definition 9.4.2.

DEFINITION 9.4-3 *An impedance is capacitive if and only if $|Z(O)| = \infty$.* ∎

An illustration of a capacitive environment is given in Figure 9.4.1c. This figure depicts a robot manipulator pushing against an object of mass h with velocity $\dot{x}$. The object is assumed to have a damping coefficient of b and a spring constant of k. The corresponding interaction force is given by

$$f = h\ddot{x} + b\dot{x} + kx;$$

therefore, utilizing (9.4.1) yields a capacitive environmental impedance of

$$Z_e(s) = hs + b + k/s. \tag{9.4.4}$$

We can easily verify that this impedance is indeed capacitive by applying Definition 9.4.3.

Position and Force Control Models

As we have seen in the preceding section, the environment can be modeled as an impedance defined by the force/velocity relationship in (9.4.1). For our impedance control formulation, we will assume that the environmental impedance is either inertial, resistive, or capacitive. The question then becomes: How does one design a controller for a given environmental impedance? The solution is obtained by formulating a manipulator impedance model, $Zm(s)$ [Anderson and Spong 1988]. In other words, a manipulator impedance (or target impedance) is selected after the environment has been modeled. The criterion for selecting the manipulator impedance is related to the dynamic performance of the manipulator. That is, the manipulator impedance is selected such that there is zero steady-state error to a step input (which may be a force or velocity command). As we will show, this performance criterion can be achieved if the manipulator impedance is the dual of the environmental impedance.

Before the concept of duality can be illustrated fully, the models for position and force control must be formulated. For position control [Anderson and Spong 1988], the relationship between force and velocity is modeled by

$$f(s) = Z_m(s)(\dot{x}_d(s) - \dot{x}(s)), \tag{9.4.5}$$

where $\dot{x}_d$ represents the input velocity of the manipulator at the environmental contact point and $Z_m(s)$ represents the manipulator impedance.

9.4 Hybrid Impedance Control

As we will show subsequently, the manipulator impedance $Z_m(s)$ is selected to "zero-out" the steady state to a step input by utilizing the dynamic relationship between $\dot{x}$ and $\dot{x}_d$. To determine the dynamic relation between $\dot{x}$ and $\dot{x}_d$, we combine (9.4.1) with (9.4.5) to yield the position control block diagram given in Figure 9.4.2. We can use this block diagram to illustrate the concept of duality. Specifically, we examine the steady-state velocity error

$$E_{ss} = \lim_{s \to 0} s \left(\dot{x}_d(s) - \dot{x}(s) \right), \tag{9.4.6}$$

where $\dot{x}_d(s) = 1/s$ for a step velocity input. Utilizing Figure 9.4.2, we can easily show that (9.4.6) can be reduced to

$$E_{ss} = \lim_{s \to 0} \frac{Z_e(s)}{Z_m(s) + Z_e(s)}. \tag{9.4.7}$$

For E_{ss} to be equal to zero in (9.4.7), $Z_e(s)$ must be a noncapacitive impedance [i.e., $Z_e(s)$ must be inertial or resistive], and $Z_m(s)$ must be a noninertial impedance. That is, zero steady-state error can be achieved for a velocity step input if inertial environments are position controlled with noninertial manipulator impedances, while resistive environments are position controlled with capacitive manipulator impedances. The aforementioned term duality is used to emphasize the fact that inertial environmental impedances can be position controlled with capacitive manipulator impedances.

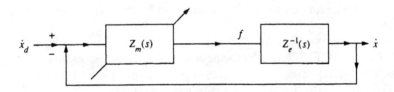

Figure 9.4.2: Position control block diagram.

From the development above, it is obvious that a capacitive environment cannot be position controlled and maintain the zero steady-state error specification. However, subsequently we will show that capacitive environments can be force controlled while maintaining the zero steady-state error specification. With regard to force control, the dynamic relationship between force and velocity is modeled [Anderson and Spong 1988] by

$$\dot{x}(s) = Z_m^{-1}(s) \left(f_d(s) - f(s) \right), \tag{9.4.8}$$

where f_d is used to represent the input force exerted at the environmental contact point.

As we will show subsequently, the manipulator impedance $Z_m(s)$ is selected to "zero-out" the steady state to a step input by utilizing the dynamic relationship between f and f_d. To determine the dynamic relation between f and f_d, we combine (9.4.1) with (9.4.8) to yield the force control block diagram given in Figure 9.4.3. We can use this block diagram to illustrate the concept of duality. Specifically, we examine the steady-state force error

$$E_{ss} = \lim_{s \to 0} s \left(f_d(s) - f(s) \right), \tag{9.4.9}$$

where $f_d(s) = 1/s$ for a step force input. Utilizing Figure 9.4.3, we can easily show that (9.4.9) can be reduced to

$$E_{ss} = \lim_{s \to 0} \frac{Z_m(s)}{Z_m(s) + Z_e(s)}. \tag{9.4.10}$$

For E_{ss} to be equal to zero in (9.4.10), $Z_e(s)$ must be a noninertial impedance, and $Z_m(s)$ must be a noncapacitive impedance. That is, zero steady-state error can be achieved for a force step input if capacitive environments are force controlled with noncapacitive manipulator impedances while resistive environments are force controlled with inertial manipulator impedances. The term "duality" is used to emphasize the fact that capacitive environmental impedances can be force controlled with inertial manipulator impedances.

The discussion above can be summarized by the following duality principle.

DUALITY PRINCIPLE *Capacitive environments are force controlled with noncapacitive manipulator impedances, inertial environments are position controlled with noninertial manipulator impedances, and resistive environments are force controlled with inertial manipulator impedances or position controlled with capacitive manipulator impedances.*

Impedance Control Formulation

Now that we have illustrated how the environment and the manipulator can be modeled as impedances, we develop an "impedance" controller based on this model. To utilize an impedance control approach, the control designer selects a task space formulation

$$x = h(q) \tag{9.4.11}$$

9.4 Hybrid Impedance Control

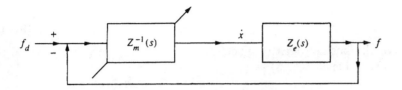

Figure 9.4.3: Force control block diagram.

for the particular position/force control application. As in Section 9.3, we can show that the torque control

$$\tau = M(q) J^{-1}(q) \left(\mathbf{a} - \dot{J}(q) \dot{q} \right) + V_m(q, \dot{q}) \dot{q} + G(q) + F(\dot{q}) + J^T(q) f \quad (9.4.12)$$

yields the linear set of equations

$$\ddot{x} = \mathbf{a}, \quad (9.4.13)$$

where $\mathbf{a}$ is an $n \times 1$ vector used to represent the impedance position and force control strategies.

As delineated by (9.4.13), the task space motion has been globally linearized; therefore, we can design separate position and force controllers for each task space degree of freedom. That is, in each task space direction represented by a component x_k, an environmental impedance relationship between $\dot{x}_k$ and the corresponding environmental force f_k is assigned. Based on this assignment of environmental impedance, the duality principle is used to determine if the corresponding element of $\mathbf{a}$ should be a force or position controller. After this determination is made, (9.4.5) and (9.4.8) are then used to obtain the specific position and force control components of $\mathbf{a}$.

As a means of separating the position control design from the force control design, we use (9.4.13) to define the equations that are position controlled in the task space directions as

$$\ddot{x}_{pi} = \mathbf{a}_{pi}, \quad (9.4.14)$$

where the subscript i denotes the ith position-controlled task space variable, and the subscript p denotes position control. The associated environmental forces in the position controlled task space directions are denoted by f_{pi}.

Assuming zero initial conditions, the Laplace transform of (9.4.14) can be written as

$$s\dot{x}_{pi}(s) = \mathbf{a}_{pi}(s). \tag{9.4.15}$$

From the position control model given by (9.4.5), we can also write the left-hand side of (9.4.15) as

$$s\dot{x}_{pi}(s) = s\left(\dot{x}_{pdi}(s) - Z_{pmi}^{-1}(s) f_{pi}(s)\right), \tag{9.4.16}$$

where Z_{pmi} is the ith position-controlled manipulator impedance. Therefore, equating (9.4.16) and (9.4.15) gives the ith position controller

$$\mathbf{a}_{pi} = L^{-1}\left\{s\left(\dot{x}_{pdi}(s) - Z_{pmi}^{-1}(s) f_{pi}(s)\right)\right\}, \tag{9.4.17}$$

where L^{-1} is used to represent the inverse Laplace transform operation.

Continuing with the separation of position and force control designs, we use (9.4.13) to define the equations that are to be force controlled in the task space directions as

$$\ddot{x}_{fj} = \mathbf{a}_{fj}, \tag{9.4.18}$$

where the subscript j denotes the jth force-controlled task space variable, and the subscript f denotes force control. The associated environmental forces in the force-controlled task space directions are denoted by f_{fj}.

Assuming zero initial conditions, the Laplace transform of (9.4.18) can be written as:

$$s\dot{x}_{fj}(s) = \mathbf{a}_{fj}(s). \tag{9.4.19}$$

From the force control model given in (9.4.8), the left-hand side of (9.4.19) can also be written as

$$\mathbf{a}_{fj} = L^{-1}\left\{sZ_{fmj}^{-1}(s)(f_{fdj}(s) - f_{fj}(s))\right\}. \tag{9.4.20}$$

$$s\dot{x}_{fj}(s) = sZ_{fmj}^{-1}(s)(f_{fdj}(s) - f_{fj}(s)), \tag{9.4.21}$$

where Z_{fmj} is the jth force-controlled manipulator impedance. Therefore, equating (9.4.21) and (9.4.19) gives the jth force controller

The overall "hybrid" impedance control strategy is obtained by using (9.4.12) in conjunction with (9.4.17) and (9.4.20). This hybrid impedance control strategy is summarized in Table 9.4.1. Note that a higher-level controller would be used to select the components of the task space which would be position or force controlled and then the appropriate manipulator impedances would be assigned. As delineated by the duality principle, the manipulator impedances Z_{fmj} in (9.4.20) are assigned to be noncapacitive

9.4 Hybrid Impedance Control

Table 9.4.1: Hybrid Impedance Controller

Torque Controller:

$$\tau = M(q) J^{-1}(q) \left(a - \dot{J}(q)\dot{q}\right) + V_m(q,\dot{q})\dot{q} + G(q)$$
$$+ F(\dot{q}) + J^T(q) f$$

where $J(q)$ is the task space Jacobian.

Position control:

$$a_{pi} = L^{-1} \left\{ s\left(\dot{x}_{pdi}(s) - Z_{pmi}^{-1}(s) f_{pi}(s)\right) \right\}$$

Force control:

$$a_{fj} = L^{-1} \left\{ sZ_{fmj}^{-1}(s) \left(f_{fdj}(s) - f_{fj}(s)\right) \right\}$$

Stability:

Zero steady-state error to a force or position step input.

Comments: Manipulator impedances Z_{pmi} and Z_{fmj} are selected by use of the duality principle.

---

and the manipulator impedances Z_{pmi} in (9.4.17) are assigned to be non-inertial. To illustrate the concept of hybrid impedance control, we now present an example.

EXAMPLE 9.4-1: Hybrid Impedance Control Along a Slanted Surface

We wish to formulate a hybrid impedance controller for the robot manipulator system given in Figure 9.2.5. The joint and surface friction may be neglected, and we assume that the tangential force (i.e., f_2) satisfies the relationship

$$f_2 = d_e \dot{v}, \tag{1}$$

and the normal force (i.e., f_1) satisfies the relationship

$$f_1 = h_e \ddot{u} + b_e \dot{u} + k_e u, \qquad (2)$$

where h_e, b_e, d_e, and k_e are all positive scalar constants.

From Table 9.4.1, the hybrid impedance controller is given by

$$\tau = MJ^{-1}\mathbf{a} + G + J^T f, \qquad (3)$$

where **a** is a 2×1 vector representing the separate position and force control strategies, and τ, J, G, and f are as defined in Example 9.2.1. The torque controller given by (3) decouples the robot dynamics in the task space as follows:

$$\ddot{x} = \begin{bmatrix} \ddot{u} \\ \ddot{v} \end{bmatrix} = \mathbf{a}; \qquad (4)$$

therefore, we can easily determine which task space directions should be force or position controlled.

Applying Definition 9.4.2 to (1) allows us to state that the environmental impedance in the task space direction given by v is a resistive impedance; therefore, by the duality principle, we will select a position controller that utilizes the capacitive manipulator impedance

$$Z_{pm1}(s) = h_m s + b_m + k_m/s, \qquad (5)$$

where h_m, b_m, and k_m are all positive scalar constants. Since the task space variable v will be position controlled, we use the notation from (9.4.14) to yield

$$\ddot{x}_{p1} = \ddot{v} = \mathbf{a}_{p1}$$

and

$$f_{p1} = f_2.$$

Now using (5) and the definition of $\mathbf{a}_{p1}$ given in Table 9.4.1, we can easily show that

$$\mathbf{a}_{p1} = \ddot{x}_{pd1} + \frac{b_m}{h_m}(\dot{x}_{pd1} - \dot{x}_{p1}) + \frac{k_m}{h_m}(x_{pd1} - x_{p1}) - \frac{1}{h_m} f_{p1}. \qquad (6)$$

9.4 Hybrid Impedance Control

Applying Definition 9.4.3 to (2) allows us to state that the environmental impedance in the task space direction given by u is capacitive; therefore, by the duality principle, we will select a force controller that utilizes the inertial manipulator impedance

$$Z_{fm1}(s) = d_m s, \qquad (7)$$

where d_m is a positive scalar constant. Since the task space variable u will be force controlled, we use the notation from (9.4.18) to yield

$$\ddot{x}_{f1} = \ddot{u} = a_{f1}$$

and

$$f_{f1} = f_1.$$

Using (7) and the definition of $\mathbf{a}_{f1}$ given in Table 9.4.1, we can easily show that

$$a_{f1} = \frac{1}{d_m}(f_{fd1} - f_{f1}). \qquad (8)$$

The overall impedance control strategy is obtained by substituting

$$\mathbf{a} = \begin{bmatrix} \mathbf{a}_{f1} \\ \mathbf{a}_{p1} \end{bmatrix} \qquad (9)$$

into (3), where $\mathbf{a}_{p1}$ and $\mathbf{a}_{f1}$ are as given by (6) and (8), respectively.

■

Implementation Issues

As mentioned earlier, the manipulator impedance $Z_m(s)$ is selected such that the duality principle is maintained. However, from a practical point of view, $Z_m(s)$ should also be selected in the expressions for $\mathbf{a}_{pi}$ and $\mathbf{a}_{fj}$ such that only measurements of f, x, and $\dot{x}$ are required. In other words, our controller should not require acceleration (i.e., $\ddot{x}$) or force derivative (i.e., $\dot{f}$) measurements. We can easily show that the expressions for $\mathbf{a}_{pi}$ and $\mathbf{a}_{fj}$ will not require measurements of $\dot{f}$ and $\ddot{x}$ if the manipulator impedance is selected as

$$Z_m = hs + Z_r,$$

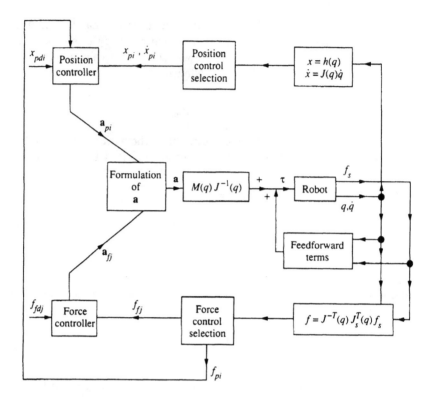

Figure 9.4.4: Hybrid impedance controller.

where h is some positive scalar constant, and Z_r is selected to be a proper transfer function [Anderson and Spong 1988].

As delineated by Table 9.4.1, the hybrid impedance controller requires task space force measurements. As stated previously for the hybrid position/force controller, a transformation can be used to obtain the task space forces for any particular force application. Specifically, the task space forces are related to the sensor forces by

$$f = J^{-T}(q) J_s^T(q) f_s,$$

where $J_s(q)$ is an $n \times n$ Jacobian sensor matrix, and f_s is an $n \times 1$ vector of sensor forces.

With these implementation concerns in mind, the overall hybrid impedance control strategy is depicted in Figure 9.4.4. The feedforward terms in the block diagram are used to represent the terms

$$-M(q)\,J^{-1}(q)\,\dot{J}(q)\,\dot{q} + V_m(q,\dot{q})\,\dot{q} + G(q) + F(\dot{q}) + J^T(q)\,f.$$

9.5 Reduced State Position/Force Control

If a rigid manipulator is constrained by a rigid environment, the degrees of freedom are reduced because the manipulator end effector cannot move through the environment; therefore, one or more degrees of freedom with regard to position are lost. Consequently, as the manipulator end effector contacts the environmental constraint, interaction forces between the end effector and the environment (sometimes referred to as constraint forces) develop. This process of "reducing" positional freedom while developing constraint forces leads one to believe that position/force controllers should be designed according to this natural phenomenon.

Recently, researchers have begun to formulate a theoretical framework [McClamroch and Wang 1988], [Kankaanranta and Koivo 1988] that incorporates the effects of the constraint forces into the robot manipulator model by utilizing classical results in dynamics. The reasoning for postulating that controllers should be designed for a rigid manipulator contacting a rigid constraint is that in most force control applications, the environment is much more rigid than the manipulator. Therefore, it seems unreasonable to assume that the manipulator is rigid while the environment is compliant.

Effects of Holonomic Constraints on the Manipulator Dynamics

For the controller given in this section, we assume that the environmental constraints are holonomic and frictionless. That is, we assume the existence of a constraint function $\overline{\psi}(q)$ (a $p \times 1$ vector function) in joint-space coordinates that satisfies

$$\overline{\psi}(q) = 0. \tag{9.5.1}$$

The relationship given by (9.5.1) illustrates that the environmental constraints are holonomic. The dimension of the constraint function is assumed to be less than the number of joint variables (i.e., $p < n$). For a specific problem, the function $\overline{\psi}(q)$ is found from the robot kinematics and the environmental configuration. To illustrate the holonomic concept, we now present an example.

EXAMPLE 9.5-1: Holonomic Constraints

We wish to formulate the constraint function $\bar{\psi}(q)$ for the robot/environmental configurations given in Figs. 9.2-5 and 9.2-6. In both of these configurations, the joint space has a dimension of 2 and the constraint function has a dimension of 1 (i.e., the constraint is a one-dimensional surface). For the manipulator system given in Figure 9.2.5, the constraint function is given by

$$\bar{\psi}(q) = q_1 - q_2 - 3 = 0.$$

For the manipulator system given in Figure 9.2.6, the constraint function is given by

$$\bar{\psi}(q) = {}^1\!/\!_4 q_1^2 + q_2^2 - 1 = 0.$$

∎

For the general n-link robot manipulator with holonomic and frictionless constraints, the constrained robot dynamics can be written in the form

$$\tau = M(q)\ddot{q} + V_m(q,\dot{q})\dot{q} + G(q) + F(\dot{q}) + A^T(q)\lambda, \qquad (9.5.2)$$

where λ is a $p \times 1$ vector that represents the generalized force multipliers associated with the constraints, and the constraint Jacobian matrix $A(q)$ is a $p \times n$ matrix defined by

$$A(q) = \partial \bar{\psi}(q)/\partial q. \qquad (9.5.3)$$

As in [Kankaanranta and Koivo 1988], we will assume that the p columns of $A^T(q)$ are linearly independent over the joint space. Also, note that the force variable λ is independent of q and $\dot{q}$.

To motivate the origin of (9.5.2), we reexamine Lagrange's equation

$$\frac{d}{dt}\left[\frac{\partial L}{\partial \dot{q}}\right] - \frac{\partial L}{\partial q} = \tau, \qquad (9.5.4)$$

where the modified Lagrangian for the constrained robot manipulator is given by

$$L = K - P - \lambda^T \bar{\psi}(q). \qquad (9.5.5)$$

Note that the Lagrangian given in (9.5.5) is really the same Lagrangian as given in Chapter 2, since

9.5 Reduced State Position/Force Control

$$\lambda^T \bar{\psi}(q) = 0,$$

as required by (9.5.1).

From (9.5.4) and (9.5.5), we can see that the structure of the robot manipulator dynamic equation is the same for the constrained manipulator as for the unconstrained manipulator with the exception of any new terms contributed by the substitution of $-\lambda^T \bar{\psi}(q)$ in Lagrange's equation. Specifically, the constrained robot manipulator equation given in (9.5.2) is obtained by substituting (9.5.5) into (9.5.4) utilizing the identity

$$\frac{\partial \left(\lambda^T \bar{\psi}(q)\right)}{\partial q} = \left[\frac{\partial \bar{\psi}(q)}{\partial q}\right]^T \lambda.$$

Before we develop the reduced-state position/force controller, we explain briefly, from a heuristic point of view, how the position and force variables have been reduced in dimension. First, note that for the constrained robot dynamics given in (9.5.2), the variable λ is used to represent the constraint forces that should be controlled. Note that the dimension of λ is p, which, by definition, is less than n. In previous formulations of the environmental forces given in this chapter, the dimension of the forces was assumed to be equal to n. In this approach we are able to reduce the number of forces that must be controlled because we have initially assumed that the constraint surface is frictionless. Of course, in reality, surface friction will exist in most robot force control applications; however, it is often treated as a disturbance because such friction is a function of the applied normal contact force.

Second, we examine the position constraints on robot motion given by (9.5.1). Because $\bar{\psi}(q) = 0$ by assumption, we can differentiate (9.5.1) with respect to time to obtain

$$A(q)\dot{q} = 0 \qquad (9.5.6)$$

where $A(q)$ is defined in (9.5.3). The expression given by (9.5.6) gives a concise form for the kinematic velocity constraints. From the position and velocity constraints given in (9.5.1) and (9.5.6), we can state that the manipulator dynamics belong to the invariant manifold C, in R^{2n}, defined by

$$C = \left\{(q, \dot{q}) : \bar{\psi}(q) = 0, \ A(q)\dot{q} = 0\right\}.$$

That is, the motion of the robot manipulator remains on the manifold defined by C. As stated in [McClamroch and Wang 1988], the manifold C is singular on R^{2n}; therefore, we can reduce the order of the motion dynamics.

Reduced State Modeling and Control

A robot system consisting of a single n-joint nonredundant manipulator constrained by a rigid environment has $n - p$ degrees of motion freedom. Note, however, that the joint variable model of a manipulator as given in (9.5.2) contains n position variables (q), which, in combination with the p force variables (λ), cause the total number of control variables (i.e., states) $n+p$ to exceed the number of control inputs n. In this section we show how a variable transformation can be used to reduce the states of the dynamical model and thereby reduce the number of control variables from $n + p$ to n.

A reduced state model can be obtained by representing the manipulator dynamics given in (9.5.2) in terms of another set of independent coordinates, called the constraint space coordinates, which will be denoted by x. It is intended that x be an $(n - p) \times 1$ vector of joint space coordinates. That is, x is a subset of q.

The constrained robot model is reduced by assuming that there exists an $n \times 1$ vector function $g(x)$ that relates the constraint space vector x [an $(n - p) \times 1$ vector] to the joint space vector q. This function is given by

$$q = g(x), \qquad (9.5.7)$$

where the function $g(x)$ must be selected such that

$$\left.\left[\frac{\partial g(x)}{\partial x}\right]^T A^T(q)\right|_{q=g(x)} = 0 \qquad (9.5.8)$$

and such that the $n \times (n - p)$ Jacobian matrix $\Sigma(x)$ defined as

$$\Sigma(x) = \partial g(x)/\partial x \qquad (9.5.9)$$

contains $n - p$ independent rows along the constraint space motion given by x.

Even though the constraint space vector x is of smaller dimension than the joint space vector q, one is usually able to find the functional mapping from x to q given in (9.5.7). For particular problems, the algebraic relations given by the holonomic equation (9.5.1) and the robot kinematics are used to find $g(x)$. Also note that the choice of $g(x)$ is nonunique and that the conditions on $g(x)$ given by (9.5.8) and on the rows of $\Sigma(x)$ are related to the reduced state model. Specifically, the condition that $\Sigma(x)$ contain $n - p$ independent rows ensures that the decoupled model represents $n - p$ independent equations. The condition on g(x) given by (9.5.8) is used to ensure that the forces represented by λ can be decoupled from the constraint space motion represented by x. It should be noted that since the constraints are assumed to be holonomic and the matrix $A^T(q)$ is assumed to have

9.5 Reduced State Position/Force Control

p linearly independent columns, we can show that both of the conditions above always hold. The reader is referred to [McClamroch and Wang 1988] for details.

To obtain the reduced state dynamics in terms of the constraint space coordinates, we first differentiate (9.5.7) with respect to time to yield

$$\dot{q} = \Sigma(x)\dot{x}, \quad (9.5.10)$$

where $\Sigma(x)$ is defined in (9.5.9). Differentiating (9.5.10) with respect to time gives

$$\ddot{q} = \Sigma(x)\ddot{x} + \dot{\Sigma}(x)\dot{x} \quad (9.5.11)$$

After substituting q, $\dot{q}$, and $\ddot{q}$ from (9.5.7), (9.5.10), and (9.5.11), respectively, into (9.5.2), we obtain the reduced-state model

$$\tau = M(x)\Sigma(x)\ddot{x} + N(x,\dot{x}) + A^T(x)\lambda, \quad (9.5.12)$$

where

$$N(x,\dot{x}) = \left(V(x,\dot{x})\Sigma(x) + M(x)\dot{\Sigma}(x)\right)\dot{x} + G(x) + F(x,\dot{x}).$$

The significance of the reduced state model given by (9.5.12) is illustrated by premultiplying (9.5.12) by $\Sigma^T(x)$ to obtain

$$\tau^* = M^*\ddot{x} + N^*, \quad (9.5.13)$$

where

$$\tau^* = \Sigma^T(x)\tau,\ M^* = \Sigma^T(x)M(x)\Sigma(x),\ \text{and}\ N^* = \Sigma^T(x)N(x,\dot{x}).$$

Note that the contact forces in (9.5.12) have been removed as a consequence of (9.5.8), which ensures that $\Sigma^T(x)A^T(x) = 0$. The model given by (9.5.13) is useful because the dynamics that govern the motion of the manipulator on the constraint surface have been reduced from n differential equations to $n - p$ differential equations; furthermore, the motion has been decoupled from the contact forces. These two results are important in the design and stability analysis of the subsequent position/force controllers.

Before the reduced state position/force controller is presented, we give some definitions with regard to position/force tracking problems. First, the constraint position tracking error is defined as

$$\tilde{x} = x_d - x. \quad (9.5.14)$$

We assume that the desired constraint space trajectory and its first two derivatives, denoted by x_d, $\dot{x}_d$, and $\ddot{x}_d$, respectively, are all bounded functions. We also assume that the desired force multiplier trajectory, λ_d, is a known bounded function, from which the corresponding force multiplier tracking error variable is defined as

$$\tilde{\lambda} = \lambda_d - \lambda. \qquad (9.5.15)$$

The reduced state position/force controller is a feedback linearizing controller; therefore, exact knowledge of the robot dynamics is required. The reduced state controller [McClamroch and Wang 1988] is

$$\tau = M(x)\Sigma(x)\left(\ddot{x}_d + K_v\dot{\tilde{x}} + K_p\tilde{x}\right) + N(x,\dot{x}) + A^T(x)\left(\lambda_d + K_f\tilde{\lambda}\right), \qquad (9.5.16)$$

where K_v and K_p are diagonal, positive-definite $(n-p) \times (n-p)$ matrices, and K_f is a diagonal, positive-definite $p \times p$ matrix.

To determine the type of stability for the position error and force tracking error, we substitute (9.5.16) into (9.5.12) to yield

$$M(x)\Sigma(x)\left(\ddot{\tilde{x}} + K_v\dot{\tilde{x}} + K_p\tilde{x}\right) + A^T(x)\left(\dot{\tilde{\lambda}} + K_f\tilde{\lambda}\right) = 0. \qquad (9.5.17)$$

Premultiplying (9.5.17) by $\Sigma^T(x)$ yields

$$M^*\left(\ddot{\tilde{x}} + K_v\dot{\tilde{x}} + K_p\tilde{x}\right) = 0 \qquad (9.5.18)$$

as a consequence of (9.5.8). Because $\Sigma^T(x)$ contains $n-p$ independent columns along the constraint space motion and $M(x)$ is a positive-definite symmetric matrix, M^* is a positive-definite symmetric matrix. Therefore, we can premultiply (9.5.18) by M^{*-1} to obtain

$$\ddot{\tilde{x}} + K_v\dot{\tilde{x}} + K_p\tilde{x} = 0 \qquad (9.5.19)$$

Because K_v and K_p are diagonal positive-definite matrices, one can apply standard linear control arguments to (9.5.19) to yield

$$\lim_{t\to\infty} \ddot{\tilde{x}}, \dot{\tilde{x}}, \tilde{x} = 0. \qquad (9.5.20)$$

To obtain the stability result for the force tracking error, we substitute (9.5.20) into (9.5.17) to yield

$$\lim_{t\to\infty} A^T(x)(I + K_f)\tilde{\lambda} = 0, \qquad (9.5.21)$$

9.5 Reduced State Position/Force Control

where I in (9.5.21) is the $p \times p$ identity matrix. Because the p columns of $A^T(x)$ are assumed to be linearly independent and the composite matrix $(I + K_f)$ is a positive-definite, diagonal matrix, we can write (9.5.21) as

$$\lim_{t \to \infty} \tilde{\lambda} = 0. \qquad (9.5.22)$$

As delineated by (9.5.20) and (9.5.22), the reduced state position/force controller given in (9.5.16) yields an asymptotic stability result for both the constraint position error and force tracking error. The reduced state position/force control strategy is summarized in Table 9.5.1. To illustrate the concept of reduced state position/force control, we now present an example.

Table 9.5.1: Reduced State Position/Force Controller

Torque Controller:

$$\tau = M(x) \Sigma(x) \left(\ddot{x}_d + K_v \dot{\tilde{x}} + K_p \tilde{x} \right) + N(x, \dot{x}) \\ + A^T(x) \left(\lambda_d + K_f \tilde{\lambda} \right)$$

where $A^T(x)$ is the constraint Jacobian matrix and

$$N(x, \dot{x}) = \left(V(x, \dot{x}) \Sigma(x) + M(x) \dot{\Sigma}(x) \right) \dot{x} + G(x) + F(x, \dot{x})$$

Stability:

Position tracking control:

$$\lim_{t \to \infty} x(t) = x_d(t)$$

Force tracking control:

$$\lim_{t \to \infty} \lambda(t) = \lambda_d(t)$$

Comments: Constraints are assumed to be holonomic and frictionless. For position/force decoupling, $A^T(q)$ must contain p independent columns in the joint space.

EXAMPLE 9.5-2: Reduced State Position/Force Control Along a Slanted Surface

We wish to formulate a reduced state position/force controller for the robot manipulator system given in Figure 9.2.5. The joint and surface friction may be neglected. As given in Example 9.5.1, the constraint function for this problem is

$$\bar{\psi}(q) = q_1 - q_2 - 3 = 0;$$

therefore, utilizing (9.5.2) and (9.5.3), the robot dynamics on the constraint surface can be written as

$$\tau = M\ddot{q} + G + A^T \lambda, \qquad (1)$$

where τ, M, q, and G are as defined in Example 9.2.1 and

$$A^T = \begin{bmatrix} 1 \\ -1 \end{bmatrix}. \qquad (2)$$

For this problem, assume that $x = q_1$; therefore, according to (9.5.7), we must find the function $g(x)$ such that

$$q = g(x).$$

From the holonomic constraints, we can easily verify from the kinematic relationships that

$$\begin{bmatrix} q_1 \\ q_2 \end{bmatrix} = \begin{bmatrix} x \\ x - 3 \end{bmatrix}. \qquad (3)$$

Utilizing (9.5.9), we can show that

$$\Sigma = \begin{bmatrix} 1 \\ 1 \end{bmatrix}; \qquad (4)$$

therefore, from Table 9.5.1, the reduced state position/force controller is given by

$$\tau = M\Sigma \left(\ddot{x}_d + K_v \dot{\tilde{x}} + K_p \tilde{x} \right) + G + A^T \left(\lambda_d + K_f \tilde{\lambda} \right), \qquad (5)$$

9.5 Reduced State Position/Force Control

where x_d represents the desired trajectory of q_1 and λ_d represents the desired force multiplier.

For this problem we can easily examine how λ is related to the normal force exerted on the surface by equating the expression for the forces in Example 9.2.1 and this example. Specifically, we have

$$A^T \lambda = J^T f, \qquad (6)$$

where J and f are as defined in Example 9.2.1. Because the surface friction has been neglected (i.e., $f_2 = 0$ in Figure 9.2.5), we can use (6) to show that

$$\lambda = f_1 / \sqrt{2}, \qquad (7)$$

where f_1 is defined as the normal force exerted on the surface in Figure 9.2.5. Similarly, from the kinematic relationships, we have

$$q_1 = \frac{\sqrt{2}}{2} v + \frac{3}{2}, \qquad (8)$$

where v is defined to be the end-effector position measured along the surface. It should be noted that the relationships given by (7) and (8) would be used for trajectory generation since the position and force control objectives would be formulated in terms of the variables f_1 and v. That is, λ_d would be obtained from the desired normal force (i.e., f_{d1}), and q_{d1} would be obtained from the desired end-effector surface position (i.e., v_d).

∎

Implementation Issues

As delineated by Table 9.5.1, the reduced state position/force controller requires measurements of the force variable λ. As stated previously for the hybrid position/force control development, a transformation can be used to obtain measurements of λ for any particular force application. That is, the force variables λ are related to the sensor forces by

$$A^T(q) \lambda = J_s^T(q) f_s, \qquad (9.5.23)$$

where $J_s(q)$ is an $n \times n$ Jacobian sensor matrix, and f_s is an $n \times 1$ vector of sensor forces. From (9.5.23), p independent equations can be obtained such that λ is given by

$$\lambda = g_1(q, f_s),$$

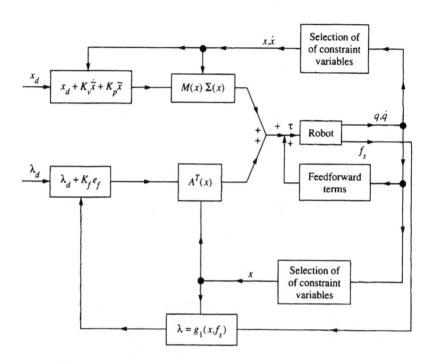

Figure 9.5.1: Reduced state position/force controller.

where $g_1(q, f_s)$ is a $p \times 1$ vector function that depends only on the measurable quantities q and f_s.

With these implementational concerns in mind, the overall reduced order position/force control strategy is depicted in Figure 9.5.1. The feedforward terms in the block diagram are used to represent the terms

$$\left(V_m(x, \dot{x})\Sigma(x) + M(x)\dot{\Sigma}(x)\right)\dot{x} + G(x) + F(x, \dot{x}).$$

9.6 Summary

In this chapter several position/force control strategies for rigid robots have been given. The intent has been to chronicle the evolution of force control development. Some new research areas, such as the effects of actuator dynamics, sensor dynamics, joint flexibilities, manipulator dynamic

9.6 Summary

uncertainty, surface uncertainty, and environmental impact instability on position/force controller performance are now being studied by many researchers. After some of these problems have been solved, one can expect to see robot manipulators used more frequently in force control applications.

REFERENCES

[Anderson and Spong 1988] Anderson, R., and M. Spong, "Hybrid impedance control of robotic manipulators," *J. Robot. Autom.*, vol. 4, no. 5, pp. 549-556, Oct. 1988.

[Chae et al. 1988] Chae, A., C. Atkeson, and J. Hollerbach, *Model-Based Control of a Robot Manipulator*. Cambridge, MA: MIT Press, 1988.

[Hogan 1987] Hogan, N., "Stable execution of contact tasks using impedance control," *Proc. IEEE Int. Conf Robot. Autom.*, pp. 595-601, Raleigh NC, Mar. 1987.

[Kankaanranta and Koivo 1988] Kankaanranta, R., and H. Koivo, "Dynamics and simulation of compliant motion of a manipulator," *IEEE Trans. Robot. Autom.*, vol. 4, pp. 163-173, Apr. 1988.

[Lipkin and Duffy 1988] Lipkin, H., and J. Duffy, "Hybrid twist and wrench control for a robot manipulator," *Trans. ASME J. Mechan. Transmissions Autom. Design*, vol. 110, pp. 138-144, June 1988.

[McClamroch and Wang 1988] McClamroch, N., and D. Wang, "Feedback stabilization and tracking of constrained robots," *IEEE Trans. Autom. Control*, vol. 33, no. 5, pp. 419-426, May 1988.

[Raibert and Craig 1981] Raibert, M., and J. Craig, "Hybrid position/force control of manipulators," *J. Dyn. Syst. Meas. Control*, vol. 102, pp. 126-132, June 1981.

[Salisbury and Craig 1980] Salisbury, J., and J. Craig, "Active stiffness control of manipulator in Cartesian coordinates," *Proc. 19th IEEE Conf. Decision Control*, Dec. 1980.

[Spong and Vidyasagar 1989] Spong, M., and M. Vidyasagar, *Robot Dynamics and Control*. New York: Wiley, 1989.

PROBLEMS

Section 9.2

9.2-1 Find the associated task space Jacobian matrix for the normal and tangent forces exerted on the environmental surface for the manipulator given in Figure 9.2.6 with the new surface function given by

$$q_1 = 1/2 q_2^2 + 5 \quad \text{for} \quad -1/2 \leq q_2 \leq 1/2$$

9.2-2 Design and simulate a stiffness controller for the robot manipulator system given in Figure 9.2.5 with the surface function given by

$$q_2 = 2q_1 - 7.$$

The control objective is to move the end effector to a desired final position of $q_{d2} = 0.3$ m while exerting a final desired normal force of 2 N. Neglect the surface and joint friction and assume that the normal force satisfies the relationship

$$f_1 = k_e (u - u_e),$$

where $k_e = 100$ N/m, and u_e is the static normal distance to the surface. The robot link masses can be assumed to be unity while the initial end-effector position is assume to be given by

$$q_1(0) = 2 \,\text{m} \quad \text{and} \quad q_2(0) = 0 \,\text{m}.$$

Section 9.3

9.3-1 Since the position and force error systems for the hybrid position/force control strategies are linear, discuss how the selection of the controller gains can be used to change the performance of the manipulator end effector.

9.3-2 Design and simulate a hybrid position/force controller for the robot manipulator system given in Figure 9.2.5 with the surface function given

$$q_2 = 2q_1 - 7.$$

The control objective is to move the end effector along the surface with trajectory

$$v_d = 1/3 \sin t \text{ m}$$

while exerting a desired normal force of

$$f_{d1} = 1 - e^{-t} \text{ N}$$

Neglect the surface and joint friction and assume that the normal force satisfies the relationship

$$f_1 = k_e (u - u_e),$$

where $k_e = 1000$ N/m, and u_e is the static normal distance to the surface. The robot link masses can be assumed to be unity while the initial end-effector position is assumed to be given by

$$q_1(0) = 2 \text{ m} \quad \text{and} \quad q_2(0) = 0 \text{ m}.$$

9.3-3 Explain why the hybrid position/force control strategy is really a positional control strategy.

Section 9.4

9.4-1 Suppose that for force or position control, the manipulator impedance is selected to be

$$Z_m(s) = hs + Z_r(s),$$

where h is positive scalar constant, and $Z_r(s)$ is a proper stable transfer function. With the manipulator impedance above, show that the hybrid impedance controller will only require measurements of the joint position, joint velocity, and contact force.

9.4-2 Design a hybrid impedance controller for the robot manipulator system given in Figure 9.2.5 with the surface function given by

$$q_2 = 2q_1 - 7.$$

Neglect the surface and joint friction and assume that the normal force and tangent force satisfies the relationship

$$f_1 = k_e (u - u_e) \quad \text{and} \quad f_2 = b_e \dot{v}$$

where $k_e = 10$ N/m, $b_e = 1$ N-s/m, and u_e is the static normal distance to the surface.

9.4-3 Explicitly show how Table 9.4.1 can be used to find equation (6) in Example 9.4.1.

Section 9.5

9.5-1 Design a reduced state controller for the robot manipulator system given in Figure 9.2.6.

9.5-2 For the controller developed in Problem 9.5-1, show how the normal force (i.e., f_1) is related to the force multiplier (i.e., λ), and how the reduced motion variable (i.e., x) is related to the motion along the surface (i.e., v).

Chapter 10

Robot Control Implementation and Software

The diversity of robot manipulator research along with the complex requirements of the associated hardware/software systems has always presented a challenge for software developers. As a result, many of the previously developed software control platforms were complex, expensive, and not very user friendly. Even though several of the previous platforms were designed to provide an open architecture-based system, very few of the previous platforms have been reused. To address previous disadvantages, this chapter describes the design and implementation of a software control platform called the Robotic Platform. The Robotic Platform is an open software development platform for robot manipulator applications. It includes hardware interfacing, servo control, trajectory generation, 3D simulation, a graphical user interface, and a math library. As opposed to distributed solutions, the Robotic Platform implements all of these components on a single hardware platform, with a single programming language (C++), and on a single operating system while guaranteeing hard real-time performance. This design leads to an open architecture that is less complex, easier to use, and easier to extend, and hence, builds on the following state-of-the-art technologies and general purpose components to further increase simplicity and reliability: i) PC technology, ii) the QNX Real-Time Operating System, iii) the Open Inventor and STL libraries, iv) object-oriented design, and v) an easy-to-use, previously-developed, low-level control environment.

10.1 Introduction

Since the construction of a software control platform requires building blocks that cover a wide range of disciplines (see Figure 10.1.1), it would be desirable to create a common generic platform that can be reused by researchers for different applications. Considering the variety of robotic applications and research areas, this task is made even more challenging due to the fact that robot manipulator vendors often supply the user with proprietary robot control languages that run on proprietary hardware components. Due to the lack of flexibility and performance of proprietary robot control languages, some of the previous work has focused on building robot control libraries on top of a commonly used programming language (RCCL [Lloyd et al. 1988] and ARCL [Corke] are examples of such libraries). While many of the previous software approaches achieved new levels of flexibility and performance by using a common programming language, many software control platforms developed in the 80's and early 90's were inherently complex due to the limitations of the operating systems and the hardware components of that time. That is, most operating systems did not support real-time programming, and hence, fostered the development of projects such as RCI [Lloyd 1990] and Chimera [Stewart 1989]. In addition, procedural programming languages, such as "C", tend to reach their limits with regard to reusability for complex projects. In addition, the limited performance of hardware components of that era forced system developers to utilize distributed architectures (see [Miller 1991], [Kapoor 1996], and [Pelich 1996]) that integrated a mix of proprietary hardware and software. Besides the obvious advantages of distributed systems (e.g., greater extensibility and more computational power), there are several disadvantages. Specifically, a distributed architecture increases the complexity of the software significantly. Additionally, hard real-time behavior over network connections often requires expensive proprietary hardware. Generally, the overall hardware cost is also higher; furthermore, users have to familiarize themselves with different hardware architectures and operating systems. Even though many software control platforms have attempted to be flexible, reconfigurable, and open, these platforms are seldom reused (compared to non-robotic platforms like Open Inventor or Corba). Apparently, engineers consider it faster and easier to develop their applications from scratch. Indeed, as many researchers have noted, the learning curve of installing, learning, and modifying previously developed software control platforms of the past is extremely steep.

Over the last ten years, many innovations in the computing area have occurred. Specifically, the advent of object-oriented software design [Stroustrup 1991] facilitated the management of more complex projects

10.1 Introduction

while also fostering code reuse and flexibility. For example, robot control libraries like RIPE [Miller 1991], MMROC+ [Zielinski 1997], OSCAR [Kapoor 1996], and ZERO++ [Pelich 1996] utilized object-oriented techniques in robot programming. The engineering community has also witnessed the proliferation of real-time Unix-like operating systems for the PC [QSSL], which facilitates the replacement of proprietary hardware components for real-time control [Costescu 1999]. Likewise, in the hardware sector, the engineering community has witnessed the advent of high-speed, low-cost PCs, fast 3D graphics video boards, and inexpensive motion control cards. Consequently, the PC platform now provides versatile functionality, and hence, makes complex software architectures and proprietary hardware components superfluous. As a result of these technological advances, the QMotor Robotic Toolkit (QMotor RTK) [Loffler 2001] was developed. The QMotor RTK was one of the first software control platforms to integrate real-time manipulator control and the graphical user interface (GUI) all on a single PC. Despite, the many advantages of the QMotor RTK, it lacked some capabilities due to the limitations of the real-time operating system (i.e., QNX4) on which it was based. For example, it lacked a 3D simulation capability, and it did not support the use of threads.

Hardware Interfacing	Real-Time Programming	Concurrency
Trajectory Generation	Real-Time Data Logging And Plotting	Interprocess Communication
Object-Oriented System Design	Complex System Design	Robotic Mathematical Functions
Utility Programs (Calibration, etc.)	Support for Different Manipulators, Sensors, and Tools	Networking
Hardware-Accelerated 3D Computer Graphics	Graphical User Interface	Programming Interface

Figure 10.1.1: Building Blocks of a Robotic Software Control Platform

In the late 1990's, QSSL, one of the leading real-time software development houses, substantially upgraded QNX4 with the release of QNX6 [QSSL] (i.e., QNX6/Neutrino operating system). This microkernel-based operating system is probably the most advanced real-time operating system currently available; furthermore, it has additional components for software development and multimedia applications. Indeed, the new attributes of QNX6 greatly facilitated the potential capabilities of an upgrade of the QMotor RTK, in fact so much so, QMotor RTK was completely transformed into a new software control platform call the Robotic Platform

[Bhargava 2001]. The rest of this chapter provides a broad overview of the Robotic Platform by examining its advantages, its design, and its operation. The word "broad" needs to be emphasized because it is really impossible to give a detailed explanation of how a software control platform should be developed for robot manipulators in one chapter of a book. With that being said, it is hope that this chapter will give the reader a better understanding of the associated complexities.

10.2 Tools and Technologies

Given recent technological advances, it now seems almost obvious that new software control platforms should integrate servo control loops, trajectory generation, task level programs, GUI programs, and 3D simulation in a homogeneous software architecture. That is, next-generation software control platforms should utilize one hardware platform (e.g., the PC), only one operating system (e.g., the QNX6 Real-Time Operating System [QSSL]), and only one programming language (e.g., C++ [Stroustrup 1998]). The advantages of this next-generation software control platform are as follows:

Simplicity. A homogeneous non-distributed architecture is much smaller and simpler than a distributed inhomogeneous architecture. It is easier to install, easier to understand, and easier to extend. Simplicity is critical with regard to motivating code reuse of the platform for different applications.

Performance. The homogeneous, non-distributed architecture also ensures high performance and hard real-time behavior, since overhead due to interprocess communication is minimal.

Flexibility at all Levels. Every component of the platform is open for extensions and modifications. Many past platforms have utilized an open architecture at some levels, but other levels had been implemented on proprietary hardware such that they could not be modified.

Cost. The Robotic Platform requires fewer hardware components than a distributed platform. Basically, a single PC with one or more input/output (I/O) boards is sufficient. Additionally, PC hardware is very cost effective as compared to other hardware platforms.

To explain the advantages in more detail and delineate how the Robotic Platform reduces software development and complexity, the general purpose tools and technologies that are used to develop the Robotic Platform

10.2 Tools and Technologies

are now examined. It seems logical to begin this discussion with the hardware engine, namely the standard desktop PC. Due to its popularity in the consumer market, the standard desktop PC has evolved dramatically over the past 20 years. While in the past, only expensive UNIX workstations provided the processing power necessary to control robotic systems, the PC has caught up or even exceeded the performance of workstations [Costescu 1999]. Compared to UNIX workstations, a PC based system allows for a greater variety of hardware and software components. Additionally, these components and the PC itself are usually cheaper than their UNIX counterparts; furthermore, the PC is a well-known platform that is comfortable for most software developers.

The second most important part of any software control platform is the operating systems. Indeed since reliability is extremely important for robot manipulation, the QNX6 Real-Time Operating Systems by QSSL [QSSL] was selected. The programming interface of QNX6 is compliant with the POSIX standards and is the first platform that combines the following features:

Hard Real-Time. QNX6 provides hard real-time response (e.g., it can execute a control loop at a certain frequency without falling behind). The real-time functionality is integrated in the microkernel.

Self-Hosted Development. The development and the execution of software are performed in the same environment and on the same PC.

Fast 3D Graphics. QNX6 provides support for hardware-accelerated 3D, as required for the graphical simulation of a robotic work cell.

Threads. QNX6 supports the concurrent execution of processes, known as threads.

Device Driver Architecture. It is very easy to develop device drivers under QNX6, as is required for integrating different robotic hardware.

Cost. QNX6 is also very cost-effective as it is free for non-commercial use and runs on low-cost standard PCs.

The third most important part or design aspect of any software control platform is the programming philosophy. With regard to developing robot manipulator control software, object-oriented programming has three main benefits over procedural programming. First, it provides language constructs that allow for a much easier programming interface. For example, a matrix multiplication can be expressed by a simple "*", similar to MATLAB programming. Second, object-oriented programming allows for a system architecture that is very flexible but yet very simple. That is, the components (classes) of a system can have a built-in default behavior and default settings. The use of this default behavior and/or default settings allows the programmer to reduce the code size. The programmer, however, can still override the default behavior for specific applications. Finally, object-oriented programming supports generic programming by facilitating the development of components that are independent from a specific implementation (e.g., a generic class "`Manipulator`" will work with different manipulator types). All of the above benefits are based on the general concepts of object-oriented programming, such as abstraction, encapsulation, polymorphism, and inheritance [Stroustrup 1991]. As will be shown later in this chapter, these object-oriented design concepts lead to an intuitive system architecture. As one might expect, the language of choice is C++, as it provides the whole spectrum of object-oriented concepts while maintaining high performance [Stroustrup 1998] (e.g., templates and inline functions can be used to create highly versatile and optimized code in the Robotic Platform). The power of C++ and object-oriented programming can be further increased when using strong object-oriented libraries such as STL [Musser 1996]. STL provides standard data structures (e.g., lists, vectors, etc.) and algorithms. These standard data structures and algorithms simplify programming effort since these tools are often required in common programming tasks. The data structures of the STL are very universal since they can be used with different data types (i.e., template programming).

The last design aspect of a software control platform has do with how the roboticist interacts with the system. Specifically, how does the roboticist animate objects for simulation purposes and how does the roboticist implement and tune control algorithms. For purposes of simulation animation, Open Inventor [Wernecke] was utilized. Open Inventor, developed by Silicon Graphics, is an object-oriented C++ library for creating 3D graphics and animation. Open Inventor minimizes development effort, as it is able to load 3D models that were created in the Virtual Reality Modeling Language (VRML) format [Hartman 1996]. A variety of software packages are available that facilitate the construction of 3D VRML models. These packages can be utilized to quickly create a graphical representation of robotic components. The Robotic Platform utilizes the built-in functionality of Open Inventor to animate these components. Open Inventor also provides advanced features like object picking, which is very useful in the programming of operator interfaces. For purposes of control implementation, QMotor [Loffler 2002] was utilized because it allows users to establish a real-time control loop, to log data, to tune control parameters, and to plot signals. QMotor control programs can be directly integrated in the class design of the Robotic Platform since QMotor supports object-oriented programming.

10.3 Design of the Robotic Platform

Overview

Each component of the Robotic Platform (e.g., the teachpendant, the trajectory generator, etc.) is modeled by a C++ class. A C++ class definition combines the data and the functions related to that component. For example, the class "Puma560" contains the data related to the Puma 560 robot (e.g., the current joint position) as well as functions related to the Puma 560 robot (e.g., enabling of the arm power). Hence, the design of the Robotic Platform results from grouping data and functions in a number of classes in a meaningful and intuitive way. A class can also use parts of the functionality and the data of another class (called the base class) by deriving this class from the base class. The classes of the Robotic Platform include GUI components and a 3D model for graphical simulation. These types of components are traditionally found in separate programs (e.g., see the RCCL robot simulator [Lloyd et al. 1988]). However, by including them in the same class, one can achieve tight integration of the user interface, 3D modeling, and other functional parts.

To provide new capabilities for different applications, the user creates new classes. Usually, new classes are derived from one of the already existing classes to minimize coding effort. This process is called inheritance, and it greatly facilitates code reuse and eliminates redundancy. To illustrate how classes are derived from each other, class hierarchy diagrams are used. The main class hierarchy diagram of the Robotic Platform is shown in Figure 10.3.1. Each arrow is drawn from the derived class to the parent class. The further to the right a class is listed, the more specific it must become. The further to the left a class is listed, the more generic it must become. The classes of the Robotic Platform can be separated into the following categories (note that some of the classes discussed below are not shown in Figure 10.3.1 since their class hierarchies will be discussed later):

The Core Classes. The classes `RoboticObject`, `FunctionalObject`, and `PhysicalObject` build the core of the Robotic Platform by defining the overall behavior of all robotic objects. These classes by themselves can be the basis for any robotic system.

Generic Robotic Classes. Derived from the core classes are a number of generic robotic classes. These classes (e.g., `ServoControl`, `Manipulator`, and `Gripper`) cannot be instantiated. Rather, these classes serve as base classes that implement common functionality

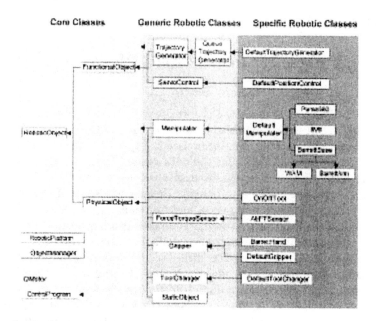

Figure 10.3.1: Class Hierarchy of the Robotic Platform.

while also presenting a generic interface to the programmer (i.e., these classes can be used to create programs that are independent from the specific hardware or the specific algorithm).

Specific Robotic Classes. Derived from the generic robotic classes are classes that implement a specific piece of hardware (e.g., the class Puma560 implements the Puma 560 robot) or a specific functional component (e.g., the class DefaultPositionControl implements a proportional integral derivative (PID) position controller).

The ControlProgram Class. This class is the basis for all real-time control loops.

Utility Classes. The utility classes manage common tasks (e.g., managing configuration files).

The Object Manager. The object manager supervises all objects and classes existing in the system.

The Math Library. The math library consists of classes for using matrices, vectors, transformations, filters, differentiators, and integrators.

10.3 Design of the Robotic Platform

The Manipulator Model Library. The manipulator model library consists of classes that model the kinematic and dynamic behavior of manipulators.

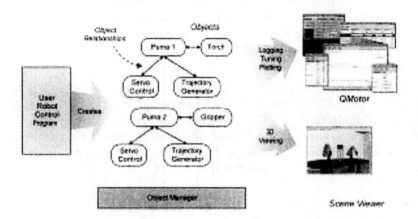

Figure 10.3.2: Run-Time Architecture of the Robotic Platform.

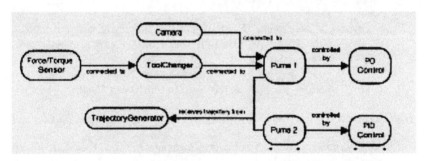

Figure 10.3.3: Object Relationships in an Example Scenario.

In a robot control program, the user instantiates objects from classes. When instantiating an object, memory for the object is reserved, and the object initializes itself. The user can create as many objects as desired from the same class. For example, two Puma robots can be operated by simply creating two objects of the class Puma560. As soon as objects are created, the user can employ their functionality. The object manager maintains a list of all currently existing objects. With the object manager, it is possible to initiate functionality on multiple objects (e.g., shutdown all objects). The Scene Viewer is the default GUI of the Robotic Platform. It contains windows to view the 3D scene of the robotic work cell and to list all objects. The overall run-time architecture is shown in Figure 10.3.2.

In a robotic system, different components are related to each other. To reflect this fact, object relationships are established between objects. For example, objects can specify their physical connection to each other. Object relationships are implemented by C++ pointers to the related object. The object relationships in an example scenario are shown in Figure 10.3.3.

Core Classes

The class RoboticObject is the base class for all classes. It defines a generic interface (i.e., a set of functions that can be used with all classes of the Robotic Platform). For example, a program can use the startShutdown() function to instruct an object of either the class Puma560 or the class Gripper to shutdown. Specifically, the following generic functionality is defined in the class RoboticObject (see also Figure 10.3.4):

Error Handling. Every object must indicate its error status.

Interactive Commands. Each object can define a set of interactive commands (e.g., "Open Gripper") that will show up in the object pop-up menu of the Scene Viewer.

Configuration Management. Each object can use the global configuration file to set itself up. Additionally, each object can also have a local object configuration file.

Shutdown Behavior. Each object is able to shutdown itself.

GUI Control Panel. Each object is able to create a control panel.

Message Handler. Each object has a message handler that can interpret custom messages.

Thread Management. Each object indicates if additional threads are required for its operation and also provides functions that execute those additional threads.

Note that the actual functionality is usually implemented in the derived class. However, the class RoboticObject also implements simple default functionality. This feature supports code reuse and simplicity by giving all classes derived from the class RoboticObject the ability to take over this default functionality if necessary. Table 10.3.1 shows all functions and their default behavior.

10.3 Design of the Robotic Platform

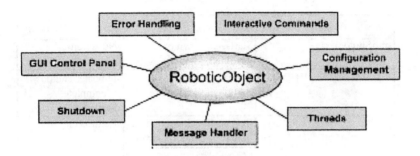

Figure 10.3.4: The Class RoboticObject.

The class `PhysicalObject` is derived from the class `RoboticObject`. It is the base class for all classes that represent physical objects (e.g., manipulators, sensors, grippers, etc.). It defines a generic interface for these classes as illustrated in Figure 10.3.5 and Table 10.3.2. Specifically, the following generic functionality is defined in `PhysicalObject`:

3D Visualization. Every physical object can create its Open Inventor 3D model. The Scene Viewer loops through all physical objects to create the entire 3D scene.

Object Connections. A physical object can specify another object as a mounting location. By using this object relationship, the Scene Viewer draws objects at the right location (e.g., the gripper being mounted on the end-effector of the manipulator).

Position and Orientation. The position/orientation information specifies the absolute location of the object in the scene (or the mounting location, if an object connection is specified).

Simulation Mode. Every physical object can be locked into simulation mode. That is, the object does not perform any hardware I/O (i.e., its behavior is only simulated).

The class `FunctionalObject` does not contain any functionality; however, it is use to build other classes. For example, functional classes, such as the class `TrajectoryGenerator`, are derived from the class `FunctionalObject`.

Robot Control Classes

The central components of any robotic work cell are manipulators. The class `Manipulator` is a generic class that defines common functionality of

Function	Description	Default Behavior
startShutdown()	Starts shutdown of the object	No action
isShutdownComplete()	Indicates when the shutdown is complete	Indicates shutdown is complete immediately
getStatus()	Returns the objects status (ok/error, including error message)	Returns ok
getCommandName()	Returns the name of one of the interactive commands.	No interactive commands are defined
doCommand()	Performs one of the interactive commands	No action
SetGlobalConfiguration()	Passes the global configuration to the object	No action
SetObjectConfiguration()	Passes the local object configuration to the object	No action
createControlPanel()	Creates the GUI of the object's control panel	No action
updateGUI()	This function is called continuously from the main GUI to allow the object to update its GUI windows (e.g., update the control panel)	No action
getNumAdditionalThreads()	Indicates how many additional threads are required	Indicates no additional threads are required
thread()	This function is executed as an additional thread	Runs control program if existing
handleMessage()	This function is called to interpret a message	No action

Table 10.3.1: The Interface of the Class RoboticObject and its Default Behavior.

manipulators with any number of joints. Derived from the class `Manipulator` is the class `DefaultManipulator`, which contains the default implementation of a manipulator. The class `DefaultManipulator` is a template class (i.e., it is parameterized with the number of joints). The generic

10.3 Design of the Robotic Platform

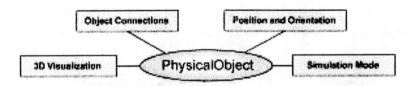

Figure 10.3.5: The Class PhysicalObject.

Function	Description	Default Behavior
get3DModel()	Returns the Open Inventor 3D model	Loads the VRML file <ClassName>.wrl
update3DModel()	Updates the Open Inventor 3D model with the current state of the object	No action
setConnection()	Specifies another object on which the object is mounted.	Sets an internal pointer
removeConnection()	Removes the connection	Sets the internal pointer to zero
setTransform() getTransform()	Sets/gets the position and orientation of the object or the mounting location	Sets/gets the root transform of the Open Inventor object tree
setDisplayOff() setDisplaySolid() setDisplayWireframe()	Disables the display of the object or selects wire frame or solid display	Sets the properties of the Open Inventor tree accordingly
setSimulationModeOn() setSimulationModeOff()	Enables/disables the simulation mode	Sets an internal variable to 1/0

Table 10.3.2: The Interface of the Class PhysicalObject and its Default Behavior.

programming interface and the default implementation are illustrated in Table 10.3.3. Derived from the class `DefaultManipulator` are the classes that implement specific manipulator types. Currently, three manipulators are supported: the Puma 560 robot, the IMI robot [Berkeley 1992], and the Barrett Whole Arm Manipulator (WAM) [Barrett] in both the 4 link and 7 link configuration. More information about the specific control implementation of these robot manipulators can be found in [Loffler 2001] and [Costescu et al. 1999]. For simulation of the manipulators, their dynamic model is required. Additionally, for Cartesian control, forward/inverse kinematics and the calculation of the Jacobian matrix are needed. All of these functions are located in the `ManipulatorModel` classes. The class hierarchy and the generic interface of the `ManipulatorModel` classes are displayed in Figure 10.3.6.

Interface (class Manipulator)	Default Implementation (class DefaultManipulator)
Determine the current manipulator joint position	Reads the motor position from the encoders of an I/O board and converts it to the joint space
Set the current manipulator joint position (position calibration)	Converts the position to the motor space, and sets the encoders of an I/O board to the motor position
Output a control torque/force to the joints	Converts the torque/force to motor space, sets the D/A channels to the control output
Convert position/torques between joint and motor space	Linear relationship, no coupling
Enable/disable the arm power	Sets/resets a digital output of the I/O board

Table 10.3.3: Functionality defined in the Classes Manipulator and DefaultManipulator.

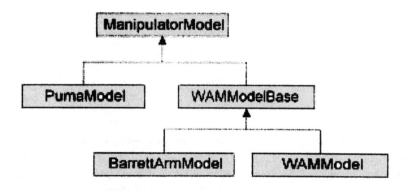

Figure 10.3.6: The ManipulatorModel Classes.

The class `DefaultManipulator` creates a real-time control loop, but it does not include the servo control algorithm. The servo control algorithm is contained in a separate object of the class `ServoControl` or of a derived class. The class `ServoControl` is a generic class that defines common

10.3 Design of the Robotic Platform

properties and functionality of a servo control. That is, a servo control obtains the current position from the manipulator object and calculates the required output control signal to achieve the control task (e.g., servo the manipulator to a desired setpoint). Derived from the class `ServoControl` is the class `DefaultPositionControl`, which contains a PID position control with friction compensation in joint and Cartesian space. The control loop in the class `DefaultManipulator` first determines the current position of the manipulator and then calls the `calculate()` function of the class `ServoControl`. The advantage of the servo control being a separate object is that the user can switch between multiple servo control algorithms, even while the manipulator is moving.

The trajectory generation is also performed in a separate class. The class `TrajectoryGenerator` defines the interface of a generic trajectory generator. A trajectory generator is any object that creates a continuous stream of setpoints and forwards this stream to a servo control object. A trajectory generator must implement its own loop such that the trajectory generator can run independently and at a different frequency than the frequency used in the servo control. The servo control calls the `getCurrentSetpoint()` function of the trajectory generator to determine the current desired position. It is also possible to define multiple trajectory generators and switch between them while the manipulator is moving. The class `QueueTrajectoryGenerator` is derived from the class `TrajectoryGenerator` to implement the generic interface and common functionality of a trajectory generator that creates the trajectory along via points and target points (i.e., motion queue management functions are implemented). Derived from the class `QueueTrajectoryGenerator` is the class `DefaultTrajectoryGenerator`. This class is the specific implementation of a trajectory generator that can interpolate both in joint space and Cartesian space, including path blending between two motion segments at the via points. Table 10.3.4 shows the interface of the default trajectory generator.

Figure 10.3.7: Object Setup for the Servo Control and the Trajectory Generation of a Manipulator.

Classes, such as `Puma560` or `WAM`, automatically create a default trajectory generator and a default servo control for the convenience of the user. However, the user can switch to a different trajectory generator or servo

Function	Description
setJointInterpolation()	Sets the joint interpolation mode
setCartesianInterpolation()	Sets the Cartesian interpolation mode
setVelocity()	Sets the velocity
move()	Queues a new target point or via point
stop()	Stops the manipulator for a certain time
setBlockingMovesOn() setBlockingMovesOff()	Specifies if move/stop commands block until they return
waitForAllMovesCompleted()	Returns after all moves have been completed
getNumRequestsInQueue()	Returns the number of via/target points that are currently in the motion queue

Table 10.3.4: Interface of the class DefaultTrajectoryGenerator.

control if desired. Figure 10.3.7 illustrates in a typical scenario how the manipulator object, the servo control object, and the trajectory generator object work together.

External Device Classes

Besides the classes related to specific manipulators, there are a several other classes that can be used for robotic applications. Specifically, these classes give the robot manipulator the ability to interface with other physical objects by means of grippers, robotic hands, and/or tools. These external device classes are given below:

Gripper. The class Gripper is the generic interface for a gripper. Basically, it defines the functions open(), close(), and relax().

DefaultGripper. The class DefaultGripper assumes that two digital output lines control the gripper, one digital line to open the gripper, and one to close it.

BarrettHand. This is a class to operate the BarrettHand [Barrett]. This class allows the user to move each of the three fingers to certain positions and also control the spread motion.

ForceTorqueSensor. This class defines the generic interface for a force/torque sensor. That is, it defines functions to read the forces and the torques.

10.3 Design of the Robotic Platform

AtiFTSensor. This class is the specific implementation of the ATI Industrial Automation Gamma 30/100 Force/Torque sensor.

ToolChanger. The class `ToolChanger` is the generic interface of a tool changer. Basically, it defines the functions `lock()`, `unlock()`, and `relax()`.

DefaultToolChanger. The class `DefaultToolChanger` assumes that one digital output line controls the lock function of the toolchanger and another line the unlock function.

StaticObject. An object of the class `StaticObject` has no functionality other than specifying its 3D model. It can be used to add objects such as a table or work pieces to the 3D scene.

Utility Classes

As with any complex software system, utility programs are needed for communication purposes and for providing the user with system related information. To this end, several separate, general-purpose utility classes are used throughout the Robotic Platform. These utility classes are given below:

Status. The class `Status` provides functionality to stack status messages. It returns detailed status reports to the user. Each class has a data member of class `Status` that indicates the status of the object.

Client. The class `Client` simplifies the sending of messages to a separate task or thread.

Server. The class `Server`, the counterpart to the class `Client`, simplifies receiving of messages and replying to messages.

Config. The class `Config` handles configuration files in ASCII format.

IOBoardClient. The class `IOBoardClient` allows communication with an I/O board (e.g., providing encoders readings, writing D/A values, reading A/D values, setting and reading digital inputs/outputs).

Lock. The class `Lock` provides mutual exclusion between two threads.

Configuration Management

The Robotic Platform utilizes a global configuration file, which is parsed by the object manager and the objects to determine the system's configuration. This file is called `rp.cfg` by default. The object manager searches for this file in the path specified by the shell variable RP_CONFIG_PATH. By starting a Robotic Platform program with the option -config <filename>, a different filename and path for the configuration file can be specified. The format of the configuration file is now discussed. For each object, the configuration file lists the object name in brackets, the class name of the object, and some additional settings (see Figure 10.3.8). Table 10.3.5 lists possible object settings and their related member functions defined by the classes RoboticObject and PhysicalObject. Derived classes can be used to define additional settings.

```
[leader]
class Puma560
config leader.cfg
position 0 0 0

[secondrobot]
class WAM
position 300 0 0
simulationMode 0
display solid

[follower]
class WAM
config /root/config/follower.cfg
position 600 0 100

[gripper]
class BarrettHand
port /dev/ser1
```

Figure 10.3-8: An Example Global Configuration File.

Object Manager

The object manager, implemented as the class ObjectManager, handles all objects in the system. Every time a new object is created, it registers itself with the object manager. Similarly, every time an object is destroyed, it is removed from the object list that is maintained by the object manager. The object manager contains the necessary functionality to loop through this list, and hence, allows one to perform operations on multiple objects. For example, the Scene Viewer retrieves a list of all objects that are derived from the class PhysicalObject. This list retrieval operation allows the Scene Viewer to render each of the objects, and thereby, render the entire 3D scene.

The object manager also maintains a list of all the classes used in the Robotic Platform. This feature is essential for fostering generic programming. The advantages of generic programming for this application are

10.3 Design of the Robotic Platform

obvious. Specifically, generic code utilizes only components that provide a generic interface; hence, generic code does not need to be changed when a component with a different implementation is used. That is, the same code can be used for a number of specific implementations, and hence, attributes to code reuse (e.g., a generic trajectory generator can be used with different manipulator types). Fortunately, C++ utilizes class inheritance and virtual functions to facilitate generic programming. To illustrate this concept, consider a section of the class hierarchy as shown in Figure 10.3.9. The following snippet of generic code demonstrates how to write a manipulator independent program that works with either the Puma robot, the WAM, the IMI, or any robot that might be added in the future:

```
Manipulator *manipulator;
ObjectManager om;

manipulator = om.createDerivedObject<Manipulator>(''leader'');
// Creates either a Puma, an IMI or a WAM object, depending on
// what is specified in the global configuration file under
// the name ''leader''

// Now, we can do generic operations
manipulator->enableArmPower();
Transform t = manipulator->getCurrentCartesianPosition();}
```

To utilize generic programming in C++, the above code creates an object of the derived class (e.g., an object of the class Puma or the class WAM), and operates the object via a pointer to the generic base class (i.e., Manipulator *manipulator). The function createDerivedObject() of the object manager is utilized to create an object of either the class Puma560, BarrettArm, WAM or IMI. To create the correct object, the createDerivedObject() function looks for the object name in the global configuration file (see Figure 10.3.8). Then, the createDerivedObject() function reads the class name of the object from the configuration file and creates an object of this class. Hence, to switch to a different manipulator type, only the class name in the global configuration file has to be changed when utilizing an existing generic program.

Name of Setting	Description	Equivalent Member Functions
class <className>	Specifies the class name of the object	-
config <configFileName>	Specifies an additional configuration file for this object. This is usually a QMotor configuration file.	RoboticObject:: setObjectConfiguration()
position <x,y,z> orientation <r,p,y>	Specifies the position and orientation of the object	PhysicalObject::setTransform()
simulationMode on simulationMode off	If "on" is specified, use simulation mode for this object	PhysicalObject::setSimulationModeOn() PhysicalObject::setSimulationModeOff()
display off display solid display wireframe	Specifies if and how the object is displayed in the Scene Viewer	PhysicalObject::setDisplayOff() PhysicalObject::setDisplaySolid() PhysicalObject::setDisplayWireframe()

Table 10.3.5: Object Settings of the Configuration File.

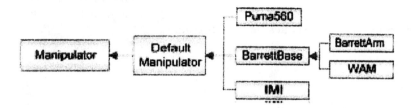

Figure 10.3.9: The Generic Class Manipulator and its Derived Classes.

10.3 Design of the Robotic Platform

Concurrency/Communication Model

An inherent complication associated with the development of a software control platform for robot manipulator systems is the issue of concurrency. That is, while it is often sufficient for many engineering systems to run as a single task, robotic systems require components, such as the servo control, to be executed concurrently with other components (e.g., the trajectory generator). The Robotic Platform addresses this issue by running all concurrent tasks on the same PC. It should be noted that since QNX6 supports symmetric multiprocessing, it is possible to distribute the tasks over multiple processors.

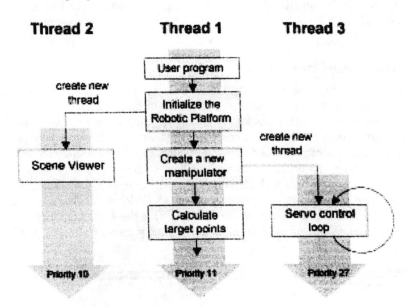

Figure 10.3.10: Creating New Threads for Concurrency.

The predecessor of the Robotic Platform, the QMotor RTK [Loffler 2001], executed multiple programs to achieve concurrency on a single processor. While this concept fosters modularity, it is inconvenient to manage the startup and termination of multiple programs. In contrast, an application that uses the Robotic Platform is compiled and linked to a single program instead. This program spawns threads if concurrent execution is required. Once the program terminates; all threads terminate automatically. Figure 10.3.10 illustrates an example of how a user program spawns multiple threads. At program start, only thread 1 is executing. At the initialization of the Robotic Platform library, a new thread is created that executes the

3D Scene Viewer. Then, the user utilizes a new object of a manipulator class. The creation of this manipulator object automatically spawns a third thread for the servo control loop. Hence, the first thread can go ahead and calculate target points for the manipulator, while the servo control loop and the Scene Viewer run in the background. To ensure real-time behavior of time critical tasks, the threads run at different priorities (e.g., the servo control loop runs at the high priority 27).

Since threads access the same address space, communication between threads can easily be performed by using this address space. However, it is important to synchronize this access to avoid corruption of the data structures. Synchronization is achieved by using the class Lock to provide mutual exclusion for accessing the data that is shared between multiple threads.

Plotting and Control Tuning Capabilities

The issue of tuning and debugging the low-level controller and trajectory planning algorithms has usually not addressed in many of the previously developed software control platforms. However, this is an extremely important issue given the level of complexity associated with accurately positioning a multi-degree of freedom system such as robot manipulator. To address this issue, the Robotic Platform utilizes QMotor for low-level algorithm development. QMotor [Loffler 2002] is a comprehensive environment for implementing and tuning control strategies. QMotor consists of: i) a client/server architecture for hardware access, ii) a C++ library to create control programs, and iii) custom GUI for control parameter tuning, data logging, and plotting.

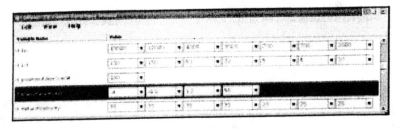

Figure 10.3.11: The QMotor Control Parameter Window.

To communicate with hardware, QMotor uses hardware servers that run in the background and perform hardware I/O at a fixed rate. Servers for different I/O boards are available (e.g., the ServoToGo board, the Quanser MultiQ board, and the ATI force/torque sensor interface board). The use of hardware servers provides an abstract client/server communication inter-

10.3 Design of the Robotic Platform

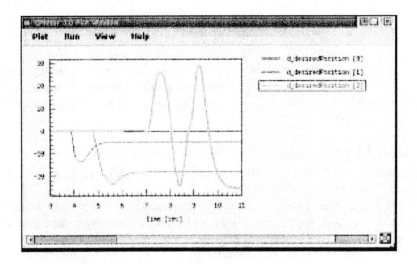

Figure 10.3.12: The QMotor Plot Window.

face such that clients can perform the same generic operations with different servers. Hence, one can quickly reconfigure the system to use different I/O boards by simply starting different servers. For writing control programs, QMotor provides a library, which defines the class ControlProgram. To implement a real-time control loop, the user derives a specific class from the class ControlProgram and defines several functions that perform the control calculation and the necessary housekeeping. Once a control program is implemented and compiled, the user can start up the QMotor GUI, load the control program, start it, and tune the control strategy from the control parameter window (see Figure 10.3.11). Furthermore, the user can open multiple plot windows (see Figure 10.3.12) and set the logging parameters.

To utilize QMotor for the Robotic Platform, classes such as the class DefaultManipulator are derived from the class ControlProgram. Hence, these classes inherit the functionality of a control program (i.e., real-time execution, data logging, and communication with the GUI). If a class is derived from the class ControlProgram, the base class RoboticObject automatically creates a new thread of execution that runs the control loop in the background. QMotor was originally designed to load and tune one control program at a time. That is, to switch to a different control program, the current control program had to be stopped and unloaded. The Robotic Platform is designed to start and run multiple control loops at the same time; hence, it became necessary to upgrade QMotor. Specifically, QMotor can now attach itself to an executing control program instead of loading and starting the control program. This QMotor modification now allows the following operational scenario in the Robotic Platform. The user first starts the control program. This control program creates several threads that execute real-time control loops. The user can then start the QMotor GUI and attach it to any of the control programs. Once the QMotor GUI is attached to a control program, control parameters can be changed and data can be plotted.

Math Library

Past robot control libraries often introduced their own specific robotic data types. Most of these data types are based on vectors or matrices (e.g., a homogeneous transformation is a 4x4 matrix). Hence, it is more feasible to use a general C++ matrix library and define robotic types on top of it. Most of the matrix libraries available for C++ use dynamic memory allocation, which risks the loss of deterministic real-time response [Loffler 2001]. Hence, the big disadvantage of libraries that use dynamic memory allocation is that they can not be used in large sections of the Robotic Platform. To overcome this problem, special real-time matrix classes were developed for the Robotic Platform that use templates for the matrix size. This means that the matrix size is known at compile time and dynamic memory allocation is not required. Besides being feasible for real-time applications, this solution also produces highly optimized code. That is, due to the use of templates and inline functions, the matrix classes can be as fast as direct programming. Specifically, with optimized implementation, the multiplication of two 2x2 matrices C = A * B is as fast as writing

```
c11 = a11 * b11 + a12 * b21;
c12 = a11 * b12 + a12 * b22; etc.
```

An additional advantage is that the compiler can check for the correct matrix sizes at compile time (e.g., a matrix multiplication of two matrices with incompatible size is detected during compilation). Figure 10.3.13, Table 10.3.6, and Table 10.3.7. Functions of the Matrix, Vector, and Transformation Classes illustrates the class hierarchy, the types classes, and the data types. The classes MatrixBase, VectorBase, Matrix, ColumnVector, and RowVector are parameterized with the data type of the elements. The default element data type is double, which is the standard floating-point data type of the Robotic Platform. The classes MatrixBase and VectorBase are pure virtual base classes that allow for manipulation of matrices and vectors of an unknown size (matrices and vectors of an unknown size are required during generic manipulator programming). Figure 10.4.1 shows an example program that uses the math library to calculate a position equation.

10.3 Design of the Robotic Platform

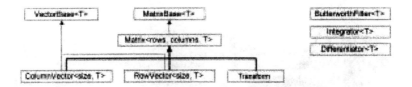

Figure 10.3.13: Class Hierarchy of the Matrix Classes.

MatrixBase<T>	Base class of all matrix and vector classes of data type <T> (e.g., T = int, float, etc.)
VectorBase<T>	Base class of all vector classes of data type <T> (e.g., T = int, float, etc.)
Matrix<row,columns,T>	Matrix of data type <T> and size <rows*columns>
ColumnVector<size,T>	Column Vector of data type <T> and <size> elements
RowVector<size,T>	Row Vector of data type <T> and <size> elements
Transform	Homogeneous transformation (4 x 4 Matrix of type double)
ButterworthFilter<T>	Low-pass 2nd order filter of data type <T>
Differentiator<T>	Numerical differentiation of data type <T>
Integrator<T>	Numerical integration of data type <T>

Table 10.3.6: Classes of the Math Library.

Matrix Functions	Vector Functions	Transformation Functions
Multiplication/division	Length (2-norm)	Translation matrix
Addition/difference	Cross-product	Rotation matrix about
Transpose	Dot-product	x, y, or z axis
Getting/setting elements	Element-by-element	Rotation matrix about an
Getting/setting sub-matrices	multiplication	arbitrary vector
Inverse		Conversion from/to
Unit/Zero matrix		Euler angles
Input/output from/to streams		Conversion from/to RPY angles

Table 10.3.7: Functions of the Matrix, Vector, and Transformation Classes.

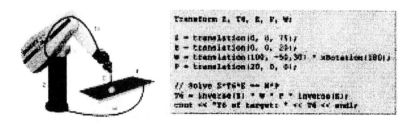

Figure 10.3.14: Example Program for the Matrix classes.

In addition to the matrix, vector, and transformation classes, the classes ButterworthFilter, Differentiator and Integrator are part of the math library. The class ButterworthFilter is used for low-pass filtering. The class Differentiator implements numerical differentiation via the backwards difference method plus low-pass filtering. Finally, the class Integrator utilizes the trapezoidal method for numerical integration. The user can derive a specific class from the class Integrator to implement more advanced integration methods. Both classes are parameterized with the data type (i.e., they work with scalars, vectors, and matrices).

Error Management and the Front-End GUI

Each object is responsible for maintaining the appropriate error status. If a fatal error occurs, any object can request the object manager to shut down the system. For example, this could be the case when a control torque exceeds its limit. For such a system shutdown, the object manager loops through all objects in the system and calls their startShutdown() function. Then, the object manager waits until all objects have completed their shutdown. The completion of the object shutdown is indicated by the is ShutdownComplete() function.

The front-end GUI components for the Robotic Platform were developed with the C++ GUI class library QWidgets++. QWidgets++ allows for object-oriented techniques in GUI programs. The GUI consists of three parts:

- The Scene Viewer is the default supervising GUI, which is opened automatically at startup of every Robotic Platform program.

- Each class can have its own control panel. The control panels are opened from the Scene Viewer.

- Several utility programs (e.g., the Manual Move program and the Teachpendant program) have a GUI.

Operation of the front-end GUI for the Robotic Platform is further explained in the next section.

10.4 Operation of the Robotic Platform

Scene Viewer and Control Panels

When a program running inside the Robotic Platform is executed, the Scene Viewer window opens up. The Scene Viewer window displays the entire 3D scene and also allows the user to open a window that displays a list of currently running objects (see Figure 10.4.2). To create a 3D scene, the Scene Viewer loops through all objects that are derived from the class `PhysicalObject` and calls the `get3DModel()` function to obtain the Open Inventor 3D data of that object. Then, the Scene Viewer uses the object connection relationships (specified by the function `setConnection()` of the class `PhysicalObject`) to display the 3D objects at the right position (e.g., to display a gripper being mounted at the end-effector of a robot manipulator). Furthermore, the Scene Viewer continuously updates the 3D scene with the current state of all objects (e.g., it uses the current joint position of a robot manipulator to display the robot joints in the correct position). Hence, the 3D scene rendered in the Scene Viewer window always represents the current state of the actual hardware (in simulation mode, the simulated state of the hardware is represented). To select the best viewing position, the user navigates in the 3D scene using a mouse. The user can rotate the view, move forward and backward, and also store/restore desired viewing positions. Additionally, the user can hide objects in the scene to increase the frame rate. Many different Scene Viewer windows can be opened to view the 3D scene from different viewing positions at the same time. The user can also open the Object Viewer window to display the list of all objects that are currently being instantiated.

Each object also has an individual pop-up menu (See Figure 10.4.3. The Object Pop-Up Menu). This pop-menu appears if the user either: i) right clicks on the object in the Scene Viewer rendering area, or ii) right clicks on an entry in the Object Viewer window. Options associated with pop-up menu can be used to hide the object in the rendering area and to select between a wire frame or a solid display. In addition, the pop-up menu displays interactive commands that are defined in the specific class of the object. For example, a gripper object has additional menu items to open, close, and relax the gripper. Finally, the user can open the control panel from the object pop-up menu. Figure 10.4.5 shows the control panel of the Barrett WAM as an example. Each control panel appears in a new window.

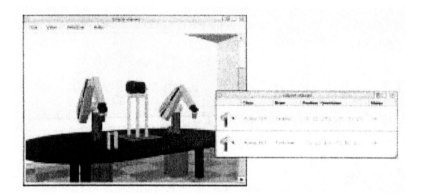

Figure 10.4.1: The Scene Viewer and Object List Windows.

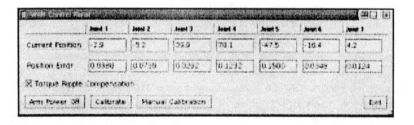

Figure 10.4.2: The Object Pop-Up Menu.

Figure 10.4.3: The Control Panel of the WAM Class.

Utility Programs for Moving the Robot

To facilitate low-level algorithm development, utility programs for moving the robot must be created. For example, the Manual Move utility (see Figure 10.4.6) is a

10.4 Operation of the Robotic Platform

program that gives the user the ability to test the low-level servo control of a robot manipulator. Specifically, the Manual Move utility contains a slider for each joint. As, the user can move the sliders with the mouse, the selected robot manipulator joint immediately follows the commanded user input.

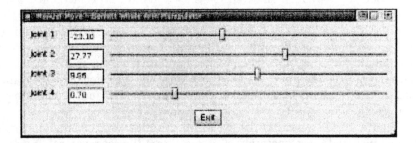

Figure 10.4.4: The Manual Move Utility.

Figure 10.4.5: The Teachpendant.

The Teachpendant (see Figure 10.5.1) implements a low-level zero gravity controller that allows the user to physically position the robot manipulator joints in any location. Once the user has moved the robot manipulator joints to a desired target position, this position can be stored as in list of robot manipulator positions. The Teachpendant utilizes the Robotic Platform's trajectory generator to move the robot manipulator to the stored positions. In this way, the user can cycle the robot manipulator through all or some of the stored positions.

Writing, Compiling, Linking, and Starting Control Programs

A control program is first compiled and then linked to the Robotic Platform library. The entire Robotic Platform (i.e., all classes and the Scene Viewer) is contained in one library. As explained earlier, the Robotic Platform can easily be extended by adding new classes. If the extension is specific to a certain control program, the classes can be added to the code of the associated control program. If an extension is used in different control programs, it would probably be more convenient to add this new functionality to the Robotic Platform library. To reflect the extensions in preexisting compiled and linked programs, the Robotic Platform library is constructed to be a dynamic library (i.e., a program loads the library whenever it is started). Therefore, after the library is extended with new functionality, programs such as the Teachpendant will take advantage of the new functionality (e.g., the teachpendant will be able to operate new manipulator types).

Figure 10.4.6 shows the listing of an example control program for a simple pick and place operation. Every Robotic Platform control program must first call RoboticPlatform::init(). This function initializes the Robotic Platform and starts up the Scene Viewer. The command line arguments are passed to RoboticPlatform::init() such that any Robotic Platform program can be started with certain default command line options (see Table 10.4.1). Additionally, these options can be controlled from C++ code as well by using the functions listed in the third column of Table 10.4.1. After RoboticPlatform::init() is called, the user's program creates all objects that are required for the associated robotic task (i.e., a gripper object, a Puma 560 object, and a trajectory generator object are created). If the option -wait was specified in the command line, no control loops are started until the user's program executes the RoboticPlatform::start() function. This function waits for the user to hit the start button. This feature is necessary since this function gives the user the ability to start the QMotor GUI, set control parameters, and select logging modes before the control loops are started. The last part of the example program utilizes the trajectory generator object and the gripper object to move the robot to the workpiece, close the gripper, pick up the workpiece, and drop it at the target position.

```
// Simple pick and place operation
#include "RoboticPlatform.hpp"

void main(int argc, char *argv[])
{
  RoboticPlatform::init(argc, argv);

  Puma560 puma;                    // Create Puma560 object
  DefaultGripper gripper;          // Create gripper object

  DefaultTrajectoryGenerator tragen;
```

10.4 Operation of the Robotic Platform

Command Line Switch	Description	Equivalent in C++
-sim -nosim	Enables/disables simulation mode for all objects	RoboticPlatform::setSimulationModeOn() RoboticPlatform::setSimulationModeOff()
-gui -nogui	Enables/disables automatic start of the Scene Viewer	RoboticPlatform::setGuiOn() RoboticPlatform::setGuiOff()
-config <filename>	Specifies the name of the global configuration file	RoboticPlatform::setConfigurationFileName()
-wait -nowait	If wait is specified, holds the program until the START button is pressed.	RoboticPlatform::setUseStartButtonOn() RoboticPlatform::setUseStartButtonOff()

Table 10.4.1: Default Command Line Options of Robotic Platform Programs.

```
    puma.setTrajectoryGenerator(tragen);

    RoboticPlatform::start();

    tragen.moveToParkPosition();
    gripper.open();
    tragen.moveTo(0, 0, 0);    // Position specified by x,y,z
    tragen.stop(1);
    gripper.close();
    tragen.stop(1);
    tragen.moveTo(50, 50, 300);
    tragen.moveTo(100, 100, 0);
    tragen.stop(1);
    gripper.open();
    tragen.stop(1);
    tragen.moveTo(50, 50, 300);
}
```

Figure 10.4-6: A Simple Pick and Place Program

10.5 Programming Examples

Comparison of Simulation and Implementation

A very interesting option is to forward the setpoints created by a trajectory generator to two manipulators. In this way, the motion of two manipulators with the same kinematics can be compared, or the behavior of a real manipulator can be compared with a dynamic simulation. The latter application is implemented in the program in Figure 10.5.1. First, two objects of the class Puma 560 are created, and one of the objects is configured into simulation mode. Then, both objects are connected to the same trajectory generator to receive the same setpoints.

```
#include "RoboticPlatform.hpp"

void main(int argc, char *argv[])
{
  RoboticPlatform::init(argc, argv);

  // Create the robot objects
  Puma560 puma;
  Puma560 pumaSimulated;
  pumaSimulated.setSimulationModeOn();

  // Connect both to the same trajectory generator
  DefaultTrajectoryGenerator tragen;
  puma.setTrajectoryGenerator(tragen);
  pumaSimulated.setTrajectoryGenerator(tragen);

  RoboticPlatform::start();

  tragen.move(0, 45, -90, 0, 0, 0);
  tragen.move(-50, 0, -70, 50, -80, 0);
}
```

Figure 10.5-1: Example to Forward the Same Trajectory to two Robots.

Virtual Walls

This is a good example on how to create a custom servo control. It also demonstrates the use of the manipulator model functions and the math library. Virtual walls are virtual planes in the manipulator's workspace that generate a reaction force once the manipulator is moved into the wall. Given a plane with the plane equation (using homogeneous coordinates):

$$\underline{w}\underline{x} = 0; \quad \underline{w} = \left[\begin{array}{c} \underline{n} \\ d_0 \end{array} \right]$$

where $\underline{n}$ is the normal vector of the plane, and d_0 is the distance from the origin. If $\underline{x}_{EndEffector}$ is the current end-effector position, then

10.5 Programming Examples

$$d = \underline{w}\underline{x}_{EndEffector} = \begin{cases} < 0 & \text{end-effector is outside the wall} \\ > 0 & \text{end-effector is inside the wall} \end{cases}$$

Using the control law

$$\tau = \begin{cases} 0 & d < 0 \quad \text{end-effector is outside the wall} \\ \underline{J}^T (k_f d\underline{n}) & d > 0 \quad \text{end-effector is inside the wall} \end{cases},$$

creates a joint torque that resists moving the robot into the wall.

To implement a new servo control algorithm, a class is derived from the class ServoControl (See line (1) of Figure 10.5.2). The above virtual wall functions are implemented in the function calculate(), which calculates the control output. In the function main(), the robot object is created as usual. In addition, an object of the virtual wall servo control class is created (See line (2) of Figure 10.5.2). Finally, the robot object is instructed to utilize the new servo control instead of the default position control, and the gravity compensation is enabled to allow the robot to be pushed around (See line (3) of Figure 10.5.2).

```
#include "RoboticPlatform.hpp"

// ----- Create a new servo control class -----

template <int numJoints>
class VirtualWallServoControl : public ServoControl              (1)
{
 public:
   virtual void calculate()
 {
     // What is the distance of the end-effector to the wall?
     Transform t = d_manipulator->getEndEffectorPosition();
     double distance =
       dotProduct(d_wallCoefficients, t.getColumn(4));

     if (distance > 0)    // No control output
       return;

     // We are inside the wall. Generate reacting force
     Vector<3> force = distance * d_wallCoefficients * d_kf;

     // Convert to joint torque
     Vector<numJoints> pos = d_manipulator->getJointPositon();
     Vector<numJoints> torque;
     d_manipulator->endEffectorForceToJointTorque(pos, force, torque);

     // Do the control output
```

```
    d_manipulator->setControlOutput(torque);
  }

  Vector<4> d_wallCoefficients;
  double d_kf;
};

// ----- Use the virtual walls servo control -----

void main(int argc, char *argv[])
{
  RoboticPlatform::init(argc, argv);

  Puma560 puma;     // Create the robot object

  // Virtual wall control for 6 joints
  VirtualWallServoControl<6> wallControl;                  (2)
  wallControl.d_kf = 0.01;
  wallControl.d_wallCoefficients = 0, 0, -1, 3;

  puma.setGravityCompensationOn();                         (3)
  puma.setServoControl(wallControl);

  RoboticPlatform::start();
}
```

Figure 10.5-2: Virtual Walls Example.

10.6 Summary

The Robotic Platform is a software framework to support the implementation of a wide range of robotic applications. As opposed to past distributed architecture-based software control platforms, the Robotic Platform presents a homogeneous, non-distributed object-oriented architecture. That is, based on PC technology and the real-time operating system QNX6, all non real-time and real-time components are integrated in a single C++ library. The architecture of the Robotic Platform provides efficient integration and extensibility of devices, control strategies, trajectory generation, and GUI components. In addition, software-based control systems implemented with the Robotic Platform are inexpensive and offer high performance. The Robotic Platform is also built on the QMotor control environment for data logging, control parameter tuning, and real-time plotting. A real-time math library simplifies operations and provides an easy-to-use programming interface. Built-in GUI components like the Scene Viewer and the control panels provide for easy operation of the Robotic Platform and a quick ramp-up-time for users that are inexperienced in C++ programming.

REFERENCES

[Lloyd et al. 1988] J. Lloyd, M. Parker and R. McClain, "Extending the RCCL Programming Environment to Multiple Robots and Processors", Proc. IEEE International Conference on Robotics & Automation, 1988, pp. 465 - 469.

[Corke] P. Corke and R. Kirkham, "The ARCL Robot Programming System", Proc. of the International Conference on Robots for Competitive Industries, Brisbane, Australia, pp. 484-493.

[Lloyd 1990] J. Lloyd, M. Parker and G. Holder, "Real Time Control Under UNIX for RCCL", Proceedings of the 3rd International Symposium on Robotics and Manufacturing (ISRAM '90).

[Stewart 1989] D.B. Stewart, D.E. Schmitz, and P.K. Khosla, "CHIMERA II: A Real-Time UNIX-Compatible Multiprocessor Operating System for Sensor-based Control Applications", Technical Report CMU-RI-TR-89-24, Robotics Institute, Carnegie Mellon University, September, 1989.

[Stroustrup 1991] B. Stroustrup, "What is 'Object-Oriented Programming'?", Proc. 1st European Software Festival. February, 1991.

[Miller 1991] D. J. Miller and R. C. Lennox, "An Object-Oriented Environment for Robot System Architectures", IEEE Control Systems Magazine, Feb. 1991, pp. 14-23.

[Zielinski 1997] C. Zielinski, "Object-Oriented Robot Programming", Robotica, Vol.15, 1997, pp. 41-48.

[Kapoor 1996] Chetan Kapoor, "A Reusable Operational Software Architecture for Advanced Robotics", Ph.D. thesis, University of Texas at Austin, December 1996.

[Pelich 1996] C. Pelich & F. M. Wahl, "A Programming Environment for a Multiprocessor-Net Based Robot Control Unit", Proc. 10th International Conference on High Performance Computing, Ottawa, Canada, 1996.

[QSSL] QSSL, Corporate Headquarters, 175 Terence Matthews Crescent, Kanata, Ontario K2M 1W8 Canada, Tel: +1 800-676-0566 or +1 613-591-0931, Fax: +1 613-591-3579, Email: info@qnx.com, http://qnx.com.

[Costescu 1999] N. Costescu, D. M. Dawson, and M. Loffler, "QMotor 2.0 - A PC Based Real-Time Multitasking Graphical Control Environment", IEEE Control Systems Magazine, Vol. 19 Number 3, Jun., 1999, pp. 68 - 76.

[Loffler 2001] M. Loffler, D. Dawson, E. Zergeroglu, and N. Costescu, "Object-Oriented Techniques in Robot Manipulator Control Software Development", Proc. of the American Control Conference, Arlington, VA, June 2001, pp. 4520-4525.

[Bhargava 2001] M. Loffler, A. Bhargava, V. Chitrakaran, and D. Dawson, "Design and Implementation of the Robotic Platform", Proc. of the IEEE Conference on Control Applications, Mexico City, Mexico, Sept., 2001, pp. 357-362.

[Stroustrup 1998] B. Stroustrup, "An Overview of the C++ Programming Language", Handbook of Object Technology, CRC Press. 1998. ISBN 0-8493-3135-8.

[Musser 1996] D. R. Musser, A. Saini, "STL Tutorial and Reference Guide", Addison-Wesley, 1996. ISBN 0-201-63398-1.

[Wernecke] Josie Wernecke, "The Inventor Mentor", Addison-Wesley, ISBN 0-201-62495-8.

[Hartman 1996] Hartman, Jed and Josie Wernecke, "The VRML 2.0 Handbook", Addison-Wesley, Reading Massachusetts, 1996.

[Loffler 2002] M. Loffler, N. Costescu, and D. Dawson, "QMotor 3.0 and the QMotor Robotic Toolkit - An Advanced PC-Based Real-Time Control Platform", IEEE Control Systems Magazine, Vol. 22, No. 3, pp. 12-26, June, 2002

[Berkeley 1992] "Direct Drive Manipulator Research and Development Package, Operations Manual", Integrated Motion Inc., Berkeley, CA, 1992.

REFERENCES

[Barrett] Barrett Technologies, 139 Main St, Kendall Square, Cambridge, MA 02142, http://www.barretttechnology.com/robot.

[Costescu et al. 1999] N. Costescu, M. Loffler, E. Zergeroglu, and D. M. Dawson, "QRobot - A Multitasking PC Based Robot Control System", Microcomputer Applications Journal Special Issue on Robotics, Vol. 18, No. 1, pp.13-22, 1999.

Appendix A

Review of Robot Kinematics and Jacobians

We review here the basic information from a first course in robotics that is needed for this book. This includes robot kinematics, the arm Jacobian, and the issue of specifying the Cartesian position. The review is detailed since those with a background in system theory and controls may not yet have seen this material. Several examples are given which are used throughout the book for design and simulation purposes.

A.1 Basic Manipulator Geometries

In this section we look at some basic arm geometries. A robot arm or manipulator is composed of a set of *joints* separated in space by the arm links. The joints are where the motion in the arm occurs (cf. our own wrist and elbow), while the links are of fixed construction (cf. our own forearm). Thus the links maintain a fixed relationship between the joints. [Although the links may be *flexible* (i.e., they may bend); we ignore flexibility effects here.]

The joints may be actuated by motors or hydraulic actuators. There are two sorts of robot joints, involving two sorts of motion. A *revolute joint* (denoted R) is one that allows rotary motion about an axis of rotation. An example is the human elbow. A *prismatic joint* (denoted P) is one that allows extension or telescopic motion. An example is a telescoping automobile antenna. There is no anthropomorphous analogy to the prismatic link. The *joint variables* of a manipulator are the variable parameters of the joints. For a revolute joint the variable is an angle, denoted θ. For a

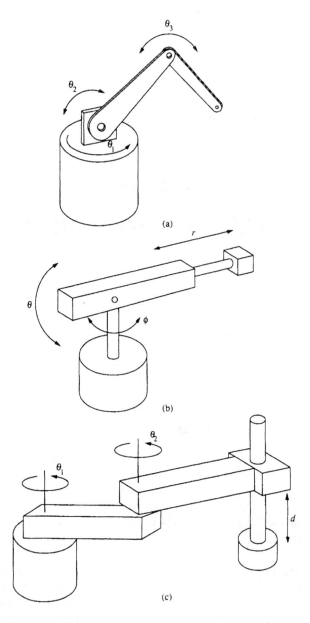

Figure A.1.1: Basic robot arm geometries: (a) articulated arm, revolute coordinates (RRR); (b) spherical coordinates (RRP); (c) SCARA arm (RRP).

A.1 Basic Manipulator Geometries

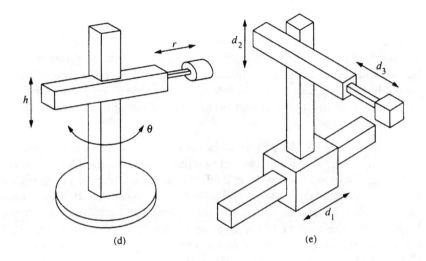

Figure A.1.1: (Cont.) (d) cylindrical coordinates (RPP); (e) Cartesian arm, rectangular coordinates (PPP).

prismatic joint it is a length, denoted d.

Some basic arm geometries are shown in Figure A.1.1. The RRR *articulated arm* in Figure A.1.1a is like the human arm, while the PPP *Cartesian arm* in Figure A.1.1e is closely tied to the coordinates used in the manipulator *workspace*, where the Cartesian coordinates (x, y, z) are often used to describe tasks to be performed. The workspace is the total volume swept out by the end effector as the robot executes all possible motions.

The *joint axis* of a revolute joint is the axis about which the rotation θ occurs. (The sense of rotation is determined using the right-handed screw rule: If the curled fingers of the right hand indicate the direction of rotation, the thumb indicates the direction of the axis of rotation.) For a prismatic joint, it is the axis along which the telescoping action d occurs. The relative orientations of the joint axes of an arm determine its fundamental properties. Figure A.1.1c shows a RRP manipulator known as the SCARA (selected compliant articulated robot for assembly) arm. It has quite a different structure than the RRP spherical arm shown in Figure A.1.1b, since its joint axes are all parallel. On the other hand, the joint axes of the spherical arm intersect at a point.

Industrial examples of the RRR arm are the PUMA and Cincinnati-Milacron T³ 735 manipulators. The Stanford manipulator is a spherical RRP arm; the AdeptOne is a SCARA RRP arm. An example of the RPP arm is the GMF M-100. The Cincinnati-Milacron T³ gantry robot is a PPP arm.

Many industrial robots are *serial link* manipulators since they consist of a series of links connected together by actuated joints. The base is called link 0, and the last link is terminated by the tool or *end effector*. Many robots have six joints, corresponding to the *six degrees of freedom* needed to obtain arbitrary position and orientation of the end effector in three-dimensional space.

Arms like the PUMA 560 have six revolute joints. In such an arm, the joints may be grouped into two sets of three joints each. The first three joints may be used to place the end effector at an arbitrary position within the three-dimensional workspace. The last three joints may be used to obtain an arbitrary orientation of the end effector at that position. In the PUMA 560, the axes of joints 4, 5, and 6 intersect at a common point and are mutually orthogonal. This makes orienting the end-effector convenient. The last three joints are known as the *wrist mechanism* (see Example A.2.4).

A.2 Robot Kinematics

Here we review the kinematics of robot manipulators, including the arm A matrices, homogeneous transformations, the T matrix, forward and inverse kinematics, and joint-space and Cartesian coordinates. Several illustrative examples are given.

A Matrices

For given values of the joint variables, it is important to be able to specify the locations of the links with respect to each other. This is accomplished by using the manipulator *kinematic equations*.

We may associate with each link i a coordinate frame (x_i, y_i, z_i) fixed to that link. See Figure A.2.1. A standard and consistent paradigm for so doing is the *Denavit-Hartenberg (D-H)* representation [Paul 1981], [Spong and Vidyasagar 1989]. The frame attached to link 0 (i.e., the base of the manipulator) is called the *base frame* or *inertial frame*.

The relation between coordinate frame $i-1$ and coordinate frame i is given by the transformation matrix

A.2 Robot Kinematics

$$A_i = \begin{bmatrix} \cos\theta_i & -\cos\alpha_i \sin\theta_i & \sin\alpha_i \sin\theta_i & a_i \cos\theta_i \\ \sin\theta_i & \cos\alpha_i \cos\theta_i & -\sin\alpha_i \cos\theta_i & a_i \sin\theta_i \\ 0 & \sin\alpha_i & \cos\alpha_i & d_i \\ \hline 0 & 0 & 0 & 1 \end{bmatrix}. \quad (A.2.1)$$

Most of the parameters in this A matrix for link i are fixed. The *link parameters* are α_i, the twist of the link i, and a_i, the length of link i. These parameters are tabulated for each link in the arm manufacturer's specifications. The *joint parameters* are the joint angle θ_i and the joint offset d_i. If joint i is revolute, the joint variable is θ_i and d_i is a constant tabulated in the specs. On the other hand, if the link is prismatic, then d_i is the joint variable and θ_i is a constant provided in the specs. The parameter a_i for a prismatic joint is defined to be zero, since the link length is variable and described by d_i.

By the D-H convention, for a revolute joint, the rotation θ_i occurs about axis z_{i-1}. For a prismatic joint d_i occurs along axis z_{i-1}. Thus the link coordinate frame is considered to be attached to the *outer end* of the link. See the examples.

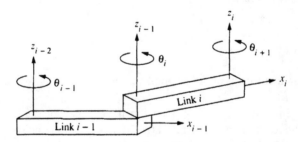

Figure A.2.1: Link kinematic relations in a manipulator.

The A matrix A_i is a function of only a single variable, namely the joint variable θ_i or d_i, since all the other parameters in A_i are fixed for a specific joint. If a manipulator has n links, *the joint-variable vector q* is an n-vector composed of the individual joint variables. Thus q is in general a combination of angles θ_i and lengths d_i. For instance, for an RRP arm,

$$q = \begin{bmatrix} \theta_1 & \theta_2 & d_3 \end{bmatrix}^T.$$

The components of q are denoted by q_i; that is, the general joint variable q_i can represent either an angle θ_i or a length d_i as appropriate.

Homogeneous Transformations

The A matrix is a *homogeneous transformation* matrix of the form

$$A_i = \begin{bmatrix} R_i & P_i \\ 0 & 1 \end{bmatrix}, \tag{A.2.2}$$

where R_i is a *rotation matrix* and p_i is a *translation vector*. Thus if $^i r$ is a point described with respect to the coordinate frame of link i, the same point has coordinates ^{i-1}r with respect to the frame of link $i-1$ given by

$$^{i-1}r = A_i {}^i r. \tag{A.2.3}$$

The homogeneous transformation is a 4×4 matrix, so that it can describe both rotations and translations; therefore, the vectors describing position in a given coordinate frame are 4-vectors. They are of the form

$$^i r = \begin{bmatrix} ^i x \\ ^i y \\ ^i z \\ 1 \end{bmatrix}, \tag{A.2.4}$$

where $(^i x, {}^i y, {}^i z)$ are the coordinates of the point in frame i. Thus, according to (A.2.3) and (A.2.2),

$$\begin{bmatrix} ^{i-1}x \\ ^{i-1}y \\ ^{i-1}z \end{bmatrix} = R_i \begin{bmatrix} ^i x \\ ^i y \\ ^i z \end{bmatrix} + p_i, \tag{A.2.5}$$

which is just a rotation of R_i applied to the coordinates in frame i plus a translation of p_i.

We may interpret frame $i-1$ as the fixed (i.e., "original") frame and frame i as the rotated and translated (i.e., "new") frame due to the following considerations. According to (A.2.5), there is an easy way to find the rotation matrix R_i that rotates a given coordinate frame $i-1$ into another given frame i. Set p_i equal to zero and $(^i x, {}^i y, {}^i z) = (1,0,0)$. Then (A.2.5) is equal to the first column of R_i. That is, the first column of R_i is nothing but *the representation of the new x-axis $^i x$ in terms of the original coordinates* $(^{i-1}x, {}^{i-1}y, {}^{i-1}z)$. Similarly, the second (respectively third) column of R_i is the representation of the rotated y-axis $^i y$ (respectively, z-axis and $^i z$) in the fixed frame $i-1$. We shall illustrate this in Example A.2.1.

Therefore, A_i may be interpreted from several points of view. It is the transformation that takes a representation $^i r$ of a vector in frame i to its

A.2 Robot Kinematics

representation ^{i-1}r in frame $i-1$. On the other hand, it is the description of frame i in terms of frame $i-1$; in fact, R_i describes the orientation of the axes of frame i in terms of frame $i-1$, while p_i describes the origin of frame i in terms of the coordinates of frame $i-1$.

A rotation matrix R enjoys the property of *orthogonality*; that is, $R^T = R^{-1}$. It has one eigenvalue at $\lambda = 1$, whose eigenvector is the axis of rotation. A rotation of θ about the x axis, for instance, looks like

$$R_{x,\theta} = \begin{bmatrix} 1 & 0 & 0 \\ 0 & \cos\theta & -\sin\theta \\ 0 & \sin\theta & \cos\theta \end{bmatrix}. \quad (A.2.6)$$

The last entry of "1" in (A.2.4) represents a *scaling factor* for the length of the vector. By using a homogeneous transformation whose (4,4) entry is not unity, vectors may be scaled. By using entries in positions, (4,1), (4,2), (4,3) of the transformation matrix, *perspective transformations* may be performed. These ideas are important, for instance, in camera-frame transformations but will not be useful in the transformations associated with the manipulator arm.

Arm T Matrix

To obtain the coordinates of a point in terms of the base (i.e., link 0) frame, we may use the matrices

$$T_i = A_1 A_2 \cdots A_i. \quad (A.2.7)$$

Then, given the coordinates $^i r$ of a point expressed in the frame attached to link i, the coordinates of the same point in the base frame are given by

$$^0 r = T_i\, ^i r. \quad (A.2.8)$$

We call T_i a *kinematic chain* of transformations.

We define the *arm T matrix* as

$$T \equiv T_n = A_1 \cdots A_n, \quad (A.2.9)$$

with n the number of links in the manipulator. Then, if $^n r$ are the coordinates of a point referred to the last link, the base coordinates of the point are

$$^0 r = T^n r. \quad (A.2.10)$$

This is an important relation since $^n r$, the coordinates of an object in the nth frame, can represent the location of the object *with respect to the tool or end effector*. It is thus important in specifying tasks to be accomplished.

On the other hand, $^0 r$ represents the location of the object with respect to the base frame, which is the object's absolute position with respect to the manipulator base.

Forward Kinematics

The position and orientation of the end effector with respect to the manipulator base frame are given by evaluating the arm T matrix. It is conventional to symbolize this homogeneous transformation as

$$T = \begin{bmatrix} n & o & a & p \\ 0 & 0 & 0 & 1 \end{bmatrix} = \begin{bmatrix} R & P \\ 0 & 1 \end{bmatrix}. \qquad \text{(A.2.11)}$$

Thus the orientations of the axes of the end-effector reference frame are described with respect to base coordinates by the rotation matrix $R = [n \; o \; a]$, and the origin of the end-effector frame has a position of p in base coordinates.

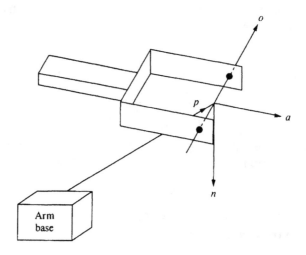

Figure A.2.2: Robot end effector, showing the definition of (n, o, a, p).

The 3-vectors n, o, a, and p are defined as in Figure A.2.2. The *approach* vector of the end effector is "a"; the *orientation vector* "o" is the direction specifying the orientation of the hand, from fingertip to fingertip. The *normal* vector "n" is chosen to complete the definition of a right-handed coordinate system using

$$n = o \times a.$$

A.2 Robot Kinematics

Thus (n, o, a) are the base coordinates of an (x, y, z) Cartesian coordinate system attached to the end effector. The *position* vector p specifies the location of the origin of the (n, o, a) frame with respect to the base frame.

The representation (n, o, a) for the orientation of the end effector is inefficient. Note that $[n \quad o \quad a]$ is a 3×3 matrix, so that it has nine entries. However, it does not take nine degrees of freedom to specify orientation. Indeed, $[n \quad o \quad a]$ is a rotation matrix, so that its columns are orthogonal; this orthogonality requirement imposes extra constraints on the elements of $[n \quad o \quad a]$, so that the nine entries of $[n \quad o \quad a]$ are not independent. Alternative more efficient methods of specifying the orientation of the end effector in base coordinates are the roll-pitch-yaw, Euler angle, quaternion, and tool-configuration vector descriptions. We discuss some of these later. The reason we use (n, o, a) is that the arm T matrix is easily computed using the A matrices in terms of the arm parameters and join variables. Then n, o, a, and p are simply read off by examining the T matrix, as we shall see in the examples.

At this point we should like to distinguish between two sets of coordinates describing the end effector. The *joint variable* coordinates of the end effector are given by the n-vector.

$$q = \begin{bmatrix} q_1 & q_2 & \cdots & q_n \end{bmatrix}^T,$$

where q_i can represent angles or lengths, depending on whether the links are revolute or prismatic. Generally, $n = 6$, so that the arm has six degrees of freedom. The end-effector *Cartesian coordinates* are the description of end effector orientation and position in terms of the arm base coordinates. According to (A.2.11), where T may be computed knowing q, the joint variable and Cartesian coordinates are equivalent, for both specify the location of the end effector.

We say that q is the *joint-space* description of the position and orientation of the end effector, while (n, o, a, p) is the *Cartesian* or *task space* description. This terminology derives from the fact that descriptions of tasks to be performed by an arm are generally given in Cartesian coordinates, not in joint coordinates.

The robot arm *kinematics problem* is as follows. Given the joint variables q, find the Cartesian position and orientation of the end of the manipulator. Thus the kinematics problem amounts to converting given joint variables into the Cartesian position and orientation of the end effector expressed in base coordinates. Let us illustrate the solution of this problem for several simple robot arms which we use as examples throughout the book. It amounts to computing T.

It should be mentioned that there are several software packages commercially available for computing the A matrices and the T matrix for a given robot arm. See, for instance, [MATMAN 1986].

EXAMPLE A.2-1: Kinematics for Three-Link Cylindrical Arm

A simple RPP manipulator is shown in Figure A.2.3a. It may be interpreted as the first three joints of an arm much as the GMF M-100. These are the joints used to position the end effector. A wrist mechanism consisting of three joints may be added to the end of the RPP arm to orient the end effector in space (see Example A.2.4). The joint variables of the three-link arm shown are θ, h, r, which correspond to the coordinates of a cylindrical coordinate system, so that the joint-variable vector is

$$q = \begin{bmatrix} \theta & h & r \end{bmatrix}^T. \tag{1}$$

a. A Matrices

Coordinate frames may be attached to links 1, 2, and 3 using any technique desired. The D-H frames, to which (A.2.1) corresponds, are shown in Figure A.2.3b. For this choice of frames, the A matrices may be determined as follows.

Frame 1 is related to the base frame 0 by a simple rotation of θ degrees about the axis z_0. A z-axis rotation, $R_{z,\theta}$, with no translation is described by

$$A_1 = \begin{bmatrix} c\theta & -s\theta & 0 & 0 \\ s\theta & c\theta & 0 & 0 \\ 0 & 0 & 1 & 0 \\ \hline 0 & 0 & 0 & 1 \end{bmatrix}, \tag{2}$$

where $c\theta$ represents $\cos\theta$ and $s\theta$ represents $\sin\theta$.

To find A_2 we may use (A.2.1). Since link 2 is prismatic with extension h, the length a_2 is zero. In this example the rotation θ_2 is also zero. The twist α_2 of link 2 is the angle of rotation about axis x_2 required to align z_1 with z_2 – that is, $-90°$. Therefore,

A.2 Robot Kinematics

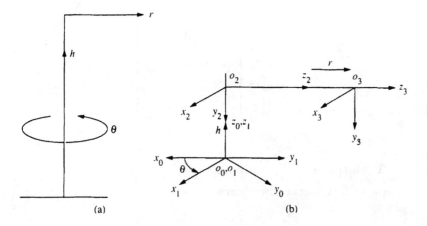

Figure A.2.3: Three-link cylindrical manipulator: (a) arm schematic; (b) D-H coordinate frames.

$$A_2 = \begin{bmatrix} 1 & 0 & 0 & 0 \\ 0 & 0 & 1 & 0 \\ 0 & -1 & 0 & h \\ \hline 0 & 0 & 0 & 1 \end{bmatrix}. \tag{3}$$

There is an attractive alternative to (A.2.1) for determining the link A matrix in simple cases. Recall that the first column of A_2 is the representation of x_2 in the coordinates (x_1, y_1, z_1). According to the figure, this is just $[1 \ 0 \ 0]^T$. The second column of A_2 is the representation of y_2 in the coordinates (x_1, y_1, z_1), which the figure shows is $[0 \ 0 \ -1]^T$. The third column of A_2 is the representation of z_2 in frame o_1, which is just $[0 \ 1 \ 0]^T$. Thus (3) is obtained by inspection.

In similar fashion we obtain,

$$A_3 = \begin{bmatrix} 1 & 0 & 0 & 0 \\ 0 & 1 & 0 & 0 \\ 0 & 0 & 1 & r \\ \hline 0 & 0 & 0 & 1 \end{bmatrix} \qquad (4)$$

b. T Matrix and Arm Kinematics

The arm T matrix is now obtained as

$$T = A_1 A_2 A_3 = \begin{bmatrix} c\theta & 0 & -s\theta & -r\,s\theta \\ s\theta & 0 & c\theta & r\,c\theta \\ 0 & -1 & 0 & h \\ \hline 0 & 0 & 0 & 1 \end{bmatrix} \qquad (5)$$

To interpret the T matrix, examine (A.2.11). The position in base coordinates of the end of the manipulator, that is, the origin of frame 3, is

$$p = \begin{bmatrix} -r\,s\theta & r\,c\theta & h \end{bmatrix}^T. \qquad (6)$$

The orientation of frame 3 described in base coordinates is expressed in (n, o, a) form by giving the coordinates of the normal, orientation, and approach vectors in terms of the base frame. These are given by

$$\begin{aligned} n &= \begin{bmatrix} c\theta & s\theta & 0 \end{bmatrix}^T \\ o &= \begin{bmatrix} 0 & 0 & -1 \end{bmatrix}^T \\ a &= \begin{bmatrix} -s\theta & c\theta & 0 \end{bmatrix}^T. \end{aligned} \qquad (7)$$

A glance at Figure A.2.3b verifies these expressions.

∎

A.2 Robot Kinematics

EXAMPLE A.2-2: Kinematics for Two-Link Planar Elbow Arm

A two-link planar RR arm is shown in Figure A.2.4a where a_1 and a_2 are the fixed and known link length parameters. The link coordinate frames are shown in Figure A.2.4b. We have taken the z-axes perpendicular to the page to conform to the convention of specifying points in a plane by (x, y) coordinates. Therefore, the frames are not defined quite as in the D-H convention.

The arm A matrices may be written by inspection as

$$A_1 = \left[\begin{array}{ccc|c} c_1 & -s_1 & 0 & a_1 c_1 \\ s_1 & c_1 & 0 & a_1 s_1 \\ 0 & 0 & 1 & 0 \\ \hline 0 & 0 & 0 & 1 \end{array}\right]. \qquad (1)$$

where c_1, s_1, represent respectively $\cos\theta_1$, $\sin\theta_1$ and

$$A_2 = \left[\begin{array}{ccc|c} c_2 & -s_2 & 0 & a_2 c_2 \\ s_2 & c_2 & 0 & a_2 s_2 \\ 0 & 0 & 1 & 0 \\ \hline 0 & 0 & 0 & 1 \end{array}\right]. \qquad (2)$$

The arm T matrix is given by

$$T = A_1 A_2 = \left[\begin{array}{ccc|c} c_{12} & -s_{12} & 0 & a_1 c_1 + a_2 c_{12} \\ s_{12} & c_{12} & 0 & a_1 s_1 + a_2 s_{12} \\ 0 & 0 & 1 & 0 \\ \hline 0 & 0 & 0 & 1 \end{array}\right]. \qquad (3)$$

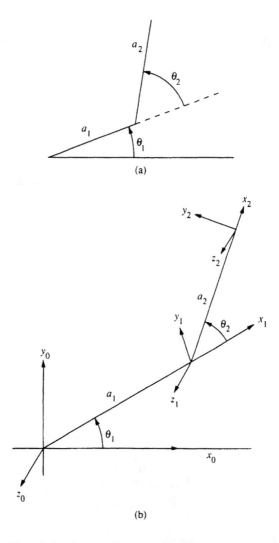

Figure A.2.4: Two-link planar RR arm: (a) arm schematic; (b) D-H coordinate frames.

where $c_{12} \equiv \cos(\theta_1 + \theta_2)$ and $s_{12} \equiv \sin(\theta_1 + \theta_2)$.

Therefore, the origin o_2 of frame 2 in terms of base coordinates is located at

$$p = \begin{bmatrix} a_1 c_1 + a_2 c_{12} & a_1 s_1 + a_2 s_{12} & 0 \end{bmatrix}^T. \tag{4}$$

A.2 Robot Kinematics

This represents the kinematic solution, which converts the joint variable coordinates (θ_1, θ_2) into the base-frame Cartesian coordinates of the end of the arm. The reader should examine Figure A.2.4a to verify this expression.

∎

EXAMPLE A.2-3: Kinematics for Two-Link Polar Arm

A two-link planar RP arm is shown in Figure A.2.5a where ℓ is the fixed known length of the base link. The link frames, with the z-axes perpendicular to the page, are shown in Figure A.2.5b. The joint vector is

$$q = [\ \theta \quad r\ ]^T, \tag{1}$$

which corresponds to polar coordinates in the plane.

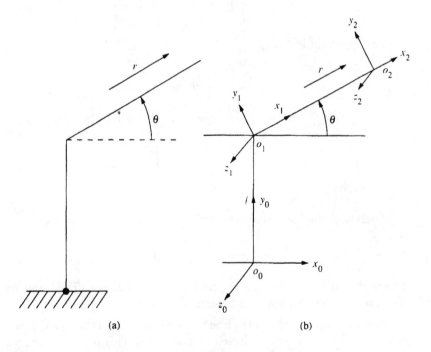

Figure A.2.5: Two-link planar RP arm: (a) arm schematic; (b) D-H coordinate frames.

By inspection, the A matrices are found to be

$$A_1 = \left[\begin{array}{ccc|c} c\theta & -s\theta & 0 & 0 \\ s\theta & c\theta & 0 & \ell \\ 0 & 0 & 1 & 0 \\ \hline 0 & 0 & 1 & 1 \end{array}\right]. \tag{2}$$

$$A_2 = \left[\begin{array}{ccc|c} 1 & 0 & 0 & r \\ 0 & 1 & 0 & 0 \\ 0 & 0 & 1 & 0 \\ \hline 0 & 0 & 0 & 1 \end{array}\right]. \tag{3}$$

The T matrix is

$$T = A_1 A_2 = \left[\begin{array}{ccc|c} c\theta & -s\theta & 0 & rc\theta \\ s\theta & c\theta & 0 & \ell + \dot{r}s\theta \\ 0 & 0 & 1 & 0 \\ \hline 0 & 0 & 0 & 1 \end{array}\right]. \tag{4}$$

Therefore, the base coordinates of the end of the arm are

$$p = \left[\begin{array}{ccc} r\,c\theta & \ell + rs\theta & 0 \end{array}\right]^T, \tag{5}$$

which should be verified by examining Figure A.2.5a. This may be interpreted as the forward kinematics solution for the arm.

This example illustrates the freedom we have to select the base frame origin o_0. We could have selected o_0 coincident with o_1. However, we have chosen to include the length ℓ of link 0 by placing o_0 at the bottom of the base link. ∎

A.2 Robot Kinematics

EXAMPLE A.2-4: Kinematics for Spherical Wrist

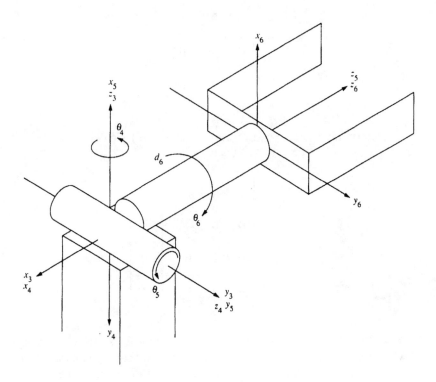

Figure A.2.6: Spherical wrist.

A spherical wrist mechanism is shown in Figure A.2.6. See [Paul 1981], [Spong and Vidyasagar 1989]. For convenient orientational control, all three joint axes intersect at a common point. Since many six-link industrial arms, including the Stanford arm, end in such a configuration, we have labeled the three joint variables $\theta_4, \theta_5, \theta_6$. The D-H frames are also shown in the figure. Recall that the rotation θ_i occurs about axis z_{i-1} in the D-H convention. The origin of frame 6 has been chosen at the base of the fingered gripper. The length d_6 is a fixed known parameter.

By expressing the axes x_4, y_4, z_4 in terms of the coordinates (x_3, y_3, z_3), we are able to directly determine the columns of A4 to obtain

$$A_4 = \begin{bmatrix} c_4 & 0 & -s_4 & 0 \\ s_4 & 0 & c_4 & 0 \\ 0 & -1 & 0 & 0 \\ \hline 0 & 0 & 0 & 1 \end{bmatrix}. \tag{1}$$

Determining x_5, y_5, z_5 in terms of (x_4, y_4, z_4) yields

$$A_5 = \begin{bmatrix} c_5 & 0 & s_5 & 0 \\ s_5 & 0 & -c_5 & 0 \\ 0 & 1 & 0 & 0 \\ \hline 0 & 0 & 0 & 1 \end{bmatrix}. \tag{2}$$

In writing down A_5, it is important to note that θ_5 is defined to be zero when the end effector is in an upright position, that is, when the arm is fully extended. We have drawn the figure with $\theta_5 = -90°$ to show more clearly the different coordinate systems.

In similar fashion, we obtain

$$A_6 = \begin{bmatrix} c_6 & -s_6 & 0 & 0 \\ s_6 & c_6 & 0 & 0 \\ 0 & 0 & 1 & d_6 \\ \hline 0 & 0 & 0 & 1 \end{bmatrix}. \tag{3}$$

The wrist T matrix is

A.2 Robot Kinematics

$$T = A_4 A_5 A_6$$

$$= \left[\begin{array}{ccc|c} c_4 c_5 c_6 - s_4 s_6 & -c_4 c_5 s_6 - s_4 c_6 & c_4 s_5 & c_4 s_5 d_6 \\ s_4 c_5 c_6 + c_4 s_6 & -s_4 c_5 s_6 + c_4 c_6 & s_4 s_5 & s_4 s_5 d_6 \\ -s_5 c_6 & s_5 s_6 & c_5 & c_5 d_6 \\ \hline 0 & 0 & 0 & 1 \end{array}\right], \quad (4)$$

where $c_4 \equiv \cos\theta_4$, and so on. It is quite interesting to note that the rotational part of this T matrix corresponds to the Euler angle transformation [Spong and Vidyasagar 1989]. Therefore, θ_4, θ_5, θ_6 may be interpreted as the Euler angles with respect to frame 3.

Suppose that we would like to determine the kinematic transformations of the 3-link cylindrical arm in Example A.2.1 terminated with a spherical wrist. Then it is only necessary to multiply the T matrix in Example A.2.1 and (4), in that order, to obtain the overall manipulator T matrix.

■

Inverse Kinematics

The location of the end effector is specified in base coordinates by the arm T matrix. On the other hand, the location of the end effector is specified in joint coordinates by giving the values of the joint variables q_i. We may compute the T matrix knowing the joint-variable vector q, as in the examples just shown. Finding T from q is the kinematics problem.

The *inverse-kinematics* problem is as follows. Given (n, o, a, p) for the end effector in base coordinates, determine the joint variables q_i in the T matrix

$$T = \begin{bmatrix} n & o & a & p \\ 0 & 0 & 0 & 1 \end{bmatrix} \quad (A.2.12)$$

that yield the specified (n, o, a, p). This problem is important in manipulator control, for the desired Cartesian orientation and position of the end effector are specified by the task. Then the solution to the inverse kinematics problems gives the joint variables q_i required to achieve that orientation and position.

Due to the functions involved in the T matrix, the relations between q_i and (n, o, a, p) are highly nonlinear; these must be inverted to obtain the inverse kinematic solution q. The solutions for q_i are generally not unique. If the last three arm axes intersect at a point (e.g., the wrist mechanism in Example A.2.4), it is common to split the inverse kinematics problem into two parts. In a six-link arm, for instance, one first determines q_1, q_2, and q_3 required to obtain the desired Cartesian position p. Then the wrist variables q_4, q_5, and q_6 that give the desired orientation (n, o, a) are determined. Some techniques for solving the inverse kinematics problem are given in [Paul 1981], [Craig 1989], [Spong and Vidyasagar 1989]. An example is now given.

EXAMPLE A.2-5: Inverse Kinematics for Two-Link Planar Elbow Arm

For the planar RR arm of Example A.2.2, the inverse kinematics problem amounts to finding the joint variable θ_1 and θ_2 given a desired Cartesian position (x, y) of the end of the arm. See Figure A.2.7.

The first thing that is evident is that, as long as $a_1^2 + a_2^2 < r \equiv x^2 + y^2$, there are two solutions. The one shown in Figure A.2.7 is the "elbow down" solution. Another solution may be determined for the "elbow up" configuration, where both links are above the vector $[x \ \ y]^T$. Thus *the inverse kinematics problem generally has a nonunique solution.* This may often be taken advantage of to obtain end-effector positioning with *collision avoidance.*

Given the T matrix from Example A.2.2, we see that θ_1 and θ_2 may be found by solving

$$a_1 c_1 + a_2 c_{12} = x \tag{1}$$

$$a_1 s_1 + a_2 s_{12} = y. \tag{2}$$

Determining θ_1 and θ_2 based on algebraic manipulations of such equations is called the *algebraic* inverse kinematics solution technique [Paul 1981]. Let us show a *geometric technique* here.

Referring to Figure A.2.7, define

$$r^2 = x^2 + y^2 \tag{3}$$

and use the law of cosines to obtain

A.2 Robot Kinematics

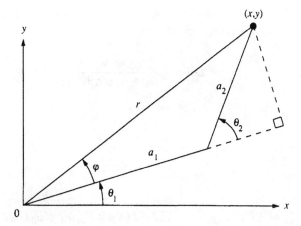

Figure A.2.7: Inverse kinematics for two-link planar RR arm.

$$\begin{aligned}r^2 &= a_1^2 + a_2^2 - 2a_1a_2\cos(\pi - \theta_2)\\ &= a_1^2 + a_2^2 + 2a_1a_2\cos\theta_2.\end{aligned} \quad (4)$$

We could now solve for θ_2 using the $\cos^{-1}$ function. However, it is better to use $\tan^{-1}$ for reasons of numerical accuracy. The FORTRAN function implementing $\tan^{-1}(b/c)$ is ATAN2(b,c). This function has a uniform accuracy over the range of its arguments, returns a unique value for the angle depending on the signs of b and c, and gives the correct solution if b and/or c is zero.

Therefore, we proceed by computing

$$\cos\theta_2 = \frac{r^2 - a_1^2 - a_2^2}{2a_1a_2} \equiv C \quad (5)$$

$$\sin\theta_2 = \pm\sqrt{1 - \cos^2\theta_2} = \pm\sqrt{1 - C^2} \equiv D \quad (6)$$

$$\theta_2 = \text{ATAN2}(D, C). \quad (7)$$

An additional advantage of using the arctangent is that the multiple solutions of the inverse kinematics problem are explicitly revealed by the choice of negative or positive sign in (6).

To determine θ_1 define the auxiliary angle φ in the figure. By inspection of the right triangle shown,

$$\tan \varphi = \frac{a_2 \sin \theta_2}{a_1 + a_2 \cos \theta_2}. \tag{8}$$

Moreover,

$$\tan (\varphi + \theta_1) = \frac{y}{x}, \tag{9}$$

so that

$$\theta_1 = \text{ATAN2}\,(y, x) - \text{ATAN2}\,(a_2 \sin \theta_2,\ a_1 + a_2 \cos \theta_2). \tag{10}$$

Note that θ_1 depends on θ_2.

■

A.3 The Manipulator Jacobian

Given a generally nonlinear transformation from the joint variable $q(t) \in R^n$ to $y(t) \in R^p$,

$$y = h(q) \tag{A.3.1}$$

we define the Jacobian associated with $h(q)$ as

$$J(q) \equiv \frac{\partial h(q)}{\partial q}. \tag{A.3.2}$$

As we have just seen, the Jacobian is useful in feedback linearization, and therefore in robot manipulator control. It is also the means by which we transform velocity, acceleration, and force between coordinate frames.

Transformation of Velocity and Acceleration
Since

$$\dot{y} = \frac{\partial h}{\partial q} \dot{q} = J \dot{q}, \tag{A.3.3}$$

the Jacobian allows us to transform velocity from joint space to "y-space." Let us discuss the special case where $\dot{y}(t)$ is the Cartesian velocity. Then $J(q)$ is called the *manipulator Jacobian*.

It is usual to define the generalized Cartesian velocity as

A.3 The Manipulator Jacobian

$$\dot{y} = \begin{bmatrix} \nu \\ \omega \end{bmatrix}, \tag{A.3.4}$$

with $\nu = [\nu_x \ \nu_y \ \nu_z]^T$ the linear velocity and $\omega = [\omega_x \ \omega_y \ \omega_z]^T$ the angular velocity. For instance, ω_x represents angular velocity about the x-axis. Thus $\dot{y}$ has six components and the arm Jacobian J is a $6 \times n$ matrix, with n the number of joints in the manipulator. If $n = 6$, the Jacobian is square. We shall soon show how to compute the manipulator Jacobian.

Using (A.3.3), we can obtain an expression for the transformation of *differential motion*. Let $dq = [dq_1 \cdots dq_n]^T$ be a differential motion in joint space, with dq_i a small rotation if joint i is revolute, and a small linear displacement if joint i is prismatic. Let

$$\mathbf{dy} = \begin{bmatrix} dx \\ dy \\ dz \\ \delta x \\ \delta y \\ \delta z \end{bmatrix} \tag{A.3.5}$$

describe the same differential motion in Cartesian coordinates, with $[d_x \ d_y \ d_z]^T$ the differential linear motion and $[\delta_x \ \delta_y \ \delta_z]^T$ representing the differential rotation. Here δ_x, for instance, means a small rotation about the x-axis. We have written the Cartesian differential motion vector **dy** in boldface to distinguish it from the small change dy in the second coordinate of Cartesian position.

According to (A.3.3), where $\dot{y} = dy/dt$ and $\dot{q} = dq/dt$, we see that

$$\mathbf{dy} = J\,dq, \tag{A.3.6}$$

with J the Jacobian relating joint space and Cartesian space.

The transformation of acceleration is found by differentiating (A.3.3) to be

$$\ddot{y} = J\ddot{q} + \dot{J}\dot{q}. \tag{A.3.7}$$

Transformation of Force

To discover the transformation of static force between coordinate frames, consider the following.

The *virtual work* resulting from the application of a generalized force/torque in joint space that results in a differential motion dq is

$$\delta W = \tau^T dq, \tag{A.3.8}$$

with τ the n-vector of arm control torques/forces, and $dq = [dq_1 \cdots dq_n]^T$ the differential change in the joint variable. If the description in another coordinate frame of the force is F and the description there of position is y, we must also have

$$\delta W = F^T dy. \quad (A.3.9)$$

Now taking (A.3.6) into account, we may write

$$\delta W = \tau^T dq = F^T J dq,$$

so that the transformation from force to torque is given by

$$\tau = J^T(q) F. \quad (A.3.10)$$

In the case that $y(t)$ is Cartesian position, we define the Cartesian generalized force to be the 6-vector

$$F = \begin{bmatrix} f_c \\ \tau_c \end{bmatrix}, \quad (A.3.11)$$

with $\mathbf{f}_c = [f_x \ f_y \ f_z]^T$ a Cartesian force 3-vector, and $\tau_c = [\tau_x \ \tau_y \ \tau_z]^T$ a 3-vector representing Cartesian torques. For instance, τ_x represents torque exerted about the x-axis.

If the arm has six links and the Jacobian is nonsingular, the transformation from generalized torque to generalized force is given by

$$F = J^{-T}(q) \tau. \quad (A.3.12)$$

The singularities in the Jacobian generally occur at the extremities of the manipulator workspace.

Specification of Cartesian Position

It is now necessary to confront an issue that is a source of confusion in robotics. Consider Equation (A.3.2). Unfortunately, when discussing the transformation $h(q)$ from joint space to Cartesian space, this equation may only be interpreted as a convenience of notation, not as a rigorous mathematical formula. The reason is that although the generalized Cartesian velocity (A.3.4) and acceleration, and the generalized Cartesian force (A.3.11) are bona fide 6-vectors, there is a problem with conveniently specifying the generalized Cartesian position $y(t)$. We now discuss several ways to specify the Cartesian position.

Representing Generalized Cartesian Position as (n,o,a,p). In our context, both the *location of the origin* of the end-effector frame as well

A.3 The Manipulator Jacobian

as its orientation must be specified in base coordinates. It is easy to specify the origin of the end-effector frame in base coordinates (x, y, z) using a 3-vector $p(t) = [p_x \; p_y \; p_z]^T$. However, specification of orientation is not so easy. This is because it takes more than three independent variables to specify uniquely the orientation of one frame with respect to another.

It should be mentioned that conventions such as the Euler angles and roll-pitch-yaw [Paul 1981], [Spong and Vidyasagar 1989] involve only three variables. However, they may not be used to uniquely specify the *absolute* orientation of one frame with respect to another, but only relative changes in orientation. We have already seen in (A.3.5) that only three variables are needed to describe *differential rotations*.

In our work we have specified the Cartesian position $y(t)$ of the end effector in base coordinates by using the (n, o, a, p) approach (see Section A.2). There we define the generalized Cartesian position as

$$y(t) \equiv \begin{bmatrix} n(t) & o(t) & a(t) & p(t) \\ 0 & 0 & 0 & 1 \end{bmatrix} = T(t), \quad (\text{A.3.13})$$

with $T(t)$ the arm T matrix, (n, o, a) the vectors needed to describe the end-effector orientation, and $p = [p_x \; p_y \; p_z]^T$ the positional portion of the Cartesian description that specifies the origin of the end-effector frame.

It is clear that the selection of the transformation $h(q)$ depends on how we wish to describe the Cartesian position $y(t)$. If we use (A.3.13), then $h(q)$ is just the arm kinematics transformation discussed in Section A.2. However, then $y(t)$ is not a 6-vector but a 4×4 matrix. Therefore, the definition of the arm Jacobian as $J(q) = \partial h / \partial q$ is a loose one purely for notational convenience. Therefore, we are faced with the problem of providing a suitable definition for $J(q)$. Before we do this, let us discuss some more techniques for representing the generalized Cartesian position.

Representing Cartesian Orientation Using Euler's Theorem. The positional portion of y is easy to specify; it is just the 3-vector $p = [p_x \; p_y \; p_z]^T$ found from the last column of the T matrix in (A.3.13). However, since $[n \; o \; a]$ is a rotation matrix having nine elements, it does not afford an efficient representation of orientation. The orthogonality of the rotation matrix imposes some constraints among its elements, meaning that it actually has only four degrees of freedom.

The orientation of one coordinate frame (frame 1) with respect to another (frame 0) may be specified uniquely using a 3-vector k representing the axis of rotation of frame 1 with respect to frame 0, and an angle φ specifying the amount of rotation about that axis. We call (k, φ) the *Euler rotation parameters*. The (k, φ) convention involves four variables, which may be found as follows. Suppose that the rotation matrix

$$R = \begin{bmatrix} n & o & a \end{bmatrix} \qquad (A.3.14)$$

describes the orientation of frame 1 with respect to frame 0. Note that R is the upper left 3 × 3 submatrix of the arm T matrix, so that it may be computed given the joint vector q.

A rotation matrix is orthogonal, so that it has one eigenvalue equal to 1. Then the axis of rotation k is the eigenvector of the eigenvalue. That is,

$$(R - I)\,k = 0. \qquad (A.3.15)$$

This is known as *Euler's theorem*. It says that $Rk = k$, or that the rotation axis is not rotated by R. There are many good routines for computing eigenvectors (e.g., [IMSL]), which may be used to find k from R. See the problems to find an explicit formula for k in terms of (n, o, a) [Paul 1981].

The angle of rotation φ about the axis k may easily be found, for the trace of a rotation matrix is equal to $1 + 2\cos\varphi$. Therefore,

$$\cos\varphi = \tfrac{1}{2}(n_x + o_y + a_z - 1), \qquad (A.3.16)$$

where n_x, for instance, denotes the x component of the normal vector n.

Using Euler's rotation parameters we may specify the generalized Cartesian position as a 7-vector, with three components for position (i.e., the last column p of the T matrix) and four components (i.e., k and φ) for orientation.

Representing Cartesian Orientation Using Quaternions. The *quaternion* representation is a 4-vector which is often used to describe the end-effector frame orientation [Ickes 1970], [Sheppard 1978], [Yuan 1988], [Stevens and Lewis 1992].

The quaternions are four numbers represented as (q_0, q), with $q = \begin{bmatrix} q_1 & q_2 & q_3 \end{bmatrix}^T$ a 3-vector. They are given in terms of the Euler parameters (k, φ) by

$$\begin{aligned} q_0 &= \cos(\varphi/2) \\ q &= k\sin(\varphi/2). \end{aligned} \qquad (A.3.17)$$

Using quaternions, the generalized Cartesian position may be represented as the 7-vector $y = \begin{bmatrix} p_x & p_y & p_z & q_0 & q_1 & q_2 & q_3 \end{bmatrix}^T$. An advantage of this representation is that no angles are involved.

A.3 The Manipulator Jacobian

Tool-Configuration Vector. A technique for specifying $y(t)$ as a 6-vector is given in [Schilling 1990]. It depends on adopting a convention for encoding orientational information.

Consider the arm T matrix in (A.3.13) whose rotational portion is R in (A.3.14). Let us select the 3-vector p to represent Cartesian linear position. The approach vector a is a 3-vector specifying the direction in which the end effector points in terms of base coordinates. Recall from the discussion on kinematics in Section A.2 that this vector specifies in base coordinates the axis z_n of the frame attached to the end of link n. Since the rotation matrix R is orthogonal, the approach vector has a length of one.

The two vectors p and a almost completely specify the Cartesian position of the end effector. The only piece of information missing is the roll angle of the end effector. However, this angle is nothing by $q_n = \theta_n$, the last joint variable of the manipulator.

To capture the information q_n, let us propose scaling the approach vector by the exponential function $e^{q_n/\pi}$ to define the *tool-configuration* vector as

$$w = \begin{bmatrix} p \\ ae^{q_n/\pi} \end{bmatrix}. \tag{A.3.18}$$

This is a 6-vector, which uniquely specifies the Cartesian position of the end effector in base coordinates.

Note that there is a unique mapping between w and (p, a, q_n), since, given w and the fact that $\|a\| = 1$, we may compute

$$q_n = \pi \ln \left(w_4^2 + w_5^2 + w_6^2\right)^{1/2} \tag{A.3.19}$$

$$a = \frac{1}{\left(w_4^2 + w_5^2 + w_6^2\right)^{1/2}} \begin{bmatrix} w_4 \\ w_5 \\ w_6 \end{bmatrix}, \tag{A.3.20}$$

with w_i the ith component of w.

Finding Cartesian Velocity from the Arm T Matrix. Note that using the definition (A.3.4), the Cartesian velocity $\dot{y}$ is a 6-vector that is not strictly speaking the derivative of $y(t)$ in (A.3.13), which is a 4×4 matrix. To compute $\dot{y}$, we may find the Jacobian $J(q)$ and then use (A.3.3). Alternatively, the Cartesian velocity may be found directly from the arm T matrix (A.2.11) as follows.

Setting $\dot{y} = [\nu^T \quad \omega^T]^T$, with ν the linear velocity and ω the angular velocity, it is clear that

$$\nu = \dot{p}, \tag{A.3.21}$$

with p found from the last column of T.

To find w, proceed as follows. Since R in (A.2.11) is orthogonal, we have

$$RR^T = I, \tag{A.3.22}$$

which on differentiation yields

$$\dot{R}R^T + R\dot{R}^T = 0.$$

Therefore, the matrix defined as

$$\Omega \equiv \dot{R}R^T \tag{A.3.23}$$

satisfies $\Omega + \Omega^T = 0$, so it is skew symmetric. It may therefore be represented as

$$\Omega = \begin{bmatrix} 0 & -\omega_z & \omega_y \\ \omega_z & 0 & -\omega_x \\ -\omega_y & \omega_x & 0 \end{bmatrix}. \tag{A.3.24}$$

The relation between the *cross-product matrix* Ω and the angular velocity vector

$$\omega = \begin{bmatrix} \omega_x \\ \omega_y \\ \omega_z \end{bmatrix} \tag{A.3.25}$$

is an interesting one, for one may easily demonstrate that

$$\omega \times w = \Omega w \tag{A.3.26}$$

for any 3-vector w. That is, the cross product may be replaced by a matrix multiplication in terms of the cross-product matrix. We denote by $\Omega(w)$ the cross-product matrix associated with a vector w.

The complete procedure for determining $\dot{y}$ from the T matrix is then as follows. First, compute ν using the p vector as in (A.3.21). Then, compute $\Omega(w)$ from the rotation matrix R using (A.3.23), and hence find the last three components w of $\dot{y}$ from the definition of $\Omega(w)$. Computing $\dot{y}$ from q thus requires a *computer subroutine*. Robot controllers generally have block diagrams in which some of the blocks are standard components like integrators, but some of the blocks are implemented in software.

Using (A.3.23), we may write the *strapdown equation*

A.3 The Manipulator Jacobian

$$\dot{R} = \Omega R, \qquad (A.3.27)$$

which is of fundamental importance in inertial navigation [Stevens and Lewis 1992].

Computing the Arm Jacobian

Let us now return to the problem of defining and computing the arm Jacobian $J(q)$. If the Cartesian position is defined by (A.3.13), then (A.3.2) does not afford an appropriate definition. Therefore, let us select (A.3.3) as the definition of $J(q)$. That is, define $J(q)$ by

$$\dot{y} = J(q)\dot{q}, \qquad (A.3.28)$$

so that it maps the joint velocity $\dot{q}$ to the Cartesian velocity $\dot{y}$ as defined by (A.3.4). Then $J(q)$ is a $6 \times n$ matrix.

Let us consider a few examples showing how to compute the arm Jacobian to demonstrate that the procedure is not complicated. A methodical technique for computing $J(q)$ will then be given.

EXAMPLE A.3-1: Arm Jacobian for Three-Link Cylindrical Arm

Let us use the kinematics derived in Example A.2.1 to compute the Jacobian for the three-link cylindrical arm. We shall call the joint variables (θ, z, r) instead of (θ, h, r) to avoid confusion with the nonlinear function $h(q)$.

Since there is no wrist on this three-link arm, the orientation of the last coordinate frame is fixed in terms of (θ, z, r), so that is not necessary to specify Cartesian orientation. This simplifies things for this example.

Denote the linear Cartesian position of the end effector in base coordinates (i.e., the first three components of y) as y_p. Then, according to Example A.2.1, y_p is the last column of the arm T matrix, so that

$$y_p = \begin{bmatrix} -r\sin\theta \\ r\cos\theta \\ z \end{bmatrix} \equiv h_p(q). \qquad (1)$$

Omitting the orientational information has allowed us to define a (positional) transformation function $h_p(q)$ simply as the last column p of the T matrix.

Using (A.3.2) and the definition of $h_p(q)$ given in (1), the Jacobian is

$$J(q) = \frac{\partial h_p}{\partial q} = \begin{bmatrix} \frac{\partial h_p}{\partial \theta} & \frac{\partial h_p}{\partial z} & \frac{\partial h_p}{\partial r} \end{bmatrix} = \begin{bmatrix} -r\cos\theta & 0 & -\sin\theta \\ -r\sin\theta & 0 & \cos\theta \\ 0 & 1 & 0 \end{bmatrix}. \quad (2)$$

Therefore, the velocities of the joints are converted into Cartesian velocities using

$$\frac{dy_p}{dt} = \begin{bmatrix} -r\cos\theta & 0 & -\sin\theta \\ -r\sin\theta & 0 & \cos\theta \\ 0 & 1 & 0 \end{bmatrix} \begin{bmatrix} \dot\theta \\ \dot z \\ \dot r \end{bmatrix}. \quad (3)$$

The determinant of J is $(r\cos^2\theta + r\sin^2\theta) = r$, so that J is nonsingular as long as $r \neq 0$.

Note that the definition (1) of the positional function $h_p(q)$ means that the Jacobian is equal to $\partial h_p / \partial q$. This is a result of neglecting the orientation portion of y.

■

EXAMPLE A.3-2: Arm Jacobian for Two-Link Planar Elbow Arm

In Example A.2.2 the joint variable was $q = [\theta_1 \ \theta_2]^T$. As in Example A.3.1, the orientation of the last coordinate frame is fixed once q is given, so that we are not concerned with the orientational portion of y, but only its linear position portion y_p.

The linear Cartesian position of the end effector in the (x_1, x_2)-plane is given by the last column of the arm T matrix as

$$y_p = \begin{bmatrix} a_1 \cos\theta_1 + a_2 \cos(\theta_1 + \theta_2) \\ a_1 \sin\theta_1 + a_2 \sin(\theta_1 + \theta_2) \end{bmatrix} \equiv h_p(q). \quad (1)$$

Since the motion is constrained to the plane, we have suppressed the x_3 component of y_p.

The Jacobian is

$$J = \frac{\partial h_p}{\partial q} = \begin{bmatrix} -a_1 \sin\theta_1 - a_2 \sin(\theta_1 + \theta_2) & -a_2 \sin(\theta_1 + \theta_2) \\ a_1 \cos\theta_1 + a_2 \cos(\theta_1 + \theta_2) & a_2 \cos(\theta_1 + \theta_2) \end{bmatrix}. \quad (2)$$

Note that the definition of $h_p(q)$ in (1) means that the Jacobian is equal to $\partial h_p / \partial q$.

A.3 The Manipulator Jacobian

The determinant of J is (verify!) equal to $a_1 a_2 \sin\theta_2$, so that J is nonsingular unless $\theta_2 = 0$ or π; that is, unless the arm is fully extended or link 2 is folded back on top of link 1. ∎

EXAMPLE A.3-3: Jacobian for Transformation to Camera Coordinates

Consider the three-link cylindrical arm of Examples A.2.1 and A.3.1 with a camera mounted vertically above the arm that measures only the position (x_1, x_2) of the end effector in the horizontal plane (in base coordinates). Call this position y. Then, according to those examples, the transformation from joint coordinates (θ, z, r) to camera coordinates (x_1, x_2) is

$$y = \begin{bmatrix} x_1 \\ x_2 \end{bmatrix} = \begin{bmatrix} -r\sin\theta \\ r\cos\theta \end{bmatrix} \equiv h(q) \tag{1}$$

and the associated Jacobian is

$$J = \begin{bmatrix} -r\cos\theta & 0 & -\sin\theta \\ -r\sin\theta & 0 & \cos\theta \end{bmatrix}. \tag{2}$$

The determinant of J with the second column deleted is equal to $-r$, so that J has full row rank as long as $r \neq 0$. The pseudo-inverse of J is

$$J^+ = \begin{bmatrix} -(\cos\theta)/r & -(\sin\theta)/r \\ 0 & 0 \\ -\sin\theta & \cos\theta \end{bmatrix}. \tag{3}$$

Since the camera cannot measure the distance x_3 perpendicular to the plane, that coordinate must be selected in any control scheme using an independent means that does not involve trajectory following control in the horizontal plane. ∎

Algorithm for Computing the Arm Jacobian. We are finally in a position to show how to compute the arm Jacobian given the joint variables q_i. The procedure follows.

Given q_i, compute the matrices T_i defined in Section A.2 as

$$T_i = A_1 A_2 \cdots A_i \equiv \begin{bmatrix} R_i & p_i \\ 0 & 1 \end{bmatrix}, \tag{A.3.29}$$

whose corresponding rotation matrices will be denoted

$$R_i = \begin{bmatrix} x_i & y_i & z_i \end{bmatrix}. \tag{A.3.30}$$

Define $T_o = I$, $R_o = I$, and the arm T matrix

$$T = T_n = \begin{bmatrix} R_n & p_n \\ 0 & 1 \end{bmatrix} \equiv \begin{bmatrix} n & o & a & p \\ 0 & 0 & 0 & 1 \end{bmatrix}. \tag{A.3.31}$$

The vector z_i represents the z-axis of frame i in base coordinates. The vector p represents the location of the origin of link frame n (the end-effector frame) in terms of base coordinates. The Jacobian is computed using the vectors p and z_i as follows.

The generalized Cartesian velocity is $\dot{y} = [\nu^T \quad \omega^T]^T$. Therefore, we may partition the Jacobian matrix into a linear and an orientational part by writing

$$\begin{bmatrix} \nu \\ \omega \end{bmatrix} = J(q)\dot{q} = \begin{bmatrix} J_p(q) \\ J_o(q) \end{bmatrix} \dot{q}, \tag{A.3.32}$$

with $J_p(q)$ the first three rows of $J(q)$ and $J_o(q)$ its last three rows.

First, consider the computation of the linear position Jacobian $J_p(q)$. Given that the linear portion of the generalized Cartesian position y is just p, we may write

$$\nu = \sum_{i=1}^{n} \frac{\partial p}{\partial q_1} \dot{q}_i = \begin{bmatrix} \frac{\partial p}{\partial q_1} & \frac{\partial p}{\partial q_2} & \cdots & \frac{\partial p}{\partial q_n} \end{bmatrix} \dot{q}. \tag{A.3.33}$$

Therefore, $J_p(q)$ is given by

$$J_p(q) = \begin{bmatrix} \frac{\partial p}{\partial q_1} & \frac{\partial p}{\partial q_2} & \cdots & \frac{\partial p}{\partial q_n} \end{bmatrix}. \tag{A.3.34}$$

It should be clearly understood that this is exactly what we used in Examples A.3.1 and A.3.2.

Let us now turn to the orientational portion $J_o(q)$ of the Jacobian. Exactly as for linear velocities, one may add angular velocities as long as they are represented in the same coordinate frame. Let us therefore add the individual angular velocities of the links in the arm to obtain the angular velocity of the end effector. A prismatic joint does not contribute to the angular velocity of the end effector.

For a revolute joint, the joint rotation $q_i = \theta_i$ occurs about joint axis z_{i-1} (see Section A.2, especially Figure A.2.1). The angular velocity for joint variable i is therefore given by $z_{i-1}\dot{q}_i$. To add the effects of all the links, it is necessary to express z_{i-1} in a common frame; we select base coordinates. However, the last column of R_{i-1} is exactly z_{i-1} in base coordinates. Therefore, we may write

A.3 The Manipulator Jacobian

$$\omega = \sum_{i=1}^{n} k_i z_{i-1} \dot{q}_i = \begin{bmatrix} k_1 z_0 & k_2 z_1 & \cdots & k_n z_{n-1} \end{bmatrix} \dot{q}. \quad (A.3.35)$$

where $z_0 = \begin{bmatrix} 0 & 0 & 1 \end{bmatrix}^T$ and the selection parameter κ_i is zero if q_i is prismatic and 1 if q_i is revolute. Thus

$$J_o(q) = \begin{bmatrix} k_1 z_0 & k_2 z_1 & \cdots & k_n z_{n-1} \end{bmatrix}. \quad (A.3.36)$$

The complete Jacobian is now given by stacking $J_i(q)$ on top of $J_o(q)$. It should now be clear that the arm Jacobian must be found from the joint vector q using a significant amount of computation. Therefore, whenever the Jacobian is required in a robot arm control scheme, it should be computed using a *computer subroutine*.

It is a good exercise to rework Examples A.3.1 and A.3.2 to determine the complete Jacobian, including the angular portion (see the problems).

EXAMPLE A.3-4: Jacobian for Spherical Wrist

In Example A.2.4 we derived the kinematics for a spherical wrist. Although the wrist would normally terminate an arm, for illustration let us take link 3 as the base link in this example. See Figure A.2.6. Thus we shall determine the Jacobian in the coordinates of the link 3 frame of reference.

To find the Jacobian, we compute the matrices

$$T_3 = I$$

$$T_4 = A_4 = \begin{bmatrix} c_4 & 0 & -s_4 & 0 \\ s_4 & 0 & c_4 & 0 \\ 0 & -1 & 0 & 0 \\ \hline 0 & 0 & 0 & 1 \end{bmatrix}$$

$$T_5 = A_4 A_5 = \begin{bmatrix} c_4 c_5 & -s_4 & c_4 s_5 & 0 \\ s_4 c_5 & c_4 & s_4 s_5 & 0 \\ -s_5 & 0 & c_5 & 0 \\ \hline 0 & 0 & 0 & 1 \end{bmatrix}$$

$$T = T_6 = A_4 A_5 A_6$$

$$= \begin{bmatrix} c_4 c_5 c_6 - s_4 s_6 & -c_4 c_5 s_6 - s_4 c_6 & c_4 s_5 & c_4 s_5 d_6 \\ s_4 c_5 c_6 - c_4 s_6 & -s_4 c_5 s_6 - c_4 c_6 & s_4 s_5 & s_4 s_5 d_6 \\ -s_5 c_6 & s_5 s_6 & c_5 & c_5 d_6 \\ \hline 0 & 0 & 0 & 1 \end{bmatrix}$$

Using the approach just derived, we compute directly that

$$J_p = \begin{bmatrix} \frac{\partial p}{\partial \theta_4} & \frac{\partial p}{\partial \theta_5} & \frac{\partial p}{\partial \theta_6} \end{bmatrix} = \begin{bmatrix} -s_4 s_5 d_6 & c_4 c_5 d_6 & 0 \\ c_4 s_5 d_6 & s_4 c_5 d_6 & 0 \\ 0 & -s_5 d_6 & 0 \end{bmatrix}$$

$$J_o = \begin{bmatrix} z_3 & z_4 & z_5 \end{bmatrix} = \begin{bmatrix} 0 & -s_4 & c_4 s_5 \\ 0 & c_4 & s_4 s_5 \\ 1 & 0 & c_5 \end{bmatrix}.$$

∎

REFERENCES

[Craig 1989] Craig, J. J., *Introduction to Robotics*. Reading, MA: Addison-Wesley, 1989.

[Ickes 1970] Ickes, B. P., "A new method for performing digital control system altitude computations using quaternions," *AIAA J.*, vol. 8, no. 1, pp. 13-17, Jan. 1970.

[IMSL] IMSL, *Library Contents Document*, 8th ed. Houston, TX: International Mathematical and Statistical Libraries.

[MATMAN 1986] MATMAN, *Symbolic Matrix Manipulation*, M. R. Driels, Pembroke, MA: Kern International, 1986.

[Paul 1981] Paul, R. P., *Robot Manipulators*, Cambridge, MA: MIT Press, 1981.

[Schilling 1990] Schilling, R. J., *Fundamentals of Robotics*, Englewood Cliffs, NJ: Prentice Hall, 1990.

[Sheppard 1978] Sheppard, S. W., "Quaternion from rotation matrix," *Eng. Notes*, pp. 223-224, May/June 1978.

[Spong and Vidyasagar 1989] Spong, M. W., and M. Vidyasagar, *Robot Dynamics and Control*. New York: Wiley, 1989.

[Stevens and Lewis 1992] Stevens, B. L., and F. L. Lewis, *Aircraft Modeling, Dynamics, and Control*. New York: Wiley, 1992.

[Yuan 1988] Yuan, J. S.-C., "Closed-loop manipulator control using quaternion feedback," *IEEE J. Robot. Autom.*, vol. 4, no. 4, pp. 434-440, Aug. 1988.

Appendix B

Software for Controller Simulation

An excellent way to gain an intuitive feel for control systems design and performance is to perform computer simulations. It is conceptually a short step from simulation to actual implementation, since the subroutines that are used on today's digital signal processors are very similar to those used for simulation. This appendix contains the software used in the text for robot controller simulation.

There are some good software packages available for design and simulation of systems, and one should be aware of them and employ them. Examples are MATLAB, MATRIX$_X$, Program CC, SIMNON, and so on. However, it is very instructive to use one's own software during the learning phase, especially since it is sometimes not clear exactly what is going on in some of these packages when dealing with digital control.

For the time-response simulation of continuous systems, a Runge-Kutta integrator works very well. In Figure B.1.1 is shown program TRESP, which uses a fourth-order Runge-Kutta subroutine to implement the simulation procedure discussed in Section 3.3. It integrates linear or nonlinear systems in the state-variable form

$$\dot{x} = f(x, u, t).$$

In Section 2.4 we showed how to place the robot dynamics into this form. Note that MATLAB's functions ode 23 and ode 45 are adaptive step size Runge-Kutta integrators that need the same form for the dynamics.

TRESP requires a subroutine F(time,x,xp), which computes $\dot{x}$ (denoted xp, or "x prime") from the current state $x(t)$ and control input $u(t)$. The control $u(t)$ and any outputs of the form

$$y = h(x, u, t)$$

are placed into COMMON storage [e.g., $u(t)$ can be computed outside F(time,x,xp), and $y(t)$ is needed for plotting in TRESP]. Samples of the use of TRESP are given in examples throughout the book.

A parameter array PAR( ) in common makes it easy to perform successive runs with different parameter values (e.g., PD gains). A time delay can be injected into the continuous dynamics if desired (e.g., using a ring buffer).

It is important to realize the following. To update $x(kT_R)$ to $x((k + 1)T_R)$, the Runge-Kutta integrator calls subroutine F(time,x,xp) four times during each Runge-Kutta integration period T_R. During these four calls, the control input should be *held constant* at $u(kT_R)$. Using subroutine SYSINP to compute $u(kT_R)$ accomplishes this.

```
C     FILE DIGCTLF.FOR
C     PROGRAM TO FIND TIME HISTORY
C        USES NONADAPTIVE RUNGE-KUTTA
C     ALLOWS SIMULATION OF SYSTEMS WITH DELAY
C     NEEDS SUBROUTINES:
C        F(TIME,X,XP) FOR CONTINUOUS DYNAMICS
C        DIG(IK,T,X) FOR DISCRETE DYNAMICS
C        SYSINP(IT,X,TIME) FOR CONTINUOUS SYSTEM INPUT (OPTIONAL)
      PROGRAM TRESP
      PARAMETER (NN=20,MM=10)
      REAL Y(0:1024,0:NN+MM),X(NN),SX(NN),XP(NN)
      INTEGER IX(NN+MM),IZ(NN+MM)
      CHARACTER *20 FILNAM,ANS,ANSC
      COMMON/CONTROL/U(MM)
      COMMON/OUTPUT/Z(MM)
      COMMON/DELAY/IT,TS
      COMMON/PARAM/PAR(10)
C
10    WRITE(*,*)'HOW MANY STATES?'
      READ(*,*) NX
      DO 15 I= 1,NX
15    SX(I)= 0.
      WRITE(*,*)'ENTER INITIAL STATES (Def= 0):'
      READ(*,*) (SX(I), I= 1,NX)
C
      I= 0
20    I= I+1
      PAR(I)= -1111
      WRITE(*,*) 'Enter parameter (Def= continue):'
      READ(*,*) PAR(I)
      IF (PAR(I) .NE. -1111) GO TO 20
C
```

```
            WRITE(*,*)'DIGITAL CONTROLLER (C) OR FILTER (F)? (DEF=N)'
            READ(*,'(A)') ANS
            IF(ANS.EQ.'F' .OR. ANS.EQ.'f') ANS= 'F'
            IF(ANS.EQ.'C' .OR. ANS.EQ.'c') ANS= 'C'
            WRITE(*,*)'CONTIN. INPUT SUBROUTINE REQUIRED? (DEF=N)'
            READ(*,'(A)') ANSC
            IF(ANSC.EQ.'Y' .OR. ANSC.EQ.'y') ANSC= 'Y'
30          WRITE(*,*) 'HOW MANY STATES TO BE PLOTTED?'
            READ(*,*) MX
            IF(MX.EQ.0) GO TO 37
            DO 35 I= 1,MX
35          IX(I)= I
            WRITE(*,*)'WHICH ONES? (DEF = IN ORDER)'
            READ(*,*) (IX(I), I= 1,MX)
37          WRITE(*,*)'PLOT HOW MANY OUTPUTS?'
            READ(*,*) MZ
            IF(MZ.EQ.0) GO TO 50
            DO 36 I= 1,MZ
36          IZ(I)= I
            WRITE(*,*)'WHICH ONES? (DEF= IN ORDER)'
            READ(*,*) (IZ(I), I= 1,MZ)
C
50          WRITE(*,*) 'RUN TIME?'
            READ(*,*) TR
            WRITE(*,*)'PRINTING TIME INTERVAL ON SCREEN?'
            READ(*,*) TPR
            IF(ANS.EQ.'F' .OR. ANS.EQ.'C') THEN
                WRITE(*,*)'SAMPLE PERIOD?'
                READ(*,*) TD
            END IF
            WRITE(*,*)'PLOTTING TIME INTERVAL?'
            READ(*,*) TPL
            IF(ANS.NE.'F' .AND. ANS.NE.'C') TD= TPL
            WRITE(*,*)'RUNGE-KUTTA INTEGRATION PERIOD?'
            READ(*,*) TS
            NPR= NINT(TR/TPR)
            NPL= NINT(TPR/TD)
            NT = NINT(TD/TPL)
            NTD= NINT(TPL/TS)
C
            TIME= 0.
            IT= 0
            IP= 0
            IK=0
            DO 60 I= 1,NX
60          X(I)= SX(I)
            Y(0,0)= TIME
            DO 70 I= 1,MX
70          Y(0,I)= X(IX(I))
            IF(MZ.GT.0) THEN
                IF(ANS.EQ.'C') CALL DIG(IK,TD,X)
                IF(ANSC.EQ.'Y') CALL SYSINP(IT,X,TIME)
                CALL F(TIME,X,XP)
```

```
              IF(ANS.EQ.'F') CALL DIG(IK,TD,X)
              DO 75 I= 1,MZ
75            Y(0,MX+I)= Z(IZ(I))
          END IF
C
          DO 110 I= 1,NPR
          DO 90 J= 1,NPL
          IF(ANS.EQ.'C') CALL DIG(IK,TD,X)
          DO 100 K= 1,NT
          DO 85 KID= 1,NTD
          IF(IT.EQ.0) THEN
              WRITE(*,*)
              WRITE(*,80) (IX(IND), IND= 1,MX)
80            FORMAT(35X,'STATES'/'    TIME',10(I12))
              WRITE(*,'(11(1PE12.3))') (Y(0,IND), IND= 0,MX+MZ)
          END IF
          IF(ANSC.EQ.'Y') CALL SYSINP(IT,X,TIME)
          CALL RUNKUT(TIME,TS,X,NX)
          IT= IT+1
85        TIME= FLOAT(IT)*TS
          IP= IP+1
          Y(IP,0)= TIME
          DO 101 Ir 1,MX
101       Y(IP,L)= X(IX(L))
          IF(MZ.LE.0) GO TO 100
          DO 105 L= 1,MZ
105       Y(IP,MX+L)= Z(IZ(L))
100       CONTINUE
          IK= IK+1
          IF(ANS.EQ.'F') CALL DIG(IK,TD,X)
90        CONTINUE
          WRITE(*,'(11(1PE12.3))') (Y(IP,L), L= 0,MX+MZ)
110       CONTINUE
C
120       WRITE(*,130)
130       FORMAT(//2X,'ENTER 0 TO FILE ANSWERS'/8X,'1 TO QUIT'
     &    /8X,'2 TO RESTART',/8X,'3 TO PICK NEW STATES',
     &    /8X,'4 TO CHANGE TIME SCALE')
          READ(*,*) I
          GO TO (150,10,30,50) 1
C
          WRITE(*,*)'OUTPUT FILE NAME?'
          READ(*,'(A)') FILNAM
          OPEN(20,FILE= FILNAM)
          REWIND 20
          WRITE(20,*) MX+MZ
          WRITE(20,*) IP+1
          DO 140 J= 1,MX+MZ
          DO 140 I= 0,IP
140       WRITE(20,'(8(1PE14.6))') Y(I,J)
          REWIND 20
          CLOSE (20)
          GO TO 120
```

```
C
150       STOP
          END
C
C
C     FOURTH-ORDER RUNGE-KUTTA INTEGRATION SUBROUTINE
C
C     REQUIRES SUBROUTINE F(TIME,X,XP) TO DESCRIBE PLANT DYNAMICS
C
          SUBROUTINE RUNKUT(TIME,TS,X,N)
C
C     TS   SAMPLE PERIOD
C     X    STATE VECTOR
C     N    NUMBER OF STATES
C     XP   DERIVATIVE OF STATE VECTOR
C
          PARAMETER (NDIM=32)
          REAL X(*), XP(NDIM), X1(NDIM), XP1(NDIM)
C
          CALL F(TIME,X,XP)
          DO 10 I= 1,N
10        X1(I)= X(I) + .5*TS*XP(I)
C
          TIME= TIME + .5*TS
          CALL F(TIME,X1,XP1)
          DO 20 I= 1,N
          XP(I)= XP(I) + 2.*XP1(I)
20        X1(I)= X(I) + .5*TS*XP1(I)
C
          CALL F(TIME,X1,XP1)
          DO 30 I= 1,N
          XP(I)= XP(I) + 2.*XP1(I)
30        X1(I)= X(I) + TS*XP1(I)
C
          TIME= TIME + .5*TS
          CALL F(TIME,X1,XP1)
          DO 40 I= 1,N
40        X(I)= X(I) + TS*( XP(I)+XP1(I) )/6.
C
          RETURN
          END
```

Figure B.1-1: Program TRESP for time response of nonlinear continuous systems.

For digital control simulation, TRESP needs subroutine DIG(IK,T,x), which contains the discrete controller equations; it is called once in every sample period T. The time T_R should be selected as an integral divisor of T. Five or 10 Runge-Kutta periods within each sample period is usually sufficient.

The program also allows digital filtering (e.g., for reconstruction of velocity estimates from joint position encoder measurements). Note that for digital controls purposes, subroutine DIG is called *before* the Runge-Kutta routine, while for digital filtering, DIG is called *after* the call to Runge-Kutta.

For some systems the Runge-Kutta integrator in the figure may not work; then an adaptive step-size Runge-Kutta routine (e.g., Runge-KuttaFehlburg) can be used [Press *et al.* 1986]. (*Note*: The program given here works for all examples in the book.)

REFERENCE

[Press *et al.* 1986] Press, W. H., Flannery, B. P., Teukolsky, S. A., and Vetterling, W. T., *Numerical Recipes*. New York: Cambridge University Press, 1986.

Appendix C

Dynamics of Some Common Robot Arms

In this appendix we give the dynamics of some common robot arms. We assume that the robot dynamics are given by

$$\tau = M(q)\ddot{q} + V_m(q,\dot{q})\dot{q} + G(q) = M(q)\ddot{q} + N(q,\dot{q}), \qquad \text{(C.1.1)}$$

where the matrix $M(q)$ is symmetric and positive definite with elements $m_{ij}(q)$, that is,

$$M(q) = [m_{ij}(q)]_{n \times n}, i,j = 1,\ldots,n$$
$$m_{ij}(q) = m_{ji}(q)$$

and $N(q,\dot{q})$ is an $n \times 1$ vector with elements n_i, that is,

$$N(q,\dot{q}) = [n_i(q,\dot{q})]_{n \times 1}.$$

Note in particular that the gravity terms are indentified in the expressions of n_i by the gravity constant $g = 9.8$ meters/s². We will also adopt the following notation:

Length of link i is L_i in meters
Mass of link i is m_i in kilograms
Mass moment of inertia of Link i about axis u is I_{uui} in kg-m-m
$S_i = \sin q_i$ and $C_i = \cos q_i$
$S_{ij} = \sin(q_i + q_j)$ and $C_{ij} = \cos(q_i + q_j)$
$S_{ijk} = \sin(q_i + q_j + q_k)$ and $C_{ijk} = \cos(q_i + q_j + q_k)$
$SS_i = \sin^2 q_i$, $CC_i = \cos^2 q_i$, and $CS_i = \cos q_i \sin q_i$;
$SS_{ij} = \sin^2(q_i + q_j)$

C.1 SCARA ARM

The first robot we consider is a general SCARA configuration robot shown in Figure C.1.1. These equations will apply to the AdeptOne and AdeptTwo robots. The dynamics include the first four degrees of freedom and are symbolically given by

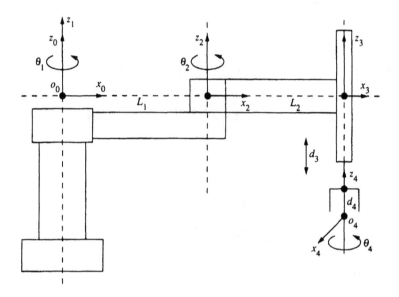

Figure C.1.1: SCARA manipulator.

$$m_{11} = (I_{zz1} + I_{zz2} + I_{zz3} + I_{zz4}) + (\frac{m_1}{4} + 2m_3 + m_4)L_1^2$$
$$+ (\frac{m_2}{2} + 3m_3 + m_4)L_1L_2C_2 + (\frac{m_2}{4} + m_3 + m_4)L_2^2$$
$$m_{12} = (I_{zz2} + I_{zz3} + I_{zz4}) + (\frac{m_2}{4} + m_3 + m_4)L_2^2$$
$$+ (2m_3 + m_4)L_1L_2C_2$$
$$m_{13} = 0$$
$$m_{14} = I_{zz4}$$
$$m_{22} = (I_{zz2} + I_{zz3} + I_{zz4}) + (\frac{m_2}{4} + m_3 + m_4)L_2^2$$
$$m_{23} = 0$$

$$m_{24} = I_{zz4}$$
$$m_{33} = m_3 m_4$$
$$m_{34} = 0$$
$$m_{44} = I_{zz4}$$

$$n_1 = (\frac{m_2}{2} - m_3)L_1 L_2 \dot{q}_1^2 - (4m_3 + 2m_4)L_1 L_2 \dot{q}_1 \dot{q}_2$$
$$n_2 = (\frac{m_2}{2} + m_3 + m_4)L_l L_2 S_2 \dot{q}_1^2$$
$$n_3 = (m_3 + m_4)g$$
$$n_4 = 0.$$

C.2 Stanford Manipulator

The Stanford manipulator shown in Figure C.2.1 has the following dynamics [Bejczy 1974], [Paul 1981]:

$$M_{11} = 1.316 - 1.056108 d_2 + 11.48 d_2^2 + 2.51 S_2^2$$
$$- 5.47995 S_2^2 d_3 + 6.47 S_2^2 d_3^2 + 0.23 S_2^2 d_3 C_5$$
$$m_{12} = -6.47 C_2 d_2 d_3$$
$$m_{13} = -6.47 S_2 d_2$$
$$m_{14} = 0$$
$$m_{15} = 0$$
$$m_{16} = 0$$
$$m_{22} = 4.721 - 5.47995 d_3 + 6.47 d_3^2 + 0.23 d_3 C_5$$
$$m_{23} = 0$$
$$m_{24} = 0$$
$$m_{25} = 0$$
$$m_{26} = 0$$
$$m_{33} = 7.252$$
$$m_{34} = 0$$
$$m_{35} = 0$$
$$m_{36} = 0$$

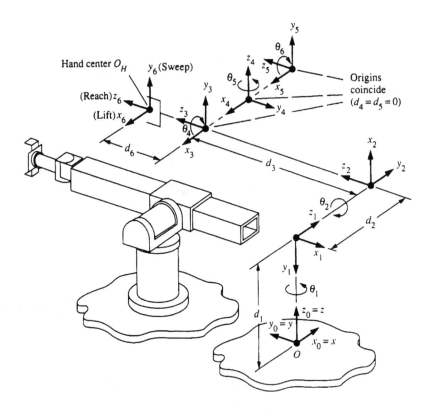

Figure C.2.1: Stanford manipulator.

$m_{44} = 0.107 + 0.203 S_5^2$
$m_{45} = 0$
$m_{46} = 0$
$m_{55} = 0.113$
$m_{56} = 0$
$m_{66} = 0.0203$

$n_1 = 0$
$n_2 = -[2.734 S_2 + 6.47 S_2 d_3 + 0.115(S_2 C_5 + C_2 C_4 S_5)]g$
$n_3 = 6.47 C_2 g$
$n_4 = 0.115(S_2 S_4 S_5)g$
$n_5 = 0.115(S_2 C_4 C_5 - C_2 S_5)g$
$n_6 = 0.$

C.3 PUMA 560 Manipulator

The PUMA 560 is shown in Figure C.3.1. Many simplifications can be made for this particular structure in order to obtain the following dynamics which appeared in [Armstrong et al. 1986].

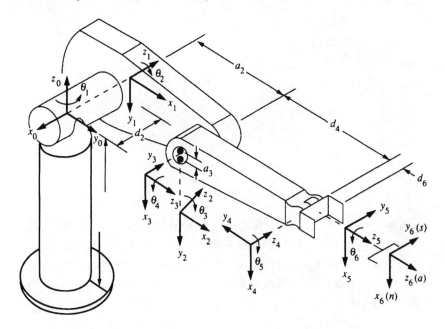

Figure C.3.1: PUMA 560 manipulator.

$$m_{11} = 2.57 + 1.38CC_2 + 0.3SS_{23} + 0.744C_2S_{23}$$
$$m_{12} = 0.69S_2 - 0.134C_{23} + 0.0238C_2$$
$$m_{13} = -0.134C_{23} - 0.00397S_{23}$$
$$m_{14} = 0$$
$$m_{15} = 0$$
$$m_{16} = 0$$
$$m_{22} = 6.79 + 0.744S_3$$
$$m_{23} = 0.333 + 0.372S_3 - 0.011C_3$$
$$m_{24} = 0$$

$m_{25} = 0$

$m_{26} = 0$

$m_{33} = 1.16$

$m_{34} = -0.00125 S_4 S_5$

$m_{35} = -0.00125 C_4 C_5$

$m_{36} = 0$

$m_{44} = 0.2$

$m_{45} = 0$

$m_{46} = 0$

$m_{55} = 0.18$

$m_{56} = 0$

$m_{66} = 0.19$

$$\begin{aligned}
n_1 =& [0.69 C_2 + 0.134 S_{23} - 0.0238 S_2] \dot{q}_2^2 + [0.1335 S_{23} - 0.00379 C_{23}] \dot{q}_3^2 \\
&+ [-2.76 S C_2 + 0.744 C_{223} + 0.6 S C_{23} - 0.0213(1 - 2 S S_{23})] \dot{q}_1 \dot{q}_2 \\
&+ [0.744 C_2 C_{23} + 0.6 S C_{23} + 0.022 C_2 S_{23} - 0.0213(1 - 2 S S_{23})] \dot{q}_1 \dot{q}_3 \\
&+ [-0.0025 S C_{23} S_4 S_5 + 0.00086 C_4 S_5 - 0.00248 C_2 C_{23} S_4 S_5] \dot{q}_1 \dot{q}_4 \\
&+ [-0.0025 (S S_{23} S_5 - S C_{23} C_4 C_5) - 0.00248 C_2 (S_{23} S_5 - C_{23} C_4 C_5) \\
&+ 0.000864 S_4 C_5] \dot{q}_1 \dot{q}_5 + [0.267 S_{23} - 0.00758 C_{23}] \dot{q}_2 \dot{q}_3
\end{aligned}$$

$$\begin{aligned}
n_2 =& -\frac{1}{2} [-2.76 S C_2 + 0.744 C_{223} + 0.6 S C_{23} - 0.0213(1 - 2 S S_{23})] \dot{q}_1^2 \\
&+ \frac{1}{2} [0.022 S_3 + 0.744 C_3] \dot{q}_3^2 \\
&+ [0.00164 S_{23} - 0.0025 C_{23} C_4 S_5 + 0.00248 S_2 C_4 S_5 + 0.00003 S_{23} \\
&\quad (1 - 2 S S_4)] \dot{q}_1 \dot{q}_4 \\
&+ [-0.00215 C_{23} S_4 C_5 + 0.00248 S_2 S_4 C_5 - 0.000642 C_{23} S_4] \dot{q}_1 \dot{q}_5 \\
&+ [0.022 S_3 + 0.744 C_3] \dot{q}_2 \dot{q}_3 - [0.00248 C_3 S_4 S_5] \dot{q}_2 \dot{q}_4 \\
&+ [-0.0025 S_5 + 0.00248 (C_3 C_4 C_5 - S_3 S_5)] \dot{q}_2 \dot{q}_5 \\
&- [0.00248 C_3 S_4 S_5] \dot{q}_3 \dot{q}_4 \\
&+ [-0.0025 S_5 + 0.00248 (C_3 C_4 C_5 - S_3 S_5)] \dot{q}_3 \dot{q}_5 \\
&- 37.2 C_2 - 8.4 S_{23} + 1.02 S_2
\end{aligned}$$

C.3 PUMA 560 Manipulator

$$n_3 = -\frac{1}{2}[-2.76SC_2 + 0.744C_{223} + 0.6SC_{23} - 0.0213(1 - 2SS_{23})]\dot{q}_1^2$$
$$- \frac{1}{2}[0.022S_3 + 0.744C_3]\dot{q}_2^2 - [0.00125C_4S_5]\dot{q}_5^2 - [0.00125C_4S_5]\dot{q}_4^2$$
$$+ [-0.0025C_{23}C_4S_5 + 0.00164S_{23} + 0.0003S_{23}(1 - 2SS_4)]\dot{q}_1\dot{q}_4$$
$$- [0.0025C_{23}S_4C_5 + 0.000642C_{23}S_4]\dot{q}_1\dot{q}_5 - [0.0025S_5]\dot{q}_2\dot{q}_5$$
$$- [0.0025S_5]\dot{q}_3\dot{q}_5 - [0.0025S_4C_5]\dot{q}_4\dot{q}_5 - 8.4S_{23} + 0.25C_{23}$$

$$n_4 = \frac{1}{2}[0.0025SC_{23}S_4S_5 - 0.00086C_4S_5 + 0.00248C_2C_{23}S_4S_5]\dot{q}_1^2$$
$$- \frac{1}{2}[0.00248C_3S_4S_5]\dot{q}_2^2$$
$$+ [0.00164S_{23} - 0.0025C_{23}C_4S_5 + 0.00248S_2C_4S_5$$
$$+ 0.0003S_{23}(1 - 2SS_4)]\dot{q}_1\dot{q}_2$$
$$+ [0.0025C_{23}C_4S_5 - 0.00164S_{23} - 0.003S_{23}(1 - 2SS_4)]\dot{q}_1\dot{q}_3$$
$$- [0.000642S_{23}C_4]\dot{q}_1\dot{q}_5 + [0.000642S_4]\dot{q}_2\dot{q}_5$$
$$+ [0.000642S_4]\dot{q}_3\dot{q}_5 + 0.028S_{23}S4S5$$

$$n_5 = \frac{1}{2}[0.0025(SS_{23}S_5 - SC_{23}C_4C_5) + 0.00248C_2(S_{23}S_5 - C_{23}C_4C_5)$$
$$- 0.000864S_4C_5]\dot{q}_1^2 - \frac{1}{2}[0.0025S_5 - 0.00248(C_3C_4C_5 - S_3S_5)]\dot{q}_2^2$$
$$+ [0.0026C_{23}S_4C_5 - 0.00248S_2S_4C_5 + 0.000642C_{23}S_4]\dot{q}_1\dot{q}_2$$
$$+ [0.0025C_{23}S_4C_5 + 0.000642C_{23}S_4]\dot{q}_1\dot{q}_3 + [0.000642S_{23}C_4]\dot{q}_1\dot{q}_4$$
$$- [0.000642S_4]\dot{q}_2\dot{q}_4 - [0.000642S_4]\dot{q}_3\dot{q}_4 - 0.028(C_{23}S5 + S_{23}C4C5)$$

$n_6 = 0.$

REFERENCES

[Armstrong et al. 1986] Armstrong, B., O. Khatib, and J. Burdick, "The explicit dynamic model and inertial parameters of the PUMA 560 arm," *Proc. 1986 IEEE Conf. Robot. Autom.*, pp. 510-518, San Francisco, Apr. 7-10, 1986.

[Bejczy 1974] Bejczy, A. K., "Robot arm dynamics and control," *NASA-JPL Technical Memorandum 33-669*, 1974.

[Paul 1981] Paul, R. P., *Robot Manipulators: Mathematics, Programming and Control.* Cambridge, MA: MIT Press, 1981.

Index

3D graphics, 519, 521

A Matrix, 558
Acceleration transformation, 576
Accuracy, 4
Activation functions, 433
Actuator
    dynamics, 152, 212, 408, 419
    saturation, 199, 217, 228
Adaptive control, 329, 399
    composite, 361
    computed-torque, 333
    desired compensation, 384
    inertia-related, 341, 384
    passivity, 349
    robust, 399
Anti-windup compensation, 239
Architecture, 518–522, 526, 539, 550
Articulated, 4
    arm, 6, 557
Assembly line, 2

Back emf, 153
Backpropagation weight tuning, 452
Barbalat's lemma, 92, 207, 338, 346, 369, 388
Barrett Whole Arm Manipulator/WAM, 527, 531, 534, 535, 545
Base frame, 558
Bellman-Gronwall lemma, 91
Bilinear transformation, 227
Boundedness, 62

global uniform, 62
global uniform ultimate, 62
uniform, 62, 270
uniform ultimate, 62, 310, 314
Bounding
    estimate update rule, 404
    function, 127, 130, 401
Brunovsky form, 147

C++, 517, 520, 522, 523, 526, 535, 539, 540, 543, 547, 550
Calculus, matrix, 129
Cartesian, 4, 6
    arm, 557
    computed-torque control, 248
    coordinates, 562
    dynamics, 148
    error, 248
    manipulator, 479
    position, 578
    velocity, 581
Centripetal force, 108, 124, 127
Class hierarchy, 523, 524, 530, 535, 540
Classical joint control, 208
Closed-loop system, 411, 419, 465, 475
Code reuse, 519, 520, 523, 526, 535
Commercial robot, 1, 3
Communication, 520, 533, 537–540
Composite adaptive control, 361

609

Computed torque, 169, 185, 205, 410, 418, 480, 482
  approximate, 203, 205, 330
Computer simulation, 591
Concurrency, 537, 538
Conservation of energy, 141, 469
Constraints, 501
Continuous-time systems, 22
Control
  loop, 520–522, 524, 530, 531, 538–540, 546
  program, 522, 525, 539, 540, 545, 546
  tuning, 539
Controllability, 94
Convergence, 52
  uniform, 59
Coordinate transformation, 560
Core classes, 523, 526
Coriolis force, 109, 124, 127
Corrective controller, 408, 417
Critical damping, 189
Cross-product matrix, 582
Cylindrical, 5

Damping
  critical, 189
  ratio, 333
DC motor, 408
Deadbeat observer, 230
Denavit-Hartenberg coordinates, 558
Dexterity, 4
Digital control, 181, 222
Discrete-time systems, 28
Discretization, 225
Disturbance, 136, 371, 400
  rejection, 372, 375
Domain of attraction, 58
Double integrator, 25
Drive train dynamics, 408
Drives
  direct, 9
  hydraulic, 9
  motor, 9
Duality principle, 494
Dynamic controllers, 278
Dynamics
  actuator, 152, 408, 419
  Cartesian, 148
  Hamiltonian, 143, 167
  independent joint, 154
  Lagrangian, 111, 120, 599
  redesign, 316
  state-variable, 142

Electrical dynamics, 408
End-effector forces, 468, 488, 509
Energy, 109
  kinetic, 109, 120
  potential, 111, 123
Environment, 464
Environmental
  forces, 468, 474, 488, 490
  stiffness, 464, 474, 481, 484
Equilibrium point, 36
Error
  Cartesian, 250
  dynamics, 203, 205
  tracking, 146
Estimate, velocity, 226
Euler parameters, 579
Euler's theorem, 580

Feedback linearization, 145, 185, 265, 483
Feedforward loop, 185
Filtered tracking error, 344, 385
Final value theorem, 466, 493, 494
Flexible joint, 156, 416
Force, 108
  centripetal, 108, 124, 127
  Coriolis, 109, 124, 127
  fictitious, 142
  transformation, 577

INDEX   611

Force control, 463
    hybrid, 479
    impedance, 489
    reduced state, 501
    stiffness, 474
Forward kinematics, 562
Friction, 125, 134
Function approximation, 434
Functions of class K, 67
    decrescent, 68
    negative definite, 49
    positive definite, 49

Gear ratio, 153
Gerschgorin theorem, 50, 340, 388, 412, 413, 420
Graphical User Interface/GUI, 519, 520, 523, 525–527, 539, 540, 543, 546, 547, 550
Gravity vector, 124, 134

Hamiltonian, 143
    dynamics, 143, 167
Hardware interfacing, 517
Hiddenlayer neurons, 433
High-mix low-volume, 2
Holonomic Constraints, 502
Homogeneous transformation, 560
Human integration, 11
Hybrid
    impedance control, 489
    position/force control, 478

Imaginary robot, 318
Impedance
    classifications, 490, 492
    control, 489
Independent joint
    control, 213
    dynamics, 154
Inertia, 108
    matrix, 122, 125
Inertial frame, 558

Infinite integral condition, 362, 365
Information integration, 11
Inheritance, 522, 523, 535
Integrator windup, 199, 228, 230
Inverse kinematics, 171, 573

Jacobian matrix, 146, 467, 480, 483, 488, 500, 504, 509, 576
    algorithm for, 585
Joint
    flexibility, 155, 417
    parameters, 559
    prismatic, 555
    revolute, 555
    variables, 555

Kinematic chain, 561
Kinematics, 468, 555
    forward, 562
    inverse, 573
Kinestatic filtering, 464
Kronecker product, 128

Lagrange's equation, 111, 123
Lagrangian dynamics, 108, 123, 501, 599
LaSalle's theorem, 206, 344, 475
Learning
    control, 392
    error, 392
    update rule, 393
Least squares
    estimation, 365
    update rule, 365
Library, 522, 525, 538–540, 543, 545, 546, 548, 550
Linear
    function with parabolic blends, 176
    quadratic control, 243
Linearity in the parameters, 136
Link parameters, 559

Local area networks, 1
Lure's problem, 85
Lyapunov designs, 268
Lyapunov function, 69, 206, 339,
 342, 344, 352, 368, 371,
 373, 376, 386, 394, 403,
 411, 419, 475

Manipulator performance, 3
Math library, 517, 524, 540, 548,
 550
Matrix
 negative definite, 49
 positive definite, 49
Meyer-Kalman-Yakubovitch (MKY)
 Lemma, 86
Minimum-time trajectories, 178
Mismatch, model, 203
Model-following design, 99
Moment of inertia, 121
Momentum, 111
Moore-Penrose matrix inverse, 148
Motion/process integration, 11
Motor dynamics, 153

Natural frequency, 189, 333
Neural net robot controller, 443
Neural networks, 431
 background, 433
 functional-link basis, 437
 gaussian, 438
 linearity in the parameters,
  435, 437
 multilayer, 433
 nonlinear-in-the-parameters, 449
 static, 433
Norms, 39
 function, 43
 induced matrix, 43
 system, 46
 vector, 40

Object manager, 524, 525, 533–
 535, 543
Object-oriented, 517–519, 521, 522,
 550
Objects, 522–527, 533, 534, 543,
 546–548
Observability, 94
One-degree-of-freedom design, 279
Open
 architecture, 517, 520
 inventor, 517, 522, 527
Operating system, 517–521, 550
Optical encoders, 16
Optimal control, 243

Parameter
 convergence, 357, 362, 368
 error, 332, 335, 385
 estimation, 365
 joint, 559
 linearity in, 136
 link, 559
 uncertainty, 329, 330
 update law, 336, 345, 349, 356,
  368, 375, 385
Passive controllers, 293
Passivity, 82, 141, 349
Passivity properties, 453
Path generation, 170
Payload weight, 4
PC, 517, 519–521, 537, 550
PD control, 197, 213, 343, 465,
 474
 gravity, 205
Persistency of excitation, 357
Perspective transformation, 561
PI adaptation, 357
PID control, 197, 215, 355
Plot window, 539, 540
Polynomial path interpolation, 173
Positive Real Systems, 84
 strictly, 85

INDEX   613

Precision, 4
Prismatic, 4
Prismatic joint, 555
Programming, 517–522, 527, 534, 535, 540, 547, 550
Proximity and distance sensors, 13
Pseudo-inertia matrix, 120
Puma, 523–527, 531, 534, 535, 546–548, 550
PUMA arm, 557, 603

QMotor, 519, 522, 537–540, 546, 550
QNX, 517, 519–521, 537, 550
Quaternions, 580
Quickness, 4

Rayleigh-Ritz theorem, 50, 338, 346, 371, 376
RCCL, 518, 523
Reach, 4
Real-time, 517–522, 524, 530, 538–540, 550
Regression matrix, 332, 348, 350, 364, 384
Repeatability, 4
Repetitive
    control law, 392
    tasks, 392
Resonant mode, 190
Revolute, 4
    joint, 555
Robot
    articulated, 8
    cartesian, 8
    controllers, 1
    cylindrical, 8
    parallel-link, 8
    SCARA, 8
    spherical, 8
Robot control languages, 518
Robotic
    classes, 523, 524
    platform, 517–520, 522–526, 533, 534, 537–540, 543, 545–547, 550
    workcells, 2
Robust control, 400
Robustness
    adaptive control, 371
Rotation matrix, 560, 579
Runge-Kutta integrator, 182, 591

Sampling period, 183, 223, 235
Saturation, actuator, 199, 217, 228
    controller, 305
    function, 310
SCARA, 4, 6
    arm, 557, 600
Scene viewer, 525–527, 534, 537, 538, 543, 545, 546, 550
Sensor data processing, 16
Sensors, 1, 12
    acceleration, 14
    force, 15
    photoelectric, 15
    position, 14
    torque, 15
    velocity, 14
Servo control, 517, 520, 530–532, 537, 538, 545, 548, 549
Set-point control, 343, 476, 479
Simulation, 517, 519, 520, 522, 523, 527, 534, 537, 543, 547, 548
Simulation, computer, 181, 591
Skew-symmetry, 131, 150
Small-gain theorem, 88
Software, 517–522, 533, 537, 539, 550
Spherical, 5
Stability
    asymptotic, 53, 294

bounded-input-bounded-output (BIBO), 80, 281
global asymptotic, 53
global exponential, 59
global uniform asymptotic, 59
in the sense of Lyapunov (SL), 51, 294
input-output, 80
total, theorem, 89
uniform, 59
uniform asymptotic, 59
Stanford arm, 601
State strict passivity, 453
State-variable model, 142
Static controllers, 274
input-output designs, 273
Steady-state
error, 466, 493, 494
force, 467, 474
position, 466, 474
Stiffness control, 464, 474
Synchronization, 538

T Matrix, 561
T_i matrices, 561
Tactile sensors, 13
Task space coordinates, 248, 467, 480, 482, 494
Teachpendant, 523, 543, 545, 546
Thread, 519, 521, 526, 527, 533, 538–540
Tool configuration vector, 581
Tracking error, 146, 185, 205, 331, 332, 335, 384, 409, 474, 480, 481, 484, 485, 505
Cartesian, 250
Trajectory
generation, 10, 171, 517, 531, 550
generation/trajectory generator, 520

generator, 531, 532, 535, 545–548
Transfer function, 350, 369, 396, 404, 466
Translation vector, 560
Two-degree-of-freedom design, 288

UNIX, 521
Unix, 519

Variable-structure controllers, 297
Velocity
estimate, 226
transformation, 576
Via points, 173
Virtual
walls, 548–550
work, 577
Vision systems, 17
VRML, 522, 527

Windup, integrator, 199, 228, 239
Workcells, 1
Workspace, 4
Wrist, 558, 571

Zero order hold, 183

CPSIA information can be obtained
at www.ICGtesting.com
Printed in the USA
LVHW081559281222
736077LV00005B/151

9 780824 740726